CATEGORICAL DATA ANALYSIS WITH SAS® AND SPSS APPLICATIONS

CATEGORICAL DATA ANALYSIS WITH SAS® AND SPSS APPLICATIONS

Bayo Lawal
St. Cloud University

LAWRENCE ERLBAUM ASSOCIATES, PUBLISHERS
2003 Mahwah, New Jersey London

Senior Editor:	Debra Riegert
Editorial Assistant:	Jason Planer
Cover Design:	Kathryn Houghtaling Lacey
Textbook Production Manager:	Paul Smolenski
Text and Cover Printer:	Sheridan Books, Inc.

Camera ready copy for this book was provided by the author.

Lawrence Erlbaum Associates, Inc., Publishers
10 Industrial Avenue
Mahwah, New Jersey 07430

Library of Congress Cataloging-in-Publication Data

Lawal, H. Bayo.
Categorical data analysis with SAS and SPSS applications / H. Bayo Lawal.
p. cm.

Includes bibliographical references and index.

ISBN 0-8058-4605-0 (cloth : alk. paper)
1. Multivariate analysis. 2. SAS (Computer file). 3. SPSS (Computer file).
I. Title.
QA278.L384 2003
519.5'3—dc21 2003044361
CIP

Books published by Lawrence Erlbaum Associates are printed on acid-free paper, and their bindings are chosen for strength and durability.

Printed in the United States of America
10 9 8 7 6 5 4 3 2 1

Contents

Preface

This book is primarily designed for a senior undergraduate class in Categorical Data Analysis and for majors in biomedical, biostatistics and statistics programs, but can also be used as reference text for researchers working in the area, and /or for an introductory text in a graduate course on the subject. A prerequisite of a one year undergraduate background in Calculus, and a basic understanding of a one-year background course in statistical methods, are therefore recommended.

I have tried to write the text in such a way that students will be at ease with the materials. Thus where necessary, concepts have been explained in details so that students can reproduce similar results for some of the problems in the exercises. Emphasis is on implementation of the models discussed in the text, with the use of the Statistical Packages SAS/STAT® Version 8e. Relevant codes and instructions are included with each example presented. Corresponding analyses with SPSS (version 11) codes are presented in the CD ROM.

Many examples are given and tables of results of analyses as well as interpretations of the results of analyses are presented. Numerous references are also given for students interested in further readings on a particular topic or topics in the bibliography listing. The students will also find exercises at the end of each chapter beginning from chapter 2. Most exercises require intensive use of PC based Statistical software. These exercises provide standard problems that have been chosen to conform to the level of the text and students. To further strengthen the students' understanding of the materials, the exercises sometimes require them to employ some of the ideas expressed in the text in a more advanced capacity.

The data for illustration and exercises have been carefully selected from various literature on the subject with one point in mind: to illustrate the concepts being discussed. Some of the data have been analyzed before in various texts and prints, but I have added relevant codes for their implementation. In some cases, I have looked at the anlysis of some of the data from a completely different perspective. I hasten to add however, that the reliability of these data depend on the honesty of the sources where these data were drawn from.

Considerable use has been made of SAS® PROC GENMOD in the text, the SAS software codes for implementing a variety of problems are presented either in the text or in the appendices which come in a CD ROM. Brief summaries of the contents of each chapter are presented below:

- Chapter 1 introduces readers to the various types of variables commonly encountered in Categorical Data analysis.
- Chapter 2 reviews the Poisson, Binomial, Multinomial and hypergeometric

distributions.It also gives an overview of the properties of generalized linear models.

- Chapter 3 discusses the one-way classification, exact and large sample tests, goodness-of-fit test statistics, local effects and goodness-of-fit tests for Poisson and Binomial data.

- Chapter 4 introduces models for the 2×2 and the $2 \times 2 \times k$ contingency tables. The chapter discusses various sampling schemes, Fisher's exact and Mid-p tests as well as the analysis of correlated 2×2 data.

- Chapter 5 gives an overview of the general $I \times J$ table. Residual analysis, quasi-independence and SAS codes for implementing all these are given.

- Chapter 6 presents a comprehensive approach to log-linear model analyses. It discusses the concepts of sufficient statistics, decomposable models, marginal and partial associations, collapsibility conditions, and problems associated with log-linear models. It also discusses weighted data as well as models for rates. All models are implemented with SAS® PROC GENMOD, and CATMOD; SPSS PROC GENLOG, HILOGLINEAR and LOGLINEAR.

- Chapter 7 discusses model selection strategies, Aitken's selection method and discusses selection criteria. Ample examples are presented.

- Chapter 8 discusses the logistic regression. Examples in bioassay are presented. The chapter also discusses analysis of cohort study data, diagnostics, over-dispersion and factor-response models, as well as parameter interpretations are discussed in this chapter. SAS® PROCs LOGISTIC, GENMOD, PHREG, PROBIT, as well as PROC CATMOD are also heavily employed in this chapter. SPSS procedures PROBIT, LOGISTIC, PLUM, NOMREG, and COXREG are similarly employed in this chapter with implementation and results in the CD ROM.

- Chapter 9 dicusses logit and multinomial logit models, Poisson regression models, multi-category response models (adjacent category, baseline, proportional odds, cumulative odds, continuation ratio) are fully discussed in this chapter with considerable number of examples together with their SAS software and SPSS implementations.

- Chapter 10 discusses association models. The chapter also discusses the general association models as well as grouping of categories in contingency tables. Extension to higher dimensional tables is provided and conditional association models are fully discussed, together with their implementation in SAS® with PROC GENMOD, and in SPSS with PROC GENLOG.

- Chapter 11 introduces the analysis of doubly classified data. A new unified approach to analyzing asymmetry, generalized independence, and skew-symmetry models using non-standard log-linear models is also presented in this chapter. The treatment of these topics in this chapter are most comprehensive and SAS® and SPSS implementations of the various models are also presented. The chapter also discusses the Bradley-Terry, and agreement models.

- Chapter 12 discusses the analysis of repeated binary response data. It also introduces readers to the generalized estimating equations (GEE) methodology. Several examples involving binary response variables are given. An example involving a non-binary response is also presented.

SPSS codes for all models discussed in this text are presented on the CD-ROM.

ACKNOWLEDGMENTS

I owe a debt of gratitude to a number of people: Dr. Raghavarao for encouraging me to write the book, Dr. Graham Upton for comments on an earlier draft of some of the chapters. I am also grateful to St. Cloud State University for the long term research grant which provided me the summer 1994 stipend that enabled me to complete chapter 11. I would also like to thank Phillip Rust from the Medical University of South Carolina, and Alexander von Eye from Michigan State University, who reviewed the manuscript for publication. Their comments have been very helpful. My sincere thanks to Debra Riegert, Jason Planer, Beth Dugger, and Paul Smolenski for their encouraging support during the acquisition and production of this text.

My thanks also go to the graduate students at Temple University for their invaluable comments on a section of this text when I taught the course in 1989. My thanks similarly go to my STAT 436 classes at St. Cloud State University in 1996, 2000 and 2002.

My gratitude also goes to the Department of Biometry and Epidemiology, MUSC, for the support provided during the final writing of this manuscript while I was on sabbatical leave at MUSC and to the graduate students in the Department for their insightful comments and input in my BIOM 711 class in the Spring of 2000.

I would also like to thank Professors Stuart Lipsitz of the Department of Biometry and Epidemiology, MUSC, Charleston, SC., for providng me some of the material in chapter 12, and Emmanuel Jolayemi of the Department of Statistics, University of Ilorin, Nigeria, whose lecture notes taken during graduate classes at the University of Michigan were very useful in some sections of this text.

I thank the SAS® Institute Inc., and SPSS Inc., for the use of their softwares for the analyses of all the data examples in this text, the results of which are presented in the text or in the appendices.

Finally, I thank my wife Nike and our children for their patience and support during the years of working on this project.

Bayo Lawal
Department of Statistics
St. Cloud State University
St. Cloud, MN

CATEGORICAL DATA ANALYSIS WITH SAS® AND SPSS APPLICATIONS

Chapter 1

Introduction

In this text, we will be dealing with *categorical data*, which consist of counts rather than measurements. A variable is often defined as the characteristic of objects or subjects that varies from one object or subject to another. Gender, for instance, is a variable as it varies from one person to another. In this chapter, because variables consist of different types, we describe the variable classification that has been adopted over the years in the following section.

1.1 Variable Classification

Stevens (1946) developed the measurement scale hierachy into four categories, namely, *nominal, ordinal, interval* and *ratio* scales. Stevens (1951) further prescribed statistical analyses that are appropriate and/or inappropriate for data that are classified according to one of the four scales above. The nominal scale is the lowest while the ratio scale variables are the highest.

However, this scale typology has seen a lot of critisisms from, namely, Lord (1953), Guttman (1968), Tukey (1961,) and Velleman and Wilkinson (1993). Most of the critisisms tend to focus on the prescription of scale types to justify statistical methods. Consequently, Velleman and Wilkinson (1993) give examples of situations where Steven's categorization failed and where statistical procedures can often not be classified by Steveen's measurement theory.

Alternative scale taxonomies have therefore been suggested. One of such was presented in Mosteller and Tukey (1977, chap. 5). The hierachy under their classification consists of *grades, ranks, counted fractions, counts, amounts*, and *balances.*

A categorical variable is one for which the measurement scale consists of a set of categories that is non-numerical. There are two kinds of categorical variables: *nominal* and *ordinal* variables. The first kind, *nominal* variables, have a set of unordered mutually exclusive categories, which according to Velleman and Wilkinson (1993) "may not even require the assignment of numerical values, but only of unique identifiers (numerals, letters, color)". This kind classifies individuals or objects into variables such as gender (male or female), marital status (married, single, widowed, divorced), and party affiliation (Republican, Democrat, Independent). Other variables of this kind are race, religious affiliation, etc. The number of occurrences in each category is referred to as the frequency count for that category. Nominal variables are invariant under any transformations that preserves the relationship between subjects (objects or individuals) and their identifiers provided we do not combine categories under such transformations. For nominal variables, the statistical analysis remains invariant under permutation of categories of the variables.

As an example, consider the data in Table 1.1 relating to the distribution 120 students in an undergraduate biostatistics class in the fall of 2001. The students were classified by their gender (males or females). Table 1.1 displays the classification of the students by gender.

	Gender		
Counts	Male	Female	Total
n_i	31	89	120

Table 1.1: Classification of 120 students in class by gender

When nominal variables such as variable gender have only two categories, we say that such variables are *dichotomous* or *binary*. This kind of variables have been sometimes referred to as the *categorical-dichotomous*. A nominal variable such as color of hair with categories {black, red, blonde, brown, gray}, which has multiple categories but again without ordering, is referred to (Clogg, 1995) as *categorical-nominal.*

The second kind of variables, *ordinal* variables, are where the categories are ordered. Using the definition in Velleman and Wilkinson (1993), if S is an ordinal scale that assigns real numbers in $\Re$ to the elements of a set P, of observed observations, then

$$\begin{aligned} &P \xrightarrow{S} \Re \quad \text{such that} \\ &\imath > \jmath \leftrightarrow S(\imath) > S(\jmath) \quad \text{for all } i, j \in P \end{aligned} \tag{1.1}$$

Such a scale S preserves the one-to-one relation between numerical order values under some transformation. The set of transformations that preserves the ordinality of the mapping in (1.1) has been described by Stevens (1946) as being *permissible.* That is, the monotonic transformation f is permissible if and only if:

$$S(\imath) > S(\jmath) \Rightarrow f[S(\imath)] > f[S(\jmath)] \tag{1.2}$$

Thus for this scale of measurements, permissible transformations are logs or square roots (nonnegative), linear, addition or multiplication by a positive constant.

Under an ordinal scale, therefore, the subjects or objects are ranked in terms of degree to which they posses a characteristic of interest. Examples are: rank in graduating class (90th percentile,80th percentile,60th percentile); social status (upper,middle,lower); type of degree (BS,MS,PhD); etc. The Likert variable such as the attitudinal response variable with four levels (strongly dissapprove, dissapprove, approve, strongly approve) can sometimes be considered as a partially ordinal variable (a variable with most levels ordered with one or more unordered levels) by the inclusion of the response level “don’t know” or “no answer” category. Ordinal variables generally indicate that some subjects are better than others but then, we can not say by how much better, because the intervals between categories are not equal. Consider again the 120 students in our biostatistics class now classified by their status at the college (freshman, sophomore, junior, or senior). Thus the status variable has four categories in this case.

Counts	Status				Total
	F	S	J	Sr	
n_i	10	45	50	15	120

Table 1.2: Distribution of 120 students in class by status

The results indicate that more than 75% of the students are either sophomores or juniors. Table 1.2 will be referred to as a one-way classification table because it was formed from the classification of a single variable, in this case, status. For the data in Table 1.2, there is an intrinsic ordering of the categories of the variable "status", with the senior category being the highest level in this case. Thus this variable can be considered ordinal.

On the other hand, a *metric variable* has all the characteristics of nominal and ordinal variables, but in addition, it is based on equal intervals. Height, weight, scores on a test, age, and temperatures are examples of interval variables. With metric variables, we not only can say that one observation is greater than another but by how much. A fuller classification of variable types is provided by Stevens (1968).

Within this grouping, that is, *nominal, ordinal,* and *metric*, the metric or interval variables are highest, followed by ordinal variables. The nominal variables are lowest. Statistical methods applied to nominal variables will also be applicable to either ordinal or interval variables but not vice versa. Similarly, statistical methods applicable to ordinal variables can also be applied to interval variables, again but not vice versa, nor can it be applied to nominal variables since these are lower in order on the measurement scale.

When subjects or objects are classified simultaneously by two or more attributes, the result of such a cross-classification can be conveniently arranged as a table of counts known as a *contingency table.* The pattern of association between the classificatory variables may be measured by computing certain *measures of association* or by fitting of *log-linear model, logit, association*, or other models.

1.2 Categorical Data Examples

Suppose the students in this class have been cross-classified by two variables, gender and status, then the resulting table will be described as a two-way table or a two-way contingency table below (see Table 1.3). Table 1.3 displays the cross-classification of students by gender and status.

Gender	**Status**				Total
	F	S	J	Sr	
Males	3	10	12	6	31
Females	7	35	38	9	89
Total	10	45	50	15	120

Table 1.3: Joint classification of students by status and gender

The frequencies in Table 1.3 above indicate that there are 31 male students in this class and 89 female students. Further, the table also reveals that of the 120 students in the class, 38 are females whose status category is junior.

In general, our data will consist of counts $\{n_i, \quad i = 1, 2, \cdots, k\}$ in the k cells (or categories) of a contingency table. For instance, these might be observations for the k levels of a single categorical variable, or for k=IJ cells of a two-way $I \times J$ contingency table. In Tables 1.1 and 1.2, k equals 2 and 4 respectively, while $k = 8$ in Table 1.3. We shall treat counts as random variables. Each observed count, n_i has a distribution that is concentrated on the non-negative integers, with expected values denoted by $m_i = E(n_i)$. The $\{m_i\}$ are called *expected* frequencies, while the $\{n_i\}$ are referred to as the observed frequencies. Observed frequencies refer to how many objects or subjects are observed in each category of the variable(s).

1.3 Analyses

For the data in Tables 1.1 and 1.2, interest for this type of data usually centers on whether the observed frequencies follow a specified distribution usually leading to what is often referred to as 'goodness-of-fit test'. For the data in Table 1.3, on the other hand, our interest in this case is often concerned with independence, that is, whether the students' status is independent of gender. This is referred to as the test of independence. An alternative form of this test is the test of "homogeneity", which postulates, for instance, that the proportion of females (or males) is the same in each of the four categories of "status". We will see that both the independence and homogeneity tests lead asymptotically to the same result in chapter 5. We may on the other hand wish to exploit the ordinal nature of the status variable (since there is an hierachy with regards to the categories of this variable). We shall consider this and other situations in chapters 6 and 9, which deals with log-linear model and association analyses of such tables. The various measure of association exhibited by such tables are discussed in chapter 5.

1.3.1 Example of a 2×2 Table

Suppose the students in the biostatistics class were asked whether they have ever smoked; this would lead to a response variable that we will call here "smoking status" with categories Yes and No. We display in Table 1.4 the two-way cross-classification of the students by gender and smoking status, leading to a 2×2 contingency table.

	Response		
Gender	Yes	No	Total
Males	18	13	31
Females	52	37	89
Total	70	50	120

Table 1.4: 2×2 contigency table of response by gender

The resulting table indicates that of the 120 students in the class, 70 indicated that they have once smoked. The 2×2 table has perhaps received more attention than many other tables. The reasons for this is related to "how the observed frequencies in the table are obtained." In this case, there are only 120 students in this class, and we may therefore consider that this value or sample size is fixed. With 120 fixed, the students are then cross-classified by the two variables, response and gender, leading to Table 1.4. This schemme leads to what has been described as the "multinomial sampling scheme." Other sampling schemes relating to the 2 × 2 table will be examined in chapter 4. Again, interests center here on whether smoking status is independent of gender or whether the response YES is uniform across gender.

1.3.2 Three-Way Tables

Assuming the students were also asked their attitudes towards abortion, this would lead to yet another variable, designated here as attitudes, with three response categories {positive, mixed, negative} designated here as 1, 2, and 3, respectively. We now cross-classified the resulting data as a three-way contingency table having variables gender, status, and attitude, resulting in a 2 × 4 × 3 three-way contingency tables having 24 cells. Table 1.5 displays the observed frequencies in this case.

		Attitudes			
Gender	Status	1	2	3	Total
	1	1	2	0	3
Males	2	4	5	1	10
	3	5	4	3	12
	4	1	3	2	6
	1	3	2	2	7
Females	2	15	14	6	35
	3	12	9	17	36
	4	2	3	4	9

Table 1.5: Three-way table of gender, status and attitudes

In Table 1.5, the status categories 1, 2, 3, and 4 refer respectively to freshman, sophomore, junior, and senior. The form of analysis here can be viewed in several ways.

(a) We may wish to consider whether the attitudes towards abortion is independent of gender. In this case, we only need to collapse Table 1.5 over the status variable (in other words we will ignore the status variable) and use the methods mentioned in the previous section to analyze the data; that is, we will conduct a test of independence, for Table 1.6.

	Attitudes			
Gender	1	2	3	Total
Males	11	14	6	31
Females	32	28	29	89
Total	43	42	35	120

Table 1.6: Two-way table of gender by attitudes collapsed over status

(b) We may also be interested in the relationship between the status category and attitudes toward abortion. This again would lead to a 4 × 3 two-way contingency table as in Table 1.7.

Again the usual test of independence will answer this question. Note here that we are collapsing over the gender variable.

Status	Attitudes 1	2	3	Total
1	4	4	2	10
2	19	19	7	45
3	17	13	20	50
4	3	6	6	15
Total	43	42	35	120

Table 1.7: Two-way table of status by attitudes collapsed over gender

(c) However, we may view collapsibility of Table 1.5 over either the status or gender variable as unsatisfactory because we do not know whether the strength and direction of the association between the resulting two variables will change when collapsed over the third variable (this is called *Simpson's paradox*).

It is therefore desirable to analyze the data in Table 1.5 like the familiar analysis of variance for the continous type data where we can study the interactions between the various variables. The equivalent analysis of variance (ANOVA) model in categorical data analysis is the log-linear models analysis, which we shall apply not only to three-way tables but to higher dimensional tables in this book. The procedure allows us to study interactions between the explanatory variables and the response variable.

If, however, we consider variable attitude as a response variabe, an even better procedure would be to consider fitting *logit* models to the data in Table 1.5. Logit models are discussed in chapter 9. We can further exploit the intrinsic ordinality of the response variable {+ve, mixed, -ve} by fitting specialized models, such as the proportional odds, continution ratio, adjacent category, and cumulative logit models to this data. These models are discussed in the latter part of chapter 9, and we also present in that chapter how to implement these models in SAS® PROC CATMOD, GENMOD, LOGISTIC and SPSS PROC GENLOG and NOMREG.

If we consider again the 4×3 table formed from the three-way table when collapased over gender in Table 1.7; we would notice here that both variables "status" and "attitudes" are ordinal in nature. We can study the associations between these two variables better by employing either the row, column or R+C association models discussed in chapter 10. These class of models, which first appeared in Goodman (1979), models the local odds-ratios in the table.

Consider again, the two-way table formed from collapsing over the "status" variable, that is, Table 1.6 where only the response variable "attitude" can be considered ordinal, gender variable being nominal in this case. Thus, we can fit the row association (R) model to this data. Another possible model is the multiplicative RC association model. These models are discussed in chapter 10 together with their SAS® implementation with example data in that chapter.

In chapter 11, we consider the analysis of doubly classified data. For the students in our class we asked the following questions: (1) What is your political party affiliation? (2) Which party did you vote for in the last Presidential election? We gave them three party choices only:Democrat, Republican, Independent. Table 1.8 displays the joint classification of these students on the two variables.
For Table 1.8, while it is a two-way 3×3 table, we cannot employ the usual two-way

Party Affiliation	Vote D	Vote R	Vote I	Total
D	35	10	8	53
R	8	39	2	49
I	4	2	12	18
Total	47	51	22	120

Table 1.8: Two-way table of party affiliation by voting pattern

table model of independence as presented earlier with these data. We would expect the analysis to be different in this case because it is obvious that the diagonal cells are highly correlated: Democrats tend to vote Democrat and Republicans likewise. Hence, the models that we would consider must take into consideration the diagonal structure of this table. In chapter 11, we shall develop diagonal parameter and similar models necessary for a proper explanation of such data. Other possible models we would want to examine are whether there is symmetry with regards to the cells or symmetry with regards to the marginal totals. Models discussed in this chapter will examine symmetry and marginal symmetry models that are appropriate for this class of data. A unified approach to fitting these models will be presented in this chapter. We shall further extend our discussion in chapter 11 to methods of analysis for agreement data arising from two and three raters. Nonstandard log-linear techniques (von Eye & Spiel, 1996) and (Lawal, 2001) will be discussed for implementing these class of models.

In the last series of questions, the students were asked regarding abortion: Should a pregnant woman be able to obtain a legal abortion if:

(1) She is married and does not want more children?

(2) The family has very low income and cannot afford any more children?

(3) She is not married and does not want to marry the man?

The form of these questions was tailored to the 1976 US General Social Survey. Since there are three questions whose responses are simply Yes or No, there are $2^3 = 8$ possible combinations of responses. The classification of our students into these eight possible combinations of responses by gender is displayed in Table 1.9.

Gender	Responses YYY	YYN	YNY	YNN	NYY	NYN	NNY	NNN	Total
Male	7	3	1	4	5	3	2	6	31
Female	18	10	7	9	11	8	5	21	89
Total	25	13	8	13	16	11	7	27	120

Table 1.9: Classification of students by gender and responses

In Table 1.9, for instance, the response NNN indicates the students responded No to questions 1, 2, and 3. In this case, there are 27 students in this category. The data in Table 1.9 fall into the category of repeated binary response data. Because the responses are repeated, there is therefore an intrinsic correlation among the three responses and models that will be employed for this would have to exploit this correlation structure within the subjects' responses. We discuss in the last chapter of this book logit, marginal, and other similar models for handling data of this nature. The discussion in this chapter also covers the case involving more than one covariates with repeated observations coming from clusters of subjects. This approach leads to the discussing of the generalized estimating equations (GEE) originally proposed by Liang and Zeger (1986). Again, SAS® programs are provided for all the examples in this chapter.

Chapter 2

Probability Models

2.1 Introduction

Categorical data analysis depends on a thorough understanding of the various sampling schemes that may give rise to either a one-way, a two-way, or a multidimensional contingency tables under consideration. These sampling schemes, which will be discussed later in the book, have various underlying probability distributions. The most common of these being the Poisson distribution. Consequently, a proper understanding of these probability distributions (or *probability models* as they are sometimes referred to) is therefore highly desirable. However, before discussing these various probability models, it would be necessary to define the concept of the moment generating functions particularly as they relate to obtaining the moments of the distributions and proving some asymptotic results.

We also introduce readers to the general theory of *generalized linear models.* The discussions here are brief and will help readers have an easy understanding of the theory behind categorical data analysis and some of the statistical softwares that have been developed to take advantage of this theory.

2.2 Moment-Generating Functions

The moment-generating function (mgf) of a random variable X is defined as

$$\mathbf{M}_X(t) = E(e^{tx}) \tag{2.1}$$

that is,

$$\mathbf{M}_X(t) = \begin{cases} \sum e^{tx} p(x), & \text{if X is discrete} \\ \int_{-\infty}^{\infty} e^{tx} p(x)\, dx & \text{if X is continuous} \end{cases}$$

where $p(x)$ is the probability density function (pdf) such that for the random variable X

$$\sum_{\text{all x}} p(x) = 1 \quad \text{if X is discrete and}$$

$$\int_{-\infty}^{\infty} p(x)\, dx = 1 \quad \text{if X is continuous}$$

The Maclurin's series expansion for e^{tx} is given by:

$$e^{tx} = 1 + tx + \frac{t^2x^2}{2!} + \cdots + \frac{t^r x^r}{r!} + \cdots \tag{2.2}$$

Hence, substituting (2.2) in the expression for M(t) in (2.1) we have for the discrete case for instance that:

$$\begin{aligned}\mathbf{M}_X(t) &= \sum(1 + tx + \frac{t^2x^2}{2!} + \cdots + \frac{t^rx^r}{r!} + \cdots)\, p(x) \\ &= 1 + \mu t + \mu_2^{'}\frac{t^2}{2!} + \cdots + \mu_r^{'}\frac{t^r}{r!} + \cdots\end{aligned} \tag{2.3}$$

since $\mu = E[\sum xp(x)]$ and $\mu_r^{'} = E[\sum x^r p(x)]$.

The case for the continuous type distribution is equivalent to the general result in (2.3) except that the summation sign is replaced by the integral sign.

The results in (2.3) indicate that if we expand $\mathbf{M}_X(t)$ as a power series in t, then the moments can be obtained as the coefficient of $\frac{t^r}{r!}$ in this expansion when evaluated at $t = 0$. That is, these moments are given by:

$$\mu_r^{'} = \frac{d^r}{dt^r}[\mathbf{M}_X(t)] \quad \text{at } t = 0 \tag{2.4}$$

Thus, for instance, $\mu = \frac{d}{dt}[\mathbf{M}_X(t)]_{t=0}$. Similar results can be obtained for other higher moments.

2.2.1 Properties of $M_X(t)$

The following results relating to $\mathbf{M}_X(t)$ are also important.

If α and β are constants, then it can be readily verified that:

(i)

$$\mathbf{M}_{(X+\alpha)}(t) = e^{\alpha t}\mathbf{M}_X(t)$$

(ii)

$$\mathbf{M}_{(\beta X)}(t) = \mathbf{M}_X(\beta t) \quad \text{or} \quad \mathbf{M}_{(X/\beta)}(t) = \mathbf{M}_X(\frac{t}{\beta})$$

(iii)

$$\mathbf{M}_{(\frac{X+\alpha}{\beta})}(t) = e^{(\frac{\alpha}{\beta})t}M_X(\frac{t}{\beta})$$

The result in **(i)** is of special importance when $\alpha = \mu$, while the result in **(iii)** is of special importance when $\alpha = -\mu$ and $\beta = \sigma$ to obtain the moment generating function of a standardized variable.

We give the proof of **(iii)** as follows:

$$\begin{aligned}\mathbf{M}_{(\frac{X+\alpha}{\beta})}(t) &= E\left[e^{t(\frac{X+\alpha}{\beta})}\right] \\ &= E\left[e^{(\frac{\alpha}{\beta})t}.e^{(\frac{t}{\beta})X}\right] \\ &= e^{(\alpha/\beta)t}E\left[e^{(\frac{t}{\beta})X}\right] \\ &= e^{(\alpha/\beta)t}M_X(t/\beta)\end{aligned}$$

If we set $\beta = 1$ in **(iii)**, we would obtain **(i)**. Similarly, setting $\alpha = 0$ in **(iii)**, we would obtain **(ii)**.

2.2.2 The Normal Probability Distribution

The normal or Gaussian probability distribution is defined as

$$p(x) = \frac{1}{\sigma\sqrt{2\pi}} e^{\left[-\frac{1}{2}\left(\frac{x-\mu}{\sigma}\right)^2\right]} \quad \text{for} \quad -\infty < x < \infty$$

where μ and σ are real constants and σ must be positive. We usually write that $X \sim N(\mu, \sigma^2)$. The random variable defined by

$$Z = \frac{X - \mu}{\sigma}$$

is referred to as the standardized normal variate and has mean zero ($\mu = 0$) and variance one ($\sigma^2 = 1$). That is, $Z \sim N(0, 1)$.

The moment generating function for the normal distribution can be obtained as:

$$\mathbf{M}_X(t) = \int_{-\infty}^{\infty} e^{tx} \frac{1}{\sigma\sqrt{2\pi}} e^{\left[-\frac{1}{2}\left(\frac{x-\mu}{\sigma}\right)^2\right]} dx$$

By completing the squares in the exponent, we have

$$\mathbf{M}_X(t) = e^{\mu t + \frac{1}{2}\sigma^2 t^2} \left\{ \frac{1}{\sigma\sqrt{2\pi}} \int_{-\infty}^{\infty} e^{-\frac{1}{2\sigma^2}[x-(\mu+\sigma^2 t)]^2} dx \right\}$$

$$= \exp[\mu t + \sigma^2 t^2/2]$$

because the quantity within the braces is the integral from $-\infty$ to ∞ of a normal density with mean $(\mu + \sigma^2 t)$, variance σ^2 and therefore equals 1.

If we set $\mu = 0$ and $\sigma^2 = 1$, we would have the corresponding moment generating function of the standardized normal variate Z. That is,

$$\mathbf{M}_Z(t) = e^{\frac{t^2}{2}} \tag{2.5}$$

Categorical data analysis deals primarily with discrete data (counts or frequencies) and hence are described by discrete type distributions. We will therefore, start by examining the various distributions that we will be considering in this book. These are discussed in the next section.

2.3 Probability Models for Discrete Data

2.3.1 The Binomial Distribution

Consider an experiment Ω consisting of a sequence of n independent trials in which the outcome at each trial is binary or dichotomous. At each trial, the probability that a particular event (A)-success occurs is π and the probability that the event does not occur (failure) is therefore $(1 - \pi)$. We shall also assume that this probability of success is constant from one trial to another. If the number of times the event A occurs in the sequence of n trials is represented by the random variable X, then, X is said to follow the binomial distribution with parameters n and π (fixed). The probability distribution for X is:

$$p[X = x] = \begin{cases} \binom{n}{x} \pi^x (1-\pi)^{n-x} & x = 0, 1, \cdots, n \\ 0 & \text{otherwise} \end{cases} \tag{2.6}$$

where $0 < \pi < 1$.

It can be shown that the moment generating function for the binomial distribution is given by: (See problem 4 in the exercises)

$$\mathbf{M}_X(t) = [1 + \pi(e^t - 1)]^n. \tag{2.7}$$

Differentiating (2.7) with respect to t, we have, (see problem 4 in the exercises)

$$\mathbf{M}_X^{'}(t) = n\pi e^t[1 + \pi(e^t - 1)]^{n-1} \quad \text{and} \tag{2.8a}$$

$$\mathbf{M}_X^{''}(t) = n\pi e^t\{1 - \pi + n\pi e^t\}[1 + \pi(e^t - 1)]^{n-2} \tag{2.8b}$$

Evaluating these at $t = 0$, give

$$\mathbf{M}_X^{'}(t)\mid_{t=0} = \mu_1^{'} \qquad = n\pi \tag{2.9a}$$

$$\mathbf{M}_X^{''}(t)\mid_{t=0} = n\pi(1 - \pi + n\pi) \tag{2.9b}$$

These lead to:

$$\mu_1 = E(X) \quad = \mu_1^{'} \quad = n\pi \tag{2.10a}$$

$$\mu_2 = \text{Var}(X) \quad = \mu_2^{'} - \mu^2 \quad = n\pi(1 - \pi) \tag{2.10b}$$

Example

Suppose the probability that a person suffering from migraine headache will obtain relief with a certain analgesic drug is 0.8. Five randomly selected sufferers from migraine headache are given this drug. We wish to find the following probabilities that the number obtaining relief will be:

(a) Exactly zero (b) Exactly two (c) More than 3

(d) Four or fewer (e) three or four (f) At least 2

Solution: Let X be the binomial random variable, with parameters $n = 5$ and $\pi = 0.8$. We can usually solve the problems in (a) to (f) by the use of the cummulative distribution Function of the binomial random variable. The cdf $F(x) = P(X \leq x)$ for this problem is generated next with the accompanying SAS software program.

```
DATA BINOMIAL;
DO X=0 TO 5;
P=PROBBNML(0.8,5,X);
CUM=ROUND(P,.0001);
OUTPUT;
END;
DROP P;
PROC PRINT NOOBS;
RUN;
```

X	CUM
0	0.0003
1	0.0067
2	0.0579
3	0.2627
4	0.6723
5	1.0000

In the program just given, we have employed the SAS software function **probbnml(p,n,X)** for generating the cummulative distribution function of a binomial variable with the specified parameters. The **ROUND** statement asked SAS® to round the probabilities to four decimal places.

(a) The number obtaining relief will be exactly 0.

$$P(X = 0) = F(0) = 0.0003$$

(b) The number obtaining relief will be exactly 2.

$$\begin{aligned} P(X = 2) &= P(X \leq 2) - P(X \leq 1) = F(2) - F(1) \\ &= 0.0579 - 0.0067 \qquad\qquad = 0.0512 \end{aligned}$$

(c) The number obtaining relief will be more than 3.

$$P(X > 3) = 1 - P(X \leq 3) = 1 - F(3) = 1 - 0.2627 = 0.7373$$

(d) The number obtaining relief will be 4 or fewer.

$$P(X \leq 4) = F(4) = 0.6723$$

(e) The number obtaining relief will be 3 or 4.

$$\begin{aligned} P(3 \leq X \leq 4) &= P(X \leq 4) - P(X \leq 2) = F(4) - F(2) \\ &= 0.6723 - 0.0579 \qquad\qquad = 0.6144 \end{aligned}$$

(f) The number obtaining relief will be at least 2.

$$\begin{aligned} P(X \geq 2) &= 1 - P(X < 2) = 1 - P(X \leq 1) \\ &= 1 - F(1) \qquad\quad = 1 - 0.0067 \\ &= 0.9933 \end{aligned}$$

The mean and variance for the above problem are $\mu = np = 5 * 0.8 = 4$ and $\sigma^2 = np(1-p) = 5 * 0.8 * 0.2 = 0.8$, respectively. Further, the moment generating function from this problem is $M_X(t) = [1 + 0.8(e^t - 1)]^5$.

2.3.2 Asymptotic Properties

All the asymptotic properties considered in this and other sections in this chapter will depend heavily on the convergence results discussed in Appendix A at the end of this chapter. Readers are advised to first familiarize themselves with the notation and definition adopted therein.

For a binomial variate X, designated as $X \sim b(n, \pi), \quad 0 < \pi < 1$ with π fixed, then as $n \to \infty$

(i)

$$Z_1 = \frac{\frac{X}{n} - \pi}{\sqrt{\frac{\pi(1-\pi)}{n}}} \tag{2.11}$$

is distributed N(0,1) approximately for large n and moderate π. That is,

$$Z_1 \xrightarrow{d} N(0,1)$$

as $n \to \infty$, and π fixed (Central Limit Theorem). This states that Z_1 converges in distribution to N(0,1).

Z_1 can also be rewritten as:

$$Z_1 = \frac{X - n\pi}{\sqrt{n\pi(1-\pi)}} = \frac{x - \mu}{\sigma}$$

That is, as $n \to \infty$, the standardized binomial distribution approaches the standard normal distribution with mean 0 and variance 1.

To prove this, we note from property **(iii)** of the moment generating functions that

$$\mathbf{M}_{\left(\frac{X+\alpha}{\beta}\right)}(t) = e^{\left(\frac{\alpha}{\beta}t\right)}\mathbf{M}_X\left(\frac{t}{\beta}\right)$$

We thus have

$$\mathbf{M}_{Z_1}(t) = e^{-\frac{\mu t}{\sigma}}\left[1 + \pi(e^{t/\sigma} - 1)\right]^n$$

where $\alpha = -\mu$, and $\beta = \sigma$ and therefore

$$\begin{aligned}
\log M_{Z_1}(t) &= -\frac{\mu t}{\sigma} + n\log[1 + \pi(e^{t/\sigma} - 1)] \\
&= -\frac{\mu t}{\sigma} + n\log\left\{1 + \pi\left[\frac{t}{\sigma} + \frac{1}{2}\left(\frac{t}{\sigma}\right)^2 + \frac{1}{6}\left(\frac{t}{\sigma}\right)^3 + \cdots\right]\right\} \\
&= -\frac{\mu t}{\sigma} + n\left\{\pi\left[\frac{t}{\sigma} + \frac{1}{2}\left(\frac{t}{\sigma}\right)^2 + \cdots\right] - \cdots\right\}
\end{aligned}$$

Expanding e^x and $\log(1+x)$ in Maclurin's series, we have

$$\log M_{Z_1}(t) = \underbrace{\left(-\frac{\mu}{\sigma} + \frac{n\pi}{\sigma}\right)}_{''0''} t + \underbrace{\left(\frac{n\pi}{2\sigma^2} - \frac{n\pi^2}{2\sigma^2}\right) t^2}_{\text{“}1/2''.t^2} + \cdots$$

$$\to \frac{1}{2}t^2, \quad \text{as n} \to \infty$$

The coefficients of $t^k, k \geq 3$, converges to zero because of σ^k in the denominator. Thus we obtain

$$\lim_{n\to\infty} M_{Z_1}(t) = e^{\left(\frac{1}{2}t^2\right)}$$

We conclude therefore that as $n \to \infty$, the standardized binomial distribution approaches the standard normal distribution with mean 0 and variance 1.

(ii)

$$\frac{X}{n} \xrightarrow{p} \pi \tag{2.12}$$

as $n \to \infty$, and π fixed. (Weak law of large numbers which indicates that $\frac{X}{n}$ converges in probability to π)

Combining (2.11) and (2.12), we have

$$Z_2 = \frac{\frac{X}{n} - \pi}{\sqrt{\frac{\frac{X}{n}(1-\frac{X}{n})}{n}}} \xrightarrow{d} N(0,1) \tag{2.13}$$

$$Z_2 \xrightarrow{d} N(0,1)$$

If p is an estimator for π, then $p = \frac{X}{n}$ and a 95% confidence interval for π is approximately given by:

$$\left(p \pm 1.96\sqrt{\frac{p(1-p)}{n}}\right)$$

(iii) If we define the odds of an event A occurring as $\frac{P(A)}{1-P(A)}$. This is sometimes referred to as the odds in favor of A. Then, the log odds of an event A occurring is

$$\phi = \log \pi - \log(1-\pi) = \log\left(\frac{\pi}{1-\pi}\right)$$

Hence,

$$\pi = e^{\phi}/(1+e^{\phi})$$

The observed log odds are given by:

$$U = \log\left(\frac{x}{n}\right) - \log\left(1-\frac{x}{n}\right) = \log\left(\frac{\frac{x}{n}}{1-\frac{x}{n}}\right) = \log\left(\frac{x}{n-x}\right).$$

Thus, $U = \log\left(\frac{x}{n-x}\right)$, and approximately,

$$Z_3 = \frac{U-\phi}{[n\pi(1-\pi)]^{\frac{1}{2}}} \sim N(0,1)$$

Since it can be shown using the *delta* method with the linearized Taylor's expansion (see Appendix A.4) that

$$\text{Var}(U) = \frac{1}{n\pi(1-\pi)}$$

Hence,

$$Z_3 \to N(0,1)$$

as $n \to \infty$ with π remaining fixed, using (2.11) and the limit theorem.

(iv)

$$\begin{aligned} Z_4 &= \frac{U-\phi}{[n\frac{x}{n}(1-\frac{x}{n})]^{\frac{1}{2}}} \\ &= \frac{U-\phi}{\left[\frac{1}{x}+\frac{1}{n-x}\right]^{\frac{1}{2}}} \end{aligned} \tag{2.14}$$

since

$$U = \log x - \log(n-x) \quad \text{and}$$

$$\text{Var}(U) = \frac{1}{x} + \frac{1}{n-x}$$

This is a general result, and using Slutsky's theorem (from Appendix A) we have

$$Z_4 \xrightarrow{d} N(0,1)$$

as $n \to \infty$, and π remains fixed. An approximate 95% confidence interval for ϕ therefore is given by

$$\left[U \pm 1.96\left(\frac{1}{x}+\frac{1}{n-x}\right)^{\frac{1}{2}}\right]$$

The results just shown are due to Cox (1970) and ϕ is defined as the logit of π. On the other hand, if ϕ is defined in terms of the inverse normal probability integral as $\phi = \Phi^{-1}(\pi)$, then ϕ is in this case referred to as the *probit* of π (Finney, 1971). Similarly, if ϕ is defined as: $\psi = \arcsin\sqrt{\pi}$, we also have the arcsin (variance stabilizing) transformation.

Other Binomial Properties

Other properties of the binomial distribution include the following:

(a) If X_1 and X_2 are independent random variables having binomial distributions with parameters n_1, π_1 and n_2, π_2, respectively, then the sum $Y = X_1 + X_2$, that is, the sum of the random variables also has the binomial distribution with parameters $(n_1 + n_2), \pi_1$ provided $\pi_1 = \pi_2$ only.

The moment generating function (mgf) of $Y = X_1 + X_2$ is:

$$[1 + \pi(e^t - 1)]^{n_1}[1 + \pi(e^t - 1)]^{n_2} = [1 + \pi(e^t - 1)]^{n_1+n_2}$$

That is, the sum of the random variables also has the binomial distribution with parameters $(n_1 + n_2)$, π. If however, $\pi_1 \neq \pi_2$, then $Y = X_1 + X_2$ does not have a binomial distribution.

(b) The distribution of X_1, conditional on $X_1 + X_2 = x$ is the hypergeometric distribution:

$$P[X_1 = r \mid x] = \frac{\binom{n_1}{r}\binom{n_2}{x-r}}{\binom{n_1+n_2}{x}} \tag{2.15}$$

subject to restrictions $r \leq n_1$; $x - r \leq n_2$ and $\max(0, x - n_2) \leq r \leq \min(n_1, r)$.

2.3.3 The Poisson Distribution

The Poisson distribution is very important in the study of categorical data and is often a realization of a rare event. If $X_i \sim P(\mu)$, $i = 1, 2, \cdots$, then the probability density function for this random variable is given by:

$$P[X = x] = \frac{e^{-\mu}\mu^x}{x!}, \quad x = 0, 1, 2, \cdots \tag{2.16}$$

It can be easily shown that:

$$\frac{P[X = x + 1]}{P[X = x]} = \frac{\mu}{(x + 1)} \tag{2.17}$$

The moment generating function for X is obtained as follows:

$$\begin{aligned}
M_X(t) &= E(e^{tx}) \\
&= \sum_{x=0}^{\infty} e^{tx}\frac{\mu^x e^{-\mu}}{x!} \\
&= \sum_{x=0}^{\infty} \frac{(\mu e^t)^x e^{-(\mu e^t)} e^{[\mu(e^t-1)]}}{x!} \\
&= e^{[\mu(e^t-1)]}\sum_{x=0}^{\infty}\frac{\theta^x e^{-\theta}}{x!} \quad \text{with} \quad \theta = \mu e^t \\
&= e^{[\mu(e^t-1)]}
\end{aligned}$$

From the above expression for $M_X(t)$, it is not too difficult to show that

$$E(X) = \mu \qquad \text{and} \qquad \text{Var(X)} = \mu$$

Hence the standard error (sd) of X is $\sqrt{\mu}$.

Sums of Independent Poissons

If X_1 and X_2 are independent random variables having Poisson distributions with parameters μ_1 and μ_2 respectively, then $X = X_1+X_2$ also has a Poisson distribution with parameters $\mu = \mu_1 + \mu_2$. (Show using mgfs.)

An Important Conditional Property

Let X_1 and X_2 be each distributed independently as Poissons with parameters μ_1 and μ_2 respectively. Then the conditional distribution of X_1 given the value of $X_1 + X_2 = N$ is binomial with parameters that are based on (the observed values) of N and $\mu_1/(\mu_1 + \mu_2)$. That is,

$$P[X_1 = x_1 \mid X_1 + X_2 = N] = \frac{N!}{x_1!(N - x_1)!}\pi^{x_1}(1 - \pi)^{N-x_1}$$

where $\pi = \mu_1/(\mu_1 + \mu_2)$.

Proof:

Let $X_1 \sim P(\mu_1)$ and $X_2 \sim P(\mu_2)$, then given $X_1 + X_2 = N$

$$\begin{aligned}
P(X_1 = x_1) &= \frac{P[X_1 = x_1; X_2 = N - x_1]}{P[X_1 + X_2 = N]} \\
&= \frac{\frac{e^{-\mu_1}\mu_1^{x_1}}{x_1!} \cdot \frac{e^{-\mu_2}\mu_2^{(N-x_1)}}{(N-x_1)!}}{\frac{e^{-(\mu_1+\mu_2)}(\mu_1+\mu_2)^N}{N!}} \\
&= \frac{e^{-\mu_1}\mu_1^{x_1}}{x_1!} \cdot \frac{e^{-\mu_2}\mu_2^{N-x_1}}{(N - x_1)} / \frac{e^{-\mu}\mu^N}{N!} \\
&= \frac{N!}{x_1!(N - x_1)!}\left(\frac{\mu_1}{\mu_1 + \mu_2}\right)^{x_1}\left(\frac{\mu_2}{\mu_1 + \mu_2}\right)^{N-x_1} \\
&= \frac{N!}{x_1!(N - x_1)!}\pi^{x_1}(1 - \pi)^{N-x_1}
\end{aligned}$$

where $\mu = \mu_1 + \mu_2$ and $\pi = \mu_1/(\mu_1 + \mu_2)$.

Moreover, if X_1 and X_2 are independent nonnegative integer-valued random variables, and the preceding expression holds, then X_1 and X_2 must be Poisson variates with parameters in the ratio μ_1 to μ_2. This is referred to as the characterization property of the Poisson distribution and the result is due to Chatterji (1963).

In general, if we have n independent Poisson variates $X_1, \cdots, X_n$ with parameters $\mu_1, \cdots, \mu_n$, respectively, then the conditional distribution of $X_1, \cdots, X_n$ given $\sum_i^n X_i = N$ is multinomial.

2.3.4 Some Asymptotic Properties

(i) Suppose $X \sim P(\mu)$ and let us consider the case when μ is assumed large. Then

$$Z_1 = \frac{X - \mu}{\sqrt{\mu}} \sim N(0, 1) \tag{2.18}$$

That is,

$$Z_1 \xrightarrow{d} N(0,1)$$

as $\mu \to \infty$. That is, $X \sim N(\mu, \mu)$.

Proof:

Again, using the third result under the moment generating functions, we have for the Poisson distribution:

$$\begin{aligned}
\mathbf{M}_{Z_1}(t) &= \exp[-(\mu/\sqrt{\mu})t]\mathbf{M}_X(t/\sqrt{\mu}) \\
&= e^{-\sqrt{\mu}t}\exp\left[\mu(e^{t/\sqrt{\mu}} - 1)\right] \\
&= e^{-\sqrt{\mu}t}\exp\left[\mu\left(\frac{t}{\sqrt{\mu}} + \frac{1}{2}\frac{t^2}{\mu} + \frac{1}{6}\frac{t^3}{\mu\sqrt{\mu}} + \cdots\right)\right] \\
&= \exp\left[\frac{1}{2}t^2 + \frac{1}{6}\frac{t^3}{\sqrt{\mu}} + \cdots\right]
\end{aligned}$$

We have employed the Maclaurin's series expansion for e^x in the above expressions. From the preceding, we therefore have

$$\lim_{\mu \to \infty} \mathbf{M}_Z(t) = e^{(\frac{t^2}{2})}$$

That is, for large μ, the Poisson distribution approaches the normal with mean μ and variance μ.

(ii)

$$\frac{X}{\mu} \xrightarrow{p} 1 \tag{2.19}$$

as $\mu \to \infty$.

Let,

$$Z_2 = \frac{X - \mu}{\sqrt{X}} \sim N(0,1) \quad \text{for large } \mu \tag{2.20}$$

then, using (2.18), (2.19), and Slutsky's theorem, we have,

$$Z_2 \xrightarrow{d} N(0,1)$$

and since

$$Z_2 = \frac{Z_1}{(X/\mu)^{\frac{1}{2}}}$$

we have

$$Pr(-1.96 < \frac{X - \mu}{\sqrt{X}} < 1.96) \simeq 0.95$$

That is, the probability that the random interval $(X - 1.96\sqrt{X},\, X + 1.96\sqrt{X})$ contains μ is $\simeq 0.95$.

We may note here that, $\dfrac{1}{\mu}$ and $\dfrac{1}{X}$ are respectively called the Fisher's expected and likelihood information. As an example, consider the random variable X having the pdf $p(x;\theta)$. Then under regularity conditions, the Fisher's Information is defined as

$$-E\left[\frac{\partial^2 \log p(x)}{\partial\theta^2}\right]$$

Thus, for the Poisson distribution, we have the log-likelihood (L) as:

$$\begin{aligned}
L &= -\mu + x \log \mu - \log x! \\
\frac{\partial l}{\partial \mu} &= -1 + \frac{x}{\mu} \\
\frac{\partial^2 l}{\partial \mu^2} &= \frac{-x}{\mu^2} \\
I(\mu) &= -E\left[\frac{-x}{\mu^2}\right] \\
&= \frac{1}{\mu}
\end{aligned}$$

which is larger for values of μ closer to zero.

(iii) Let

$$Z_3 = \frac{\log X - \log \mu}{\mu^{\frac{1}{2}}} \xrightarrow{d} N(0,1) \quad \text{as } \mu \to \infty \tag{2.21}$$

This result follows from (2.18) and the limit theorem.

That is,

$$\log X \sim N\,(\log \mu,\, \mu^{-1})$$

approximately for large μ, and hence,

$$Z_4 = \frac{\log X - \log \mu}{X^{\frac{1}{2}}} \xrightarrow{d} N(0,1) \quad \text{as } \mu \to \infty \tag{2.22}$$

The preceding result follows from combining (2.19), (2.21) and certain limiting distribution theorem.

Thus,

$$(\log (X) - 1.96\sqrt{X},\, \log (X) + 1.96\sqrt{X})$$

is the approximate 95% confidence interval for large μ.

We note here that the Poisson probability model for counts $\{n_i\}$ assumes that they are independent Poisson variates. The joint probability function for $\{n_i\}$ is then the product of the probabilities for the k cells. The total sample size $n = \sum n_i$ also has a Poisson distribution with parameter $\sum \mu_i$.

2.4 The Multinomial Distribution

Here, a sample of n observations is classified into k mutually and exhaustive categories according to the specified underlying probabilities $\mathbf{\Pi} = (\pi_1, \pi_2, \pi_3, \cdots, \pi_k)$. Let n_i be the resulting observed frequencies in each of the k classes. We display this in Table 2.1.

The joint distribution for the random variables $n_1, n_2, ..., n_k$, the observed frequencies, is given by the *multinomial* distribution:

	Response Categories					
	1	2	3	$\cdots$	k	Totals
Probabilities	π_1	π_2	π_3	$\cdots$	π_k	1
Obs. Freq.	n_1	n_2	n_3	$\cdots$	n_k	n

Table 2.1: Table of observed frequencies with the underlying probability values

$$\begin{aligned} P(\mathbf{n}, \mathbf{\Pi}) &= \frac{n!\pi_1^{n_1}\pi_2^{n_2}\cdots\pi_k^{n_k}}{n_1!n_2!\cdots n_k!} \\ &= n!\prod_{i=1}^{k}\frac{\pi_i^{n_i}}{n_i!} \end{aligned} \tag{2.23}$$

where $\pi_i > 0$, $\sum \pi_i = 1$ and $\sum_{i=1}^k n_i = n$. $\mathbf{n} = \{n_1, n_2, \cdots, n_k\}$ is a random vector which can take on any value for which

(i) $0 \leq n_i \leq n$ for $i = 1, 2, \cdots, k$

(ii) $\sum n_i = n$. Thus we have linear dependence as a result of

(iii) $n_r = n - \sum_{i=1 \neq r}^{k} n_i$

The expression in (2.23) is referred to as the multinomial distribution.

2.4.1 Proof:

n_1 objects can be selected into n_1 positions in $\binom{n}{n_1}$ ways, after which we select the n_2 from the remaining $n - n_1$ into positions $\binom{n-n_1}{n_2}$ ways, and so on. Using the multiplication rule, we have therefore the number of distinguishable permutations of the n objects comprising of n_1 of one kind, n_2 of another kind, etc. as:

$$\begin{aligned} \binom{n}{n_1} \times \binom{n-n_1}{n_2} \times \binom{n-n_1-n_2}{n_3} \cdots &= \binom{n}{n_1, n_2, \cdots, n_k} \\ &= \frac{n}{n_1!n_2!\cdots n_k!} \end{aligned} \tag{2.24}$$

The expression in (2.24) is sometimes referred to as the *multinomial coefficient.* The joint probability density function of the linearly dependent random variables $(n_1, n_2, \cdots, n_k)$ is given by:

$$f(n_1, n_2, \cdots, n_{k-1}) = P(n_1, n_2, \cdots, n_{k-1})$$

where $n_k = n - \sum_{i=1}^{k-1} n_i$. Each particular arrangement of the n_is is from independent trials and has probability

$$\pi_1^{n_1}\pi_2^{n_2}\cdots\pi_k^{n_k}$$

Since the number of such arrangement is given by (2.24), hence it follows that the product of these expressions gives the joint pdf of the ns. That is,

$$f(n_1, n_2, \cdots, n_k) = \frac{n!\pi_1^{n_1}\pi_2^{n_2}\cdots\pi_k^{n_k}}{n_1!n_2!\cdots n_k!} \tag{2.25}$$

2.4.2 Moments of the Multinomial Distribution

The factorial moment generating function (fmgf) of a random variable X can be defined as $K(t) = E(t^X)$ when it exists for all real values of t in an open interval that includes the point $t = 1$. Then $K^m(1)$ is equal to the *mth factorial moment* defined as

$$K^m(1) = E[X(X-1)(X-2)\cdots(X-m+1)].$$

With this definition of factorial moments, it can be shown that the (fmgf) for the multinomial distribution is

$$\eta(t) = (\pi_1 t_1 + \pi_2 t_2 + \cdots + \pi_k t_k)^n \tag{2.26}$$

We can obtain the means, variances and covariances for each element and pairs of elements of the vector $\mathbf{n}$ from the first and second partial derivatives of $\eta(t)$ evaluated at $t_1 = t_2 = \cdots = t_k = 1$. Differentiating $\eta(t)$ in (2.26) with respect to t, we have

$$\frac{\partial \eta(t)}{\partial t_j} = n\pi_j(\pi_1 t_1 + \pi_2 t_2 + \cdots + \pi_k t_k)^{n-1} \tag{2.27a}$$

$$\frac{\partial^2 \eta(t)}{\partial t_j t_{j'}} = n(n-1)\pi_j \pi_{j'}(\pi_1 t_1 + \pi_2 t_2 + \cdots + \pi_k t_k)^{n-2} \tag{2.27b}$$

For the means, we evaluate (2.27a) at $t_1 = t_2 = \cdots = t_k = 1$, to have

$$\frac{\partial \eta(t)}{\partial t_j} \,|_{\{t_1=t_2=\cdots=t_k=1\}} = n\pi_j \quad j = 1, 2, \cdots, k \tag{2.28}$$

Similarly for the variances and covariances, we evaluate (2.27b) at $t_1 = t_2 = \cdots = t_k = 1$ to have

$$\frac{\partial^2 \eta(t)}{\partial t_j t_{j'}} \,|_{\{t_1=t_2=\cdots=t_k=1\}} = n(n-1)\pi_j \pi_{j'} \tag{2.29}$$

From the expression in (2.29), therefore, we have

(i) For $j \neq j'$

$$E(n_j n_{j'}) = n(n-1)\ \pi_j \pi_{j'}$$

so that

$$\begin{aligned} \text{COV}(n_j, n_{j'}) &= n(n-1)\pi_j\pi_{j'} - n^2\pi_j\pi_{j'} \\ &= -n\pi_j\pi_{j'} \end{aligned}$$

(ii) Similarly, for $j = j'$

$$E(n_j n_{j'}) = n(n-1)\pi_j^2$$

so that

$$\begin{aligned} \text{Var}(n_j) &= E[n_j(n_j-1)] + E(n_j) - [E(n_j)]^2 \\ &= n(n-1)\pi_j^2 + n\pi_j - (n\pi_j)^2 \\ &= n\pi_j(1-\pi_j) \end{aligned}$$

Note: Var(X) $= E[X(X-1) + E(X)] - [E(X)]^2$

Hence,

(a)

$$E(n_i) = n\pi_i \qquad i = 1, 2, \cdots, k$$

(b)

$$\begin{aligned} E(n_i n_j) &= n(n-1)\pi_i\pi_j, \quad \text{for} \quad i \neq j \\ \text{Cov}(n_i, n_j) &= -n\pi_i\pi_j \end{aligned}$$

(c) For $i = j$,

$$\text{Var}\{n_i\} = n\pi_i(1-\pi_i)$$

2.4.3 Special Case when $k = 2$

When $k = 2$, the multinomial reduces to the binomial case, and in this case,

- $E(n_1) = n\pi_1$, where $n = n_1 + n_2$
- $E(n_2) = n\pi_2 = n(1 - \pi_1)$, since $\pi_2 = 1 - \pi_1$ in this case
- $\text{Var}(n_1) = n\pi_1(1 - \pi_1)$
- $\text{Var}(n_2) = n\pi_1(1 - \pi_1)$
- $\text{Cov}(n_1, n_2) = -n\pi_1(1 - \pi_1) = -\text{Var}(n_1)$

2.4.4 Maximum Likelihood Estimation of Parameters

Let $N_1, N_2, ..., N_n$ denote a random sample of size n from a probability density function $p(\mathbf{n}; \pi)$, $\pi \in \Omega$, where the parameter space Ω is an interval. Then, the *likelihood function* (the joint pdf of $N_1, N_2, ..., N_n$) is

$$L(\pi) = p(n_1; \pi)p(n_2; \pi)...p(n_n; \pi)$$

Using (2.23), the likelihood function for the multinomial distribution becomes:

$$L = \frac{n!}{n_1!n_2! \cdots n_k!} \prod_{j=1}^{k} \pi_j^{n_j}$$

so that

$$\ell = \log n! - \sum \log n_j! + \sum n_j \log \pi_j$$

We seek to maximize the likelihood function of the given observed sample $n_1, n_2, ...$, n_n. However, since the same value of π would maximize both $L(\theta)$ and its logarithm, we therefore wish to minimize the above log-likelihood subject to the constraint $\sum \pi_j = 1$. The Lagrange's multipliers method is most appropriate for this situation. The method seeks to find relative maxima and minima for a function on which certain *constraints* are imposed. For this method, suppose we have a function $p(n, y)$ subject to the constraint $h(n, y) = 0$. Then, we construct a new function P of three variables defined by the following:

$$P(n, y, \lambda) = p(n, y) - \lambda h(n, y)$$

Partial derivatives of $P(n, y, \lambda)$ are then obtained, set to zeros, resulting in simultaneous equations whose solution (n_0, y_0, λ_0) correspond to the solution of p(n) with the given constraint.

If we now apply this method to the problem at hand, if we let

$$G = \sum_{j=1}^{k} n_j \ln \pi_j - \lambda(\sum \pi_j - 1)$$

then

$$\frac{\partial G}{\partial \pi_j} = \frac{n_j}{\pi_j} - \lambda$$

$$\frac{\partial G}{\partial \lambda} = \sum \pi_j - 1$$

Setting these to zero, we have

$$\frac{n_j}{\hat{\pi}_j} - \hat{\lambda} = 0$$
$$\Longrightarrow n_j = \hat{\lambda}\hat{\pi}_j$$

But $\sum \hat{\pi}_j = 1$, so that $n = \sum n_j = \hat{\lambda}$.

Hence,

$$\hat{\pi}_j = \frac{n_j}{\hat{\lambda}} = \frac{n_j}{n} = p_j.$$

and therefore, we note here that

$$\begin{aligned} E(p_j) &= E(\hat{\pi}_j) &&= \pi_j \\ \text{Var}(p_j) &= \frac{1}{n^2}\text{Var}(n_j) &&= \frac{\pi_j(1-\pi_j)}{n} \quad \text{and} \\ \text{Cov}(p_i, p_j) &= \frac{1}{n^2}\text{Cov}(n_i, n_j) &&= -\frac{\pi_i \pi_j}{n} \end{aligned}$$

These can be summarized in matrix form as: $E(\mathbf{P}) = \mathbf{\Pi}$

$$V_{\mathbf{P}} = \begin{bmatrix} \frac{\pi_1(1-\pi_1)}{n} & -\frac{\pi_1\pi_2}{n} & \cdots & -\frac{\pi_1\pi_k}{n} \\ \vdots & \frac{\pi_2(1-\pi_2)}{n} & \cdots & -\frac{\pi_2\pi_k}{n} \\ -\frac{\pi_1\pi_k}{n} & \cdots & \ddots & \frac{\pi_k(1-\pi_k)}{n} \end{bmatrix}$$

That is,

$$\mathbf{V_P} = \frac{1}{n}\begin{bmatrix} \pi_1(1-\pi_1) & -\pi_1\pi_2 & \cdots & -\pi_1\pi_k \\ \vdots & \pi_2(1-\pi_2) & \cdots & -\pi_2\pi_k \\ -\pi_1\pi_k & \cdots & \ddots & \pi_k(1-\pi_k) \end{bmatrix}$$

$$\mathbf{V_P} = \frac{1}{n}[\mathbf{D_\Pi} - \mathbf{\Pi}\,\mathbf{\Pi}']$$

where

$$\mathbf{D_\Pi} = \text{diag}[\pi_1, \pi_2, \cdots, \pi_k], \text{a diagonal matrix}$$

Estimator for V_P

p_j is an unbiased estimator for π_j with

$$\text{Var}(p_j) = \frac{\pi_j(1-\pi_j)}{n}$$

This implies that p_j is a consistent estimator for π_j. That is,

$$\text{Var}(p_j) = \frac{p_j(1-p_j)}{n}$$
$$\text{Cov}(p_j, p_{j'}) = -\frac{p_j p_{j'}}{n}$$

Hence,

$$\mathbf{V}_{\underline{\mathbf{P}}} = \frac{1}{n}[\mathbf{D}_{\underline{\mathbf{P}}} - \underline{\mathbf{P}}\,\underline{\mathbf{P}}']$$

with

$$\mathbf{D}_{\underline{\mathbf{P}}} = \text{diag}[p_1, p_2, \cdots, p_k], \text{a diagonal matrix}$$

2.5 The Hypergeometric Distribution

Consider a finite population of N subjects, where each subject is classified as either S (success) or F (failure). Suppose there are R of type S and therefore, $(N-R)$ of type F. Suppose we wish to draw a sample of size n from this population of size N (equivalent to sampling without replacement). Our aim is to find the probability of obtaining exactly r S's (successes) in the sample of size n.

	Sample		
Population	S	F	Totals
S	r	$R-r$	R
F	$n-r$	$N-n-R+r$	N-R
Totals	n	$N-n$	N

Table 2.2: The sampling procedure arranged as a 2×2 table

Let us denote the probability that $X = r$ successes in the sample of size n from the finite population containing R successes and N minus R failures by

$$P[X = r \mid n, N, R] = \frac{\binom{R}{r}\binom{N-R}{n-r}}{\binom{N}{n}}$$

where the values or r satisfy the constraint

$$\max\{0, n-(N-R)\} \leq r \leq \min\{n, R\} \quad .$$

2.5.1 Proof

The number of different samples of size n which can be drawn from the finite population of N subjects is

$$\binom{N}{n} = \frac{n!}{n!(N-n)!}.$$

Of these, the number of different ways of drawing (without replacement) the r sample successes out of the possible R population successes is $\binom{R}{r}$, and for each of these ways, there are $\binom{N-R}{n-r}$ different ways of selecting $(n-r)$ failures from the $(N-R)$ population failures.

Thus the proportion of samples of size n having exactly r successes is given by:

$$Pr\{X = r \mid n, N, R\} = \frac{\binom{R}{r}\binom{N-R}{n-r}}{\binom{N}{n}}$$

Example

If $N = 10$, $R = 5$ and $n = 6$, then $N - R = 5$. The extreme samples are SSSSSF and FFFFFS. That is, min $= 1$ and max $= 5$.

As discussed earlier in section 2.3.2, if X_1 and X_2 are two independent binomial variates with parameters (n_1, π) and (n_2, π) respectively, then $X_1 + X_2$ has a binomial distribution with parameters $N = n_1 + n_2$ and π. The conditional distribution of X_1 given $X_1 + X_2 = n$ is the *hypergeometric distribution* and is derived as follows: From Table 2.3, $n_2 = N - n_1$ and if we let $n = X_1 + X_2$, then we have

Outcome	Population I	Population II	Totals
S	x	$n-x$	n
F	n_1-x	n_2-n+x	$N-n$
Totals	n_1	n_2	N

Table 2.3: The distribution arranged as a 2×2 table

$$\begin{aligned} P[X_1 = x \mid n] &= \frac{P[X_1 = x; X_2 = n-x]}{P[X_1 + X_2 = n]} \\ &= \frac{\left\{\binom{n_1}{x}\pi^x(1-\pi)^{(n_1-x)}\right\}\left\{\binom{n_2}{n-x}\pi^{(n-x)}(1-\pi)^{n_2-(n-x)}\right\}}{\binom{n_1+n_2}{n}\pi^n(1-\pi)^{n_1+n_2-n}} \\ &= \frac{\binom{n_1}{x}\binom{n_2}{n-x}}{\binom{N}{n}} \end{aligned} \tag{2.30}$$

since $N = n_1 + n_2$. Notice that all the π and $(1-\pi)$ terms cancel in both the numerator and denominator in the above expression.

The above is the hypergeometric distribution with constraints $x \leq n_1$, $n - x \leq n_2$. This distribution is used to test therefore the equality of two binomial probabilities, and as observed above, this test does not depend on their common value. Rearranging (2.30), we have

$$P[X_1 = x \mid X_1 + X_2 = n] = P[X_1 = x \mid N, n_1, n] = \frac{\binom{n}{x}\binom{N-n}{n_1-x}}{\binom{N}{n_1}} \tag{2.31}$$

with the constraints $\max(0, n_1+n-N) \leq X \leq \min(n_1, n)$. The expression in (2.31) can be rewritten using Table (2.3) as:

$$P[X_1 = x \mid N, n_1, n] = \frac{n!(N-n)!n_1!n_2!}{x!(n-x)!(n_1-x)!(n_2-n+x)!N!}$$

From (2.30) and (2.31), the mean of X are respectively given by:

$$E(X) = \frac{Rn}{N} \text{ and } \frac{nn_1}{N}$$

The corresponding variances are also respectively given by:

$$\text{Var}(X) = \frac{Rn}{N}\frac{(N-R)(N-n)}{N(N-1)} \quad \text{and} \quad \frac{nn_1}{N}\frac{(N-n_1)(N-n_1)}{N(N-1)}$$

2.5.2 Generalizations of the Hypergeometric

1. Our first generalization of the hypergeometric distribution relates to the case when the assumption that the two binomial probabilities are equal is no longer applicable. This leads to the *extended hypergeometric distribution.* Here again, as in the previous section, if X_1 and X_2 are two independent binomial variates with parameters (n_1, π_1) and (n_2, π_2) respectively such that $\pi_1 \neq \pi_2$, the distribution of $X_1 + X_2$ no longer has a binomial distribution. In this case, the conditional distribution of X_1 given $X_1 + X_2 = n$ is derived as follows:

From the above, $n_2 = N - n_1$ and again if we let $n = X_1 + X_2$, we have

$$\begin{aligned} P[X_1 = x \mid n] &= \frac{P[X_1 = x; X_2 = n - x]}{P[X_1 + X_2 = n]} \\ &= \frac{\left\{\binom{n_1}{x}\pi_1^x q_1^{(n_1-x)}\right\}\left\{\binom{N-n_1}{n-x}\pi_2^{(n-x)}q_2^{N-n_1-(n-x)}\right\}}{\sum_j \left\{\binom{n_1}{j}\pi_1^j q_1^{n_1-j}\right\}\left\{\binom{N-n_1}{n-j}\pi_2^{n-j}q_2^{N-n_1-n+j}\right\}} \end{aligned} \tag{2.32}$$

where $q_1 = (1 - \pi_1)$, $q_2 = (1 - \pi_2)$, and the denominator is the sum of the numerator over all possible values of x, which are constrained by:

$$\max(0; n_1 + n - N) \leq x \leq \min(n_1, n)$$

If we denote the log-odds ratio of the two binomial probabilities by

$$\lambda = \log\left\{\frac{\pi_1(1 - \pi_2)}{\pi_2(1 - \pi_1)}\right\}$$

then the probability expression in (2.32) can be written as:

$$P[X_1 = x \mid N, n_1, n, \lambda] = \frac{\binom{n_1}{x}\binom{N-n}{n-x}e^{\lambda x}}{\sum_j \binom{n_1}{j}\binom{N-n_1}{n-j}e^{\lambda j}} \tag{2.33}$$

When $\lambda = 0$, the expression in (2.33) reduces to those in (2.30). The expression in (2.33) is sometimes written as:

$$P[X_1 = x \mid N, n_1, n, \lambda] = \frac{Ce^{\lambda x}}{x!(n_1 - x)!(n - x)!(N - n_1 - n + x)!}$$

The factorials are those in the 2×2 table cells in Table 2.3. C is a function of (λ, N, n_1, n), which is independent of x. The probability function in (2.33) has been referred to as the *extended* or *noncentral* hypergeometric distribution with noncentrality parameter λ. If we instead write $\psi = e^{\lambda}$ representing the odds, rather than the log odds of the binomial probabilities, then the probability distribution in (2.33) becomes

$$P[X_1 = x \mid N, n_1, n, \psi] = \frac{C\psi^x}{x!(n_1 - x)!(n - x)!(N - n_1 - n + x)!}$$

Example

Consider again the above example in which $N = 10, n = 5$, and $n_1 = 6$; then $N - n = 5$.

	Population		
Outcomes	I	II	Totals
S	x	$5 - x$	5
F	$6 - x$	$x - 1$	5
Totals	6	4	10

Table 2.4: The distribution arranged as a 2×2 table

Here again, min = 1 and max = 5. $n_{11} = 2$ and $\lambda = \log(2/12)$. We give an SAS® program and its corresponding output for generating the hypergeometric distribution.

```
data hyper;
n=10; k=5; m=6; i2=min(k,m);
i1=max(0,(k+m-n));
sum=0.0;
do i=i1 to i2;
if i-1 lt i1 then prob=probhypr(n,k,m,i);
    else prob=probhypr(n,k,m,i)-probhypr(n,k,m,i-1);
sum=sum+prob;
output; end;
proc print data=lab2 noobs;
var i prob sum;
format prob sum 8.4;
run; quit;
```

i	prob	sum
1	0.0238	0.0238
2	0.2381	0.2619
3	0.4762	0.7381
4	0.2381	0.9762
5	0.0238	1.0000

The corresponding extended hypergeometric distribution when we assume that $x = 2$, that is, the odds ratio in this case being $2/12$, is similarly generated with the SAS® program below.

```
set hyper;
u=2/12;
do i=i1 to i2;
if i-1 lt i1 then prob=probhypr(n,k,m,i,u);
    else prob=probhypr(n,k,m,i,u)-probhypr(n,k,m,i-1,u);
sum=sum+prob;
output;
end;
proc print data=hyper noobs;
var i prob sum;
format prob sum 8.4;
run;
quit;
```

Extended Hypergeometric distribution

i	prob	sum
1	0.3059	0.3059
2	0.5098	0.8157
3	0.1699	0.9856
4	0.0142	0.9998
5	0.0002	1.0000

The extended hypergeometric is most useful for power calculations and sample size determination.

2. Another extension of the hypergeometric distribution relates its extension to the general $I \times J$ contingency tables: that is, the multivariate case. This can further be extended to cases involving more than two dimensions.

2.6 Generalized Linear Models

The generalized linear models (GLM) are a restricted class of the usual linear models used for regression analysis and the analysis of variance. This class of models is well treated in the book by McCullagh and Nelder (1989). SAS® PROC GENMOD, SPSS PROC GENLOG and the statistical package GLIM (generalized linear interactive modeling) is based on the principles of GLM. The term *generalized linear model* is due to Nelder and Wedderburn (1972) while applying Fisher's scoring method to obtain maximum likelihood estimates for exponentially distributed variables.

The general linear model assumes having an observation y_i, with $\mathbf{x}'_{i1}$ being the vector of explanatory variables, and if $\boldsymbol{\beta} = (\beta_1, \cdots, \beta_p)$ represents the p-column vector of unknown parameters, then the linear model is written in the form:

$$\mathbf{y} = \mathbf{x}'\boldsymbol{\beta}$$

We know from the assumptions of linear models that the y_i are independently distributed normal with means $x_i\beta'_i$ and variance σ^2. That is,

$$y_i \sim N(x'_i\beta, \sigma^2), \quad i = 1, 2, \cdots, n$$

with $E(y_i) = x'_i\beta$.

The GLM extends the classical linear model:

- to include a broader class of distributions and gives a
- more general relationship between $E(\mathbf{Y})$ and the linear combination of predictors $\mathbf{X}\boldsymbol{\beta}$.

The GLM is described by three components, namely:

1. a *random* component: which describes the distribution of the response variable (continuous or categorical);
2. a *systemic* component: which describes how the covariates enter into the model for the mean; and
3. a *link* between the random and systematic components: which describes the functional relationship between the mean response and the systematic component.

2.6.1 Random Component

The responses, $y_1, y_2, ...y_n$ can be a random sample from any distribution within the *exponential family*. Distributions belonging to this category are the normal, the binomial, the Poisson, and the gamma, among many others. The exponential family distributions all depend on a vector of parameters $\boldsymbol{\theta}$ whose log-likelihood can be written in the form:

$$L = -\frac{r(\boldsymbol{\theta})}{a(\phi)} + \frac{\boldsymbol{\theta} y}{a(\phi)} + \log[h(\phi, y)] \tag{2.34}$$

where

- $\boldsymbol{\theta}$=natural (*canonical*) parameters of the distribution (more explicitly, $\boldsymbol{\theta}(\mu)$)
- ϕ = (nuisance) scale parameter
- a(), r() and h() characterize the particular member of the exponential family

Note: Often $a(\phi_i)$ is of the form ϕ/ω_i, where the weights (or prior weights) (ω_i) are known and vary from one observations to another. $\boldsymbol{\theta}$ is a scalar (for the one-parameter exponential family) and ϕ is a constant parameter. The GLM belongs to the one-parameter exponential family of distributions and consequently, the pdf is of the form

$$f(y \mid \theta; \phi) = \exp\left[-\frac{r(\theta)}{a(\phi)}\right] \exp\left[\frac{\theta}{a(\phi)} y\right] h(\phi, y) \tag{2.35}$$

The particular form of the probability density distribution $f(y \mid \theta, \phi)$ in (2.35) is chosen so that the maximum likelihood estimate of θ does not depend on ϕ.

The family of distribution in (2.35) has an expected value of Y depending only on θ but not on ϕ.

From the log-likelihood function of $f(y \mid \theta; \phi)$ given in (2.34), we have:

$$\frac{\partial L}{\partial \theta} = -\frac{r'(\theta)}{a(\phi)} + \frac{y}{a(\phi)} \quad \text{and} \tag{2.36}$$

$$\frac{\partial^2 L}{\partial \theta^2} = -r''(\theta)/a(\phi). \tag{2.37}$$

Let us show first that the following well known results apply to exponential family distributions. Consider for example, the Poisson distribution.

$$\begin{aligned} L(y, \mu) &= -\mu + y \ln \mu - \ln(y!) \\ \frac{\partial L}{\partial \mu} &= -1 + \frac{y}{\mu} \\ \frac{\partial^2 L}{\partial \mu^2} &= -\frac{y}{\mu^2} \end{aligned}$$

Hence,

$$E\left(\frac{\partial L}{\partial \mu}\right) = -1 + \frac{\mu}{\mu} = 0$$

Similarly,

$$E\left(\frac{\partial^2 L}{\partial \mu^2}\right) + E\left(\frac{\partial L}{\partial \mu}\right)^2 = -\frac{\mu}{\mu^2} + \frac{1}{\mu^2} E[y^2 - \mu^2] = -\frac{1}{\mu} + \frac{1}{\mu^2}[\mu + \mu^2 - \mu^2] = 0.$$

In general, therefore, for an exponential family of distributions, we have

$$E\left(\frac{\partial L}{\partial \theta}\right) = 0 \tag{2.38}$$

$$E\left(\frac{\partial^2 L}{\partial \theta^2}\right) + E\left(\frac{\partial L}{\partial \theta}\right)^2 = 0 \tag{2.39}$$

Applying these to our problem, we have:

$$E(Y) = \mu \qquad = r'(\theta), \quad \text{and} \tag{2.40a}$$

$$\operatorname{Var}(Y) = V(\mu)\, a(\phi) \qquad = r''(\theta)\, a(\phi) \tag{2.40b}$$

where $r'(\theta)$ and $r''(\theta)$ are, respectively, the first and second derivatives (with respect to θ) and $V(\mu)$ is often called the *variance function*, which describes how the variance is related to the mean.

2.6.2 Systematic Component

The linear predictor: $\boldsymbol{\eta} = \mathbf{X}\boldsymbol{\beta}$ is a linear function of covariates:

$$\eta_i = \beta_0 + \beta_1 x_{1i} + \ldots + \beta_k x_{pi} + \epsilon_i$$

$$\boldsymbol{\eta} = \begin{bmatrix} \eta_1 \\ \eta_2 \\ \vdots \\ \eta_n \end{bmatrix}$$

2.6.3 Link Functions

The link function g(.), describes the relationship between $E(y_i) = \mu_i$ and the linear predictor:

$$\eta_i = \mathbf{x}_i^{'}\boldsymbol{\beta} = g(\mu_i)$$

(Any monotonic differentiable function can be a link.)

Example: Binomial proportion:

$$g(\mu_i) = g(\pi_i) = \log\left(\frac{\pi_i}{1-\pi_i}\right)$$

$$g(.) = \text{logit function}$$

The canonical link is the link that relates μ_i to the canonical parameter $\theta_i(\mu_i)$ and is often used in practice. That is, g(.) is called the *canonical link* function. The density $f(y \mid \theta, \phi)$ is called the error function (as used in GENMOD) and the parameter ϕ is called the *dispersion parameter.*

We consider four examples of the error functions, namely, the normal, the Poisson, the binomial, and the gamma, in the next section.

The Normal Error Function

If $y \sim N(\mu, \sigma^2)$, then

$$f(y \mid \mu; \sigma^2) = \frac{1}{\sqrt{2\pi\sigma^2}}\exp\left[-\frac{(y-\mu)^2}{2\sigma^2}\right]$$
$$= \exp(-\mu^2/2\sigma^2)\exp(\mu y/\sigma^2)\left(\frac{1}{\sqrt{2\pi\sigma}}e^{-y^2/2\sigma^2}\right)$$

Consequently,

$$L = \frac{-\mu^2}{2\sigma^2} + \frac{\mu y}{\sigma^2} + \log\{(1/\sqrt{2\pi\sigma})e^{-y^2/2\sigma^2}\} \tag{2.41}$$

Comparing the above with (2.34), we have

$$\theta = \mu; \qquad \phi = \sigma^2; \quad h(\phi, y) = (1/\sqrt{2\pi\sigma})e^{-y^2/2\sigma^2}$$
$$a(\phi) = \phi; \quad r(\theta) = \frac{\mu^2}{2}$$

Since E(Y) $= \mu$, the canonical linear structure is $\theta = \mu = \mathbf{x}'\boldsymbol{\beta}$, that is, a standard linear model. This is often referred to as the *identity link* because $g(\mu) = \mu$.

The Poisson Error Function

For Poisson data, $Y \sim \text{Pois}(\mu)$ and the pdf is given by:

$$\begin{aligned} f(y \mid \mu) &= \frac{\mu^y e^{-\mu}}{y!}, \quad y = 0, 1, 2, \cdots \\ &= \exp(-\mu)\exp(y\ln(\mu))(y!)^{-1} \end{aligned}$$

Hence,

$$L = -\mu + y\log(\mu) - \log(y!) \tag{2.42}$$

Here again, comparing with (2.34), we have

$$\theta = \log(\mu); \qquad \phi = 1; \quad h(\phi, y) = (y!)^{-1}$$
$$a(\phi) = 1; \qquad r(\theta) = \mu$$

We know from distribution theory that if $Y \sim \text{Pois}(\mu)$, then $\text{E(Y)} = \mu$ and $\text{Var(Y)} = \mu$. From the above, $\mu = \text{E(Y)} = r'(\theta)$ and with $a(\phi) = 1$, $r(\theta) = \mu = e^\theta$ and thus $r'(\theta) = e^\theta$ and, $r''(\theta) = e^\theta$ and the canonical linear structure is $\theta = \log(\mu) = \mathbf{x}'\boldsymbol{\beta}$. The canonical link therefore is the *log* and leads to a standard *log-linear model* for Poisson data.

The Binomial Error Function

If $Y \sim b(n, \pi)$, with n known, then the pdf is given by

$$\begin{aligned} f(y \mid \pi) &= \binom{n}{y}\pi^y(1-\pi)^{n-y} \\ &= \binom{n}{y}(1-\pi)^n\left(\frac{\pi}{1-\pi}\right)^y \\ &= (1-\pi)^n\exp\left[y\log\left(\frac{\pi}{1-\pi}\right)\right]\binom{n}{y} \end{aligned}$$

This again leads to

$$L = n\log(1-\pi) + y\log\left(\frac{\pi}{1-\pi}\right) + \log\binom{n}{y} \tag{2.43}$$

from which again, on comparing with (2.34), we have

$$\theta = \log\left(\frac{\pi}{1-\pi}\right) \qquad \phi = 1 \qquad h(\phi, y) = \binom{n}{y}$$
$$a(\phi) = 1 \qquad r(\theta) = -n\log(1-\pi)$$

Here again, $\text{E(Y)} = \mu = n\pi$, so that the canonical linear structure is $\theta = \log\left(\frac{\pi}{1-\pi}\right) = \log\left(\frac{\mu}{n-\mu}\right) = \mathbf{x}'\boldsymbol{\beta}$. The canonical link therefore is the *logit*.

The Gamma Error Function

The gamma distribution with parameters α and λ, written $G(\alpha, \lambda)$, has the pdf given by

$$\begin{aligned} f(y \mid \alpha, \lambda) &= \frac{\lambda^\alpha}{\Gamma(\alpha)}e^{-\lambda y}y^{\alpha-1}, \quad \text{for } y > 0 \quad \text{and } \alpha > 0 \\ &= \left(\frac{\lambda}{\alpha}\right)^\alpha \exp\left[\alpha(\frac{-\lambda}{\alpha})y\right]\left(\frac{\alpha^\alpha}{\Gamma(\alpha)}y^{\alpha-1}\right) \end{aligned}$$

Therefore,

$$L = \alpha \log\left(\frac{\lambda}{\alpha}\right) + \alpha(\frac{-\lambda}{\alpha})y + \log\left(\frac{\alpha^\alpha}{\Gamma(\alpha)}y^{\alpha-1}\right) \tag{2.44}$$

Comparing the above with (2.34) again, we have

$$\begin{array}{lll} \theta = -\lambda/\alpha & \phi = \alpha & h(\phi, y) = \alpha^\alpha y^{\alpha-1}/\Gamma(\alpha) \\ a(\phi) = 1/\alpha & r(\theta) = -\log(\lambda/\alpha) & \end{array}$$

with $E(Y) = \mu = \alpha/\lambda = -1/\theta$ so that the canonical link structure is $\theta = \frac{-1}{\mu} = \mu^{-1} = \mathbf{x}'\boldsymbol{\beta}$. The distribution is only defined for $y > 0$. Thus for any gamma distribution, $\mu > 0$, an adequate restriction must therefore be placed on the parameter vector $\boldsymbol{\beta}$ to ensure that the expected value is positive.

The canonical generalized linear model for n independent observations with gamma distribution is (Christensen, 1991) given by:

$$\begin{aligned} y_i &\sim \Gamma(\alpha, \lambda_i) \\ E(y_i) = \mu_i &= \frac{\alpha}{\lambda_i} \\ \frac{-1}{\mu} &= \mu^{-1}\,\mathbf{x}_i'\boldsymbol{\beta} \end{aligned}$$

where $\boldsymbol{\beta}$ is restricted so that $-\mathbf{x}_i'\boldsymbol{\beta} > 0$ for all i. The gamma distribution regression has been found to be very useful in modeling situations in which the coefficient of variation is constant: that is, when

$$\frac{\sqrt{\text{Var}(y_i)}}{E(y_i)} = \frac{\sqrt{\alpha/\lambda_i^2}}{\alpha/\lambda_i} = \frac{1}{\sqrt{\alpha}}.$$

As pointed out by Christensen (1987) and McCullagh and Nelder (1989), when the coefficient of variation is constant in a set of data, such data are often adequately fitted by using the gamma distribution on the logs of the data.

2.6.4 Summary of Canonical Links

The distributions discussed in the preceeding section each have distinct link functions $g(\mu)$. These functions are called the *canonical links* when $g(\mu) = \theta$, where again,

$$g(\mu_i) = \mathbf{x}_i'\boldsymbol{\beta} = \sum x_i\beta_i$$

From the preceding results, we can summarize the canonical link functions for the distributions considered in the preceeding section in Table 2.5.

2.6.5 Parameter Estimation for the GLM

Again, consider a single observation (the i-th) from an exponential family of distributions with a random variable Y given in (2.35) with the log-likelihood given as:

$$L = -\frac{r(\theta_i)}{a(\phi)} + \frac{\theta_i y_i}{a(\phi)} + \ln[h(\phi, y_i)]$$

and

Distributions	Canonical link	Name	ϕ	$\mu(\theta)$
Normal	$g(\mu)=\mu$	*Identity*	σ^2	θ
Poisson	$g(\mu)=\log\,\mu$	*Log*	1	e^{θ}
Binomial	$g(\mu)=\log\left(\frac{\mu}{n-\mu}\right)$	*Logit*	1	$e^{\theta}/(1+e^{\theta})$
Gamma	$g(\mu)=\mu^{-1}$	*Reciprocal*	$1/\lambda$	$-1/\theta$

Table 2.5: Canonical links for some distributions

$$\frac{\partial L}{\partial \theta_i} = [-r'(\theta_i) + y_i]/a(\phi) = (y_i - \mu_i)/a(\phi)$$

since $\mu = r'(\theta_i)$. Also,

$$\begin{aligned} \frac{\partial L}{\partial \mu_i} &= \frac{\partial L}{\partial \theta_i} / \frac{\partial \theta_i}{\partial \mu_i} \\ &= \frac{(y_i - \mu_i)/a(\phi)}{r''(\theta_i)} \qquad = \frac{(y_i - \mu_i)/a(\phi)}{V(Y_i)/a(\phi)} \\ &= \frac{y_i - \mu_i}{V} \end{aligned}$$

since

$$\frac{\partial \mu_i}{\partial \theta_i} = r''(\theta_i) - V/a(\phi)$$

Similarly,

$$\frac{\partial L}{\partial g} = \frac{\partial L}{\partial \mu_i} \frac{d\mu_i}{dg}$$

But

$$g(\mu_i) = x_i \beta_i$$

hence, $\dfrac{\partial g}{\partial \beta_i} = x_i$. And finally, we can write

$$\frac{\partial L}{\partial \beta_i} = \frac{\partial L}{\partial g} \frac{\partial g}{\partial \beta_i} = \frac{\partial L}{\partial \mu_i} \frac{\partial \mu_i}{\partial g} x_i = \left(\frac{y_i - \mu_i}{V} \right) \frac{\partial \mu_i}{\partial g} x_i$$

For n independent observations therefore , the maximum likelihood equations for the β_i terms are given by:

$$\sum \left(\frac{y_i - \mu_i}{V} \right) \frac{\partial \mu_i}{\partial g} x_i = \sum W (y_i - \mu_i) \frac{dg}{d\mu_i} x_i \quad = 0 \tag{2.45}$$

where

$$W = \frac{1}{V} \left(\frac{\partial \mu}{\partial g} \right)^2 = V^{-1} \left(\frac{\partial \mu}{\partial g} \right)^2 \tag{2.46}$$

The above is generally referred to as the *likelihood equation*, and these equations are nonlinear functions in the parameters $\boldsymbol{\beta}$.

2.6.6 Fisher's Scoring Algorithm

Consider the Hessian matrix,

$$
\begin{aligned}
E\left(\frac{\partial^2 l}{\partial\beta_i\partial\theta_j}\right) &= E\left(\frac{\partial}{\partial\beta_j}\left[\sum(Y-\mu_i)V^{-1}\frac{\partial\mu_i}{\partial g}x_i\right]\right)\\
&= E\left(\sum(Y-\mu_i)\frac{\partial}{\beta_j}\left[V^{-1}\frac{\partial\mu_i}{\partial g}x_i\right]+\sum\frac{\partial}{\partial\beta_j}(Y-\mu_i)V^{-1}\frac{\partial\mu_i}{\partial g}x_i\right)\\
&= -\sum V^{-1}\left(\frac{\partial\mu_i}{\partial g}\right)^2 x_i x_j\\
&= -\sum W x_i x_j
\end{aligned}
$$

The above can be written in matrix notation as $-\mathbf{X}'\mathbf{W}\mathbf{X}$. Thus, the Fisher's information matrix $\mathbf{I}(\boldsymbol{\beta})$ defined by $E\left(-\frac{\partial^2 l}{\partial\beta_i\partial\beta_j}\right)$ is given by

$$\mathbf{I}(\boldsymbol{\beta}) = \mathbf{X}'\mathbf{W}\mathbf{X}$$

where $\mathbf{W}$, a diagonal matrix, is as previously defined in (2.46). The matrix $\mathbf{I}(\boldsymbol{\beta})$ is particularly important in ML estimation, since its inverse gives the asymptotic variance-covariance matrix of the ML parameters.

As observed before, the likelihood equations are non-linear functions of the $\beta's$, consequently, iterative methods are usually employed for the solutions of the equations. The Newton-Raphson iterative method is usually employed in which we shall assume that $\beta_i^{(t)}$ is the t-th approximation for the ML estimate of $\hat{\beta}_i$. Then, the N-R method based on Fisher (1935) scoring is for the $\mathbf{p}$ $\boldsymbol{\beta}$ parameters in matrix form is:

$$\mathbf{IM}^{(t)}\boldsymbol{\beta}^{(t+1)} = \mathbf{IM}^{(t)}\boldsymbol{\beta}^{(t)} + \mathbf{q}^{(t)} \tag{2.47}$$

where $\mathbf{IM}^{(t)}$ is the t-th approximation for the estimated information matrix evaluated at $\boldsymbol{\beta}^{(t)}$, $\mathbf{q}^{(t)}$ is the vector having elements $\partial l/\partial\beta_j$ also evaluated at $\boldsymbol{\beta}^{(t)}$.

McCullagh and Nelder (1989), and Agresti (1900) have shown that the RHS of equation (3.18) can be written in the form:

$$\mathbf{IM}^{(t)}\boldsymbol{\beta}^{(t)} + \mathbf{q}^{(t)} = \mathbf{X}'\mathbf{W}^{(t)}\mathbf{z}^{(t)}$$

where $\mathbf{W}^{(t)}$ is $\mathbf{W}$ estimated at $\boldsymbol{\beta}^{(t)}$ and where $\mathbf{z}^{(t)}$ is the dependent variate,

$$
\begin{aligned}
z_i^{(t)} &= \sum_i x_{ij}\beta_j^{(t)} + (y_i - \mu_i^{(t)})\left(\frac{\partial g_i^{(t)}}{\partial\mu_i^{(t)}}\right)\\
&= g_i^{(t)} + (y_i - \mu_i^{(t)})\left(\frac{\partial g_i^{(t)}}{\partial\mu_i^{(t)}}\right)
\end{aligned}
$$

The Fisher scoring algorithm reduces therefore to

$$(\mathbf{X}'\mathbf{W}^{(t)}\mathbf{X})\boldsymbol{\beta}^{(t+1)} = \mathbf{X}'\mathbf{W}^{(t)}\mathbf{z}^{(t)}$$

The above are the normal equations for a weighted least squares for fitting a linear model with a response variable $\mathbf{z}^{(t)}$, with design matrix of constants $\mathbf{X}$ and weight matrix $\mathbf{W}^{(t)}$. The corresponding weighted least squares solutions are:

$$\underline{\beta}^{(t+1)} = (\mathbf{X}'\mathbf{W}^{(t)}\mathbf{X})^{-1}\mathbf{X}'\mathbf{W}^{(t)}\mathbf{z}^{(t)} \tag{2.48}$$

where we regress for the t-th iteration cycle, $z_i^{(t)}$ on $x_1, x_2, \cdots, x_p$ with weights $\mathbf{W}^{(t)}$ to obtain a new estimate $\boldsymbol{\beta}^{(t+1)}$. These new estimates are then used to obtain a new *linear predictor* value $g^{(t+1)} = \mathbf{X}\boldsymbol{\beta}^{(t+1)}$ and a corresponding new adjusted dependent variable value $\mathbf{z}^{(t+1)}$, which will in turn, be used in the next cycle of iteration. The process is repeated until changes are small. This process has been described as the *iterative re weighted least squares.* We note here that the vector $\mathbf{z}$ is a linearized form of the link function applied to the data at $\boldsymbol{\mu}$ evaluated at $\mathbf{y}$. Hence, to first order approximation, we have

$$\begin{aligned} h(y_i) &\simeq h(\mu_i) + (y_i - \mu_i)h'(\mu_i) \\ &\simeq g_i + (y_i - \mu_i)(\partial g_i/\partial \mu_i) \end{aligned}$$

As pointed out in McCullagh and Nelder, the iterative process is usually started by using the data values $y_i, i = 1, 2, ..., n$ as the first estimate of $\hat{\mu}_0$, which in turn determines $\hat{g}_0$, $(\partial g/\partial \mu)\,|_0$, and V_0. The iterative process continues until differences in estimates between successive cycles are very small. Note that both the adjusted dependent variable $\mathbf{z}$ (it is sometimes referred to as the "working" dependent variable) and the weights $\mathbf{W}$ depend on the fitted values based on current estimates.

A more comprehensive literature on Fisher's scoring algorithm can be found in Finney (1971) or Green (1984).

For *canonical* links, both the Fisher scoring algorithm and the Newton-Raphson method are equivalent and if $a(\phi)$ is identical for all y_i, the like-lihood equations reduces to

$$\sum_i x_i y_i = \sum_i x_i \mu_i$$

Examples

1. For binomial data with parameters n and π, we have $E(Y) = n\pi_i = \mu_i$ and

$$V(Y_i) = n\pi_i(1 - \pi_i) = \mu_i(n - \mu_i)/n$$

The link is:

$$\mathrm{g}_i = \log\left(\frac{\mu_i}{n - \mu_i}\right) = \log(\mu_i) - \log(n - \mu_i)$$

Hence, $\partial g_i/\partial \mu_i = \dfrac{n}{\mu_i(n - \mu_i)}$ and the maximum likelihood equations in (2.45) become

$$\begin{aligned} \sum_i \left(\frac{(y_i - \mu_i)}{V}\right)\frac{\partial \mu_i}{\partial g_i} x_i &= \sum_i \frac{y_i - \mu_i}{\mu_i(n - \mu_i)/n} \times \frac{\mu_i(n - \mu_i)}{n} x_i \\ &= \sum_i (y_i - \mu_i)x_i \\ &= 0 \end{aligned}$$

Similarly, for Poisson data, we have shown that $E(Y_i) = \mu_i = V(Y_i)$ and the link $\mathrm{g}_i = \log \mu_i$, that is,

$$\log(\mu) = \mathbf{X}\beta$$

Hence, $\mu_i = e^{g_i}$ and $\partial \mu_i/\partial g_i = e^{g_i} = \mu_i$. Consequently, the likelihood equations become

$$\sum_i (y_i - \mu_i)x_i = 0.$$

In general, for exponential-family models, the likelihood equations reduce to the form given above for both binomial and Poisson cases. That is,

$$\sum_i (y_i - \mu_i)x_i = 0 \tag{2.49}$$

In this case, the Fisher's scoring algorithm is equivalent to the well known Newton's Raphson Iterative procedure. Most statistical packages, GLIM, SAS®, SPSS, EGRET have capabilities for fitting models to binary data and other data arising from exponentially distributed variables. Collett (1991) discussed the fitting of a generalized linear model to binomial data in its appendix.

2.7 Exercises

1. Given the multinomial probability model $P(\underline{n}) = \dfrac{n!\pi_1^{n_1}\pi_2^{n_2}\cdots\pi_k^{n_k}}{n_1!n_2!\cdots_k!}$ for the one-way classification having k categories, find the MLE of π_j, its variance and covariance.

2. Using results in (2.40a) and (2.40b), obtain the mean and variance for the binomial distribution with parameters n and θ.

3. Repeat the previous question for the Poisson distribution with parameter μ.

4. Show that the moment generating function for the binomial distribution with parameters n and θ is as given in equation (2.7).

5. (i) Give the name of the distribution of X, (ii) find the values of μ and σ^2 and calculate (iii) $P(2 \leq X \leq 4)$ when the mgf of X is given by:

   (a) $M(t) = (0.4 + 0.6e^t)^{10}$

   (b) $M(t) = e^{8(e^t - 1)}$

6. A prescription drug company claims that 15% of all new drugs that are shown to be effective in animal tests ever got marketed. Suppose the company currently has 10 new drugs that have been shown to be effective in animal tests:

   (a) Find the probability that none of the 10 drugs will ever be marketed.

   (b) Find the probability that at most four of the drugs get marketed.

7. In a random sample of 1500 adult Americans, the question was asked as to how many support the presidents proposals on the national health bill, 920 say they favor the proposals. Obtain a 95% confidence interval for the proportion of adult Americans who support the bill. Interpret your result. Also, use the procedure developed in section 1.2.2 to obtain a 95% confidence interval for U. Compare your two results.

8. A jar contains 25 pieces of candy, of which 11 are yogurt-covered nuts and 14 are yogurt-covered raisins. Let X equal the number of nuts in a random sample of 7 pieces of candy that are selected without replacement. (This example is from Hogg & Tanis, 1997) Find
   (a) $P(X = 3)$
   (b) $P(X = 6)$
   (c) The mean and variance of X.

9. Let X have the hypergeometric distribution given in (2.31). Find the probability of $X = 2$ given $N = 80, n = 4$, and $n_1 = 40$. Compare this answer to a binomial approximation for X with $n = 4$ and $\pi = 40/80$.

10. For the binomial distribution with parameters n and θ, show, using moment-generating function techniques that $Z_1 = \frac{X-n\theta}{\sqrt{n\theta(1-\theta)}}$ has asymptotically a standard normal distribution.

11. Use SAS® software for this problem. Let X have the hypergeometric distribution. Find the probability density function of X given that $N = 80, n = 4$, and $n_1 = 40$. Compare your answer to a binomial approximation for X with $n = 4$ and $\pi = 40/80$. Find $P(X = 2)$ in both cases.

12. Obtain the hypergeometric pdf of X, given that $N = 75, n = 3$, and $n_1 = 5$. Compare your results here to a Poisson approximation distribution with a mean $\lambda = \dfrac{(3 \times 5)}{75}$. Find $P(X = 1)$ again in both cases.

13. Find all possible outcomes that are consistent with this set of marginal totals:

		18
		9
17	10	27

14. Suppose that an electronic equipment contains eight transistors, three of which are defective. Four transistors are selected at random and inspected. Let X be the number of defective transistors observed in the sample, where $X = 0, 1, 2, 3, 4$. Find the probability distribution for X. Obtain E(X).

15. For the geometric distribution with probability density function

$$f(x|\theta) = \theta^x(1-\theta), \quad x \geq 0$$

Show that the geometric distribution belongs to the exponential family of distributions. Find its canonical parameter.

16. Consider the multinomial distribution (trinomial really) with probability parameters π_1, π_2, and π_3. Suppose we wish to test the hypothesis

$$H_0 : \frac{\pi_1}{\pi_2} = \frac{\pi_2}{\pi_3} = 2$$

Show that H_0 is equivalent to

$$\pi_1 = \frac{4}{7}, \quad \pi_2 = \frac{2}{7}, \quad \pi_3 = \frac{1}{7}$$

Chapter 3

One-Way Classification

3.1 The Multinomial Distribution

Suppose we are sampling from a population Ω which contains k types of objects for which π_i = P{an object selected at random is of type i} for $i = 1, 2, ..., k$. Now, suppose we draw a simple random sample of size n from Ω and classify the n objects into the k mutually and exhaustive classes or categories according to the specified underlying probabilities $\mathbf{\Pi} = (\pi_1, \pi_2, \pi_3, \cdots, \pi_k)$. Let n_i be the resulting observations in each of the k classes. These observed frequencies and the underlying probabilities are displayed in Table 3.1. Refer to discussion in Chapter 2 for more details.

	Response Categories					Totals
	1	2	3	$\cdots$	k	
Π	π_1	π_2	π_3	$\cdots$	π_k	1
Obs. Freq.	n_1	n_2	n_3	$\cdots$	n_k	n

Table 3.1: Table of observed frequencies with the underlying probability values

Since the categories are mutually exclusive and exhaustive, we have $\sum \pi_i = 1$ and $\sum n_i = n$. Thus, it can be shown that the probability mass function (called *the multinomial distribution*) for $\mathbf{n} = (n_1, n_2, \cdots, n_k)$ with $\mathbf{\Pi}$ as defined above is given by:

$$\begin{aligned} P(\mathbf{n}, \mathbf{\Pi}) &= \frac{n! \pi_1^{n_1} \pi_2^{n_2} \cdots \pi_k^{n_k}}{n_1! n_2! \cdots n_k!} \\ &= n! \prod_{i=1}^{k} \frac{\pi_i^{n_i}}{n_i!} \end{aligned} \tag{3.1}$$

where $\pi_i > 0$, $\sum \pi_i = 1$ and $\sum_{i=1}^{k} n_i = n$. Thus $\mathbf{n}$ is a random vector which can take on any value for which

(i) $0 \leq n_i \leq n$ for $i = 1, 2, ..., k$, and

(ii) $\sum n_i = n$.

Now suppose we have a null hypothesis of the form:

$$H_0 : \mathbf{\Pi}_0 = (\pi_{01}, \pi_{02}, \cdots, \pi_{0k})$$

with $\sum \pi_{0i} = 1$. The *exact multinomial probability* of the observed configuration is from (3.1) given by:

$$\begin{aligned} P(\mathbf{n}, \mid H_0) &= \frac{n! \pi_{01}^{n_1} \pi_{02}^{n_2} \cdots \pi_{0k}^{n_k}}{n_1! n_2! \cdots n_k!} \\ &= n! \prod_{i=1}^{k} \frac{\pi_{0i}^{n_i}}{n_i!} \end{aligned} \tag{3.2}$$

To test at an 100α % significance level, if the observed outcome $\mathbf{n}^0$ came from a population distributed according to H_0, the following steps are performed:

(a) For every other possible outcome $\mathbf{n}$, calculate the exact probability as in (3.2) above.

(b) Rank the probabilities from smallest to largest.

(c) Starting from the smallest rank, add the consecutive probabilities up to and including that associated with the observed configuration $\mathbf{n}^0$. This cumulative probability α', gives the chance of obtaining an outcome that is no more likely than $\mathbf{n}^0$ (that is, outcomes that are as extreme or more extreme than $\mathbf{n}^0$).

(d) Reject H_0 if $\alpha' \leq \alpha$ where α is specified a priori.

The above procedure was originally due to Roscoe and Byars (1971) and is known as *the exact test procedure.* This procedure, as we saw, is based on probability ranking. We shall explore other exact tests that are not based on probability ranking later in this chapter.

3.1.1 An Example

Consider the simple case when $k = 3$ and $n = 3$. Suppose we observed a one-way table with observed counts given by $\mathbf{n}^0 = \{2, 0, 1\}$ and we wish to test whether such an observed table could have come from an underlying population with probabilities {0.1,0.2,0.7}. That is, we are interested in testing the null hypothesis $H_0 : \mathbf{\Pi}_0 = (0.1, 0.2, 0.7)$ against a two-sided alternative. Then, the number of possible vectors, $\mathbf{M}$, such as {3, 0, 0} that are consistent with $k = 3$ and $n = 3$ are given in the one-way multinomial by $\mathbf{M} = \binom{n+k-1}{k-1}$ (Lawal, 1984). In our simple case, $\mathbf{M} = 10$. These possible configurations (vectors) that are consistent with k and n with their probabilities computed under H_0 are displayed below from a SAS® software program implementation. The program can be found in appendix B.1.

Vectors					
n1	n2	n3	Prob	Cum	
3	0	0	0.0010	0.0010	
2	1	0	0.0060	0.0070	
0	3	0	0.0080	0.0150	
1	2	0	0.0120	0.0270	
2	0	1	0.0210	0.0480	**
0	2	1	0.0840	0.1320	
1	1	1	0.0840	0.2160	
1	0	2	0.1470	0.3630	
0	1	2	0.2940	0.6570	
0	0	3	0.3430	1.0000	

For the observed configuration $\mathbf{n}^0 = \{2, 0, 1\}$, α' equals 0.048, and based on this value, we would reject H_o at a specified α level of 0.05 for example. We note here that this attained α' is sometimes referred to as the P-value.

The above test has been referred to (Tate and Hyer 1970) as the *exact multinomial test* or **EMT** for brevity. There are certain problems associated with implementing the EMT which basically have to do with the

(i) enumeration of the **M** vectors, and the

(ii) computation of probabilities given in (3.2) for each vector.

Since the number of configurations is given by $\mathbf{M} = \binom{n+k-1}{k-1}$, with moderate n and k therefore, the EMT may be feasible but with n and k large, the sheer number of these configurations as n or k (or both) increase can make the exact test cumbersome. Table 3.2 for instance, gives values of **M** for some n and k.

	Sample size (n)			
k	5	10	20	50
3	21	66	231	1326
5	126	1001	10,626	316,251
10	2002	92,378	100015005	$> 10^9$
20	42504	20030010	$> 6 \times 10^{10}$	

Table 3.2: Number of possible configurations (**M**) consistent with given n and k

From the preceeding table of possible configurations, it is realized that in order to conduct the exact multinomial test (EMT), one would need to generate all such configurations that are consistent with n and k which we have seen from the above table, that, it can be really tasking for k and n large. Additionally, the null multinomial probabilities given in (3.2) would need to be computed for each of these possible configurations, ranked, and the probabilities accumulated. No wonder, then, that there are good reasons to look for alternative procedures (or test statistics) to the exact multinomial test.

3.2 Test Statistics for Testing H_0

Under the null hypothesis (H_0) model specified earlier, interest centers on whether the observed data fit the hypothesized model. This fit is usually assessed by comparing the observed cell frequencies $\mathbf{n}^0$ with the *expected* frequencies $\mathbf{m}$, which are given by $n\pi_{0i}$. The hypothesized model would be rejected if there is considerable discrepancy between the observed frequencies and those expected from the model. Below, we discuss some of the test statistics that are employed to test this discrepancy between the observed and expected frequencies.

3.2.1 Goodness-of-Fit Test Statistics

One of the foremost goodness-of-fit test statistics developed is the classical χ^2 test statistic introduced by Karl Pearson (1900) and given by:

$$X^2 = \sum_{i=1}^{k} \frac{(n_i - n\pi_{0i})^2}{n\pi_{0i}} \tag{3.3}$$

where $n\pi_{0i} = m_i$ is the expected frequency under H_0 for category i and n_i is the corresponding observed frequency. This test rejects H_0 when X^2 exceeds some appropriate upper percentage point of a χ^2 distribution with $(k-1)$ degrees of freedom. When there is perfect fit between the null model and the data, $m_i = n_i$ for $i = 1, 2, ..., k$ and $X^2 = 0$ in this case. However, as the discrepancy between the observed n_i and the corresponding expected m_i increases, so will X^2. In this case, the true upper tail probability $P(X^2 \geq c)$ is approximated by the corresponding upper tail probability of the χ^2 distribution with $(k-1)$ degrees of freedom. This is referred to as the χ^2 approximation.

We give in the next sections other goodness-of-fit test statistics that have also received considerable attention in recent years.

The Likelihood Ratio Test Statistic

The likelihood ratio test (also known as the log-likelihood ratio test) is defined (Wilks, 1938) as:

$$G^2 = 2\sum_i n_i \log\left(\frac{n_i}{m_i}\right) \tag{3.4}$$

To illustrate the derivation of the above, consider for example independent response variables $Y_1, Y_2, \cdots, Y_n$ having Poisson distributions with parameters μ_i. The maximum likelihood estimator of the parameter $\boldsymbol{\theta}$ is the value $\hat{\boldsymbol{\theta}}$ which maximizes the likelihood function (LF) or the log-likelihood function (ℓ). That is, $L(\hat{\boldsymbol{\theta}}; \mathbf{y}) \geq L(\boldsymbol{\theta}; \mathbf{y})$ for all $\boldsymbol{\theta}$ in the parameter space. The log likelihood ratio statistic is

$$G^2 = 2\log\lambda = 2[\ell(\hat{\boldsymbol{\theta}}_{\max}; \mathbf{y}) - \ell(\boldsymbol{\theta}; \mathbf{y})] \tag{3.5}$$

where $\hat{\boldsymbol{\theta}}_{\max}$ is the maximal model likelihood estimate.

Returning to our example above, the log-likelihood function is:

$$\ell(\boldsymbol{\theta}; \mathbf{y}) = \sum_{i=1}^{n} y_i \log\mu_i - \sum_{i=1}^{n} \mu_i - \sum_{i=1}^{n} \log y_i! \tag{3.6}$$

In the case where all the Y_i terms have the same parameter μ, the MLE is given by

$$\hat{\mu} = \sum y_i/n = \bar{y}$$

and so

$$\ell(\hat{\boldsymbol{\theta}}; \mathbf{y}) = \sum y_i \log\bar{y} - n\bar{y} - \sum \log y_i!$$

For the maximal model, the MLE are $\hat{\mu}_i = y_i$ (a perfect fit), and so

$$\ell(\hat{\boldsymbol{\theta}}_{\max}; \mathbf{y}) = \sum y_i \log y_i - \sum y_i - \sum \log y_i!$$

G^2 in equation (3.5) therefore becomes:

$$\begin{aligned} G^2 &= 2[\ell(\hat{\boldsymbol{\theta}}_{\max}; \mathbf{y}) - \ell(\boldsymbol{\theta}; \mathbf{y})] \\ &= 2[\sum y_i \log y_i - \sum y_i \log\bar{y}] \\ &= 2\sum y_i \log\left(\frac{y_i}{\bar{y}}\right) \end{aligned}$$

Hence, the definition of G^2.

The Deviance

The deviance D was introduced by Nelder and Wedderburn (1972) and is written as

$$\mathrm{D} = 2[\ell(\hat{\boldsymbol{\theta}}_{\max}; \mathbf{y}) - \ell(\boldsymbol{\theta}; \mathbf{y})] \tag{3.7}$$

We notice that the deviance is equivalent to the log-likelihood by definition. That is, $\mathrm{D} \equiv G^2$. A model's adequacy is easily assessed by computing D or G^2 and comparing the result with a χ^2 distribution with the relevant degrees of freedom. In general, a model that is based on the estimation of p parameters from a data set with n observations would have its computed test statistic distributed as χ^2_{n-p}. In particular, for D, we have

$$\mathrm{D} \cong n - p \tag{3.8}$$

As we demonstrate from the Poisson example above, D can be calculated from the observed and fitted values respectively. However, for some other distributions, D cannot be calculated directly because of some nuisance parameters. For the normal distribution for example, it can be shown that

$$\sigma^2 D = \sum (y_i - \hat{\mu}_i)^2$$

where $\hat{\mu}_i$ denotes the MLE of μ_i. PROC GENMOD in SAS® computes $\sigma^2 D$ and also gives the **scale parameter** which is an estimate of σ^2. That is,

$$\hat{\sigma}^2 = \frac{D}{n - p} \tag{3.9}$$

For the one way multinomial, both G^2 and D have asymptotic χ^2 distribution with $(k - 1)$ degrees of freedom in case of specified null probabilities.

The Freeman-Tukey Test

The Freeman-Tukey test was introduced by Freeman and Tukey (1950). It is given by

$$T^2 = 4 \sum_i (\sqrt{n_i} - \sqrt{m_i})^2 \tag{3.10}$$

while Bishop et al. (1975) introduced the improved or modified Freeman Tukey test which is given by

$$\mathtt{FT} = \sum_i \left\{ \sqrt{n_i} - \sqrt{(n_i + 1)} + \sqrt{(4m_i + 1)} \right\}^2 \tag{3.11}$$

The Modified Log-Likelihood Ratio Test Statistic

Kullback (1959) gives the minimum-discriminant information statistic for the external constraints problem as:

$$\mathtt{GM}^2 = 2 \sum_i m_i \log \left(\frac{m_i}{n_i} \right) \tag{3.12}$$

The Neyman Modified X^2 Statistic

Neyman (1949) proposed

$$\mathtt{NM}^2 = \sum_i (n_i - m_i)^2 / n_i \tag{3.13}$$

The Cressie-Read Power Divergence Test Statistic

Cressie and Read (1984) proposed the power divergent test statistic

$$\mathtt{I}(\lambda) = \frac{2}{\lambda(\lambda+1)} \sum_i n_i \left\{ (\frac{n_i}{m_i})^\lambda - 1 \right\} \tag{3.14}$$

where λ is a real parameter. It can be shown that:

(i) For $\lambda = 1$, $I(1) = X^2$

(ii) As $\lambda \to 0$, $\lim_{\lambda \to 0} I(\lambda) = G^2$.

(iii) As $\lambda \to -1$, $\lim_{\lambda \to -1} I(\lambda) = \mathtt{GM}^2$.

(iv) For $\lambda = -2$, $I(-2) = \mathtt{NM}^2$.

(v) For $\lambda = -\frac{1}{2}$, $I(-\frac{1}{2}) = T^2$.

Cressie and Read recommend $\lambda = 2/3$ because of its superiority over the other values of λ in terms of attained significance levels and small sample power properties.

Under H_0 and certain conditions on $\mathbf{\Pi}_0$ and k, all the statistics discussed above all have asymptotically, the χ^2 distribution with $(k-1)$ degrees of freedom. Pearson (1900) derived the asymptotic (i.e, as the sample size n increases) distribution for X^2, while Wilks (1938) derived that of G^2. Cramer (1946) had also derived the limiting distribution of X^2.

Lawal (1984), Larntz (1978), Koehler and Larntz (1980), and West and Kempthorne (1972) have, among others, studied the comparative accuracies of some of the above test statistics. One advantage of G^2 over other test statistics however, relates to its partioning property, which allows us to decompose the overall G^2 into small components. A typical example of this is based on the Lancaster (1949a) partioning principle, which will be discussed in chapter 5. Another advantage of G^2 is that it simplifies the process of comparing one model against another. For this reasons, the likelihood ratio test statistic G^2 is the choice test statistic in this text.

3.3 An Alternative Test to EMT

Earlier, we discuss the exact multinomial test (EMT) in the context of ranking probabilities. An alternative exact test suggested by Radlow and Alf (1975) ranks the possible outcomes in terms of a test criterion (say) for example X^2. We give below the result of this approach based on the X^2 test criterion. As in the previous case, we again perform the following:

a For every other possible outcome $\mathbf{n}^0$, calculate the exact probability as in (3.2) and the corresponding test statistic X^2 under H_0.

b Rank the test statistic X^2 in the order of magnitude from largest to smallest and allow each ranked value to carry its corresponding multinomial probability.

c Starting from the largest ranked, add the consecutive corresponding probabilities up to and including that associated with the observed configuration X_0^2. This cumulative probability gives the chance of obtaining an outcome that is at least as unlikely as X_0^2 (i.e., outcomes that are as extreme as or more extreme than X_0^2).

d Reject H_0 if this cumulative probability α' is less or equal to α: that is, if $\alpha' \leq \alpha$.

Again, let us implement this procedure by referring to our simple example. I shall use the X^2 criterion for illustration here, although any other test statistic could have been used. In the table 3.3 are the ordered values of the X^2 statistic: starting from the largest to the smallest under the null hypothesis $H_0 = \mathbf{\Pi}^0 = (0.1, 0.2, 0.7)$. We have also presented a SAS software program for implementing this procedure in appendix B.2.

X-squard	n1	n2	n3	Prob	Cum	
27.0000	3	0	0	0.0010	0.0010	
12.0000	0	3	0	0.0080	0.0090	
12.0000	2	1	0	0.0060	0.0150	
10.8095	2	0	1	0.0210	0.0360	**
7.0000	1	2	0	0.0120	0.0480	
4.1429	0	2	1	0.0840	0.1320	
2.4762	1	1	1	0.0840	0.2160	
2.2381	1	0	2	0.1470	0.3630	
1.2857	0	0	3	0.3430	0.7060	
0.5714	0	1	2	0.2940	1.0000	

Table 3.3: Ordered values of X^2 with accompanying probabilities

Here, $X_0^2 = 10.8095$, corresponding to the observed configuration $\mathbf{n}^0 = \{2, 0, 1\}$, and therefore the pvalue equals 0.036 and we notice that the vectors $\mathbf{n}^0 = \{n_1, n_2, n_3\}$ have been reordered from those given under the EMT. We further note that for this case at least, there is stronger evidence for the rejection of H_0 in this example. When this result is compared with that obtained from the EMT (i.e., where outcomes are ranked by probabilities), the corresponding pvalue is 0.048, indicating that the two procedures in many cases do give different results, and this is the main reason for the controversy regarding whether to conduct exact test by using ranked probabilities or ranked test statistics.

The number of distinct values of the statistic X^2, say, $\mathbf{S}$ from table 3.3, is 9 (instead of 10), because there are two configurations that yield the same X^2 of 12.000. These are {0,3,0} and {2,1,0} with multinomial probabilities of 0.0080 and 0.0060, respectively. This value of X^2 therefore carries a total probability of 0.0140, which is the sum of the two probabilities. The above table should have therefore been displayed with only one value of $X^2 = 12.0000$ with a corresponding probability 0.0140.

In general, $\mathbf{S} \leq \mathbf{M}$, as some vectors sometimes yield the same value of the test statistic, as is indeed the case for the two configurations above.

As mentioned earlier, not only do the EMT and exact test based on X^2 give different results, but this problem is further complicated by the fact that different test criteria rank outcomes differently (Lawal, 1984). For example, the likelihood test statistic G^2 ranks the outcomes differently as seen from the results in Table 3.4. In this case, the number of distinct test statistics, $\mathbf{S}$, equals the number of possible configurations **M**. The pvalue based on the G^2 criterion for our observed outcome is 0.048 in this case. We have presented again an SAS® software program that implements this test in appendix B.3.

Obs	GG2	n1	n2	n3	PROB	CUM
1	13.8155	3	0	0	0.0010	0.0010
2	9.6566	0	3	0	0.0080	0.0090
3	8.6101	2	1	0	0.0060	0.0150
4	7.2238	1	2	0	0.0120	0.0270
5	6.1046	2	0	1	0.0210	0.0480
6	3.3320	0	2	1	0.0840	0.1320
7	2.2128	1	0	2	0.1470	0.2790
8	2.1400	0	0	3	0.3430	0.6220
9	1.9457	1	1	1	0.0840	0.7060
10	0.8265	0	1	2	0.2940	1.0000

Table 3.4: Ordered values of G^2 with accompanying probabilities

As another illustration, consider again the null hypothesis, $\mathbf{\Pi}_{0i} = (0.3, 0.3, 0.4)$ and $n = 3$; we still have $\mathbf{M} = 10$ but $\mathbf{S} = 6$ for the X^2 criterion in this case. We list the values of this statistic in Table 3.5.

X^2	Probability	Cumulative Probability
7.000	0.054	0.054
4.500	0.064	0.118
2.556	0.162	0.280
2.278	0.216	0.496
1.444	0.288	0.784
0.056	0.216	1.000

Table 3.5: Ordering of all distinct vectors consistent with $k = 3, n = 3$ by X^2

In Table 3.5, vectors {0, 3, 0} and {3, 0, 0} with probabilities 0.0270 each give a computed X^2 value of 7.000 and therefore a combined probability of $0.0270 + 0.0270 = 0.054$. We give in Table 3.6 the vectors and their corresponding probabilities for all the $M = 10$ possible vectors in this case.

X^2	Vector	Probability
7.000	0 3 0	0.0270
	3 0 0	0.0270
4.500	0 0 3	0.0640
2.556	1 2 0	0.0810
	2 1 0	0.0810
2.278	0 2 1	0.1080
	2 0 1	0.1080
1.444	0 1 2	0.1440
	1 0 2	0.1440
0.056	1 1 1	0.2160

Table 3.6: Ordered vectors for X^2, under H_0= (0.3,0.3,0.4)

Suppose $\mathbf{n}^0 = (2,1,0)$; then the corresponding pvalue would be 0.2800 for the X^2 criterion under the above null hypothesis. Similar results can be obtained for the test statistics $I(\frac{2}{3})$ and the G^2. Table 3.7 gives some results for some null hypotheses for three goodness-of-fit statistics.

H_o	Test Criterion	Sample Size	Observed Vector	Observed test value	S	P-Value	Tail Probability
.1 .2 .7	X^2	3	2 0 1	10.8095	9	0.0368	0.0045
	G^2	3	2 0 1	6.1046	10	0.048	0.0473
	$I(\frac{2}{3})$	3	2 0 1	8.4495	10	0.0360	0.0146
.1 .2 .7	X^2	10	3 0 7	6.000	53	0.0481	0.0498
	G^2	10	3 0 7	6.592	66	0.0345	0.0370
	$I(\frac{2}{3})$	10	3 0 7	5.832	66	0.0345	0.0541
.3 .3 .4	X^2	10	3 0 7	5.250	27	0.0691	0.0724
	G^2	10	3 0 7	7.835	35	0.0374	0.0199
	$I(\frac{2}{3})$	10	3 0 7	5.698	36	0.0559	0.0579
.2 .2 .6	X^2	10	3 0 7	2.667	24	0.3986	0.2636
	G^2	10	3 0 7	4.591	36	0.2395	0.1007
	$I(\frac{2}{3})$	10	3 0 7	3.039	36	0.2836	0.2188

Table 3.7: Comparative results for various null hypotheses, sample sizes and three test statistics X^2, G^2 , $I(2/3)$

In Table 3.7, the tail probabilities are the corresponding upper tail probability of the approximating χ^2 distribution with two degrees of freedom. For instance, for the very first line in Table 3.7 above, the tail probability in the last column is obtained by evaluating $P(\chi^2_2 > 10.8095) = 0.0045$.

We see that the χ^2 distribution gives a very poor approximation to the exact pvalues in most of the cases. The reason for this is that the three statistics have discrete type distributions and these are being approximated by a continuous type distribution. Further, the different test statistics order the vectors differently (Lawal, 1984; West & Kempthorne, 1972). We also observe that the number of distinct values of the test statistics varies from one test criterion to another and from one null hypothesis to another. Our results in Table 3.7 further show that the distribution of X^2 is better approximated by the χ^2 distribution than any of the other test criteria in terms of attained nominal level. The above conclusion has been supported by various works (e.g. Lawal 1984; Koehler and Larntz 1980 among others).

Returning to the underlying χ^2 approximation to each of these statistics, the results in the preceeding table indicate that this approximation is not well suited for the situations displayed in the above table. Consequently, it has been suggested that this approximation is only valid when the expected values are large and that the approximation ceases to be appropriate if any of the expected cell frequencies $m_i = n\pi_{0i}$ becomes too small, since the χ^2 approximation to X^2 was derived under the assumption that with k fixed, m_i becomes increasingly large as the sample size n also becomes large. That is,

$$m_i \to \infty, \quad \text{as} \quad n \to \infty, \quad \text{for} \quad i = 1, 2, \cdots, k$$

How small is "too small" has been the subject of controversy for quite some time. Suggested minimum expected values range from 1 to 20. Good et. al. (1970) provide an overview of the historical recommendations. Earlier recommendations are Fisher (1924) $m_i \geq 5$, Cramer (1946) $m_i \geq 10$, and Kendall (1952) $m_i \geq 20$, while Cochran (1954) states that in goodness-of-fit tests of unimodal distributions (e.g., Poisson or Normal), tail expectations should be at least 1, that is, $m_i \geq 1$. In an earlier article, Cochran (1952) has advocated that satisfactory approximation may be obtained in tests of distributions such as the normal when a single expectation is as low as 0.5, the others "being above the conventional limits of 5 to 10." Yarnold (1970) has suggested that the following criteria be used:

> If the number of classes $k \geq 3$ and if r is the number of expectations less than 5, then the minimum expectation may be as small as $\frac{5r}{k}$.

However, strict application of this rule generally involves collapsing together of one or more categories, and in most cases this may detract greatly from the interest and usefulness of the study. For instance in Table 3.8 (taken from Lawal, 1980), $k = 6$ and $r = 4$.

Classes	1	2	3	4	5	6	Total
m_i	0.1	0.2	0.3	0.4	10.0	10.0	21

Table 3.8: Expected values under some model

Hence, the minimum expected value by Yarnold's rule is 3.33. Thus, it would be necessary to collapse the first five cells, leaving us with only 1 degree of freedom for our test of significance! A less restrictive use of Yarnold's rule was proposed by Lawal (1980), and this is examined later in this chapter.

Because small expectations affect the validity of the χ^2 approximation to the discrete distribution of X^2 (and indeed to all the test statistics mentioned above) and because X^2 has received more attention than any other test criteria, we give below some of the alternative approximations that have been proposed over the years.

3.4 Approximations to the Distribution of X^2

3.4.1 The Continuity Correction

Because of the discreteness of X^2, Cochran (1942) proposed the continuity corrected X^2 whose upper tail area is given by

$$P[X_0^2 \geq c] \approx H_{k-1}(\frac{c+d}{2})$$

where $\approx$ means "approximately," H_{k-1} is the upper tail probability of a χ^2 distribution with $(k-1)$ d.f., c is the observed value of X^2, and d is the next smallest possible value of X^2.

As an example, consider the observed configuration {2,0,1} in the case when $k = 3$ and $n = 10$ under the null $\mathbf{\Pi}^0 = \{0.1, 0.2, 0.7\}$ discussed earlier. There, $X_0^2 = c = 10.810$, $d = 7.000$, and hence $P[X^2 \geq 10.810] \approx P[\chi_2^2 \geq 8.905] = 0.012$. This value, while not exactly equaling the exact probability of 0.036, it is nonetheless closer to the exact value than the usual χ^2 approximation pvalue of 0.0045 obtained again earlier.

3.4.2 The C(m) Approximation

The C(**m**) distribution was first proposed by Cochran (1942) and is derived under the assumption that as $n \to \infty$, some expected values will remain finite, while the remaining expected values will be large. The limiting distribution of X^2, called the $C(\mathbf{m})$ distribution with parameters s, r, and $\mathbf{m}$, is defined as:

$$C(\mathbf{m}) = \sum_{i=1}^{r} \frac{(u_i - m_i)^2}{m_i} + \chi^2_{(s-r)}$$

where $s = (k - 1)$ and U_i is a Poisson variate with mean m_i. Cochran's definition of the above distribution was limited because it was defined only for cases when $r = 1$ and $r = 2$. Yarnold (1970) extended Cochran's results and showed that the approximation works very well in the region of the parameter space (upper tail area) in which the χ^2 approximation fails. Lawal (1980) obtained the 5% and 1% points of the C(**m**) distribution.

3.4.3 The Gamma Approximation

Nass (1959) suggested approximating cX^2 by a χ^2_d distribution, where c and d are chosen by matching the means and variances of both cX^2 and a χ^2_d. That is, $E(cX^2) = d$ and $\text{Var}(cX^2) = 2d$ since $E(\chi^2_d) = d$ and $\text{Var}(\chi^2_d) = 2d$. From the above,

$$\Rightarrow c = \frac{2E(X^2)}{\text{Var}(X^2)} \quad \text{and} \qquad d = \frac{2E(X^2)^2}{\text{Var}(X^2)} \tag{3.15}$$

Haldane (1937) showed that:

$$E(X^2) = k - 1 \quad \text{and} \qquad \text{Var}(X^2) = 2(k-1) + (R - k^2 - 2k + 2)/n \tag{3.16}$$

where $R = \sum_i \pi_i^{-1}$. Consequently,

$$Pr[cX^2 \geq \chi^2_d] = Pr[X^2 \geq \frac{1}{c}\chi^2_d]$$

where c and d are given by (3.15).

It is evident that practical use of this approximation necessarily involves *fractional* degrees of freedom. Earlier attempts to use this approximation (Nass, 1959; Yarnold, 1970) have sought to obtain the critical value corresponding to the fractional degree of freedom by interpolation in standard χ^2 tables. This procedure has limited utility because it can lead to inaccurate critical values and it is labor intensive.

The **cube** approximation in Abramowitz and Stegun (1970, pg. 941) for approximating the percentage points of the χ^2 distribution when there are either integral or fractional degrees of freedom has been used to overcome this difficulty. The Best and Roberts (1975) algorithm along with the necessary auxiliary routines is an excellent program for computing the lower tail area of the χ^2 distribution with d degrees of freedom for a given percentile. Lawal (1986) produced excellent results from this approximation.

3.4.4 The Lognormal Approximation

Lawal and Upton (1980) employed the two-parameter lognormal distribution to approximate the upper tail distribution of X^2. From the expressions for the mean and variance of X^2 in (3.16) we observe that $E(X^2) = k-1$, which is also the mean of the approximating χ^2 distribution with $(k-1)$ degrees of freedom. Similarly, when $R = \sum_i \pi_{0i}^{-1}$ and k are small in comparison to n, then in this case, $\text{Var}(X^2) \approx 2(k-1)$, again the variance of the approximating χ^2. In this situation, we would hope the approximation would work well. Unfortunately, if R/n is large, as it would be if there are some very small expected values, then the variance of X^2 is greater than the variance of the approximating distribution and, consequently, the tail probabilities of X^2 may greatly exceed the nominal values.

The Lawal and Upton (1980) approximation sought to correct this by once again matching the first two moments of X^2 with those of the two parameter lognormal distribution. Basically, the approximation amounts to writing

$$P[X^2 > z] = P[Z > z] \tag{3.17}$$

where Z has a lognormal distribution with parameters μ and σ^2, which are related by the method of moments to the mean and variance of X^2 given in (3.16) by

$$\begin{aligned} \mu &= \theta - \frac{1}{2}\psi \\ \sigma^2 &= \psi - \theta \end{aligned} \tag{3.18}$$

which translates to

$$\begin{aligned} \theta &= 2 \log\,[E(X^2)] \\ &= 2 \log\,(k-1) \qquad \text{and} \\ \psi &= \log\,[E(X^2)^2 + \text{Var}(X^2)] \\ &= \log\,[k^2 - 1 + (R - k^2 - 2k + 2)/n] \end{aligned}$$

If u_α is the upper α point of a unit normal random variable, the above implies that the upper α point for X^2 is obtained to be

$$\exp\{\mu + \sigma u_\alpha\}$$

where $P[X^2 > z] = \Phi[(\log z - \mu)/\sigma]$, and $\Phi(.)$ is the unit normal distribution function.

Lawal and Upton showed that the lognormal approximates well the upper tail areas of X^2 and that the approximation seems impervious to the number of small cell expectations. They further suggested that the approximation will usually work well if we allow minimum expected cell value not to be less than $r/d^{\frac{3}{2}}$ where r is the number of expected values less than 3.0 and d is the degrees of freedom. Based on this, Lawal (1980) gives the following slightly restricted rule for the use of the χ^2 distribution.

> If the degree of freedom on which fit is based is d and r is the number of expectations less than 3, then the minimum expectation may be as small as $r/d^{3/2}$.

We summarize these various rules in Table 3.9.

If Yarnold and Lawal's rules are satisfied, use the usual χ^2 as the approximating distribution. Both the lognormal and the $C(\mathbf{m})$ require the use of computed critical levels based on the expressions above and those in Lawal (1980).

Rules	Number of small expectations (r)	minimum m_i
Yarnold	r < 5	$\frac{5r}{k}$
Lawal	r < 3	$\frac{r}{(k-1)^{3/2}}$
Lognormal	r < 3	$\frac{r}{(k-1)^{3/2}}$
C_m	r < 3	$\frac{r}{(k-1)^{3/2}}$

Table 3.9: Summary of rules for the case when $k \geq 3$

3.4.5 The Modified X^2 Statistic

Because the variance of X^2 is highly inflated when some expected values are small, Lawal (1992a) has proposed a variance stabilizing test statistic: the modified X^2 test statistic, which is defined as:

$$T^2 = \sum_{i=1}^{k}[(n_i - \frac{1}{2}) - m_i]^2/m_i \tag{3.19}$$

T^2 has variance $2(k - 1)$ if the n_i follow the Poisson distribution and variance $2(k-1)(1-\frac{1}{n})$ if the n_i follow the multinomial distribution. This test is also shown to belong to the family of the power divergence test statistic proposed by Cressie and Read (1988), which includes

$$h^\lambda(n_i + c, m_i + d)$$

where

$$h^\lambda(u,v) = \frac{2}{\lambda(\lambda+1)}\left\{u\left[(\frac{u}{v})^\lambda - 1\right] + \lambda[u - v]\right\}$$

Then T^2 is equivalent to the family with $\lambda = 1$, $c = -\frac{1}{2}$ and $d = 0$. The T^2 defined in (3.19) above is related to the statistic D^2 proposed by Zelterman (1987) by:

$$T^2 = D^2 + k + \frac{1}{4}\sum m_i^{-1}$$

where $D^2 = \sum[(n_i - m_i)^2 - n_i]/m_i$.

3.4.6 Applications

We now give examples of the use of these approximations. The first example is concerned with gene heredity. Peas may be yellow or green, round or wrinkled, short or tall. Within each of these classifications, the dominant gene is believed to be the first-named category, and a theory suggests that these characteristics are mutually independent with, for each characteristic, the ratio of dominant to recessive being 3 to 1. Suppose that we have a random sample of 64 peas whose characteristics are given in Table 3.10 below.

	Yellow				Green			
	round		wrinkled		round		wrinkled	
	Short	long	short	long	short	long	short	long
Observed Count	36	12	8	4	2	1	1	0
Expected Count	27	9	9	3	9	3	3	1

Table 3.10: Example of a nominal data: Pea types

The expected frequencies to the genetic theory are included in the table and we wish to enquire whether the observed data are consistent with the theory, where, for instance, the expected values of 27 and 1 are obtained as:

$$27 = \frac{3}{4} \times \frac{3}{4} \times \frac{3}{4} \times 64$$

$$1 = \frac{1}{4} \times \frac{1}{4} \times \frac{1}{4} \times 64$$

The expected frequency of 1 in the final category rules out the use of the χ^2 approximation since Yarnold's rule is violated. The common practice of combining this category with some adjacent category does not make sense here because the data are nominal and not ordinal. There is therefore no category that can be said to be adjacent and arbitrary clumping of categories seems unsatisfactory.

To use the lognormal approximation, we note that for these data $n = 64, k = 8$ and $R = 2.37$, giving $\theta = 3.89$, $\psi = 4.12$ and hence $\mu = 1.83$ and $\sigma^2 = 0.23$. Consequently, the upper 1% point of the distribution of X^2 is estimated as 19.13, compared with the tabulated value of 18.48. The corresponding 1% point using the C(**m**) approximation is 19.35 (from Lawal, 1980). Since the observed value of X^2 is 13.56, there is clearly no cause to doubt the theory in the present case.

3.5 Goodness-of-Fit (GOF) Tests

The standard situation involves n observations obtained as a simple random sample from a population, which are classified into k mutually exclusive categories. There is some theory or null hypothesis which gives the probability π_j that an observation will be classified into the j-th category. The π_j are sometimes completely specified by the theory, e.g., genetic characteristics, or sometimes they are functions of other parameters $\alpha_1, \alpha_2, \cdots, \alpha_k$ whose actual values are unknown. An example here is, for instance, an underlying normal distribution with unknown mean and variance. The quantities m_j are called the expected frequencies. Under the null hypothesis, interest centers on estimating the expected frequencies and consequently using any of the well known statistics above to conduct the test.

We have shown in chapter 1 that the maximum likelihood estimate (MLE) of π_i from the multinomial one-way distribution in (3.1) is given by $\hat{\pi}_i = \dfrac{n_i}{n}$. We shall use this result and the concept of MLE to find solutions to the examples in the next section.

3.5.1 Examples: Nominal Variable with Many Categories

The data below relate to five methods of child delivery in eclamptic (eclampsia) patients (Jolayemi, 1990b). The variable "mode of delivery" in this case has five categories. A variables with many categories is sometimes referred to as a *polytomous variable.*

Mode of Delivery	Number of Cases	%
CS	313	39.72
Normal	258	32.74
Vacuum	95	12.06
Forceps	81	10.28
ABD	41	5.20

Table 3.11: Number of deliveries in eclamptic patients

for CS (Caesarian section); Normal (SVD) delivery; Vacuum (Vacuum extraction); Forceps (Forceps delivery); and ABD (Assisted breech delivery).
Here, we have five classes or categories and the following null hypotheses are of interest:

(i) $H_{01} : \pi_i = \pi$, $i = 1, 2, \cdots, 5$, that is, the equi-probable case

(ii) $H_{02} : \pi_1 = \pi_2; \pi_3 = \pi_4 = 2\pi_5$

(iii) $H_{03} := \frac{\pi_1}{7} = \frac{\pi_2}{7} = \frac{\pi_3}{2} = \frac{\pi_4}{2} = \pi_5$

against the general alternatives.

Under H_{01}, the likelihood function becomes:

$$L(n, H_0) = C\pi^{n_1+n_2+n_3+n_4+n_5}$$

Thus, minimizing $\log L$ subject to the constraint $\sum_{i=1}^{5} \pi_i = 1$ gives MLE estimates $\hat{\pi} = \frac{1}{k}$. The expected values are n/k (=157.60) and the corresponding G^2 and X^2 are 369.4453 and 365.5533, respectively on 4 d.f This model gives a poor fit to the data with a pvalue = 0.0000.

For the second null hypothesis, if we let $\pi_1 = \pi_2 = \pi_A$, and $\pi_3 = \pi_4 = 2\pi_5 = \pi_B$, then the likelihood function becomes:

$$L(n, H_0) = C\pi_A^{n_1+n_2}\pi_B^{n_3+n_4}\left(\frac{\pi_B}{2}\right)^{n_5}$$

Again, maximizing $\log L$ subject to the constraint $2\pi_A + \dfrac{5\pi_B}{2} = 1$ gives MLE estimates $\hat{\pi}_A = \dfrac{n_1 + n_2}{2n} = 0.3623$ and $\hat{\pi}_B = \dfrac{n_3 + n_4 + n_5}{\frac{5}{2}n} = 0.1102$ and estimated cell probabilities {0.3623, 0.3623, 0.1102, 0.1102, 0.0551}. The corresponding expected values for each cell which are the product of corresponding probability estimates and the total sample size n are respectively {285.4924, 285.4924, 86.8376, 86.8376, 43.4188}. The corresponding G^2 and X^2 are 6.4315 and 6.5944, respectively, and these are each based on $(5-3) = 2$ degrees of freedom since two parameters (namely, π_A and π_B) are estimated with corresponding pvalue of 0.0401.

For the third null hypothesis, let $\pi_5 = \pi$. Then, $\pi_1 = \pi_2 = 7\pi$, $\pi_3 = \pi_4 = 2\pi$, and

$$L = C(7\pi)^{n_1+n_2}(2\pi)^{n_3+n_4}\pi^{n_5}$$

Maximizing $\log L$ subject to the constraint $7\pi + 7\pi + 2\pi + 2\pi + \pi = 1$ gives MLE estimate $\hat{\pi} = 1/19 = 0.0526$, and the corresponding expected values are $\{290.3158,$ $290.3158, 82.9474, 82.9474, 41.4737\}$. Computed G^2 and X^2 under this model are 7.1900 and 7.1766, respectively, and are again based on $(5 - 2) = 3$ degrees of freedom since only one parameter was estimated. The corresponding pvalue here is 0.0681.

Based on the above results, while both hypotheses H_{02} and H_{03} are tenable at $\alpha = 0.01$ nominal level, only H_{03} is tenable at $\alpha = .05$ nominal level and is considered more parsimonious than H_{02}. We would therefore prefer H_{03}.

We would thus conclude that the occurrence rate of the mode of delivery by cesarian section and SVD is seven times the occurrence rate of birth by ABD and that the occurrence rates of delivery by vacuum and forceps are twice that of delivery by ABD. Thus in the population, the categories are not uniformly distributed, with the cesarean mode being as likely as the SVD mode. Similarly, the vacuum model is as likely as the forceps mode of delivery. We present in appendix B.4 the SAS® software program and corresponding output employed in implementing some of the results above.

In the SAS® software program in appendix B.4, we first fit the equiprobable model using PROC GENMOD. In the **model** statement, we specify that the data is Poisson distributed with **dist=poisson** and that **link=log**. The make obstats **obstats** option asks for the output to be written to a file named **aa**. The print command requests specific variables to be printed with a format statement indicating output to four decimal places. The GENMOD output includes predicted values, lower and upper confidence intervals for each predicted observation, and residuals and other statistics.

The corresponding SAS® software output gives the **estimate**, **std Err**, and the relevant **ChiSquare** for the test of statistical significance of the estimated parameter. The column headed **pred** gives the estimated m_i, while **xbeta** gives the corresponding estimated log-means. The **Resraw** gives the raw residuals $n_i - m_i$, while **Reschi** gives the standardized Pearson's residual, which is the square root of the individual Pearson's X^2. The model gives a deviance or G^2 value of 369.4458 on 4 d.f.

PROC FREQ can also be used to obtain some of the results in the preceeding section. The first **TABLES** statement in this appendix fits the equi-probable model. The second TABLES statement computes X^2 for the model based on the second hypothesis using **TESTP**, where the last **P** proportions are inputed. The third TABLES statement also computes X^2 statistic for the model based on the third hypothesis using **TESTF**, with the last **F** indicating that frequencies are being inputed. In all the cases, the X^2 values are correct but the degrees of freedom in TESTP and TESTF should be 2 and 3, respectively.

Finally, PROC CATMOD is also employed to obtain some of the results as in the previous case. This is implemented in CATMOD by stating in the **RESTRICT** line $\log\left(\frac{m_i}{m_I}\right)$. The results are again similar to those obtained under PROC FREQ.

Another way of analyzing the data in the last example is to consider the variable of interest *mode of delivery* as a nominal polychotomous variable (Clogg & Shockey, 1988) where if we denote the variable as A, then the saturated log-linear model (this will be fully discussed in Chapter 6) is given by:

$$\log(m_i) = \mu + \lambda_i^A$$

If we let $L_i = \log(m_i)$, then the MLE estimates are given by $\hat{\mu} = L_+/I$ and $\hat{\lambda}_i^A = L_i - \hat{\mu}$, with $\sum_i \lambda_i^A = 0$. As pointed out by Clogg and Shockey (1988), it is sometimes useful to examine a set of logits for this type of data, where one category is used as a reference point. In this case, the logit $\phi_i = L_i - L_I = \log(n_i/n_I) = \lambda_i^A - \lambda_I^A$. This is the log odds that $A = i$ rather than $A = I$.

In the example above, if we use the first category as the reference, then we can compute the logits $\hat{\phi}_i = \log(n_i/n_1)$, $i = 2, \cdots, 5$. These are respectively {-0.213, -1.192, -1.352, -2.033}.

3.5.2 Analysis When Variable Categories Are Ordered

When the variable of interest is ordinal in nature, then we may exploit this ordinal nature of the variable in our modeling techniques. Again, both the saturated model and the equiprobability model ($H_0 : \pi = 1/k$) discussed above are also relevant but various other models are quite possible. Suppose the ordered categories are assigned scores $\{\nu_i\}$, $i = 1, 2, \cdots, I$. Various forms of the scoring can be used. Scores can be obtained that are based on distributional assumptions or scores can be assigned based on prior knowledge. Popular scores are the *integer scores*, $\nu_i = i$, which are equivalent to the assumption of equal spacing or $\nu_i = i - (I+1)/2$, which centers the scores. Whichever scores we adopt, inferences derived from subsequent analyses are dependent on the scoring system adopted. We shall elaborate further on this in Chapter 10. Four main models have received wide attention in relation to the analysis of an ordered polytomous variable. These are the *the equiprobable model*, *the linear effect model*, the *the quadratic effect model*, and the *symmetry effect model*. The log-linear model formulations for these models are described below with the relevant identifiability constraints.

(i) The equiprobable model:

$$\ell_i = \lambda, \qquad i = 2, \cdots, I$$

(ii) The linear effect model:

$$\ell_i = \lambda + \beta\nu_i, \quad i = 1, 2, \cdots, I$$

(iii) The quadratic model:

$$\ell_i = \lambda + \beta\nu_i + \gamma\nu_i^2$$

(iv) The symmetry model:

$$\ell_i = \lambda + \lambda_i^A$$

The symmetry model in (iv) is subject to the following constraints: $\lambda_1^A = \lambda_I^A$, $\lambda_2^A = \lambda_{I-1}^A$, $\lambda_3^A = \lambda_{I-2}^A$, $\cdots$. We note here that since the model of symmetry above implies that, $\ell_1 = \ell_I$, $\ell_2 = \ell_{I-1}$, $\ell_3 = \ell_{I-2}$, and so on, the symmetry model actually implies a composite of separate equiprobability models. That is, categories 1 and I are equiprobable, categories 2 and $(I-1)$ are equiprobable, and so forth. This is sometimes referred to in nonparametric statistics as the "umbrella alternatives."

Example

We now apply some of the models discussed in the preceeding section to the following data taken from Haberman (1978, p. 85), which concern the political views of subjects interviewed during the U.S. 1975 General Social Survey. The data are displayed in Table 3.12 with seven categories: (1) extremely liberal, (2) liberal, (3) slightly liberal, (4) moderate, (5) slightly conservative, (6) conservative, (7) extremely conservative.

	Extremely liberal (1)	(2)	(3)	Mode-rate (4)	(5)	(6)	Extremely conservative (7)
n_i	46	179	196	559	232	150	35
m_i	40.5	164.5	214.0	559	214.0	164.5	40.5

Table 3.12: Ordinal data: Political views

In Table 3.13 are the results of fitting some of the models (under integer scores) discussed above to our data.

Model	d.f.	G^2	pvalue
Equiprobable	6	832.82	0.0000
Linear	5	832.28	0.0000
Quadratic	4	133.94	0.0000
Symmetry	3	7.09	0.0691

Table 3.13: G^2 and pvalues based on the above models

In this example, the model of symmetry fits and has MLE given by the following expressions. These expected frequencies are displayed in Table 3.12.

$$\hat{m}_1 = \hat{m}_7 = \frac{1}{2}(n_1 + n_7)$$

$$\hat{m}_2 = \hat{m}_6 = \frac{1}{2}(n_2 + n_6)$$

$$\hat{m}_3 = \hat{m}_5 = \frac{1}{2}(n_3 + n_5)$$

$$\hat{m}_4 = n_4$$

The above models can be implemented in SAS® with the program in appendix B.5. The symmetry model for instance, can be decomposed into its components in terms of their contributions to the overall G^2 value of 7.09 as follows. This is accomplished with the contrasts statements in the SAS® software program in appendix B.5.

Type of symmetry	df	G^2
1 & 7	1	1.4985
2 & 6	1	2.5596
3 & 5	1	3.0316
Total	3	7.0897

Table 3.14: Results from symmetry analysis

The results are given in Table 3.14, where for instance the $G^2 = 2\sum n_i \log\left(\frac{n_i}{\hat{m}_i}\right)$ $= 1.4985$ contribution is computed as:

$$1.4985 = 2[46 \times \log\left(\frac{46}{40.5}\right) + 35 \times \log\left(\frac{35}{40.5}\right)]$$

since $n_1 = 46, n_7 = 35$, and $\hat{m}_1 = \hat{m}_7 = 40.5$. The SAS® software results from the symmetry model implementation are displayed below.

```
          SAS CONTRAST Statement Results

Contrast    DF    ChiSquare   Pr>Chi   Type
--------------------------------------------
SYM          3       7.0896   0.0691   LR
1 & 7        1       1.4985   0.2209   LR
2 & 6        1       2.5596   0.1096   LR
3 & 5        1       3.0316   0.0817   LR
```

Haberman (1978) gives some further examples of analyses of data of this type. Plackett (1981) discusses the general analysis of Poisson data, while McCullagh (1980) discusses the various types of regression models that may be used for ordinal data.

3.6 Goodness of Fit for Poisson Data

The theoretical distribution for counts that is often considered first is the Poisson distribution, which represents events in either time or space as random processes and where the probabilities of counts of 0,1,2,... are given by:

$$p(x) = P\{X = x\} = \frac{e^{-\lambda}\lambda^x}{x!} \quad \text{for} \quad x = 0, 1, ...$$

where $p(x)$ is completely defined by the one parameter λ, which is the average count.

The following cases for goodness-of-fit tests for the Poisson data are considered: The parameter λ is known a priori, or unknown. When λ is known and interest centers on whether the observed counts follow the Poisson distribution with the given parameter, then, specifically, we wish to test $H_0 : X \sim P(\lambda)$ where λ is known. This is illustrated with Table 3.15, where we consider a one-way table (with k classes) having the observed frequencies n_i and corresponding expected frequencies under a Poisson distribution with parameter λ assumption.

x	0	1	2	...	$\geq k-1$	Total
Observed	n_0	n_1	n_2	$\cdots$	n_{k-1}	n
Expected	$\hat{n}_0$	$\hat{n}_1$	$\hat{n}_2$	$\cdots$	$\hat{n}_{k-1}$	n

Table 3.15: Table of observed and expected counts

With λ known, the expected values $m_x = n\pi_x$ for $x = 0, 1, \cdots, (k-1)$, where $\pi_x = \dfrac{e^{-\lambda}\lambda^x}{x!}$.

The corresponding Pearson's statistic is computed as:

$$X^2_{GOF} = \sum_{i=0}^{k-1} (n_i - m_i)^2 / m_i$$

and is distributed χ^2 with $(k-1)$ degrees of freedom.

Rarely is there any theoretical reason for expecting a particular value of λ. More often than not, λ has to be estimated from the sample. The maximum likelihood estimate of λ was shown from Chapter 1 to be

$$\hat{\lambda} = n^{-1} \sum y_i = \bar{y}.$$

Consequently,

$$\hat{\pi_x} = \frac{e^{-\hat{\lambda}}\hat{\lambda}^x}{x!} = \frac{e^{-\bar{y}}\bar{y}^x}{x!},$$

and $m_x = n\hat{\pi}_x$ for $x = 0, 1, ..., (k-1)$.

It follows that $X^2 = \sum_{i=0}^{k-1} (n_i - m_i)^2/m_i$ is distributed χ^2 with $(k-t-1)$ degrees of freedom where t is the number of parameters estimated from the sample. In this case, the d.f. equals $(k-2)$, since only one parameter, λ is estimated from the data.

Example 3.5: Horse Kicks in the Prussian Army

As an example to illustrate the fitting of a Poisson model to data, let us consider the frequency of death by horsekicks in the Prussian corps as recorded by Bortkiewicz (1898) for each of 14 corps for each of 20 years, giving a total of 280 observations.

Death per corps per year	0	1	2	3	4+
Frequency of occurrence	144	91	32	11	2

Table 3.16: Number of deaths from Prussian horse kicks

The hypothesis of interest centers on whether the observed data can be modeled by a Poisson distribution with parameter λ. An estimate of the mean death rate (that is, λ) is

$$\hat{\lambda} = \frac{0(144) + 1(91) + 2(32) + 3(11) + 4(2)}{280} = \frac{196}{280} = 0.70$$

and hence, $P[X = 0] = e^{-\hat{\lambda}} = 0.4966$. The remaining probabilities are obtained from the recursive formula given in chapter 2. (*Show this.*) Hence, $P[X = 1] = \frac{0.70}{1} P[X = 0] = 0.7 \times 0.4966 = 0.3476$ and similarly for probabilities for when $X = 2, 3$, and 4, respectively. These results are displayed in the next table.

Deaths	Probability	Expected frequency	Observed frequency
0	0.4966	139.05	144
1	0.3476	97.33	91
2	0.1217	34.08	32
3	0.0284	7.95	11
4+	0.0057	1.60	2

The computed G^2 and X^2 are 1.9978 and 2.1266, respectively, which are clearly not significant (pvalue = 0.5466) when compared with χ^2_3 (d.f. $3 = 5 - 1 - 1$) at $\alpha = 0.05$, and we therefore have no reason to doubt the data follows a Poisson distribution.

We can implement the above fit with the SAS® software statements in appendix B.6 with a partial output below from the program. Note that **FAC** variable in the program generates the factorials of 0 to 4. A more detailed output is provided in appendix B.6.

```
              Criteria For Assessing Goodness Of Fit

Criterion                  DF           Value          Value/DF
---------------------------------------------------------------
Deviance                    3          1.9978            0.6659
Pearson Chi-Square          3          2.1266            0.7089


                 Analysis Of Parameter Estimates

                             Standard       Wald 95%            Chi-
Parameter  DF  Estimate       Error    Confidence Limits      Square  Pr > ChiSq
---------------------------------------------------------------------------------
Intercept   1   -0.7028      0.0782   -0.8560    -0.5495      80.77      <.0001
DEATH       1   -0.3516      0.0720   -0.4928    -0.2104      23.82      <.0001
Scale       0    1.0000      0.0000    1.0000     1.0000
```

The estimate of λ is given by the negative of the intercept (0.7028), where $X = 0, 1, 2, 3, 4$. This estimate does not exactly match the estimate of $\hat{\lambda} = 0.70$ that we obtained earlier but is very close, and the model fits the data well. Similarly, the logarithm of the mean also gives the log estimate of the slope. That is, $-0.3516 = \log(0.7028)$.

Notice that one of the expected values above is less than 3. Here $r = 1$ and $d = 3$ and hence the minimum expected value, in order to use either the $C(\mathbf{m})$ or the lognormal approximation, cannot be lower than $r/d^{\frac{3}{2}} = 0.19$. The computed smallest expectation of 1.4 clearly satisfies this condition. Hence, the corresponding critical value from the $C(\mathbf{m})$ table at the 5% nominal level is 8.02, which indicates that the null hypothesis is tenable.

In the above analysis, we have assumed that more than four deaths per corps per year are impossible hence our estimate of λ from the Poisson model is not exactly 0.7. The Poisson model is based on the assumption that the sum of the probabilities $\sum_{i=0}^{4} p(x) = 1$ when $\hat{\lambda} = 0.7$. As seen below from the SAS® software output, these probabilities do not sum to 1 until the number of deaths gets to 6.

```
data horse;
sum=0.0;
do i=0 to 7;
if i-1 lt 0 then prob=poisson(.7,0);
    else prob=poisson(.7,i)-poisson(.7,i-1);
sum=sum+prob;
output;
end;
proc print data=horse noobs;
var i prob sum;
format prob sum 8.4;
run;
```

i	prob	sum
0	0.4966	0.4966
1	0.3476	0.8442
2	0.1217	0.9659
3	0.0284	0.9942
4	0.0050	0.9992
5	0.0007	0.9999
6	0.0001	1.0000

It is therefore desirable to consider the case where the number of deaths per corps per year is more than 4 to have zero frequencies. This can be implemented by setting the frequencies to zero for categories 5 and 6. When this new model is implemented, we now have the estimate of λ being equal to the negative of the intercept (0.7000), while the log of the mean is also given by the slope. That is, $0.7000 = e^{-0.3566}$. This model gives a $G^2 = 2.4378$ on 5 d.f., which is a better fit than the earlier model (pvalue = 0.7858). The corresponding SAS® software program is in appendix B.7 and again a partial output from the implementing the new model is displayed below.

Criteria For Assessing Goodness Of Fit

Criterion	DF	Value	Value/DF
Deviance	5	2.4378	0.4876
Pearson Chi-Square	5	2.3687	0.4737

Analysis Of Parameter Estimates

Parameter	DF	Estimate	Standard Error	Wald 95% Confidence Limits		Chi-Square	Pr > ChiSq
Intercept	1	-0.7000	0.0779	-0.8528	-0.5473	80.70	<.0001
DEATH	1	-0.3566	0.0714	-0.4966	-0.2166	24.91	<.0001
Scale	0	1.0000	0.0000	1.0000	1.0000		

Let us now consider some variations of the above assumptions. First, assume the data follow another single parameter distribution such as the negative binomial distributions instead of the Poisson distribution. The second possible violation of the above analyses is if, for instance, the means of the observations n_i follow some systemic pattern. For example, for data gathered over several days, the means might be constant within a day but vary over days (e.g., traffic accidents across different seasons).

In such situations, a more powerful test than the X^2_{GOF} against the alternatives is the *variance* test. The rationale for this procedure is based on the fact that both the mean and the variance of a Poisson variate are equal to λ. That is, $\hat{\mu}_x = \bar{x}$, $\hat{\sigma}^2_x = \bar{x}$.

Hence,

$$\frac{(x_i - \bar{x})^2}{\bar{x}} = z^2 \sim N(0,1)$$

Thus,

$$\begin{aligned} X^2_\nu &= \frac{\sum_{i=1}^n (x_i - \bar{x})^2}{\bar{x}} \\ &= \bar{x}^{-1} \sum_{i=1}^n (x_i - \bar{x})^2 \end{aligned}$$

The above is known as the variance test and under H_0, $X^2_\nu \sim \chi^2_{n-1}$. The degree of freedom $(n-1)$ is due to the fact that the summation is over n observations and only one parameter, namely, $\bar{x}$, was estimated from the data.

We note here that this statistic is calculated from the individual observations and does not require the construction of k classes as does the X^2_{GOF}. For this reason, X^2_ν can be used in small samples where X^2_{GOF} would be inappropriate.

Example

In the preceding example, $\hat{\lambda} = 0.70$ and we give in the next table the number of deaths and their corresponding observed values:

i	Observed frequencies
0	144
1	91
2	32
3	11
4+	2
Totals	280

The variance test is now computed as follows:

$$\begin{aligned} X_\nu^2 &= \frac{144(0-0.7)^2}{0.7} + \frac{91(1-0.7)^2}{0.7} + \frac{32(2-0.7)^2}{0.7} + \frac{11(3-0.7)^2}{0.7} + \frac{2(4-0.7)^2}{0.7} \\ &= 100.80 + 11.70 + 77.257 + 83.129 + 31.114 \\ &= 304.00 \end{aligned}$$

because the grouped data can be visualized as the following:

$$\underbrace{0,0,...,0,}_{144}\ \underbrace{1,1,...,1,}_{91}\ \cdots,\ \underbrace{3,3,...,3,}_{11}\ \underbrace{4,4}_{2}$$

The above is based on 279 (280−1) degrees of freedom with a pvalue 0.1454. That is, we have no reason to doubt that the data actually came from a Poisson distribution.

3.6.1 Tests For Change in Level of Poisson Data

Consider a situation in which the first n_1 observations have a common mean λ_1 and the additional $n_2 = (n - n_1)$ observations have a common mean λ_2. Such a case usually arises in situations when we are interested in testing that two Poisson distributions (the first having sample size n_1 and parameter λ_1 and the second with sample size n_2 and parameter λ_2) have the same parameter λ. Then it is of interest to test

$$H_0 : \lambda_1 = \lambda_2 \quad \text{versus} \quad H_a : \lambda_1 \neq \lambda_2$$

Under H_0,

$$Z = \frac{\bar{x}_1 - \bar{x}_2}{\sigma\sqrt{\left(\frac{1}{n_1} + \frac{1}{n_2}\right)}} \sim N(0,1)$$

where σ^2 is the common variance. From this we have

$$\begin{aligned} Z^2 &= \frac{(\bar{x}_1 - \bar{x}_2)^2}{\sigma^2\left(\frac{1}{n_1} + \frac{1}{n_2}\right)} \\ &= \frac{n_1 n_2}{n_1 + n_2} \frac{(\bar{x}_1 - \bar{x}_2)^2}{\sigma^2} \end{aligned}$$

and is distributed χ^2 with *one* degree of freedom. Under H_0, we can estimate the common variance as

$$\hat{\sigma}^2 = \bar{x}_p \quad = \frac{(n_1\bar{x}_1 + n_2\bar{x}_2)}{n}$$

Thus we have, under H_0,

$$X^2 = \frac{n_1 n_2}{n_1 + n_2} \frac{(\bar{x}_1 - \bar{x}_2)^2}{\bar{x}_p} \quad \sim \chi_1^2$$

Example

Consider as an example the data below, which relate to the number of accidents per working day in the months of December and January at a certain busy university road in Minnesota.

```
December
                    3 3 3 4 9 7 6 4 3 4 8 4 2 5 5
January
                    4 1 4 3 1 3 3 5 0 2 5 2 3 3 4 5 3 5 3 8
```

For this data, $n_1 = 15, n_2 = 20$, and $n = n_1 + n_2 = 35$; $\bar{x}_1 = 4.6667$, $\bar{x}_2 = 3.350$, and $\bar{x}_p = 3.9143$; hence

$$X^2 = \frac{15 \times 20}{35} \frac{(4.6667 - 3.500)^2}{3.9143} = 2.981$$

The corresponding pvalue equals 0.0842. Thus we do not have sufficient evidence to reject the null hypothesis that the rates of accidents for the two months are equal.

3.6.2 Linear Trend Models

We shall illustrate the implementation of a linear trend model with the data in the following example.

The data in Table 3.17 relate to parity (number of children born prior to present condition) distribution of 186 infertile patients that reported in the Maternity Wing of the University of Ilorin (Nigeria) teaching hospital between 1982 and 1987.

Parity (Y)	No. of cases	Proportions
i	n_i	p_i
0	86	0.462
1	41	0.220
2	29	0.156
3	14	0.076
4	9	0.048
5	5	0.027
6+	2	0.011

Table 3.17: Parity distribution of 186 patients

In the constant-parity (or equiprobable) model, the expected parity rates m_i are all equal to some constant μ. Then one obtains the model in the form

$$\log(m_i) = \lambda \quad \text{where } \lambda = \log \mu$$

Using PROC GENMOD, the MLE is $\hat{\lambda} = 3.280$ with an estimated asymptotic standard error (a.s.e $= 0.07318$ and $G^2 = 178.18$ on 6 degrees of freedom, which clearly indicates a lack of fit of the model.

A good guess on the adequacy of the model is also provided by the following (Lindsey, 1995). If

$$| \text{Parameter estimate} | \geq 2 \text{ a.s.e.}$$

then the estimate is significantly different from zero at the 5% nominal level. Clearly for the above model $| 3.280 | > 2(0.07318) = 0.1464$. The Lindsey result is motivated by the fact that under the null hypothesis H_0 : parameter = 0, and the standardized z score = $\frac{\text{Parameter estimate}}{ASE}$ is distributed normally with mean 0 and variance 1. Since, for normal distributions, the interval $\mu \pm 2\sigma$ contains 95% of the data, it is therefore reasonable to assume that any parameter value whose standardized z score is more than 2 in absolute value must belong either to the bottom 2.5% or the upper 2.5%.

The SAS® software program and a modified output for fitting the equiprobability model are displayed next. The estimated model is $\log{(n_k)} = 3.2798$.

```
DATA EXAMPLE;
INPUT PARITY COUNT INT @@;
DATALINES;
0 86 1 1 41 1 2 29 1 3 14 1 4 9 1 5 5 1 6 2 1
;
PROC GENMOD DATA=EXAMPLE order=data;
make 'obstats' out=aa;
model count=/DIST=POI LINK=LOG OBSTATS;
run;
proc print data=aa noobs;
var count pred resraw reschi;
format pred resraw reschi 10.4;
run;
```

Criteria For Assessing Goodness Of Fit

Criterion	DF	Value	Value/DF
Deviance	6	178.1752	29.6959
Pearson Chi-Square	6	198.7742	33.1290

Algorithm converged.

Analysis Of Parameter Estimates

Parameter	DF	Estimate	Standard Error	Wald 95% Confidence Limits		Chi-Square	Pr > ChiSq
Intercept	1	3.2798	0.0733	3.1361	3.4235	2000.86	<.0001
Scale	0	1.0000	0.0000	1.0000	1.0000		

Summary Data set

COUNT	Pred	Resraw	Reschi
86	26.5714	59.4286	11.5289
41	26.5714	14.4286	2.7991
29	26.5714	2.4286	0.4711
14	26.5714	-12.5714	-2.4388
9	26.5714	-17.5714	-3.4088
5	26.5714	-21.5714	-4.1848
2	26.5714	-24.5714	-4.7668

Looking at the raw residuals above, we observe that the first three are positive and the last four are negative, indicating that the probability of infertility is more than average in those patients with high parity values and less than average in patients with fewer parities.

Suppose that the probability of infertility diminishes in the same proportion between any two consecutive parity values. That is,

$$\frac{\pi_k}{\pi_{k-1}} = \phi$$

Then,

$$\frac{\pi_k}{\pi_1} = \phi^{k-1}$$

and this implies that

$$\log\left(\frac{\pi_k}{\pi_1}\right) = (k-1)\log(\phi)$$

But $\pi_k = m_k/n$ where m_k is the expected frequency in category k and n is the sample size such that $\sum m_k = n$. Thus,

$$\begin{aligned}\log\left(\frac{\pi_k}{\pi_1}\right) &= \log\left(\frac{m_k}{m_1}\right) \\ \Rightarrow \log(m_k) &= \log(m_1) + (k-1)\log(\phi) \\ &= \log(m_1/\phi) + k\log(\phi)\end{aligned}$$

which is of the form

$$\log(m_k) = \beta_0 + \beta_1\, k \tag{3.20}$$

where

$$\beta_0 = \log(m_1/\phi) \quad \text{and} \quad \beta_1 = \log(\phi)$$

Fitting this log linear linear-trend model using GENMOD, we have

$$\hat{\beta}_0 = 4.4217 \quad \hat{\beta}_1 = -0.5794 \quad \text{and}$$

$$\log(m_k) = 4.4217 - 0.5794\, k$$

with fitted $G^2 = 1.3828$ on 5 d.f. The model clearly fits the data well (pvalue = 0.9262), and the negative value of β_1 indicates a decrease in infertility with previous parity. Since $\hat{\beta}_1 = \log(\phi)$, $\phi = 0.5602$ is the proportional decline in probability per parity.

Hence,

$$\pi_k = \pi_1\phi^{k-1} = \pi_1 e^{\hat{\beta}_1(k-1)}$$

Because $\hat{\beta}_1$ is negative, this is a model of exponential decay. If $\beta_1 > 0$, it would have been a model of exponential growth. Below are the SAS® software program and the accompanying modified output.

```
set example;
PROC GENMOD DATA=EXAMPLE order=data;
make 'obstats' out=bb;
model count=PARITY/DIST=POI LINK=LOG OBSTATS;
run;
proc print data=bb noobs;
var count pred resraw reschi;
format pred resraw reschi 10.4;
run;
```

Criteria For Assessing Goodness Of Fit

Criterion	DF	Value	Value/DF
Deviance	5	1.3828	0.2766
Pearson Chi-Square	5	1.3579	0.2716

Analysis Of Parameter Estimates

Parameter	DF	Estimate	Standard Error	Wald 95% Confidence Limits		Chi-Square	Pr > ChiSq
Intercept	1	4.4217	0.0944	4.2368	4.6066	2196.01	<.0001
PARITY	1	-0.5794	0.0516	-0.6806	-0.4782	126.00	<.0001
Scale	0	1.0000	0.0000	1.0000	1.0000		

Summary Data set

COUNT	Pred	Resraw	Reschi
86	83.2388	2.7612	0.3026
41	46.6330	-5.6330	-0.8249
29	26.1252	2.8748	0.5624
14	14.6362	-0.6362	-0.1663
9	8.1996	0.8004	0.2795
5	4.5937	0.4063	0.1896
2	2.5735	-0.5735	-0.3575

We also consider fitting a gamma-type model to the data. This is accomplished by defining a new variable, which is the logarithm of X, and then fitting the two-parameter gamma distribution. The results of such a fit is reproduced below from a GENMOD output. We added 1 to parity values to avoid taking the logarithm of zero. The values of X now ranges from 1 to 7.

```
set example;
PP=PARITY+1;
PLOG=LOG(PP);
PROC GENMOD DATA=EXAMPLE order=data;
make 'obstats' out=aa;
model count=PP plog/dist=poi link=log obstats;
run;
proc print data=aa noobs;
var count pred reschi streschi;
format pred reschi streschi 7.4;
run;
```

Criteria For Assessing Goodness Of Fit

Criterion	DF	Value	Value/DF
Deviance	4	1.2758	0.3190
Pearson Chi-Square	4	1.2586	0.3147

Analysis Of Parameter Estimates

Parameter	DF	Estimate	Standard Error	Wald 95% Confidence Limits		Chi-Square	Pr > ChiSq
Intercept	1	4.9596	0.1827	4.6016	5.3176	737.10	<.0001
PP	1	-0.5208	0.1857	-0.8846	-0.1569	7.87	0.0050
PLOG	1	-0.1466	0.4474	-1.0235	0.7303	0.11	0.7432
Scale	0	1.0000	0.0000	1.0000	1.0000		

COUNT	Pred	Reschi	Streschi
86	84.6778	0.1437	0.9051
41	45.4451	-0.6594	-0.9666
29	25.4400	0.7058	0.8803
14	14.4892	-0.1285	-0.1457
9	8.3307	0.2319	0.2656
5	4.8185	0.0827	0.0995
2	2.7986	-0.4774	-0.5987

As expected, the model fits the data very well, being based on a G^2 value of 1.2758 on 4 d.f. (pvalue = 0.8655) as compared to 1.3828 on 5 d.f. for the exponential.

The exponential is certainly the more parsimonious in this case. Of course, the exponential is a special case of the gamma distribution with α being 1 in the two-parameter gamma, $\Gamma(\alpha, \beta)$.

3.7 Local Effects Models

Lindsey (1995) describes fitting local effects models to certain data, where models are fitted to only some observations in the data, the rest being ignored. The ignored data values are often described as outliers. We consider below the following data which relate to intervals between explosions in British mines between 1851 and 1962 (Jarret, 1979). The distribution of the number of accidents on each day of the week during the period covered is presented in Table 3.18.

Sunday	5
Monday	19
Tuesday	34
Wednesday	33
Thursday	36
Friday	35
Saturday	29

Table 3.18: Weekly distribution of number of accidents: 1851 - 1962

Interest centers on how the accidents vary over the week. We first consider fitting the equiprobability model to the data. This model gives a $G^2 = 37.705$ on 6 d.f. The GENMOD output for this model is given below. Examination of the residuals indicate that in the first two weekdays, the model overestimates the rate and underestimates the rate the last five days (Tuesday to Saturday). Sunday has a significant residual of -4.266. The SAS® software program for implementing the models discussed here is presented in appendix B.8.

```
Obs    DAYS    COUNT    cex    weight    LL
-------------------------------------------
1      1       5        0      1         1
2      2       19       0      2         4
3      3       34       1      3         9
4      4       33       1      3         16
5      5       36       1      3         25
6      6       35       1      3         36
7      7       29       1      3         49

COUNT       Pred       Reschi
-----------------------------
    5    27.2857      -4.2664
   19    27.2857      -1.5862
   34    27.2857       1.2854
   33    27.2857       1.0939
   36    27.2857       1.6683
   35    27.2857       1.4768
   29    27.2857       0.3282
```

A revised model, which assigns 0 to Sunday and Monday and 1 to Tuesday through Saturday, was next fitted. This dichotomous variable is the variable CEX in the output. That is,

$$\text{CEX} = \begin{cases} 0 & \text{if days=1, } \quad 2 \\ 1 & \text{elsewhere} \end{cases}$$

This model is implemented with the SAS® software program below together with a partial output.

```
set local;
if 0 le days le 2 then cex=0;
else cex=1;
          Criteria For Assessing Goodness Of Fit

Criterion                  DF           Value          Value/DF
----------------------------------------------------------------
Deviance                    5          9.6028            1.9206
----------------------------------------------------------------

                         COUNT        Pred      Reschi
                         ---------------------------
                             5     12.0000     -2.0207
                            19     12.0000      2.0207
                            34     33.4000      0.1038
                            33     33.4000     -0.0692
                            36     33.4000      0.4499
                            35     33.4000      0.2769
                            29     33.4000     -0.7613
```

The model fits the data with a $G^2 = 9.603$ on 5 d.f., but the residuals for Sunday and Monday are again significant. A look at the plot in Figure 3.1 of the number of accidents versus day of the week indicates that perhaps a quadratic model might be appropriate.

We next therefore fit a quadratic model to the data, and the model gives $G^2 = 4.4578$ on 4 d.f. (pvalue = 0.3476). This model fits the data and the residuals are now not significant. The parameter estimates for this model are $\hat{\beta}_0 = 1.2388$, $\hat{\beta}_1 = 0.9848$, and $\hat{\beta}_2 = -0.0990$. A partial SAS® software output is also displayed for this model.

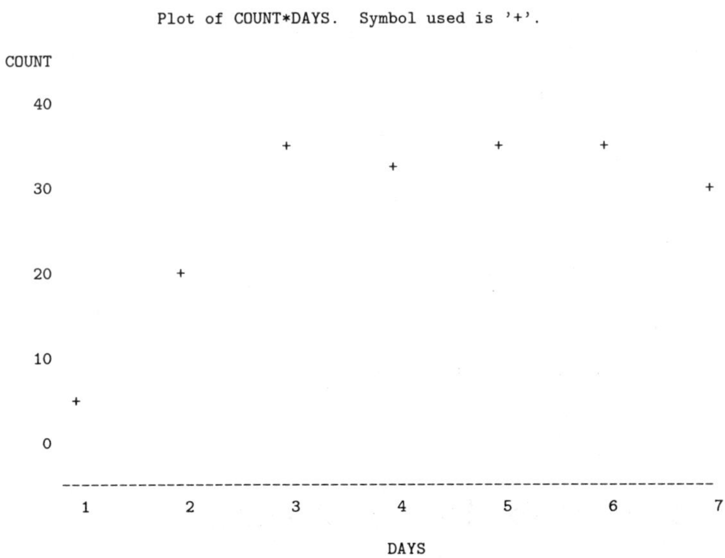

Figure 3.1: Plot of count by days

```
              Criteria For Assessing Goodness Of Fit

Criterion                    DF            Value           Value/DF
-------------------------------------------------------------------
Deviance                      4           4.4578             1.1145
Pearson Chi-Square            4           4.3530             1.0883
-------------------------------------------------------------------

                            COUNT        Pred      Reschi
                            ---------------------------
                                5      8.3692     -1.1646
                               19     16.6474      0.5766
                               34     27.1640      1.3116
                               33     36.3602     -0.5573
                               36     39.9249     -0.6212
                               35     35.9621     -0.1604
                               29     26.5724      0.4709
```

But then, if one may ask, is this the most parsimonious model? To answer this question, we next fit another model, which involves a new factor variable we call weight (WT), which takes values 1 and 2, respectively, on Sunday and Monday, and value 3, Tuesday through Saturday. That is,

$$\text{wt} = \begin{cases} 1 & \text{if days} = 1 \\ 2 & \text{if days} = 2 \\ 3 & \text{elsewhere} \end{cases}$$

This model ignores the first two weekdays (Sunday and Monday) and then fits the equiprobability model to the remaining days. This model fits with a $G^2 = 0.8952$ on 4 d.f. (pvalue $= 0.9252$). Thus we can conclude that the number of accidents between Tuesday and Saturday is equally probable but varies between Sunday and Monday. This latter model is implemented in SAS® with the following statements, and again with a partial output.

```
set local;
if days eq 1 then weight=1;
else if days eq 2 then weight=2;
else weight=3;
proc genmod order=data;
class weight;
model count=weight/dist=poi;
run;
```

```
            Criteria For Assessing Goodness Of Fit

Criterion                    DF            Value           Value/DF
-------------------------------------------------------------------
Deviance                      4           0.8952             0.2238
Pearson Chi-Square            4           0.8743             0.2186
-------------------------------------------------------------------

                          COUNT        Pred      Reschi
                          ---------------------------
                              5      5.0000      0.0000
                             19     19.0000      0.0000
                             34     33.4000      0.1038
                             33     33.4000     -0.0692
                             36     33.4000      0.4499
                             35     33.4000      0.2769
                             29     33.4000     -0.7613
```

3.8 Goodness of Fit for Binomial Data

Suppose there are k experiments, each of which consists of n independent trials that result in either success (S) with probability π or failure (F) with probability $1 - \pi$.

The resulting data may be displayed as in Table 3.19.

Experiment number	Response S	Response F	Total
1	n_{11}	n_{12}	n
2	n_{21}	n_{22}	n
⋮	⋮	⋮	⋮
k	n_{k1}	n_{k2}	n
Totals	n_{+1}	n_{+2}	$N = kn$

Table 3.19: Binomial experiment data display

If $x_i = n_{i1}$ is the number of successes in the i-th experiment, we wish to test

$$H_0 : x_i \quad \sim \quad \mathrm{b}\,(n, \pi) \qquad i = 1, 2, ..., k$$

and the corresponding categories are:

	Number of successes					
Frequency	0	1	2	$\cdots$	n	Totals
Observed	n_0	n_1	n_2	$\cdots$	n_n	k
Expected	m_0	m_1	m_2	$\cdots$	m_n	k

where n_s denotes the number of times that

$$x_i = s \quad \text{for} \quad s = 0, 1, 2, \cdots, n$$

The expected frequencies are determined according to whether or not the success rate π is specified (CASE I) or estimated from the data (CASE II).

Below is an example of 100 simulated experiments ($k = 100$) from a binomial distribution with $n = 5$ and $\pi = 0.3$. The numbers of successes for each of the experiments are also displayed. For instance, the first experiment results in 0 successes, the second in 2, the third in 3, etc. These values correspond respectively to $n_{11}, n_{21}, n_{31}, ...$ The frequency distributions of these are then tallied, resulting in a distribution to the table of observed and expected frequencies above.

```
0  2  3  2  1  1  1  2  1  1  0  3  3  1  0
1  4  1  1  2  1  0  2  0  3  1  1  1  3  1
2  1  2  1  1  1  3  2  2  1  0  2  2  0  1
5  1  2  2  2  1  0  4  0  3  3  1  1  1  0
1  2  3  1  1  1  0  0  0  2  2  1  2  2  1
0  2  3  1  0  2  1  3  1  0  3  0  4  0  0
2  1  2  1  2  1  1  0  2  2

    X  COUNT
    --------
    0    20
    1    38
    2    26
    3    12
    4     3
    5     1
   N=   100
```

3.8.1 Case I: π Specified

If π is specified, then

$$m_s = k\binom{n}{s}\pi^s(1-\pi)^{n-s} \quad \text{for} \quad s = 0, 1, 2, ..., n \tag{3.21}$$

since there are k repeated binomial trials. The resulting X^2_{GOF} statistic has an approximate χ^2 distribution with $(n+1) - 1 = n$ degrees of freedom.

Case II: π Estimated from Data

For a $b(n, \pi)$ the likelihood can be written as

$$\begin{aligned} L(x;\pi) &= \prod_{i=1}^{k}\binom{n}{n_{k1}}\pi^{n_{k1}}(1-\pi)^{n-n_{k1}} \\ &= C\pi^{\sum n_{k1}}(1-\pi)^{kn-\sum n_{k1}} \\ &= C\pi^{n_{+1}}(1-\pi)^{N-n_{+1}} \end{aligned}$$

where C is a constant. Then

$$\log L = \log C + n_{+1}\log\pi + N\log(1-\pi) - n_{+1}\log(1-\pi)$$

and the MLE of π is

$$\hat{\pi} = \frac{n_{+1}}{N},$$

$$m_s = k\binom{n}{s}(\frac{n_{+1}}{N})^s(1-\frac{n_{+1}}{N})^{n-s}, \quad \text{for } s = 0, 1, ..., n \tag{3.22}$$

and

$$X^2_{GOF} = \sum_{s=0}^{n}(n_s - m_s)^2/m_s$$

is approximately distributed as a χ^2 with $n-1 = (n+1)-1-1$ degrees of freedom.

Example

The following example in Table 3.20 is taken from Mead and Curnow (1983). It is claimed that housewives cannot tell product A from product B. To test this assertion, 100 housewives are each given eight pairs of samples, each pair containing one sample of product A and one of product B, and are asked to identify the A sample. The number of times (out of eight) that each housewife is correct is recorded as follows:

In this example, $k = 100$, $n = 8$, and hence $N = 800$. The following two possible null hypotheses are considered:

(a) The housewives have no discriminatory power, in which case the null hypothesis is of the form $H_{01} : \pi = \frac{1}{2}$ against the alternative $H_{11} : \pi \neq \frac{1}{2}$.

(b) The housewives have discriminatory ability and the null hypothesis will be of the form $H_{02} : \pi = \pi_0 > \frac{1}{2}$.

For case (a), we are interested in testing the null hypothesis

$$H_{01} : \pi = \frac{1}{2}$$

Time correct	Number of housewives	Expected values under H_{01}	Expected values under H_{02}
0	2	0.39	0.11
1	0	3.13	1.16
2	5	10.94	5.46
3	13	21.88	14.76
4	36	27.34	24.96
5	13	21.88	27.02
6	18	10.94	18.28
7	10	3.13	7.07
8	3	0.39	1.19
Totals	100	100.02	100.01

Table 3.20: Results from binomial fits

against the alternative

$$H_{11} : \pi \neq \frac{1}{2}$$

Under H_0, π is known and the problem reduces to Case I in subsection (3.8.1). The expected values based on the null hypothesis are computed using (3.21) as:

$$\hat{m}_0 = 100\binom{8}{0}(\frac{1}{2})^8$$

$$\vdots = \quad \vdots$$

$$\hat{m}_8 = 100\binom{8}{8}(\frac{1}{2})^8$$

These expected values are given in column 3 of Table 3.20. Observe that two expected values are less than 3. The minimum expected values by the Lawal (1980) rule should be $\hat{m} = r/d^{\frac{3}{2}}$ where $r = 2$ in this problem and $d = 8$. Hence, the minimum expected value equals 0.08. Because the computed expected values are all greater than 0.08, we can justifiably employ either the lognormal approximation or the critical values obtained from the $C(\mathbf{m})$ tables.

To use the $C(\mathbf{m})$ table in Lawal (1980), first compute an effective small expectation as:

$$\hat{m}_G = (0.39 \times 0.39)^{\frac{1}{2}} = 0.39$$

Hence, at the 5% and 1% levels, the computed X^2 will be compared with the $C(\mathbf{m})$ value of 16.30 and 23.62, respectively.

The computed Pearson's $X^2 = 60.06$, which clearly indicates that H_0 should be rejected, concluding that the housewives have discriminatory ability.

To test for these discriminating powers, it is required to test whether these powers are the same for different housewives and whether the results within each set of eight are independent. What this implies is that we are interested in testing whether the data are compatible with a binomial distribution with probability π, the proportion of a correct identification, greater than $\frac{1}{2}$. Thus we wish to test the null hypothesis

$$H_{02} : \pi = \pi_0 > \frac{1}{2}$$

against the alternative

$$H_{22} : \pi \leq \frac{1}{2}$$

Based on the data, an estimate of π is computed from Table 3.20 as follows:

$$\hat{\pi} = \frac{0(2) + 1(0) + 2(5) + 3(13) + \cdots + 8(3)}{8 \times 100} = \frac{460}{800} = 0.575$$

That is, $\hat{\pi} = 0.575$ and $(1 - \hat{\pi}) = 0.425$.

The expected values based on the null hypothesis are computed using (3.21) or (3.22) and are displayed in column 4 of Table 3.20. They are again computed as follows:

$$\hat{m}_0 = 100\binom{8}{0}(0.575)^0(0.425)^8$$

$$\vdots = \quad \vdots$$

$$\hat{m}_8 = 100\binom{8}{8}(0.575)^8(0.425)^0$$

Here, there are $r = 3$ expected values less than 3, and with $d = 7$ in this case, the required minimum expected value is again computed as $3/7^{\frac{3}{2}} = 0.16$. Since one of the expected cell frequencies 0.11 is less than this minimum allowable value, it is necessary to collapse this cell with the adjacent second cell to give observed and expected values of 2 and 1.27, respectively, with the corresponding decrease in the degree of freedom from 7 to 6. With this new setup, the minimum expected frequency for $r = 2$ and $d = 6$ is now $2/6^{\frac{3}{2}} = 0.13$. The expected values satisfy this minimum requirement and consequently

$$\hat{m}_G = (1.27 \times 1.19)^{0.5} = 1.23$$

The corresponding tabulated values with $k = 6, r = 2$ are, by interpolation, $C(\mathbf{m}) = 12.91$ and $C(\mathbf{m}) = 18.32$ at the 5% and 1% levels, respectively. The computed Pearson's $X^2 = 16.17$. Hence, the null is rejected at the 5% level but we would fail to reject the null at the 1% level. Observe that the results would be based on 8 classes instead of 6 if we were to use the usual χ^2 approximation.

Thus, at the 1% point, there is no reason to doubt that the binomial model fits these data. An estimate of the variance of the estimator is given by Var $= \dfrac{\hat{\pi}(1-\hat{\pi})}{N} = \dfrac{(0.575)(0.425)}{800} = 0.0003055$, and an approximate 99% confidence interval for the discriminating probability is

$$0.575 \pm 2.58\sqrt{0.0003055} \quad \text{or} \quad (0.530, 0.620)$$

3.8.2 Variance Test for the Binomial Distribution

(a) Suppose that rather than the binomial distribution, the data follow a different discrete distribution (usually with a larger variance) or

(b) The probabilities of success π_i follow some systemic source of variation.

Situation (a) usually arises if the binary responses are correlated or if the wrong explanatory-response model is assumed when the actual relationship is different from that assumed (e.g., linear model versus quadratic model). Similarly, (b) usually arises in situations when one or more explanatory variables or interactions have been omitted from the model.

In both situations like these, the variance test is more powerful than the X^2_{GOF} test. The resulting data can be displayed in Table 3.21. Note that the sample sizes n_i may differ.

Experiment number	Response S	F	Total
1	n_{11}	n_{12}	n_{1+}
2	n_{21}	n_{22}	n_{2+}
$\vdots$	$\vdots$	$\vdots$	$\vdots$
k	n_{k1}	n_{k2}	n_{k+}
Totals	n_{+1}	n_{+2}	N

Table 3.21: Resulting data arranged in a table

Let $\pi_i = \dfrac{n_{i1}}{n_{i+}}$ for $i = 1, 2, ..., k$. Then

$$\hat{\pi} = \frac{n_{+1}}{N}$$

and the variance test can be expressed as:

(i)

$$\begin{aligned} X^2_\nu &= \sum_{i=1}^{k} \frac{(n_{i1} - n_{i+}\pi_0)^2}{n_{i+}\pi_0(1-\pi_0)} \\ &= \sum_{i=1}^{k} \frac{n_{i+}(\pi_i - \pi_0)^2}{\pi_0(1-\pi_0)} \end{aligned} \tag{3.23}$$

where π_0 is specified a priori. Under H_0, the statistic X^2_ν is distributed χ^2 with k degrees of freedom.

(ii) When π is unknown,

$$\begin{aligned} X^2_\nu &= \sum_{i=1}^{k} \frac{(n_{i1} - n_{i+}\hat{\pi})^2}{n_{i+}\hat{\pi}(1-\hat{\pi})} \\ &= \sum_{i=1}^{k} \frac{n_{i+}(\hat{\pi}_i - \hat{\pi})^2}{\hat{\pi}(1-\hat{\pi})}. \end{aligned} \tag{3.24}$$

This variance test is equivalent to the *overdispersion* test that will be discussed in chapter 8 in the context of the linear logistic modeling of binary data. The statistic is based on $k - 1$ degrees of freedom, since one parameter is estimated from the data.

Example

The variance test for the data in Table 3.20 can be conducted as follows:

$$\begin{aligned}X_\nu^2 &= \sum_{i=1}^{100} n_{i+}(\hat{\hat{\pi}}_i - \hat{\pi})^2/\hat{\pi}(1-\hat{\pi}) \\ &= \frac{8}{(0.575)(0.425)} \sum_{i=1}^{100} (\hat{\hat{\pi}}_i - 0.575)^2 \\ &= 32.73657 \left\{ \sum_{i=1}^{100} \pi_i^2 - 100(0.575)^2 \right\} \\ &= 128.900\end{aligned}$$

on 99 degrees of freedom and where

$$\begin{aligned}\sum_{i=1}^{100} \hat{\hat{\pi}}^2 &= 2(\frac{0}{8})^2 + 0(\frac{1}{8})^2 + 5(\frac{2}{8})^2 + \\ &13(\frac{3}{8})^2 + 36(\frac{4}{8})^2 + 13(\frac{5}{8})^2 + 18(\frac{6}{8})^2 + 10(\frac{7}{8})^2 + 3(\frac{8}{8})^2 \\ &= 37.000\end{aligned}$$

The above variance test is implemented in SAS® with either PROC LOGISTIC or PROC GENMOD. The following program along with its partial output gives the relevant results.

```
data new;
input r count @@;
n=8;
datalines;
0 2 1 0 2 5 3 13 4 36 5 13 6 18 7 10 8 3
;
run;
proc logistic data=new; freq count;
model r/n=/scale=none; run;

        Deviance and Pearson Goodness-of-Fit Statistics

Criterion          DF         Value      Value/DF      Pr > ChiSq
-----------------------------------------------------------------
Deviance           99      149.2817        1.5079          0.0008
Pearson            99      128.9003        1.3020          0.0234

Number of events/trials observations: 100

proc genmod;
model r/n= /dist=b;
freq count;
run;
                    Criteria For Assessing Goodness Of Fit

          Criterion                      DF           Value         Value/DF
          ------------------------------------------------------------------
          Deviance                       99        149.2817           1.5079
          Pearson Chi-Square             99        128.9003           1.3020
```

An alternative approximation suggested by Cochran (1954) utilizes the normal distribution and is given by

$$z = \frac{X_\nu^2 - E(X_\nu^2)}{\sqrt{\text{Var}(X_\nu^2)}}$$

where

$$E(X_\nu^2) = 99$$

and

$$\text{Var}(X^2_\nu) = 2(k-1).\left(\frac{n-1}{n}\right) = (2 \times 99)\frac{7}{8} = 173.250$$

Hence,

$$z = \frac{128.900 - 99}{\sqrt{173.250}} = 2.27$$

We observe that the pvalue is < 0.0001. This test has been shown to be more powerful than the X^2 test.

3.9 Exercises

1. The power divergence goodness-of-fit test statistic is defined as:

$$I(\lambda) = \frac{2}{\lambda(\lambda+1)} \sum_i n_i \left\{ \left(\frac{n_i}{m_i}\right)^\lambda - 1 \right\}$$

   where $-\infty < \lambda < \infty$.
   Show that:
   (i) When $\lambda = 1$, $I(\lambda)$ equals the Pearsons test statistic X^2.
   (ii) As $\lambda \to -1$ or 0, $I(\lambda)$ converges respectively to $2\sum_i m_i \ln\left(\frac{m_i}{n_i}\right)$, and $2\sum_i n_i \ln\left(\frac{n_i}{m_i}\right)$.

2. For the data relating to methods of child delivery in eclamptic patients, estimate the probabilities if the null hypothesis is of the form: $H_o : \frac{3}{4}\pi_1 = \pi_2 = 3\pi_3 = \pi_4 = 7\pi_5$. Test this hypothesis and compare your results with those obtained in the text.

3. The 1982 General Social Survey of political opinions of subjects interviewed has seven categories, which are displayed in the next table as

(1)	(2)	(3)	(4)	(5)	(6)	(7)
31	136	217	583	206	201	55

   where (1) extremely liberal, (2) liberal, (3) slightly liberal, (4) moderate, (5) slightly moderate, (6) conservative, (7) extremely conservative. Fit the equiprobable model, and the linear and quadratic models to these data. Why does the symmetry model fail to fit this data (partition the symmetry G^2 into its components)?

4. The number of shutdowns per day caused by a breaking of the thread was noted for a nylon spinning process over a period of 10 days (Ott, 1984). The data collected are displayed below:

y_i	0	1	2	3	4	≥ 5
n_i	20	28	15	8	7	12

   Fit a Poisson distribution to the above data and test for goodness of fit at $\alpha = 0.05$ level of significance. Also do a variance test for the data.

5. The progeny in a qualitative experiment were expected to segregate into classes A, B, C, and D in the ratios 9:3:3:1. The numbers actually observed in the respective classes were 267, 86, 122, and 25. Were these data significantly compatible with the theoretical expected numbers? (From Box, 1987.)

6. The data below (Haberman, 1978) relate to individuals randomly selected California responding to one stressful event and time to illness.

Individuals reporting one stressful event.

Months before	Number
1	15
2	11
3	14
4	17
5	5
6	11
7	10
8	4
9	8
10	10
11	7
12	9
13	11
14	15
15	6
16	1
17	1
18	4

Fit an exponential decay model to the above data.

7. Jarret (1979) also gave the following data, relating to the distribution of the mine explosions in British mines analyzed in section (3.6).

Distribution of explosions by month

Months	Number
January	14
February	20
March	20
April	13
May	14
June	10
July	18
August	15
September	11
October	16
November	16
December	24

Study to see if there is any systemic change over the year.

8. A class of 100 students took an exam consisting of 15 questions, with results as summarized below. Assume these data behave as a random sample. Fit the binomial distribution to this data and state your conclusion.

Number of questions missed	Number of students
0	1
1	6
2	16
3	23
4	23
5	17
6	9
7	4
8	1
≥ 9	0

9. The numbers of tomato plants attacked by spotted wilt disease were counted in each of 160 areas of 9 plants. The results are displayed below (Snedecor & Cochran, 1973):

No.of diseased plants	0	1	2	3	4	5	6	7	Total
n_i	36	48	38	23	10	3	1	1	160

Fit a binomial distribution to this data and perform the relevant goodness-of-fit test.

Chapter 4

Models for 2 × 2 Contingency Tables

4.1 Introduction

In this chapter, I shall consider the case where subjects are jointly classified by two dichotomous classificatory variables **A** and **B** indexed by $i = 1, 2$ and $j = 1, 2$ respectively, providing four category combinations. The joint frequency distribution of the two variables is frequently displayed in an array called a 2×2 contingency table. If we let n_{ij} denote the number of subjects in the sample who are jointly classified as belonging to the i-th level of A and the j-th level of B, then these data can be summarized as shown in Table 4.1.

Observed Frequencies

A	B: 1	B: 2	Total
1	n_{11}	n_{12}	n_{1+}
2	n_{21}	n_{22}	n_{2+}
Total	n_{+1}	n_{+2}	N

Population Probabilities

A	B: 1	B: 2	Total
1	π_{11}	π_{12}	π_{1+}
2	π_{21}	π_{22}	π_{2+}
Total	π_{+1}	π_{+2}	1

Table 4.1: Notation for the (2×2) contingency table

Each entry in the body of the table is said to refer to a cell of the table.

In this formulation, $n_{i+} = \sum_j n_{ij}$ denotes the marginal number of subjects in the i-th level of A. Similarly, $n_{+j} = \sum_i n_{ij}$ denotes the marginal total number of subjects classified in the j-th level of B and $N = \sum_i \sum_j n_{ij}$ denotes the total sample size. If the row margin is assumed to be fixed, these totals will be denoted by $M_1 = \{n_{1+}, n_{2+}\}$, and similarly, if the column margin is assumed to be fixed, these totals will be denoted as $M_2 = \{n_{+1}, n_{+2}\}$.

Let us consider a simple example of a data set that give rise to a 2×2 contingency table. In Table 4.2 are 13 males between the ages of 11 and 30 who were operated on for knee injures using arthroscopic surgery. The patients were classified by type of injury: direct blow (D), or both twisted knee and direct blow (B). The results of the surgery were also classified as excellent (E) or good (G). The resulting 2×2 contingency table is displayed in Table 4.2.

	Result		
Injury	Excellent	Good	Total
Direct	3	2	5
Both	7	1	8
Total	10	3	13

Table 4.2: Surgery results

We would be interested in whether the two classificatory variables are independent of one another. That is, whether the result of a patient's surgery is independent on the type of injury sustained. When there is such an independence, then the conditional proportion of the result of a patient's surgery being excellent would be the same regardless of whether the patient sustains a direct or both injuries. That is:

> The probability that the result of a patient's surgery will be excellent given that he or she sustained a direct injury would be equal to the probability that the result of a patient's surgery is excellent given that he or she sustained both injuries.

There are at least four situations that might give rise to the observed 2×2 table in Table 4.2 above. These are:

1. The cross-classification was based on the information that 5 patients had sustained direct and 8 had sustained both. Further, we were informed that of these 13 patients, 10 had excellent surgery result and 3 had good surgery result.

2. The cross-classification was solely based on the knowledge that 5 patients had sustained direct and 8 patients had similarly sustained both injuries.

3. Here, all the 13 patients were cross-classified according to type of injury and the result of the surgery. The classification is random.

4. Patients were cross-classified each at random until 3 were said to have sustained direct injury.

The differences in the situations lie in the numbers fixed by the schemes. In situation 1, each of n_{1+}, n_{2+}, n_{+1}, and n_{+2} are fixed. In situation 2, only $M_1 = \{n_{1+}, n_{2+}\}$ is fixed. In situation 3, only N, the total sample size, is fixed, while in situation 4, only n_{11} is fixed. Of course for situation 2 it is also possible to have fixed only n_{+1} and n_{+2}. A particularly clear discussion of the conceptual differences in these sampling schemes has been given by Pearson (1947).

The underlying distributions for the four situations that could give rise to the 2×2 contingency table displayed in Table 4.2 are respectively the hypergeometric, the product binomial, the full multinomial, and the negative binomial (Kudo & Tarumi, 1978). The hypergeometric model, which assumes that both margins are fixed, has had most publicity, though genuine examples in which both sets of marginal totals are fixed are very rare. The first three sampling models are discussed in turn in the following sections. The fourth scheme has not gained much popularity and will therefore not be discussed in this text.

4.2 The Hypergeometric Probability Model

For observational data obtained from restricted populations or for experimental design data viewed within the framework of a strict randomization model, both marginal distributions M_1 and M_2 are assumed to be fixed (either by design or by conditional distribution arguments). In this context, the null hypothesis of "randomness" can be stated as:

> H_0: The classificatory variable B is randomly distributed with respect to the other classificatory variable A.

In other words, the data in the first row of the 2×2 table can be regarded as a simple random sample of size n_{1+} from a fixed population consisting of the column marginal distribution $\mathbf{M}_2$. Under H_0, it can be shown using conditional arguments (see Chapter 2, section 2.4) that the vector $\mathbf{n}$ where $\mathbf{n}' = \{n_{11}, n_{12}, n_{21}, n_{22}\}$ follows the hypergeometric distribution given by the probability model:

$$P[\mathbf{n} \mid \mathbf{M}_1, \mathbf{M}_2; H_0] = \frac{n_{1+}!n_{2+}!n_{+1}!n_{+2}!}{N!n_{11}!n_{12}!n_{21}!n_{22}!} \tag{4.1}$$

The above can also be expressed in the form

$$\begin{aligned} P[\mathbf{n} \mid \mathbf{M}_1, \mathbf{M}_2; H_0] &= \frac{\binom{n_{1+}}{x}\binom{n_{2+}}{n_{+1}-x}}{\binom{N}{n_{+1}}}, \quad \text{for } x = 1, 2, \cdots, n_{+1} \\ &= P[n_{11} \mid n_{11} + n_{21} = n_{+1}] \end{aligned} \tag{4.2}$$

For brevity, we usually write $P[n_{11} \mid n_{11} + n_{21} = n_{+1}]$ succinctly as simply $P[n_{11}]$. Note that under the assumption of both marginal distributions $\mathbf{M}_1$ and $\mathbf{M}_2$ being fixed, the entire distributions of the vector $\mathbf{n}$ can be characterized by the distribution of any one of the internal cell frequencies n_{ij}. Specifically, once the value of one of the cell frequencies is fixed, the other three can be determined immediately from the fixed margins $\mathbf{M}_1$ and $\mathbf{M}_2$. Thus the distribution of $\mathbf{n}$ can be determined completely from the distribution of one of the "pivot" cells, say n_{11} as illustrated below in Table 4.3.

	B		
A	1	2	Total
1	n_{11}	$n_{1+} - n_{11}$	n_{1+}
2	$n_{+1} - n_{11}$	$n_{2+} - n_{+1} + n_{11}$	n_{2+}
Total	n_{+1}	n_{+2}	N

Table 4.3: Observed frequencies as functions of pivot cell and fixed marginal totals

As was discussed in Chapter 2, the range of n_{11} is from $L_1 = \max(0, n_{1+} - n_{+2})$ to $L_2 = \min(n_{1+}, n_{+1})$. In other words, the number of jointly classified subjects in the first row and first column, n_{11} cannot be less than either 0 or $(n_{1+} - n_{+2})$, whichever is larger, and cannot be greater than n_{1+} or n_{+1}, whichever is smaller. As a result, the probability density for $\mathbf{n}$ can be relabeled as

$$P[n_{11}] = P\{n_{11} \mid \mathbf{M}_1, \mathbf{M}_2; H_0\} \tag{4.3}$$

and it can be shown that

$$\sum_{n_{11}=L_1}^{L_2} P[n_{11}] = 1$$

which would indicate that the probabilities in (4.3) do constitute a probability density function for the pivot cell frequency. We illustrate this below with the data from Table 4.2.

For the data in Table 4.2, $n_{11} = 3$ and $L_1 = \max(0, 5-3) = 2$ and $L_2 = \min(5, 10) = 5$. The probability for the observed configuration $\mathbf{n} = \{3, 2, 7, 1\}$ therefore is given based on (4.1) as:

$$\mathrm{P}(3, 2, 7, 1) = \frac{5!8!10!3!}{13!3!2!7!1!} = 0.27972$$

Similarly, the probability for the possible outcome $\mathbf{n} = \{5, 0, 5, 3\}$ is given by:

$$\mathrm{P}(5, 0, 5, 3) = \frac{5!8!10!3!}{13!5!0!5!3!} = 0.19580$$

The collection of all possible tables consistent with the marginal totals is displayed below.

$$\begin{array}{|cc|}\hline 2 & 3 \\ 8 & 0 \\ \hline\end{array} \qquad \begin{array}{|cc|}\hline 3 & 2 \\ 7 & 1 \\ \hline\end{array} \qquad \begin{array}{|cc|}\hline 4 & 1 \\ 6 & 2 \\ \hline\end{array} \qquad \begin{array}{|cc|}\hline 5 & 0 \\ 5 & 3 \\ \hline\end{array}$$

A short-cut method of evaluating the probabilities for each of the table is described by Feldman and Klinger (1963). This method involves the application of the following recursive formula:

$$P(a+1) = P(a) \times \frac{bc}{(a+1)(d+1)}$$

$$\begin{array}{|cc|}\hline \text{a} & \text{b} \\ \hline \text{c} & \text{d} \\ \hline\end{array} \qquad \begin{array}{|cc|}\hline \text{a+1} & \text{b-1} \\ \hline \text{c-1} & \text{d+1} \\ \hline\end{array}$$

where we are starting with the $a = 2$ table. Here, a is the smallest of all possible values of the pivot cell n_{11}. Hence,

$$\begin{aligned}
P(2) &= P(2) \times 1 &&= 1P(2) \\
P(3) &= P(2)\,\frac{8 \times 3}{3 \times 1} &&= 8P(2) \\
P(4) &= P(3)\,\frac{7 \times 2}{4 \times 2} &&= 14P(2) \\
P(5) &= P(4)\,\frac{6 \times 1}{5 \times 3} &&= 5.6P(2)
\end{aligned}$$

Adding these probabilities and since they must sum to 1, we have

$$P(2) + 8P(2) + 14P(2) + 5.6P(2) = 1,$$

Hence,

$$P(2) = \frac{1}{28.6} = 0.03497$$

The other probabilities can now be obtained in terms of $P(2)$ and these probabilities are displayed in the next table. The table therefore gives the distribution of the pivot cell (for all the possible configurations) for the data in Table 4.2.

n_{11}	Vector ($\mathbf{n}'$)	Probability
2	{2,3,8,0}	0.03497
3	{3,2,7,1}	0.27972
4	{4,1,6,2}	0.48951
5	{5,0,5,3}	0.19580
Total		1.00000

As expected, the probabilities based on the distribution of the pivot cell do add up to 1.00.

4.2.1 Mean and Variance of the n_{ij} Terms

The moments for the hypergeometric distribution are most readily obtained (Johnson & Kotz 1982) by computing the factorial moments. If we define

$$\begin{aligned}(a)^k &= \frac{a!}{(a-k)!} \\ &= a(a-1)(a-2)\cdots(a-k+1)\end{aligned}$$

then the k-th factorial moment can be expressed as

$$E\{(n_{11} \mid H_0)^k\} = \frac{(n_{1+})^k (n_{+1})^k}{(N)^k} \tag{4.4}$$

As a result of (4.4), it follows that the expected value of n_{11} under H_0 (that is, the first factorial moment when $k = 1$) is

$$\begin{aligned}m_{11} &= E\{n_{11} \mid H_0\} \\ &= \frac{n_{1+}n_{+1}}{N}\end{aligned} \tag{4.5}$$

Similarly, the second factorial moment ($k = 2$) is also given by:

$$E\{n_{11}(n_{11}-1) \mid H_0\} = \frac{n_{1+}(n_{1+}-1)n_{+1}(n_{+1}-1)}{N(N-1)} \tag{4.6}$$

Consequently, the variance of n_{11} under H_0 is obtained directly from (4.5) and (4.6) as

$$\begin{aligned}V_{11} &= \text{Var}(n_{11} \mid H_0) \\ &= E\{n_{11}(n_{11}-1) \mid H_0\} + m_{11} - m_{11}^2 \\ &= \frac{n_{1+}n_{2+}n_{+1}n_{+2}}{N^2(N-1)}\end{aligned} \tag{4.7}$$

We are using here the fact that the variance of X can be defined as $\text{Var}(X) = E[X(X-1)] + E(X) - E(X)^2$.

The central moments in (4.5) and (4.7) will be used later to develop a large sample test statistic for H_0. We may note here that in general, for the covariance structure of $\mathbf{n}$, each pairwise calculations can be obtained directly from the relationship shown in Table 4.3. In particular, for n_{11} and n_{22}, we obtain

$$\begin{aligned}\text{COV}(n_{11}, n_{22}) &= COV(n_{11}, n_{2+} - n_{+1} + n_{11}) \\ &= \text{COV}(n_{11}, n_{11}) \\ &= V_{11}\end{aligned} \tag{4.8}$$

because n_{11} determines n_{22} and n_{2+}, n_{+1} are constants. These covariances can be summarized in a variance - covariance matrix as:

$$\text{VAR}\{\underline{n}\} = V_{11} \begin{bmatrix} 1 & -1 & -1 & 1 \\ -1 & 1 & 1 & -1 \\ -1 & 1 & 1 & -1 \\ 1 & -1 & -1 & 1 \end{bmatrix} = \text{COV} \begin{bmatrix} n_{11} \\ n_{12} \\ n_{21} \\ n_{22} \end{bmatrix}$$

4.2.2 Example 3.2

The data in Table 4.4 which appeared in Freeman (1987), were originally published in Correa et al. (1983) to study the effect of passive smoking on lung cancer. Each of 155 nonsmokers is tabulated according to whether the spouse smoked. For the moment, we will focus only on the females. Freeman (1987) had considered the case for the males. There were only 22 cancer cases among ever-married nonsmoking females, and only 75 spouses in the sample have ever smoked. In strict sense, both margins of the table were not fixed in the actual sampling scheme. However, if we assume that the "spouse-smoked" sample sizes are fixed by design, then the proportion of being a case (having lung cancer), which equals 22/155, is of no interest in this situation, since our data are a nonrandom sample from the population. This proportion is referred to as the nuisance parameter and we would accept this observed value as given. The test of interest therefore is whether the proportions of cases reported for both the spousal smoking and nonspousal smoking groups are the same, that is, a conditional homogeneity test: see section 10.11 for details of the homogeneity test.

	Lung cancer status?		
Spouse smoked?	Case	Control	Total
Yes	14	61	75
No	8	72	80
Totals	22	133	155

Table 4.4: Spouse smoking by case-control study by sex: Nonsmokers who have been married (females)

In this example, $N = 155$, $\mathbf{M}_1 = \{75, 80\}$, and $\mathbf{M}_2 = \{22, 133\}$.

From the above, we would like to answer questions as to how likely it is that exactly 14 cases are observed? Since there are 22 cases in a sample of 155, there are therefore $\binom{155}{22}$ possible samples. However, 14 of the 75 smoking spouses had diagnosed as cases, and this can occur in $\binom{75}{14}$ ways. Similarly, 8 of the 80 nonsmoking spouses are also diagnosed as cases, and this can also occur in $\binom{80}{8}$ ways. Thus both the 22 smoking cases and the 8 nonsmoking cases can occur jointly in $\binom{75}{14} \times \binom{80}{8}$. Hence, the probability of observed outcome is given by:

$$\frac{\binom{75}{14}\binom{80}{8}}{\binom{155}{22}} = \frac{22!133!75!80!}{155!14!8!61!72!} = 0.05673.$$

We notice that the arguments lead to the hypergeometric distribution of the pivot cell $n_{11} = 14$. With the pivot cell being the observed value 14, we have $L_1 = \max(0, 22-80) = 0$ and $L_2 = \min(22, 75) = 22$. Under H_0 and the assumption that

$\mathbf{M}_1$ and $\mathbf{M}_2$ are fixed, the range of n_{11} is restricted from $L_1 \leq n_{11} \leq L_2$: that is, from 0 to 22 with probabilities obtained from (4.1), which are summarized in Table 4.5. All other possible values of the pivot cell with the corresponding probabilities are tabulated in the table below. There is a total of $D = 1+\min(n_{1+}, n_{2+}, n_{+1}, n_{+2})$ tables that are consistent with the fixed marginals. As in chapter 2, these probabilities are generated with the following SAS software program.

```
data hyper;
n=155; k=75; m=22; i2=min(k,m); i1=max(0,(k+m-n));
sum=0.0;
do i=i1 to i2;
if i-1 lt i1 then prob=probhypr(n,k,m,i);
    else prob=probhypr(n,k,m,i)-probhypr(n,k,m,i-1);
output; end;
proc print data=hyper noobs;
var i prob;
format prob 8.4; run;
```

n_{11}	$P(n_{11})$	Z	X^2
0	0.000000	-4.89	23.88
1	0.000003	-4.43	19.61
2	0.000034	-3.97	15.75
3	0.000273	-3.51	12.32
4	0.001507	-3.05	9.31
5	0.006114	-2.59	6.72
6	0.018948	-2.13	4.55
7	0.045975	-1.67	2.80
8	0.088815	-1.21	1.47
9	0.138157	-0.76	0.57
10	0.174321	-0.30	0.09
11	0.179144	0.16	0.03
12	0.150140	0.62	0.39
13	0.102479	1.08	1.17
14	0.056729	1.54	2.37
15	0.025282	2.00	4.00
16	0.008968	2.46	6.04
17	0.002490	2.92	8.51
18	0.000528	3.38	11.40
19	0.000082	3.84	14.71
20	0.000009	4.29	18.44
21	0.000001	4.75	22.59
22	0.000000	5.21	27.17
Total	1.000000		

Table 4.5: pdf for the pivot cell in Table 4.4

The observed table is enclosed between the two lines, and X^2 is the Pearson's test statistic defined in Chapter 3 and Z is defined as,

$$Z = \sum_i \sum_j \frac{(n_{ij} - m_{ij})}{\sqrt{m_{ij}}}$$

The pivot cell $n_{11} = 0$ leads to the configuration $\mathbf{n}' = \{0, 75, 22, 58\}$. The null hypothesis of homogeneity $H_0 : \pi_1 = \pi_2$ versus $\pi_1 \neq \pi_2$ can be tested in terms of the probabilities in Table 4.5 using the well-known Fisher's exact test, which is discussed in the next section, or a large sample test that will be developed in subsection 4.3.1 by using the moments summarized in the previous section.

4.3 Fisher's Exact Test

Fisher's exact test consists first of generating all tables that are consistent with the given marginal totals M_1, M_2. Then for each such table, we calculate the sums of the individual probabilities associated with tables that are as extreme as or more extreme than the observed table. This sum depends on what one considers to be an extreme or more extreme table. Possible definitions of extreme or more extreme tables are any other table satisfying the marginal total constraints having cell counts $\{n'_{11}, n'_{12}, n'_{21}, n'_{22}\}$ n'_{11} such that

$$P(n'_{11}) \leq P(n_{11}) \tag{4.9a}$$

$$\prod_{j=1}^{2}\prod_{i=1}^{2} n'_{ij}! \geq \prod_{j=1}^{2}\prod_{i=1}^{2} n_{ij}! \tag{4.9b}$$

$$\mid n'_{11}n'_{22} - n'_{12}n'_{21} \mid \geq \mid n_{11}n_{22} - n_{12}n_{21} \mid \tag{4.9c}$$

where $n_{ij} = \{n_{11}, n_{12}, n_{21}, n_{22}\}$ are the cell counts of the observed table. Equations (4.9a) to (4.9c) refer respectively to the *probability*, *log-likelihood*, and *absolute difference of cross-products (CP)* orderings of events. We shall conduct Fisher's exact test based on these methods later in this section.

If we assume that π_1 and π_2 are the underlying probabilities of successes ($A_2 = 1$) for each respective rows in the 2×2 setup in Table 4.1, then the null hypothesis can be conceptualized in terms of these underlying probabilities as

$$H_0 : \pi_1 = \pi_2 = \pi$$

Then the pvalue for the test of H_0 against the one-sided alternative

$$H_1 : \pi_1 > \pi_2$$

is obtained as:

$$\begin{aligned} F_1(n_{11}) = \sum_{a \geq n_{11}}^{L_2} P(a) &= \sum_{a \geq n_{11}}^{L_2} \frac{\binom{n_{1+}}{a}\binom{n_{2+}}{n_{+1}-a}}{\binom{N}{n_{+1}}} \\ &= P(n_{11}) + P(n_{11}+1) + \cdots + P(L_2) \end{aligned} \tag{4.10}$$

In our example, we have $L_2 = 22$; thus,

$$\begin{aligned} F_1(n_{11}) = \sum_{14}^{22} P(u) &= P(14) + P(15) + \cdots + P(22) \\ &= 0.094089 \end{aligned}$$

Similarly, the pvalue for the test of H_0 against the one-sided alternative

$$H_2 : \pi_1 < \pi_2$$

is given by

$$\begin{aligned} F_2(n_{11}) = \sum_{a=L_1}^{n_{11}} P(a) &= \sum_{a=L_1}^{n_{11}} \frac{\binom{n_{1+}}{a}\binom{n_{2+}}{n_{+1}-a}}{\binom{N}{n_{+1}}} \\ &= P(L_1) + P(L_1+1) + \cdots + P(n_{11}) \end{aligned} \tag{4.11}$$

Again for our example, we have

$$\begin{aligned} F_2(n_{11}) = \sum_{0}^{14} P(a) &= P(0) + P(1) + \cdots + P(14) \\ &= 0.96260 \end{aligned}$$

In many situations, one is interested in the general two-sided alternative hypothesis

$$H_1 : \pi_1 \neq \pi_2$$

which reflects departure from the null hypothesis of homogeneity in either direction. For this two-tailed test, a natural choice for the pvalue calculations is the cumulation of the point probabilities for all extreme tables in either direction. However, unless $n_{1+} = n_{2+}$ or $n_{+1} = n_{+2}$ (in which case, the distribution is symmetric), the two-tail sums are not equal. Thus the two-tailed pvalue is computed in this case as the sum of all table probabilities that are as extreme as or more extreme than the observed probability of 0.0567. That is,

$$\text{pvalue} = \underbrace{P(0) + \cdots + P(7)}_{ST} + \underbrace{P(14) + \cdots + P(22)}_{PT} \tag{4.12}$$

where PT is sometimes referred to as the *primary tail* probability, which is the sum of the set of tables $A_P = \{n'_{11}\}$ in the same tail of the distribution as n'_{11}, that is, $n'_{11} \geq n_{11}$ such that $P(n'_{11}) \leq P(n_{11})$.

Similarly, we can define the *secondary tail* (ST) as the set of tables $A_S = \{n'_{11}\}$ in the opposite tail of the distribution from the observed table, that is, $n'_{11} \leq n_{11}$, such that $P(n'_{11}) \leq P(n_{11})$. In this context, the primary tail is

$$PT(n_{11}) = \sum_{n'_{11} \in A_P} P(n'_{11})$$

Similarly,

$$ST(n_{11}) = \sum_{n'_{11} \in A_S} P(n'_{11}), \quad \text{and therefore}$$

$$PT(n_{11}) = \min[PT(n_{11}), ST(n_{11})]$$

and the two-tailed pvalue can be obtained as

$$P(n_{11}) = \text{PT}(n_{11}) + \text{ST}(n_{11})$$

This result agrees with the recommendation in Bradly (1968). Cox (1970a), however, suggested using

$$P^* = 2\ PT(n_{11})$$

Again in our example, the primary tail probability is $PT(14) = 0.094089$ while the secondary tail probability is

$$ST(14) = \sum_{n'_{11}=0}^{7} Pr(n_{11}) = 0.0728854$$

As a result, the two-tailed pvalue is given as either

$$P(14) = 0.094089 + 0.072854 = 0.1669 \tag{4.13}$$

or

$$P^*(14) = 2(0.094089) = 0.1882$$

The null hypothesis H_0 is rejected whenever the computed pvalue $\leq \alpha^*$ where α^* is usually specified. For instance, if we choose α^* to be 0.05 in our example, then we would conclude that we do not have sufficient evidence to reject H_0 and the null hypothesis of randomness will be tenable in this case.

We can implement Fisher's exact test in SAS® version 8, for the data in Table 4.4, as follows:

```
DATA FISHER;
DO STATUS=1 TO 2;
   DO SMOKE=1 TO 2;
INPUT COUNT @@;
OUTPUT;
END; END;
```

```
DATALINES;
14 61 8 72
;
TITLE 'FISHERS EXACT TEST';
PROC FREQ; WEIGHT COUNT;
TABLES STATUS*SMOKE/EXACT;
RUN;
                    Statistics for Table of STATUS by SMOKE

          Statistic                        DF        Value       Prob
          ------------------------------------------------------------
          Chi-Square                        1       2.3873     0.1223
          Likelihood Ratio Chi-Square       1       2.4068     0.1208
          Continuity Adj. Chi-Square        1       1.7288     0.1886
          Mantel-Haenszel Chi-Square        1       2.3719     0.1235
          Phi Coefficient                           0.1241
          Contingency Coefficient                   0.1232
          Cramer's V                                0.1241

                           Fisher's Exact Test
                    ----------------------------------
                    Cell (1,1) Frequency (F)        14
                    Left-sided Pr <= F          0.9626
                    Right-sided Pr >= F         0.0941

                    Table Probability (P)       0.0567
                    Two-sided Pr <= P           0.1669
                             Sample Size = 155
```

The SAS software output above gives results that agree with our results to four decimal places. Fisher's exact test can be summarized in terms of the alternatives as follows:

1. If the hypothesis is of the form
$$H_0 : \pi_1 = \pi_2 \quad \text{versus} \quad H_1 : \pi_1 > \pi_2$$
then the one-tailed pvalue equals the primary tail, which in our example equals 0.094089.

2. If the hypothesis is of the form
$$H_0 : \pi_1 = \pi_2 \quad \text{versus} \quad H_1 : \pi_1 < \pi_2$$
then the one-tailed pvalue equals the sum of one minus the primary tail probability and $P(n_{11})$, which in our example equals (1 - 0.094089) + 0.056729 = 0.96260.

3. If the hypothesis is of the form
$$H_0 : \pi_1 = \pi_2 \quad \text{versus} \quad H_1 : \pi_1 \neq \pi_2$$
then the two-tailed pvalue equals the sum of the primary and secondary tail probabilities, which in our example equals $0.094089 + 0.072854 = 0.1669$. We can show that an exact test based on the X^2 criterion for the data above yields 0.166944, exactly the same as the result obtained for the two-sided alternative in Fisher's test.

The pvalues based on the other ranking methods discussed earlier can also be obtained for the data in Table 4.4. For this table, we have presented in the next SAS software output, the values of $L(n'_{11})$, $CP(n'_{11})$, and $P(n'_{11})$. For the observed $n_{11} = 14$, therefore, P(14)=0.0567, L(14)=467.5133 and CP(14)=520.00. Extreme or more extremme tables are those indicating $P(n'_{11}) \leq 0.0567$, $L(n'_{11}) \geq 467.5133$, and $CP(n'_{11}) \geq 520$ for the three ranking procedures, respectively.

i	prob	LHOOD	CP	sum
0	0.0000	480.8179	1650.000	0.0000
1	0.0000	477.4869	1495.000	0.0000
2	0.0000	474.9258	1340.000	0.0000
3	0.0003	472.8491	1185.000	0.0003
4	0.0015	471.1414	1030.000	0.0018
5	0.0061	469.7409	875.0000	0.0079
6	0.0189	468.6099	720.0000	0.0269
7	0.0460	467.7235	565.0000	0.0729
8	0.0888	467.0650	410.0000	0.1617
9	0.1382	466.6232	255.0000	0.2998
10	0.1743	466.3907	100.0000	0.4741
11	0.1791	466.3634	55.0000	0.6533
12	0.1501	466.5400	210.0000	0.8034
13	0.1025	466.9219	365.0000	0.9059
14	0.0567	467.5133	520.0000	0.9626
15	0.0253	468.3215	675.0000	0.9879
16	0.0090	469.3579	830.0000	0.9969
17	0.0025	470.6393	985.0000	0.9994
18	0.0005	472.1905	1140.000	0.9999
19	0.0001	474.0494	1295.000	1.0000
20	0.0000	476.2779	1450.000	1.0000
21	0.0000	478.9913	1605.000	1.0000
22	0.0000	482.4754	1760.000	1.0000

The pvalues for the test of H_0 against a two-sided alternative are computed for each case therefore as:

(i) Probability ranking method:

$$\text{pvalue} = P(0) + \cdots + P(7) + P(14) + \cdots + P(22) = 0.1642$$

(ii) Likelihood ranking method:

$$\text{pvalue} = P(0) + \cdots + P(7) + P(14) + \cdots + P(22) = 0.1642$$

(iii) CP ranking method:

$$\text{pvalue} = P(0) + \cdots + P(7) + P(14) + \cdots + P(22) = 0.1642$$

In the above cases, all the three ranking methods give the same pvalue for the data in Table 4.4. While it is true that the three cases usually lead to the same ordering of events and consequently identical results, for some situations the three methods may not necessarily lead to the same results. While this does not create much problem for the one tailed test, it does create a problem for two-tailed tests where the concept of *primary* and *secondary* tail probabilities are important. Therefore, in some cases, the three ordering results may not lead to the same conclusion. Consider, for example, a table whose observed cell counts are $\mathbf{n} = \{1, 10, 12, 13\}$.

i	prob	LHOOD	CP	sum
0	0.0023	60.0417	143.0000	0.0023
1	0.0248	57.6438	107.0000	0.0270
2	0.1061	56.1885	71.0000	0.1331
3	0.2334	55.4000	35.0000	0.3665
4	0.2918	55.1769	1.0000	0.6583
5	0.2162	55.4764	37.0000	0.8745
6	0.0961	56.2873	73.0000	0.9706
7	0.0253	57.6224	109.0000	0.9959
8	0.0038	59.5195	145.0000	0.9997
9	0.0003	62.0532	181.0000	1.0000
10	0.0000	65.3674	217.0000	1.0000
11	0.0000	69.8021	253.0000	1.0000

For this table, the corresponding SAS software output is displayed above, and the pvalues computed for a two-sided alternative using the three methods are:

(a) Probability and likelihood ranking methods:

$$\text{pvalue} = \underbrace{P(0) + P(1)}_{PT} + \underbrace{P(8) + P(9) + P(10) + P(11)}_{ST} = 0.0311$$

(b) CP ranking method:

$$\text{pvalue} = \underbrace{P(0) + P(1)}_{PT} + \underbrace{P(7) + P(8) + P(9) + P(10) + P(11)}_{ST} = 0.0564$$

Notice that the (CP) ordering includes $n'_{11} = \{7\}$ in its secondary tail probability; hence the pvalues are not equal for this data set and at an $\alpha' = 0.05$, the results will lead to different conclusions.

4.3.1 Large Sample Test

With larger frequencies, the exact calculation is awkward as a result of the factorial calculations, hence, it is simpler to use the excellent χ_1^2 approximation due to Yates (1934).

We may appeal to standard asymptotic (that is, as n tends to ∞) or large sample size results concerning the convergence of hyper geometric distribution to the normal distribution. That is,

$$n_{11} \sim AN(m_{11}, V_{11})$$

where $m_{11} = \dfrac{n_{1+}n_{+1}}{N}$ and $V_{11} = \dfrac{n_{1+}n_{2+}n_{+1}n_{+2}}{N^2(N-1)}$.

A reasonable test suggested is the Wald (1943) test defined as:

$$Q = \frac{(n_{11} - m_{11})^2}{V_{11}} = \frac{(N-1)(n_{11}n_{22} - n_{12}n_{21})^2}{n_{1+}n_{2+}n_{+1}n_{+2}}$$

and the test asymptotically follows a χ^2 distribution with one degree of freedom.

We may also compute the expected values for each of the observed cells n_{ij} by noting that under H_0, the expected values for each of the n_{ij} terms are (from the results in chapter 4)

$$m_{ij} = E\{n_{ij} \mid H_0\} = \frac{n_{i+}n_{+j}}{N}$$

Then Pearson's X^2 test statistic is

$$X^2 = \sum\sum (n_{ij} - m_{ij})^2 / m_{ij} = \frac{N(n_{11}n_{22} - n_{12}n_{21})^2}{n_{1+}n_{2+}n_{+1}n_{+2}}$$

Under H_0, X^2 is asymptotically distributed χ^2 with one degree of freedom.

Note here that

$$Q = \frac{N-1}{N} X^2$$

Thus, Q is asymptotically equivalent to the X^2 criterion. Further, Q is equivalent to the empirically obtained X_U^2 proposed by Upton (1982) where

$$X_U^2 = \frac{N-1}{N} X^2$$

We shall discuss this and other statistics further later in this chapter.

4.3.2 The Yates Continuity Correction

The Yates correction to X^2 involves calculating

$$X_Y^2 = \frac{N(\mid n_{11}n_{22} - n_{12}n_{21} \mid - \frac{N}{2})^2}{n_{1+}n_{2+}n_{+1}n_{+2}}$$

The correction was based on the concept that a discrete distribution was being approximated by a continuous distribution. The Yates corrected X^2 has been shown to give a better approximation to the hyper geometric probabilities than the uncorrected X^2 (Yates, 1934; Mantel & Greenhouse, 1968; Grizzle, 1967; Conover, 1974). For the hypergeometric sampling scheme, there are a possible number of outcomes $\frac{1}{2}(N+2)$ that are consistent with the given marginals. Ideally, the exact test is usually employed for situations in which one or more of the expected values m_{ij} is small. In the example above, the expected values are reasonably large and one would therefore be expected to use the χ^2 approximation. Consequently, for the data in Table 4.4, $X^2 = 2.3873$ and $X_Y^2 = 1.7288$, which when both are compared with the tabulated χ_1^2 gives pvalues of 0.1223 and 0.1886, respectively. Thus, again at $\alpha = 0.05$, the results indicate that we will fail to reject the null hypothesis of independence, which agrees with the earlier results obtained using Fisher's exact test. It may also be noted here that the above asymptotic tests involving the test criteria are two-tailed tests. These results agree with those given in the SAS software output earlier displayed.

The following is another example taken from Goldstein (1965). There is reason to believe that a certain drug might have value in particular kinds of cancer. However, the drug is quite toxic, so it is not desirable to embark on large scale clinical trials without some good preliminary evidence of efficacy. In this, randomized, double-blind clinical trial, 10 patients are assigned at random to control and treatment groups and their status is evaluated at the end of 6 months, with the following results displayed in Table 4.6.

	Improved	Not improved	Total
Treated	4	1	5
Control	0	5	5
Total	4	6	10

Table 4.6: Goldstein's table

Here the expected values in all the cells are very small and SAS® software warns us of this problem with a warning below the output (see below). The exact test therefore would be most appropriate for the analysis of this data set. The two-sided and right-sided pvalues for the data, based on Fisher's, are 0.0476 and 0.024, respectively. The results imply that we would have to reject the null hypothesis at the 5% significance level that the treatment is ineffective. The corresponding computed values of X^2 and X_Y^2 are respectively 6.667 and 3.75 (see SAS software output below) with corresponding (large sample approximation based) pvalues of 0.010 and 0.053, respectively. Had we used these, the adjusted X_Y^2 would have resulted in a contradictory conclusion (since pvalue > 0.05) in this case. Although the result from the unadjusted X^2 agrees even more strongly with our conclusion, this test is very unreliable in view of the very small expected values. In fact, the continuity corrected X^2 is more reliable in this case. Thus, the use of X_Y^2 for the two-sided alternative gives close to similar result to the exact test. We shall give some guidelines for the use of these tests in a later section.

```
                 STATISTICS FOR THE GOLDSTEIN DATA

Statistic                        DF      Value        Prob
------------------------------------------------------
Chi-Square                        1      6.667       0.010
Likelihood Ratio Chi-Square       1      8.456       0.004
Continuity Adj. Chi-Square        1      3.750       0.053
Mantel-Haenszel Chi-Square        1      6.000       0.014
Fisher's Exact Test (Left)                           1.000
                    (Right)                          0.024
                    (2-Tail)                         0.048

Sample Size = 10
WARNING: 100% of the cells have expected counts less
         than 5. Chi-Square may not be a valid test.

                 STATISTICS FOR TABLE 2

Statistic                        DF      Value        Prob
------------------------------------------------------
Chi-Square                        1      1.311       0.252
Likelihood Ratio Chi-Square       1      1.287       0.257
Continuity Adj. Chi-Square        1      0.219       0.640
Mantel-Haenszel Chi-Square        1      1.210       0.271
Fisher's Exact Test (Left)                           0.315
                    (Right)                          0.965
                    (2-Tail)                         0.510

Sample Size = 13
WARNING: 75% of the cells have expected counts less
         than 5. Chi-Square may not be a valid test.
```

I also note here that the one-tailed Fisher's exact test, when significant, indicates a departure from the null hypothesis in a specific direction, while the X^2 test indicated departure from the hypothesis in either direction. In the example above, for instance, the one-tailed exact test is used to determine whether the proportions of patients in the two groups having improved condition are equal or whether the proportion of treated patients with improved condition is less than the proportion of the control patients. The X^2 test, on the other hand, tests whether these proportions are equal or unequal regardless of the direction of inequality.

4.4 The Mid-P Test

The mid-P tests can be considered as alternatives to Fisher's exact test. To conduct the mid-P test, let us consider $G(y) = P(Y \geq y)$, where y is the observed value of a random variable Y having a continuous type distribution. Then, in this case it can be shown (Upton, 1992) that $E[G(y)] = 1/2$. However, when Y is discrete as it would be in the 2×2 table with small sample sizes, it can be shown (Barnard, 1989) that $E[G(y)] > 1/2$. In this case, Fisher's exact test would be biased in an upward direction. In order to correct this, Lancaster (1961), Barnard (1989, 1990), Hirji et al. (1991) among others in recent years have suggested the use of the mid-P test, which is defined as follows:

Under H_0, and the given marginal totals $\{M_1, M_2\}$, let us define

$$\begin{aligned} f(n'_{11}, \mid H_0) &= P(n'_{11} = n_{11}) \quad \text{and} \\ g(n'_{11}, \mid H_0) &= P(n'_{11} > n_{11}) \end{aligned} \tag{4.14}$$

where n_{11} is the observed value of the pivot cell and n'_{11} (a random variable) denotes any other observed value of n_{11} that are consistent with $\{M_1, M_2\}$ under H_0. For brevity, we simply write (4.14) above as $f(n'_{11})$ and $g(n'_{11})$ respectively.

With the above definitions, the one-sided Fisher's exact test for example is now equivalent to rejecting H_0 if

$$f(n'_{11}) + g(n'_{11}) \leq \alpha^*$$

For the data in Table 4.4, for example, $f(n'_{11}) = 0.056729$ and $g(n'_{11}) = 0.0374$. Fisher's pvalue equals $0.056729 + 0.0374 = 0.0941$, which as expected agrees with the primary tail probability and our result above.

Based on the above definitions of the functions f and g, the one-sided mid-P test proposed by Hirji et al. rejects H_0 if:

$$M_0 = \frac{1}{2} f(n'_{11}) + g(n'_{11}) \leq \alpha^*$$

and for the data in Table 4.4, for example, the mid-P based on Hirji's formulation equals: $0.5(.056729) + 0.0374 = 0.0658$. Hirji et al. also considered two other methods for obtaining pvalues for the two-sided hypothesis. These are the methods proposed by Cox (1970b) and that by Gibbons and Pratt (1975). The Gibbons and Pratt method (M_A), called *the minimum likelihood* method, rejects H_0 if

$$P\{n'_{11} : f(n'_{11}) \leq f(n'_{11} = n_{11})\} \leq \alpha^*$$

while the Cox method (M_B), also called *twice the smallest tail* method, rejects H_0 if

$$2 \min \{f(n'_{11}) + g(n'_{11}), 1 - g(n'_{11})\} \leq \alpha^*$$

Again in the example in Table 4.4, M_A gives a pvalue of $0.0728854 + 0.056729 + 0.0374 = 0.1670$. Similarly, $M_B = 2 \min(0.0941, 0.9626) = 0.1882$. In both cases, these results agree with the results given earlier and both support the null hypothesis.

4.5 Product Binomial Sampling Scheme

Consider for example a binary risk factor (e.g, smoking) and a disease state (e.g., lung cancer). The distribution of the number of subjects in a reference population can compactly be written as $\mathbf{n}' = (n_{11}, n_{12}, n_{21}, n_{22})$ and the classification by the risk factor and disease status is presented in Table 4.7 together with the accompanying underlying probability distribution, which can also be compactly written in vector form as $\mathbf{\Pi}' = (\pi_{11}, \pi_{12}, \pi_{21}, \pi_{22})$.

	Disease		
R factor	Cases	Controls	Total
Exposed	n_{11}	n_{12}	n_{1+}
Unexposed	n_{21}	n_{22}	n_{2+}
Total	n_{+1}	n_{+2}	N

	Disease		
R factor	Cases	Controls	Total
Exposed	π_{11}	π_{12}	π_{1+}
Unexposed	π_{21}	π_{22}	π_{2+}
Total	π_{+1}	π_{+2}	

Table 4.7: Observed and underlying probability tables

There are two possible situations that could give rise to the sampling scheme above. It is not uncommon to also refer to this design as arising from *observational studies*. In the first case, called *prospective studies*, the marginal totals $M_1 = n_{i+}, i = 1, 2$, are often fixed. Other names for this scheme are *cohort* studies or *follow-up* studies. The sampling scheme involves characterizing or identifying one of the samples with a predetermined number n_{1+} of subjects that have the presence (the exposed) of the suspected antecedent or risk factor (e.g., smoking) and the other sample also of a predetermined number n_{2+} of subjects do not have the suspected antecedent factor (the unexposed). The subjects are then followed up over time into the future (that is, prospectively) and the proportions of subjects developing the disease (outcome variable or response, e.g., lung cancer) at some point in time are then estimated for both samples and inference made. As an example, we might identify one cohort (a group of individuals who have a common characteristic) of smokers and the other cohort of nonsmokers, and then observe them for a period of time to determine the rate of incidence of lung cancer in the two groups. This kind of study is often very expensive and time-consuming. Further, the study may sometimes not be feasible if the disease of interest is a rare one such as particular types of cancer. Under this scheme, therefore, the observed frequencies and underlying population probabilities are displayed in Table 4.8.

	Disease		
R factor	Cases	Controls	Total
Exposed	n_{11}	n_{12}	n_{1+}
Unexposed	n_{21}	n_{22}	n_{2+}
Total	n_{+1}	n_{+2}	N

	Disease		
R factor	Cases	Controls	Total
Exposed	π_{11}	π_{12}	1
Unexposed	π_{21}	π_{22}	1

Table 4.8: Observed and probability structure with M_1 fixed

In Table 4.8, n_{+1} and n_{+2} are random variables, with n_{1+} and n_{2+} fixed by design. Under this scheme, the π_{ij} are unknown parameters that satisfy the constraints

$$\sum_j \pi_{ij} = 1 \quad \text{for} \quad i = 1, 2$$

In the second case, called *retrospective studies*, the marginal totals $M_2 = n_{+j}, j = 1, 2$ are often fixed. Retrospective study or the *case-control* study is an alternative to the prospective study and is characterized by selecting or identifying subjects that fall into the categories of the outcome variable (disease of interest) and accordingly predetermining these numbers based on the presence (cases) or absence (controls) of the disease and then estimating the proportions possessing the antecedent factor retrospectively. They are retrospective because we are *looking back* for possible causes. Examples of both studies can be found in Fleiss (1981). Again under this scheme, the observed frequencies and underlying population probabilities are displayed in Table 4.9.

In the above, n_{1+} and n_{2+} are random variables, with n_{+1} and n_{+2} fixed by design. Under this scheme, the π_{ij} are unknown parameters that satisfy the constraints

$$\sum_i \pi_{ij} = 1 \quad \text{for} \quad j = 1, 2$$

Thus in the cohort studies, individuals (study units) are selected who are initially free of disease, and the rate of occurrence of the disease is observed over some period

Rfactor	**Disease** Cases	Controls	Total
Exposed	n_{11}	n_{12}	n_{1+}
Unexposed	n_{21}	n_{22}	n_{2+}
Total	n_{+1}	n_{+2}	N

Rfactor	**Disease** Cases	Controls
Exposed	π_{11}	π_{12}
Unexposed	π_{21}	π_{22}
Total	1	1

Table 4.9: Observerd and probability Structure with M_2 fixed

of time, in the presence and absence of factors suspected of being associated with causing the disease. On the other hand, in the case-control study, individuals are selected on the basis of presence or absence of the disease, and we determine the rate of possible causative factors by looking back at their past history.

In general:

- In prospective studies, the marginal totals n_{i+} are often fixed.
- In retrospective studies, the marginal totals n_{+j} are often fixed.
- In cross-sectional studies, the sample size N is assumed fixed.

Now returning to the problem, under the first (prospective studies) sampling scheme, n_{11} follows the binomial distribution with parameters n_{1+}, π_{11}. Similarly, n_{21} follows the binomial distribution with parameters n_{2+}, π_{21}, respectively. That is,

$$n_{11} \sim b(n_{1+}, \pi_{11}) \quad n_{21} \sim b(n_{2+}, \pi_{21})$$

Since these are independent samples, it follows that the vector

$$\mathbf{n}' = (n_{11}, n_{12}, n_{21}, n_{22})$$

follows the product binomial probability model

$$P\{\mathbf{n} \mid M_1, \mathbf{\Pi}\} = \binom{n_{1+}}{n_{11}} \pi_{11}^{n_{11}} \pi_{12}^{n_{12}} \binom{n_{2+}}{n_{21}} \pi_{21}^{n_{21}} \pi_{22}^{n_{22}} \tag{4.15}$$

with the constraints $\sum_j n_{ij} = n_{i+}$ and $\sum_j \pi_{ij} = 1$ for $i = 1, 2$.

4.5.1 Homogeneity Hypothesis

In the above framework, the usual hypothesis of interest involves testing for the equality of the "cases" rates π_{11} and π_{21} for the two independent samples or subpopulations (exposed and unexposed). That is,

$$H_0 : \pi_{11} = \pi_{21} \tag{4.16}$$

The above hypothesis is equivalent to the following:

$$\pi_{11} - \pi_{21} = 0 \quad \text{or}$$

$$\frac{\pi_{11}}{\pi_{21}} = 1$$

The ratio π_{11}/π_{21} is often referred to as the "relative risk' of the event under consideration.

Under H_0, let the common parameter for the two populations be π_a and its compliment π_b so that

$$\pi_a + \pi_b = 1 \tag{4.17}$$

Thus under H_0 the probability model in (4.15) becomes

$$P\{n_{+1} \mid M_1, \pi_a, H_0\} = \binom{n_{1+}}{n_{11}} \pi_a^{n_{11}} \pi_b^{n_{12}} \binom{n_{2+}}{n_{21}} \pi_a^{n_{21}} \pi_b^{n_{22}}$$

The above simplifies to:

$$P\{n_{+1} \mid M_1, \pi_a, H_0\} = \binom{N}{n_{+1}} \pi_a^{n_{+1}} \pi_b^{n_{+2}}$$

where $\sum_j n_{ij} = n_{+j}$ is the column totals for $j = 1, 2$.

4.5.2 Maximum Likelihood Estimates of π_a

From the above expression, the likelihood equation is

$$\begin{aligned} L(\pi, \mathbf{n}, H_0) &= \binom{N}{n_{+1}} \pi_a^{n_{+1}} \pi_b^{n_{+2}} \\ \ell &= \log N! - \sum\sum \log(n_{ij}) + n_{+1}\log(\pi_a) + n_{+2}\log(\pi_b) \end{aligned} \tag{4.18}$$

where ℓ is the log-likelihood. Let $\text{G} = n_{+1}\log(\pi_a) + n_{+2}\log(\pi_b)$, the kernel of the log-likelihood; then, since the first two terms on the R.H.S. of (4.18) do not involve the πs, we will therefore maximize G subject to the constraints in (4.17). Using Lagrange multipliers, we can write G^* as:

$$\text{G}^* = n_{+1}\log(\pi_a) + n_{+2}\log(\pi_b) - \lambda(\pi_a + \pi_b - 1)$$

then

$$\frac{\partial \text{G}^*}{\partial \pi_a} = \frac{n_{+1}}{\pi_a} - \lambda \tag{4.19a}$$

$$\frac{\partial \text{G}^*}{\partial \pi_b} = \frac{n_{+2}}{\pi_b} - \lambda \tag{4.19b}$$

$$\frac{\partial \text{G}^*}{\partial \lambda} = -(\pi_a + \pi_b - 1) \tag{4.19c}$$

Setting the equations in (4.19a) through (4.19c) to zero, we have

$$\hat{\pi}_a \hat{\lambda} = n_{+1}$$
$$\hat{\pi}_b \hat{\lambda} = n_{+2}$$

Adding these, and noting that from equation (4.19c) we have $\hat{\lambda} = \text{N}$, consequently,

$$\begin{aligned} \hat{\pi}_a &= \frac{n_{+1}}{N} \quad \text{and} \\ \hat{\pi}_b &= \frac{n_{+2}}{N} \end{aligned} \tag{4.20}$$

The corresponding expected frequencies under H_0 become

$$m_{ij} = n_{i+}\,\hat{\pi}_a \quad \text{if } i = 1, 2 \text{ and } j = 1$$
$$m_{ij} = n_{i+}\,\hat{\pi}_b \quad \text{if } i = 1, 2 \text{ and } j = 2$$

4.5.3 Parameter Estimates

If we let the estimates of the success (or cases) rates for the response or outcome variable be given by

$$p_{11} = \frac{n_{11}}{n_{1+}}, \quad \text{and} \quad p_{21} = \frac{n_{21}}{n_{2+}}$$

and if we further let

$$d = p_{11} - p_{21} = \frac{n_{11}}{n_{1+}} - \frac{n_{21}}{n_{2+}} = \frac{(n_{11}n_{22} - n_{12}n_{21})}{n_{1+}n_{2+}}$$

be the estimate of the observed difference of the success rates in the two subpopulations, then

$$\mathrm{E}\{d\} = \pi_{11} - \pi_{21}$$

so that under H_0,

$$\mathrm{E}\{d\} = 0$$

and

$$\mathrm{Var}(d) = \mathrm{Var}(p_{11} - p_{21}) = \frac{\pi_{11}\pi_{12}}{n_{1+}} + \frac{\pi_{21}\pi_{22}}{n_{2+}}$$

so that

$$\begin{aligned}\mathrm{Var}[d \mid H_0] &= \frac{\pi_a\pi_b}{n_{1+}} + \frac{\pi_a\pi_b}{n_{2+}} &&= \pi_a\pi_b\left(\frac{1}{n_{1+}} + \frac{1}{n_{2+}}\right)\\ &= \frac{N\pi_a\pi_b}{n_{1+}n_{2+}} &&= \frac{N\pi_a(1-\pi_a)}{n_{1+}n_{2+}}\end{aligned}$$

Thus, the variance of d under the null hypothesis depends on the nuisance parameter π_a.

Let us consider two possible estimates of the variance of d under H_0. The first is to simply substitute the MLE of π_a obtained above into the expression for the variance to have an estimate of the variance of d under H_0 as

$$\mathrm{Est}\left(V_{d,H_0}\right) = \frac{N\hat{\pi}_a\hat{\pi}_b}{n_{1+}n_{2+}}$$

An a large sample test statistic for H_0 based on the MLE for the variance of d is

$$X_P^2 = d^2/\mathrm{est}(V_{d,H_0}) = \frac{N(n_{11}n_{22} - n_{12}n_{21})^2}{n_{1+}n_{2+}n_{+1}n_{+2}} = X^2$$

The Pearson's X^2 criterion can be viewed as the large sample test statistic based on the MLE for the variance of d.

The second alternative approach to estimating the variance of d, which is based on the difference $\pi_{11} - \pi_{21}$, is given in appendix C.1. In this case, the variance estimate leads to the Wald large sample test statistic against the two-sided alternative and is given by

$$Q = \frac{(N-1)(n_{11}n_{22} - n_{12}n_{21})^2}{n_{1+}n_{2+}n_{+1}n_{+2}} = \frac{N-1}{N}X^2$$

which again is equivalent to the Q statistic obtained for the hypergeometric model. The test statistic based on the above is generated from an unbiased estimator for the variance of d under H_0.

4.5.4 Example 4.4: HIV Testing

The data in Table 4.10, adapted from Samuels and Witmer (1999), concern a survey conducted to test the relative attitude of males and females to HIV testing in a certain college. Of 61 females, 9 had taken the HIV test, while of the 59 men, 8 had taken the HIV test. The data are summarized in Table 4.10.

	HIV testing status?		
Gender	Yes	No	Total
Female	9	52	61
Male	8	51	59
Totals	17	103	120

Table 4.10: HIV testing data

We note in this example that the marginal total {61, 59} are fixed by the design of this study. The marginal totals {17, 103} in this case can be considered as random variables. Thus in this case, for each level of gender, HIV testing status (Yes or No) can be assumed to have arisen from binomial random vector. With this study design, we would then be interested in the hypothesis of homogeneity. That is,

$$\pi_{11} = \pi_{21} = \pi_a$$

The maximum likelihood estimate of $\pi_a = n_{+1}/N = 17/120 = 0.14167$. The expected frequencies under H_0 are $\hat{m}_{11} = 61 \times (0.14167) = 8.6417$ and $\hat{m}_{21} = 59 \times (0.14167) = 8.3583$. Note that $\hat{m}_{11} + \hat{m}_{21} = 17.00$. Also, $\hat{m}_{12} = n_{1+} - \hat{m}_{11} =$ 61-8.6417 =52.3583. Similarly $\hat{m}_{22} = n_{2+} - \hat{m}_{21} = 59 - 8.3583 = 50.6417$.

The Pearson's X^2 test statistic equals:

$$\begin{aligned} X^2 &= \frac{(9-8.6417)^2}{8.6417} + \frac{(8-8.3583)^2}{8.3583} \\ &+ \frac{(52-52.3583)^2}{52.3583} + \frac{(51-50.6417)^2}{50.6417} \\ &= 0.0352 \end{aligned}$$

or

$$X^2 = \frac{120(9 \times 51 - 8 \times 52)^2}{17 \times 103 \times 61 \times 59} = 0.0352$$

Alternatively, X^2 can also be obtained as:

$$\begin{aligned} X^2 &= W \sum_{j=1}^{2} \sum_{i=1}^{2} \frac{1}{\hat{m}_{ij}} \\ &= 0.1284 \left(\frac{1}{8.6417} + \frac{1}{52.3583} + \frac{1}{8.3583} + \frac{1}{50.6417} \right) \\ &= 0.0352 \end{aligned}$$

where $W = (n_{11} - \hat{m}_{11})^2 = (9 - 8.6417)^2 = 0.1284$.

Similarly,

$$\begin{aligned} G^2 &= 2 \left[9 \log\left(\frac{9}{8.6417}\right) + 52 \log\left(\frac{52}{52.3583}\right) \right. \\ &+ \left. 8 \log\left(\frac{8}{8.3583}\right) + 51 \log\left(\frac{51}{50.6417}\right) \right] \\ &= 2(0.3656 - 0.3571 - 0.3505 + 0.3596) \\ &= 0.0353 \end{aligned}$$

The value of X^2 computed above when compared with the tabulated χ^2 with one degree of freedom is not significant (pvalue = 0.8512). This result therefore indicates that the proportions of females who had been tested for HIV seem to be the same as the proportion of males who had similarly been tested for HIV. Ideally, Fisher's exact test is most appropriate for the case when at least one of the expected values $\hat{m}_{ij}$ are small, that is, less than 3.

The product binomial (product multinomial in higher dimensional tables) sampling scheme has certain implications for models that can be fitted to data arising from such a scheme. In the HIV example above, the marginal totals n_{i+} are fixed. Any model that we would employ for these data must therefore accomodate these constraints and be such that the expected values under such models must satisfy

$$\hat{m}_{i+} = n_{i+} \quad \text{for } i = 1, 2$$

Thus a log-linear model (chapter 6) must include the term λ_i^G, where G stands for gender. We shall discuss these implications more in chapter 6.

4.6 The Full Multinomial Model Scheme

Suppose we randomly sampled subjects according to whether they belong to either of two groups (risk factor present and risk factor not present). We next determine the number of subjects or proportion of subjects in each group that have the outcome variable (case or no case). Here, the subjects are observed at only one point in time, and this represents an *independent-sample* design. Such a design is referred to as a cross-section study. For cross-sectional studies, only the sample size N is fixed and we are interested in the relationship between the two risk and outcome variables. In this situation, individuals sampled are classified simultaneously on both variables and thus only the sample size N is fixed here. In this context, the observed counts and the corresponding underlying probability structure for these counts are summarized in Table 4.11.

Observed Frequencies

	Response		
Popl	S	F	Total
I	n_{11}	n_{12}	n_{1+}
II	n_{21}	n_{22}	n_{2+}
Total	n_{+1}	n_{+2}	N

Population Probabilities

	Response		
Popl	S	F	Total
I	π_{11}	π_{12}	π_{1+}
II	π_{21}	π_{22}	π_{2+}
Total	π_{+1}	π_{+2}	1

Table 4.11: Observed counts and corresponding probality structure

The π_{ij} in Table 4.11 are unknown parameters that satisfy the constraint

$$\sum_{ij} \pi_{ij} = 1$$

Under this sampling scheme, the observed frequencies $\mathbf{n}$ follow the multinomial distribution with parameters N and $\mathbf{\Pi}$, that is,

$$P[\mathbf{n} \mid N, \underline{\Pi}]\} = \binom{N}{n} \pi_{11}^{n_{11}} \pi_{12}^{n_{12}} \pi_{21}^{n_{21}} \pi_{22}^{n_{22}}$$

where
$$\binom{N}{n} = \frac{N!}{n_{11}!n_{12}!n_{21}!n_{22}!}$$
is the standard multinomial coefficient.

4.6.1 The Independence Hypothesis

The primary focus is on the relationship between the two classificatory variables. Specifically, the null hypothesis of independence can be stated formally as the conditional proportion of being in column 1, given that an individual belongs to a known row is the same for both rows, that is,
$$\frac{\pi_{11}}{\pi_{1+}} = \frac{\pi_{21}}{\pi_{2+}}$$
The above can generally be written in the form
$$H_0 : \pi_{ij} = \pi_{i+}\pi_{+j} \quad \text{for } (i,j) = 1,2 \tag{4.21}$$
That is, under H_0, the joint probabilities can be obtained directly as products of the corresponding marginal probabilities. From the above, it follows that
$$\pi_{11}\pi_{22} = \pi_{12}\pi_{21}, \quad \Longleftrightarrow \quad \pi_{11}\pi_{22} - \pi_{12}\pi_{21} = 0$$

The latter is sometimes called the *cross-product difference*. Similarly, H_0 implies that the *cross-product ratio* or *odds-ratio*, θ, is
$$\theta = \frac{\pi_{11}\pi_{22}}{\pi_{12}\pi_{21}} = 1$$
so that H_0 can be stated as $H_0 : \theta = 1$.

We observe here that under H_0, n_{1+} and n_{+1} are independent in spite of sharing n_{11}. This result can be demonstrated by using conditional arguments and is presented in appendix C.2.

4.6.2 MLE Estimates of the π Terms

Under H_0, the π_{ij} are estimated by
$$\hat{\pi}_{ij} = \pi_{i+}\pi_{+j} \tag{4.22}$$

Consequently, the expected values under H_0 are given by
$$\hat{m}_{ij} = N\hat{\pi}_{ij} = \frac{n_{i+}n_{+j}}{N}$$

Substituting these in X^2, we have again,
$$X^2 = \frac{N(n_{11}n_{22} - n_{12}n_{21})^2}{n_{1+}n_{2+}n_{+1}n_{+2}}$$

4.6.3 Example 4.5: Longitudinal Study Data

Christensen (1990) gives the following example from a longitudinal study. A sample of 3,182 individuals without cardiovascular disease were cross-classified by two factors: personality type and exercise. Personality type has two categories: type A and type B, where a type A individual shows signs of stress, uneasiness, and hyperactivity. On the other hand, a type B individual is relaxed, easygoing, and

	Personality		
Exercise	A	B	Total
Regular	483	477	960
Other	1101	1121	2222
Totals	1584	1598	3182

Table 4.12: Longitudinal study data

normally active. The exercise variable is categorized as individuals who exercise regularly and those who do not. The data are summarized in Table 4.12.
In this example, only the sample size $N = 3182$ is assumed fixed. Consequently, $\pi_{11} + \pi_{12} + \pi_{21} + \pi_{22} = 1$ under this sampling scheme. The hypotheses of interest here are:

$$H_0 = \text{Exercise is independent of personality type}$$
$$H_a = \text{Exercise is dependent on personality type}$$

For these data under H_0, therefore, $\hat{m}_{11} = (960)(1584)/3182 = 477.8881$. Similarly, $\hat{m}_{12} = 482.1119; \hat{m}_{21} = 1106.1119$ and $\hat{m}_{22} = 1115.8881$, and the Pearson's $X^2 = 0.1559$ (pvalue $= 0.6929$), which is not significant when compared with the χ^2 distribution with one degree of freedom. Thus there is no evidence against independence of exercise and personality type. Corresopondingly, $G^2 = 0.1559$ for this data.

4.6.4 Alternative Formulation

We mentioned earlier that a test of independence in the 2×2 contingency table under the multinomial sampling scheme is equivalent to testing that the cross-product ratio (or the odds ratio) is

$$\theta = \frac{\pi_{11}\pi_{22}}{\pi_{12}\pi_{21}} = 1$$

An estimate of θ can be obtained as

$$\hat{\theta} = \frac{\hat{m}_{11}\hat{m}_{22}}{\hat{m}_{12}\hat{m}_{21}}$$

If we define η to be

$$\eta = \ln(\hat{\theta}) = \ln(\hat{m}_{11}) + \ln(\hat{m}_{22}) - \ln(\hat{m}_{12}) - \ln(\hat{m}_{21})$$

Then η can be viewed as a *log-contrast* where

$$\eta = \sum_{i=1}^{2}\sum_{j=1}^{2} a_{ij} \log\left(\hat{m}_{ij}\right) \quad \text{with} \sum_{ij} a_{ij} = 0 \tag{4.23}$$

Let the corresponding observed log-odds be given by

$$\ell = \ln(n_{11}) + \ln(n_{22}) - \ln(n_{12}) - \ln(n_{21})$$

Again, we see that ℓ can be considered as a log-contrast. The asymptotic variance of ℓ using the *delta method* equals

$$\text{Var}_A(\ell) = \sum_i \sum_j n_{ij}^{-1}$$

A large sample Wald statistic $\dfrac{\ell}{\sqrt{\text{var}(\ell)}}$ converges in distribution to $N(0, 1)$ and a test of the hypothesis that $\theta = 1$ is equivalent to the hypothesis that

$$H_0 : \eta = 0 \qquad \text{versus} \qquad H_a : \eta \neq 0$$

Under H_0, we note that

$$\frac{\ell}{(\sum_{ij} \frac{1}{n_{ij}})^{\frac{1}{2}}} \sim N(0,1)$$

which implies that

$$\frac{\ell^2}{\sum_{ij} \frac{1}{n_{ij}}} \sim \chi_1^2 \tag{4.24}$$

The above is the test statistic suggested by Lindley (1964) and can be presented in the form:

$$B^2 = \frac{\ell^2}{V} = \frac{[\ln(n_{11}) + \ln(n_{22}) - \ln(n_{12}) - \ln(n_{21})]^2}{n_{11}^{-1} + n_{22}^{-1} + n_{12}^{-1} + n_{21}^{-1}} \tag{4.25}$$

where V is the variance of ℓ and we would reject H_0 at 5% significance level if $B^2 > (1.96)^2 = 3.84$, or at the 1% level if $B^2 > (2.58)^2 = 6.66$. The corresponding pvalue is given approximately by

$$\alpha_p \approx 2\Phi(-\frac{|\,\ell\,|}{V^{\frac{1}{2}}}) \quad \text{where } V = \sum_{ij} n_{ij}^{-1} \tag{4.26}$$

For the data in Table 4.12,

$$\hat{\ell} = \ln(483) + \ln(1121) - \ln(477) - \ln(1101) = 0.0305$$

and the asymptotic standard error is

$$\hat{V}^{\frac{1}{2}} = \sqrt{\frac{1}{483} + \frac{1}{1121} + \frac{1}{477} + \frac{1}{1101}} = 0.0772$$

Hence,

$$B = \frac{\hat{\ell}}{\hat{V}^{\frac{1}{2}}} = 0.3955 \quad \text{and} \quad B^2 = 0.1564$$

The corresponding pvalue $= 2\Phi(-0.3955) = 2\ (0.3462) = 0.6924$. There is no sufficient evidence to reject H_0 either by comparing B^2 with a χ^2 distribution with one degree of freedom or by the use of the pvalue computed above.

The odds ratio, $\theta = e^{\hat{\ell}}$ is estimated to be $e^{0.0305} = 1.0310$ and a 95% confidence interval for log θ is given by

$$(\hat{\ell} \pm 1.96\hat{V}^{\frac{1}{2}}) = 0.0305 \pm 0.1513 = (-0.1208, 0.1818)$$

Since the interval includes 0, i.e., the null hypothesis is plausible, we would again fail to reject H_0 in this example.

4.7 Summary Recommendations

For the three sampling schemes designated as A, B, and C in Table 4.13, we give the number of possible outcomes that could be consistent with a given sample size N, when both margins, one margin and only the sample size is fixed.

We recognize from the above table that for a given N, there are more possible configurations for C, followed respectively by B and A. In other words, the distribution of a test statistic, say, X^2, under sampling scheme A is more discrete than those

Sampling scheme	Number of outcomes (M)
Hypergeometric: A	$\frac{1}{2}(N+2)$
Product Binomial: B	$\frac{1}{4}(N+2)^2$
Multinomial: C	$\frac{1}{6}(N+1)(N+2)(N+3)$

Table 4.13: Possible number of configurations (M) under the sampling schemes

for schemes B or C. Thus X^2 would be better approximated by a continuous χ^2 distribution for those schemes than for scheme A. The Yates continuity correction is therefore meant to correct this for the case when the sampling scheme is of type A. It should be pointed out here that X^2 will be based on 1 d.f. for each of the sampling schemes.

Another continuity-corrected X^2 criterion similar to the one discussed in chapter 3 for the one-way classification that has been found to be very successful for the general 2×2 table is Cochran's (1942) correction, which is given by

$$P(X^2 > \chi_1^2) \sim P(X_C^2 > \chi_1^2)$$

where $X_C^2 = \dfrac{1}{2}(X_0^2 + X_{-1}^2)$, X_0^2 is the observed X^2 statistic, and X_{-1}^2 is the next lower value (when ranked) under the null distribution of X^2. The correction is of course based on the discreteness of the distribution of X^2.

Conover (1974) has advocated this procedure. The argument for this is based on the fact that the correction is an adaptive correction with respect to the sampling models, by the fact that its importance will decrease as the empirical discrete distribution of the X^2 criterion becomes less discrete as we go from sampling scheme A to C. Although, this represents a considerable improvement over Yate's (fixed) correction, the search for the next lower value of the test criterion can be computationally expensive, especially for sampling schemes B and C.

For the data in Table 4.4, for instance, we list in Table 4.14, the configurations as well as the corresponding X^2 for each configuration, ranked in terms of the test statistic X^2, from largest to smallest. If we chose to use any other test statistics (not X^2), the result would be different. Cochran's procedure involves evaluation of

$$P(\chi_1^2 > \frac{2.37 + 1.47}{2}) = P(\chi_1^2 > 1.92) = 0.1659$$

However, it should be pointed out here that regardless of the sampling scheme, under the null hypothesis of independence or homogeneity, the expected values are given by $\hat{m}_{ij} = n_{i+}n_{+j}/N$, the Pearsons's test statistic reduces in each case to $X^2 = N(n_{11}n_{22} - n_{12}n_{21})^2/n_{1+}n_{2+}n_{+1}n_{+2}$ and is based on 1 degree of freedom. The corresponding likelihood ratio test statistic is similarly computed from

$$G^2 = 2\sum_{i=1}^{2}\sum_{j=1}^{2} n_{ij} \log\left(\frac{n_{ij}}{\hat{m}_{ij}}\right)$$

4.7.1 Equivalent χ^2-Based GOF Test Statistics

The following test statistics have all been employed in conducting tests for the general 2×2 contingency table.

Configuration number	Ranked X^2
22, 0,	27.17
0, 22,	23.88
⋮	⋮
14, 8	2.37
8, 14	1.47
⋮	⋮
11, 12	0.09
11, 11	0.03

Table 4.14: Relevant list of configurations for the data in Table 4.4

Upton's (1982) modified X^2 statistic is

$$X_U^2 = \frac{N-1}{N} X^2$$

We notice that Upton's scaled statistic is equal to the statistic Q developed earlier under the three sampling schemes.

Other test statistics that have also been considered are the following:

$$X_{+1}^2 = \frac{N(\mid n_{11}n_{22} - n_{12}n_{21} \mid - \frac{1}{10}N)^2}{n_{1+}n_{2+}n_{+1}n_{+2}}$$

$$X_{+25}^2 = \frac{N(\mid n_{11}n_{22} - n_{12}n_{21} \mid - \frac{1}{4}N)^2}{n_{1+}n_{2+}n_{+1}n_{+2}}$$

where X_{+1}^2 and X_{+25}^2 are continuity corrected statistics introduced by Burstein (1981) to approximate two-tailed exact binomial significance levels for cases when $n_{1+} \neq n_{2+}$ and $n_{1+} = n_{2+}$ respectively for the sampling scheme B.

Berchtold (1972) also introduced the test statistic with continuity correction:

$$X_B^2 = \frac{(N-1)(n_{11}n_{22} - n_{12}n_{21} + S)^2}{n_{1+}n_{2+}n_{+1}n_{+2}}$$

where $S = \dfrac{1}{6}(z_\alpha^2 - 1)(2q-1)(n_{1+} - n_{2+})$; z_α is the α-quantile of a standard normal distribution and q equals $\frac{n_{+1}}{N}$. Following Sachs (1986), the one-sided X_B^2 statistic will be designated as b^*. For this case, the z_α quartile would be taken as 1.645 (Dozzi & Riedwyl, 1984).

Obviously, when $n_{1+} = n_{2+}$, then X_B^2 reduces to Upton's test statistic discussed above, which was labeled as X_U^2 by Rice (1988).

And yet another test statistic proposed by Schouten et al. (1980) is defined here as:

$$X_S^2 = \frac{(N-1)\{\mid n_{11}n_{22} - n_{12}n_{21} \mid - \frac{1}{2}\min(n_{1+}, n_{2+})\}^2}{n_{1+}n_{2+}n_{+1}n_{+2}} \tag{4.27}$$

Again as pointed out in Schouten et al. (1980), this result extends those of Pirie and Hamden (1972) for in the case when $n_{1+} = n_{2+}$, the correction factor in (4.27) becomes $\frac{N}{4}$, which asymptotically is equivalent to $X_{.25}^2$.

The computation of χ^2 tail probabilities in SAS® for each of the above test statistics can be accomplished by using **probchi** and **probnorm** functions for the upper tail of a χ^2 with one degree of freedom and normal distributions respectively.

We give below some summary recommendation for the use of the continuity corrected X^2 suggested earlier.

4.7.2 For Observational Data

1. If the sample size N is very small, use Fisher's exact test or the mid-P test. The latter is recommended.

2. For moderate N, use $Q = \frac{N-1}{N} X^2$ with Yate's continuity correction.

4.7.3 If the Sampling Scheme Is the Product Binomial

1. If $n_{1+} = n_{2+}$, use uncorrected Q, or Burstein's X^2_{+25} statistic or simply resort to the exact binomial test, which is discussed below.

2. For $n_{1+} = kn_{2+}$, where k is an integer, use Q with continuity correction of $\dfrac{n_{2+}}{2}$ where n_{2+} is the smaller sample. The exact binomial test or the X^2_{+1} will provide suitable alternatives. We would, however, recommend X^2_{+1} if the exact binomial test is not immediately feasible.

3. If n_{1+} is not a multiple of n_{2+}, use X^2 with $\dfrac{k}{2}$, where k is the largest common divisor of n_{1+} and n_{2+}. Again, Burstein's X^2_{+1} will be highly recommended in lieu of the exact binomial test.

4. If the sampling scheme is multinomial (cross-classifying) with N fixed, use Q.

4.7.4 The Exact Binomial Test

Burstein (1981) obtained for the test of H_0 versus H_1, significance levels based on the joint probabilities:

$$B = \sum_i \sum_j \binom{n_{1+}}{i} \theta^i (1-\theta)^{n_{1+}-i} \binom{n_{2+}}{j} \theta^j (1-\theta)^{n_{2+}-j}$$

where $0 \leq i \leq n_{1+}$; $0 \leq j \leq n_{2+}$, with Δ as defined in the Burstein (1981) expressions for the upper-tailed probability and the two-tailed probability.

Thus, B_1, the upper tail probability, included events are those permitting $(\dfrac{i}{n_{1+}} - \dfrac{j}{n_{2+}}) \geq \Delta$ while B_2 = Tail-2 probability, included events are those permitting $(\dfrac{j}{n_{2+}} - \dfrac{i}{n_{1+}}) \geq \Delta$ and B, the two-tailed significance level will be given by $B_1 + B_2$. Burstein also provided simpler expressions and a computer program for calculating B_1, B_2, and hence B.

The one-tailed probability B_1 so obtained is equivalent to the CBET result of Rice (1988) case 0.

We give in Table 4.15 computed results for the exact binomial, mid-P, and Fisher's exact test for the data in Tables 4.4 and 4.6.

	Mid-P		Fisher's		Binomial	
	1-Tail	2-Tail	1-Tail	2-Tail	1-Tail	2-Tail
Table 4.4	0.0658	0.1386	0.0941	0.1669	0.0637	0.1281
Table 4.6	0.0119	0.0238	0.0240	0.0476	0.0094	0.0188

Table 4.15: Summary of results for the data in Tables 4.4 and 4.6

4.8 The 2×2 Tables with Correlated Data

The procedures that we have employed so far for the analysis of 2×2 table assume that the samples are independent. Sometimes the data arising from two binomial populations can be correlated. Our interest still centers on testing for homogeneity in the two populations. For instance, 2×2 table arising from husband and wife pair cannot be considered to be independent. Also not independent are data arising from matched-pair experiments. An example of such a matched-pair experiment is the data in Table 4.16, which relate to the side effect of pump therapy in the control of blood glucose levels in diabetic patients (Mecklenburg et al., 1984). The data is on the occurrence of diabetic ketoacidosis (DKA) in patients before and after the onset of pump therapy. Here, individuals provide a pair of measurements before and after diabetic pump therapy.

Before pump therapy	**After pump therapy** DKA	No DKA	Total
DKA	7	7	14
No DKA	19	128	147
Total	26	135	161

Table 4.16: Matched-pair data for DKA patients

The *concordant* pairs (that is, pairs in which outcomes are the same) equal $128 + 7 = 135$. The *discordant* pairs (cases in which outcomes are different for each member of the pair) usually designated as n_D equals $19 + 7 = 26$. The concordant pairs do not provide any information regarding the treatments effects. The discordant pairs, however, do provide this information, and these pairs can further be subdivided into pairs:

(a) A **type A** discordant pair is one in which a before treatment (pump therapy) member has the event (DKA) and the after pump therapy member does not. This pair is designated as $n_A = n_{12}$.

(b) A **type B** discordant pair is one in which an after pump therapy member has the event (DKA) and the before therapy member does not. Again, we shall designate this pair by $n_B = n_{21}$.

The number of counts in each of these two discordant pairs for the data above are $n_A = 7$ and $n_B = 19$, respectively. The test of the hypotheses

$$H_0 : \pi_{12} = \pi_{21} \quad \text{versus} \quad H_1 : \pi_{12} \neq \pi_{21}$$

is provided by the McNemar's test statistic, given by (see exercise 4.3)

$$X_N^2 = \frac{(n_{12} - n_{21})^2}{n_{12} + n_{21}}$$

The statistic is distributed χ_1^2 and the pvalue is given by $P(X_N^2 > \chi_1^2)$. A continuity-corrected version of the test statistic is defined as:

$$X_{Nc}^2 = \frac{(\mid n_{12} - n_{21} \mid -1)^2}{n_{12} + n_{21}}$$

For the data in Table 4.16, we have

$$X_N^2 = \frac{(19 - 7)^2}{19 + 7} = \frac{144}{26} = 5.5385 \quad \text{and}$$

$$X_{Nc}^2 = \frac{(\mid 19 - 7 \mid -1)^2}{19 + 7} = \frac{121}{26} = 4.6538$$

The pvalues for both statistics are respectively, 0.0186 and 0.0310. Thus there is significant statistical DKA proportional difference between the before and after pump therapies. The after pump therapy would be preferred. The SAS software program and output for this analysis is provided below. The statistic used by SAS software is clearly the X_N^2.

```
data corr;
input before after count @@; datalines;
1  1 7 1 2 7 2 1 19 2 2 128
;
TITLE 'McNema's Test';
proc freq order=data;
weight count; tables before*after/agree; run;
```

```
                    STATISTICS FOR TABLE OF BEFORE BY AFTER

                               McNemar's Test
                               --------------
          Statistic = 5.538          DF = 1          Prob = 0.019

                          Simple Kappa Coefficient
                          ------------------------
                                              95% Confidence Bounds
          Kappa = 0.267      ASE = 0.102          0.067      0.468
          Sample Size = 161
```

The X_{NC}^2 can be alternatively written in terms of n_D and n_A as follows:

$$X_{NC}^2 = \left(\mid n_A - \frac{n_D}{2} \mid - \frac{1}{2} \right)^2 / \left(\frac{n_D}{4} \right)$$

Substituting the relevant values into the above expression leads to a value of 4.6538 as expected.

The above test is valid if $n_D \geq 20$. For cases when $n_D < 20$, we shall use the exact test, which is based on the binomial distribution. The exact pvalues can be computed from the expressions below,

$$p = \begin{cases} 2 \times \sum_{j=0}^{n_A} \binom{n_D}{j} \left(\frac{1}{2} \right)^{n_D} & \text{if } n_A < n_D/2 \\ 2 \times \sum_{j=n_A}^{n_D} \binom{n_D}{j} \left(\frac{1}{2} \right)^{n_D} & \text{if } n_A > n_D/2 \\ 1 & \text{if } n_A = n_D/2 \end{cases}$$

4.8.1 Another Example

A recent phenomenon in the recording of blood pressure is the development of the automated blood-pressure machine, where for a small fee a person can sit in a booth and have his or her blood pressure measured by a computer device. A study is conducted to compare the computer device with the standard methods of measuring blood pressure. Twenty patients are recruited, and their hypertensive status is assessed by both the computer device and a trained observer. Hypertensive status is defined as either hypertensive (+), if either systolic bp ≥ 160 or diastolic bp ≥ 95, or normotensive (−) otherwise. The data are presented in Table 4.17, (taken from Rosner, 2000). Assess the statistical significance of these findings.

	Hypertensive status	
Person	Computer device	Trained observer
1	−	−
2	−	−
3	+	−
4	+	+
5	−	−
6	+	−
7	−	−
8	+	+
9	+	+
10	−	−
11	+	−
12	+	−
13	−	−
14	+	−
15	−	+
16	+	−
17	+	−
18	−	−
19	−	−
20	−	−

Table 4.17: Hypertensive status of 20 women

The 2 × 2 table arising from this problem is presented in a SAS software output below.

```
data matched;
input  computer $ obs$@@; datalines;
- - - - + - + + - - + - - - + + + + - - + - + - - - + - - + + -
+ - - - - - - -
;
proc freq; tables computer*obs/agree; run;
data exact; do i=0 to 8;
if i-1=-1 then prob=probbnml(0.5,8,0);
    else prob=probbnml(0.5,8,i)-probbnml(0.5,8,i-1);
output; end; run;
proc print noobs; run;
proc means data=exact noprint;
where i le 6; var prob;
output out=aa sum=p; run;
data result; set aa;
exact=2*(1-p); proc print data=result;
```

```
var exact; format exact 10.4; run;
                          The FREQ Procedure

                    Table of computer by observer

                          observer
            Computer      +          -          Total
            ------------------------------------------
            +             3          7          10
            ------------------------------------------
            -             1          9          10
            ------------------------------------------
            Total         4         16          20

            Statistics for Table of computer by obs

                        McNemar's Test
                  ------------------------
                  Statistic (S)     4.5000
                  DF                     1
                  Pr > S            0.0339

                     Obs          exact
                      1          0.0703
```

The McNemar's test above indicates a pvalue of 0.0339, which indicate that we would have to reject the null hypothesis. On the other hand, the exact result based on the binomial gives a pvalue 0.0703, indicating that we would fail to reject the null hypothesis in this case. That is, there is agreement between the trained observer and the computer device. The exact result is more accurate in this case.

4.9 Measures of Association in 2 × 2 Contingency Tables

In this section, we shall introduce some measures of association that have received considerable attention for the 2 × 2 contingency tables and later generalize this to higher dimensional tables. The structure of measures of association in larger tables can be easily constructed from a proper understanding of the 2 × 2 case (Goodman, 1979a, 1985).

For the 2 × 2 table, let the underlying probability model under the multinomial sampling scheme (cross-sectional study design) be displayed as in the next table.

	Response		
Popl	S	F	Total
I	π_{11}	π_{12}	π_{1+}
II	π_{21}	π_{22}	π_{2+}
Total	π_{+1}	π_{+2}	1

Let π_{i+} be the probability that a randomly selected individual is in the i-th row. Similarly, let π_{+j} be the corresponding probability of a randomly selected individual being in the j-th column. Also let π_{ij} be the unconditional probability that an individual selected at random is in row i and column j, $i, j = 1, 2$.

A measure of association between the row categories and the column categories is a population parameter which summarizes the relationship between the joint variation of the two classificatory variables.

For the 2×2 table, interest is centered on the null hypothesis of independence that is, the hypothesis when

$$H_0 : \pi_{ij} = \pi_{i+}\pi_{+j}$$

Suitable measures of association that we would be considering here are those that assume zero under the null hypothesis of independence or homogeneity. We shall commence our discussion on these measures by first distinguishing between a measure of association and a test of independence. A test of independence is used to determine whether a relationship exists between the two classificatory variables. On the other hand, a measure of association indicates the particular *type* and *extent* of this relationship. Thus tests of independence are existence tests, but measures of association reflect the strength of that relationship.

For the 2×2 table with the underlying probabilities above, the odds that a randomly selected individual will be classified in column 1 to being in column 2 given that he or she is classified in row 1 is given by

$$\left[\frac{\pi_{11}}{\pi_{1+}}\right] / \left[\frac{\pi_{12}}{\pi_{1+}}\right] = \frac{\pi_{11}}{\pi_{12}}$$

Similarly, the odds of being classified in column 1 to being in column 2 given that the individual is classified in row 2 are given by

$$\left[\frac{\pi_{21}}{\pi_{2+}}\right] / \left[\frac{\pi_{22}}{\pi_{2+}}\right] = \frac{\pi_{21}}{\pi_{22}}$$

The odds ratio is defined as the ratio of these two odds. That is,

$$\theta = \frac{\pi_{11}\pi_{22}}{\pi_{12}\pi_{21}} \tag{4.28}$$

θ is referred to as the odds ratio or cross-product ratio. Under the model of independence, $\theta = 1$. If $\theta > 1$, then there is said to be *positive association* between the two variables. Similarly, if $\theta < 1$, there is said to be *negative association* between the two variables.

4.9.1 Properties of θ

1. Under $H_0 : \theta = 1$.

2. θ is invariant under interchange of rows and columns. However, an interchange of only rows or columns will change θ to $1/\theta$.

3. θ is invariant under row and column multiplication by a constant, say, $\beta > 0$.

4. θ ranges from 0 to ∞ but the logarithm of θ, that is, the log odds ratio, is symmetrical about 0. Thus, $0 < \theta < \infty$ which implies that $-\infty < \log(\theta) < \infty$ and under $H_0 : \log(\theta) = 0$.

5. If we let $r_1 = \dfrac{\pi_{11}}{\pi_{1+}}$ and $r_2 = \dfrac{\pi_{21}}{\pi_{2+}}$, be functions of the conditional row probabilities and similarly let $c_1 = \dfrac{\pi_{11}}{\pi_{+1}}$ and $c_2 = \dfrac{\pi_{12}}{\pi_{+2}}$ be functions of the conditional column probabilities, then we see that

   (a)

$$\theta = \frac{\pi_{11}\pi_{22}}{\pi_{12}\pi_{21}} = \left[\frac{r_1}{(1-r_1)}\right]\left[\frac{(1-r_2)}{r_2}\right] = \left[\frac{c_1}{(1-c_1)}\right]\left[\frac{(1-c_2)}{c_2}\right]$$

and we can show that X^2 reduces in this case to

$$X^2 = N[(r_1 + c_1)(r_2 + c_2) - 1]$$

that is, a function estimated by X^2.

(b) If we define, $\gamma = \log(\theta) = \log(\pi_{11}) + \log(\pi_{22}) - \log(\pi_{12}) - \log(\pi_{21})$, then we can reexpress γ as:

$$\gamma = \log\left(\frac{r_1}{1-r_1}\right) - \log\left(\frac{r_2}{1-r_2}\right) = \theta_1 - \theta_2$$

where $\theta_i = \log\left(\frac{r_i}{1-r_i}\right)$, $\quad i = 1, 2$, is the difference between two logits $r_i = \dfrac{\pi_{i1}}{\pi_{i1} + \pi_{i2}}$ based upon conditional probabilities r_1, r_2.

4.9.2 MLE and Asymptotic Variance

Under either the multinomial sampling scheme or the product binomial scheme, the MLE estimator for θ is given by

$$\hat{\theta} = \frac{n_{11}n_{22}}{n_{12}n_{21}}$$

It is sometimes advocated that for those situations where there are zero cell frequencies, the estimate can be computed as

$$\hat{\theta}^* = \frac{(n_{11} + 0.5)(n_{22} + 0.5)}{(n_{12} + 0.5)(n_{21} + 0.5)}$$

We have shown earlier using the delta method that an estimate of the asymptotic variance of $\hat{\theta}$ is given by

$$\widehat{\text{Var}}_A(\hat{\theta}) = \hat{\theta}^2 \left\{ \frac{1}{n_{11}} + \frac{1}{n_{12}} + \frac{1}{n_{21}} + \frac{1}{n_{22}} \right\}$$

For the data of Table 4.4 we have

$$\hat{\theta} = \frac{14 \times 72}{8 \times 61} = 2.066$$

and

$$\hat{V}(\hat{\theta}) = 2.066^2(0.2267) = 0.9676$$

The approximate 95% confidence interval for the true value of association, as measured by θ, is therefore

$$2.066 \pm 1.96\sqrt{.9676} \text{ that is, } (0.1380, 3.9940)$$

The confidence interval includes the value 1, the independence value, so that the data do not rule out the possibility of independence (as measured by θ) or homogeneity.

A $100(1-\alpha)\%$ confidence interval is sometimes obtained first in terms of the log of the odds ratio as:

$$e^{\ell \pm z_{1-\alpha/2}\sqrt{V}}$$

where ℓ is the log of the estimated odds ratio and

$$V = \left(\frac{1}{n_{11}} + \frac{1}{n_{12}} + \frac{1}{n_{21}} + \frac{1}{n_{22}} \right)$$

is the asymptotic variance of ℓ. We have therefore, for our data,

$$e^{\ln(2.066)\pm 1.96\sqrt{0.2267}} = e^{0.7256\pm 0.9332} = e^{(-0.2076,1.6588)}$$
$$= (0.8125, 5.2530)$$

The latter is the confidence intervals computed by SAS®.

4.9.3 General Functions of θ

Several measures of association are monotonically increasing or decreasing functions of θ. Let $f(\theta)$ be a positive monotonically increasing function of θ such that $f(1) = 1$. Then a normalized measure of association based on $f(\theta)$ whose maximum absolute value is +1 is obtained as

$$g(\theta) = \frac{f(\theta) - 1}{f(\theta) + 1}$$

The asymptotic variance of $g(\theta)$ is given by

$$\text{Var}_A\{g(\theta)\} = \frac{[1 - g(\theta)]^4 [f^{'}(\theta)]^2}{4} \text{Var}_A(\hat{\theta})$$

where $f^{'}(\theta)$ is the derivative of $f(\theta)$ with respect to θ.

Proof

We have shown using the linearized Taylor's series method, which is otherwise known as the δ method, that the variance of a function $g(\theta)$ can be expressed as

$$\text{Var}_A\{g(\theta)\} = g^{'}(\theta)^2 \text{Var}(\theta)$$

In the present case,

$$g(\theta) = \frac{f(\theta) - 1}{f(\theta) + 1}$$

$$g^{'}(\theta) = \frac{2f^{'}(\theta)}{[1 + f(\theta)]^2}$$

Hence,

$$\text{Var}_A\{g(\theta)\} = \frac{4f^{'}(\theta)^2}{[1 + f(\theta)]^4} \text{Var}_A(\hat{\theta})$$

but

$$1 - g(\theta) = \frac{2}{1 + f(\theta)},$$

Hence,

$$\text{Var}_A\{g(\theta)\} = \frac{[1 - g(\theta)]^4 [f^{'}(\theta)]^2}{4} \text{Var}_A(\hat{\theta})$$

Specific Functions of θ

Yules's (1900) Q measure of association: Let $f(\theta) = \theta$; then

$$Q = \frac{\theta - 1}{\theta + 1} = \frac{\pi_{11}\pi_{22} - \pi_{12}\pi_{21}}{\pi_{11}\pi_{22} + \pi_{12}\pi_{21}}$$

$Q = +1$ if either $\pi_{21} = 0$ or $\pi_{12} = 0$; that is, as the value of the classificatory variable Popl increases, the value for Response either remains the same or increases. Similarly, $Q = -1$ if either $\pi_{22} = 0$ or $\pi_{11} = 0$.

A sample estimate of Q is given by

$$\hat{Q} = \frac{\hat{\theta} - 1}{\hat{\theta} + 1} = \frac{n_{11}n_{22} - n_{12}n_{21}}{n_{11}n_{22} + n_{12}n_{21}}$$

with an estimate of the asymptotic variance given by

$$\widehat{\text{Var}}_A\{\hat{Q}\} = \frac{(1 - \hat{Q}^2)^2}{4}\left\{\frac{1}{n_{11}} + \frac{1}{n_{12}} + \frac{1}{n_{21}} + \frac{1}{n_{22}}\right\}$$

The hypothesis of independence is rejected when $Q = 0$. Referring again to the data in Table 4.4, we have $\hat{Q} = \frac{2.066-1}{2.066+1} = 0.3477$ with estimated asymptotic variance $\widehat{\text{Var}}(Q) = 0.1932 \times 0.2267 = 0.0438$. An approximate 95% CI for Q is therefore given by:

$$0.3477 \pm 1.96\sqrt{0.0438} \quad \text{that is,} \quad (-0.0625, 0.7579)$$

Again, this interval contains zero; hence, we will again fail to reject the null hypothesis of independence.

4.9.4 Measure of Relative Risk

In situations in which one variable is assumed to be *antecedent* to another, but both are measured at the same point in time, e.g., retrospective, or case-control studies in epidemiology, it is of interest to investigate the *relative risk* of developing the *consequent condition* based on the *antecedent characteristic.*

Notation

Let D denote the condition of interest (that is, the disease variable) e.g Cancer of the lung, and let E denote the characteristic, factor, or explanatory variable of interest, e.g. smoking, often called the exposure variable.

Let + denote presence and − denote absence. In this framework, the corresponding 2 × 2 (exposure disease) table is

	D		
E	+	−	Totals
+	π_{11}	π_{12}	π_{1+}
−	π_{21}	π_{22}	π_{2+}
Totals	π_{+1}	π_{+2}	1

Among subjects with the factor present or exposed, the risk of D is

$$P[D = + \mid E = +] = \frac{\pi_{11}}{\pi_{1+}}$$

and among subjects with factor absent or unexposed, the risk of D is

$$P[D = + \mid E = -] = \frac{\pi_{21}}{\pi_{2+}}$$

The *relative risk*, **RR** is then defined as

$$\text{RR} = \frac{\pi_{11}/\pi_{1+}}{\pi_{21}/\pi_{2+}} = \frac{P(\text{disease} \mid \text{exposed})}{P(\text{disease} \mid \text{unexposed})} = \frac{\pi_{11}\pi_{2+}}{\pi_{21}\pi_{1+}} \tag{4.29}$$

In words, we can define relative risk as the ratio of developing a disease among exposed (or at risk) subjects to the risk of developing the disease among un exposed subjects. The *relative risk* and the *odds ratio* are two different measures that sought to explain the same phenomenon. The relative risk can be expressed in terms of the odds ratio by writing

$$\text{RR} = \frac{\pi_{11}\{\pi_{21}+\pi_{22}\}}{\pi_{21}\{\pi_{11}+\pi_{12}\}} = \frac{\pi_{11}\pi_{22}\{1+\pi_{21}/\pi_{22}\}}{\pi_{12}\pi_{21}\{1+\pi_{11}/\pi_{12}\}} = \theta\left\{\frac{1+\pi_{21}/\pi_{22}}{1+\pi_{11}/\pi_{12}}\right\}$$

Now if D is a "rare disease" in the population, $\frac{\pi_{21}}{\pi_{22}}$ and $\frac{\pi_{11}}{\pi_{12}}$ will be essentially 0 (often referred to as the rare outcome assumption). Thus in this case (usually when outcome << 10%),

$$\hat{\theta} \simeq \widehat{\text{RR}}$$

For prospective data, $\text{RR} = \frac{\pi_{11}/n_{1+}}{\pi_{21}/n_{2+}}$. Similarly, for cross-sectional studies the ratio $\frac{\pi_{11}}{\pi_{21}}$ is often referred to as the *prevalence ratio.* However, this ratio does not indicate risk since the disease and risk factors are assessed at the same time, but gives a comparison of the prevalence of the disease for the "at risk" group versus the other group.

Example

The example below relates to the relationship between aspirin use and heart attack (Agresti, 1990). During the Physician's Health Study, 11,037 physicians were randomly assigned to take 325 mg of aspirin every other day. In another group, 11,034 physicians were randomly assigned to take a placebo. The resulting number of heart attacks in each group is displayed in the next table as a 2 × 2 contingency table.

	Myocordial infarction		
	Heart attack	No heart attack	Totals
Placebo	189	10,845	11,034
Aspirin	104	10,933	11,037

This was a randomized clinical trial testing whether aspirin regularly taken reduces mortality from cardiovascular disease. Let

π_1 = P(Heart attack taking placebo)

π_2 =P(Heart attack taking aspirin)

The estimates of these probabilities from the data are:

$$\hat{\pi}_1 = 189/11034 = 0.0171$$
$$\hat{\pi}_2 = 104/11037 = 0.0094$$

Sample difference of proportions is $0.0171 - 0.0094 = 0.0077$ and the relative risk is

$$\hat{\text{RR}} = \frac{0.0171}{0.0094} = 1.82$$

The proportion suffering heart attacks was 1.82 times higher for those taking placebo than for those taking aspirin. The sample odds ratio is $\hat{\theta} = \frac{(189\times 10933)}{(104\times 10845)} = 1.83$, which is very close to the estimate of the relative risk. The above data are from a prospective or cohort study, which allows us to estimate π_1 and π_2. In a retrospective or case-control study, however, we would not be able to estimate these probabilities because in that case, the $n_{+1} = 293$ and $n_{+2} = 21,778$ were fixed by design and the data therefore contain no information about the π values. We can, however, use the result established earlier to show that regardless, an estimate of θ is 1.83.

4.9.5 Sensitivity, Specificity and Predictive Values

Other measures for testing the effectiveness of a test procedure (screening test or set of symptoms), such as a medical test to diagonose a disease, are *sensitivity, specificity,* and *predictive values.*

For screening tests, it must be realized that these tests themselves can sometimes be wrong. That is, a testing procedure may yield either a *false positive* or a *false negative* result.

1. A false positive results when a test indicates a positive status when the true status is negative.

2. A false negative results when a test indicates a negative status when the true status is positive.

Suppose we have for a sample of n subjects (n always large) the information in the next table:

	Disease		
Test result	Present (D)	Absent (D)	Total
Positive (T)	a	b	a+b
Negative ($\bar{T}$)	c	d	c+d
Total	a+c	b+d	n

(i) The **sensitivity** of a test (or symptom) measures how well it detects disease or set of symptoms. It is the proportion among those with the disease who give a positive result. That is, it gives the probability of a positive test result (or presence of the symptom) given the presence of the disease. In the context of the above table, then, it is given by:

$$P(T \mid D) = \frac{a}{a+c}$$

(ii) The **specificity** of a test (or symptom) measures how well it detects absence of disease. It is the proportion of those without the disease that are correctly diagnosed (i.e., negative result). Again, this is the probability of a negative test result (or absence of the symptom) given the absence of the disease. That is, it is

$$P(\bar{T} \mid \bar{D}) = \frac{d}{b+d}$$

(iii) The **predictive value positive** of a screening test or symptom is the probability that subject has the disease given that the subject has a positive screening test result (or has the symptom):

$$P(D \mid T+) = \frac{P(T \mid D)\, P(D)}{P(T \mid D)\, P(D) + P(T \mid \bar{D})\, P(\bar{D})}$$

(iv) The **predictive value negative** of a screening test (or symptom) is the probability that subject does not have the disease given that the subject has a negative screening test result (or does not have the symptom).

$$P(\bar{D} \mid \bar{T}) = \frac{P(\bar{T} \mid \bar{D})\, P(\bar{D})}{P(\bar{T} \mid \bar{D})\, P(\bar{D}) + P(\bar{T} \mid D)\, P(D)}$$

Example

As an example, consider the data here from Pagano and Gauvreau (1993). Among the 1,820 subjects in a study, 30 suffered from tuberculosis and 1,790 did not. Chest x-rays were administered to all individuals; 73 had a positive x-ray, indicating the significance presence of inflammatory disease, and 1,747 had a negative x-ray. The data for the study are displayed in Table 4.18.

	Disease status		
X-ray result	Present(D)	Absent($\bar{D}$)	Total
Positive (T)	22	51	73
Negative ($\bar{T}$)	8	1739	1747
Total	30	1790	1,820

Table 4.18: Study data on Tuberculosis

We note that the prevalence of the disease (tuberculosis) in the population is 30/1820, or 1.65%.

(a) The sensitivity of the test is given by:

$$P(T \mid D) = \frac{22}{30} = 0.7333$$

A *false negative* occurs when the test of an individual who has tuberculosis incorrectly indicates that the individual does not. In our example this is given by:

P[test negative | tuberculosis] $= \frac{8}{30} = 0.2667$. That is,
P[*false negative*] $= 1 - P$[*sensitivity*]

(b) The specificity of the test is given by:

$$P(\bar{T} \mid \bar{D}) = \frac{1739}{1790} = 0.9715$$

Since not all individuals tested are actually carrying tuberculosis, 2.85% of the tests were *false positive* outcomes, which implies that

P[test positive | no tuberculosis] $= \frac{51}{1790} = 0.0285$
That is, P[*false positive*] = 1-P[*specificity*]

(c) The predictive value positive of the test that is, the probability that a subject who is positive on the x-ray test has tuberculosis is:

$$\begin{aligned} P(D \mid T+) &= \frac{P(T \mid D)\, P(D)}{P(T \mid D)\, P(D) + P(T \mid \bar{D})\, P(\bar{D})} \\ &= \frac{(0.7333)(0.0165)}{(0.7333)(0.0165) + (0.0285)(1 - 0.0165)} \\ &= 0.3015 \end{aligned}$$

since, $P(T \mid \bar{D}) = 51/1790 = 0.0285$. We see that the predictive value of the test is low.

(d) The predictive value negative of the test that is, the probability that a subject who is negative on the x-ray test is tuberculosis free, is:

$$\begin{aligned}
P(\bar{D} \mid \bar{T}) &= \frac{P(\bar{T} \mid \bar{D})\, P(\bar{D})}{P(\bar{T} \mid \bar{D})\, P(\bar{D}) + P(\bar{T} \mid D)\, P(D)} \\
&= \frac{(1 - 0.0285)(1 - 0.0165)}{1 - 0.0285)(1 - 0.0165) + (1 - 0.7333)(0.0165)} \\
&= \frac{(0.9715)(0.9835)}{(0.9715)(0.9835) + (0.2667)(0.0165)} \\
&= 0.9954
\end{aligned}$$

We observe here that both the predictive value positive and negative can be easily computed from the above table respectively as:

$$\begin{aligned}
P[D \mid T] &= \frac{22}{73} &= 0.3014 \\
P[\bar{D} \mid \bar{T}] &= \frac{1739}{1747} &= 0.9954
\end{aligned}$$

It should be noted that for rare diseases, high values of specificity or sensitivity are not necessarily sufficient to ensure that a large proportion of those testing positive actually have the disease.

4.9.6 Product Moment Correlation Coefficient

Consider a population with the following underlying probability structure.

	B		
A	0	1	Totals
0	π_{11}	π_{12}	π_{1+}
1	π_{21}	π_{22}	π_{2+}
Totals	π_{+1}	π_{+2}	1

Means of rows are:

$$E(\text{Row}) = 0(\pi_{11}) + 0(\pi_{12}) + 1(\pi_{21}) + 1(\pi_{22}) = \pi_{21} + \pi_{22} = \pi_{2+}$$

Similarly,

$$E(\text{Column}) = 0(\pi_{11}) + 0(\pi_{21}) + 1(\pi_{12}) + 1(\pi_{22}) = \pi_{12} + \pi_{22} = \pi_{+2}$$

and

$$\rho = \frac{\text{Cov}(A, B)}{\sqrt{\text{Var}(A)\text{Var}(B)}}$$

From the above specified probability structure, we have

$$\begin{aligned}
E(A) &= \pi_{2+} \\
E(B) &= \pi_{+2} \\
E(AB) &= \pi_{22}
\end{aligned}$$

so that

$$\begin{aligned}\text{COV}(A,B) &= \pi_{22} - \pi_{2+}\pi_{+2} \\ &= \pi_{22} - (\pi_{21} + \pi_{22})(\pi_{12} + \pi_{22}) \\ &= \pi_{11}\pi_{22} - \pi_{12}\pi_{21} \\ &= \sigma_{12}\end{aligned}$$

Thus the variances are obtained as:

$$\begin{aligned}\text{Var}(A) &= \pi_{2+}(1 - \pi_{2+}) = \pi_{1+}\pi_{2+} = \sigma_1^2 \\ \text{Var}(B) &= \pi_{+2}(1 - \pi_{+2}) = \pi_{+1}\pi_{+2} = \sigma_2^2\end{aligned}$$

and hence,

$$\rho = \frac{\sigma_{12}}{\sigma_1\sigma_2} = \frac{\pi_{11}\pi_{22} - \pi_{12}\pi_{21}}{\{\pi_{1+}\pi_{2+}\pi_{+1}\pi_{+2}\}^{\frac{1}{2}}}$$

In the above formulation, we are assuming that the response variables A and B are each measured on a dichotomous (or binary) scale for each of a sample of N subjects, and that the measurement scale can be coded as 0 (absent) and 1 (present) so that the usual arithmetic means represent the proportion present for each variable.

Properties of ρ

1. ρ is invariant under a change of both rows and columns. ρ changes only its sign if either rows or columns are interchanged, but not both.

2. Under independence, $\rho = 0$.

3. $\rho = +1$ if and only if $\pi_{12} = \pi_{21} = 0$.

4. $\rho =$ -1 if and only if $\pi_{11} = \pi_{22} = 0$.

5. ρ is invariant under positive linear transformations, so that the same value of ρ is obtained under scoring the rows and columns by any monotonic increasing functions of 0 and 1.

An MLE for ρ is given by:

$$\hat{\rho} = r = \frac{n_{11}n_{22} - n_{12}n_{21}}{(n_{1+}n_{2+}n_{+1}n_{+2})^{\frac{1}{2}}}$$

Under $H_0 : \rho = 0$ and $\text{Var}(\hat{\rho}) = 1/N$.

In most statistical literature,

$$\hat{\rho} = \pm\sqrt{\frac{X^2}{N}}$$

The above is also known as the *phi* (ϕ) coefficient. For the data in Table 4.4, we have:

$$\hat{\rho} = \frac{14 \times 72 - 8 \times 61}{\sqrt{(22 \times 133 \times 75 \times 80)^{\frac{1}{2}}}} = \frac{520}{4189.9881} = 0.1241$$

$\widehat{\text{Var}}(\hat{\rho}) = \frac{1}{N} = \frac{1}{155} = 0.0065$, and an approximate 95% CI for ρ equals

$$0.1241 \pm 1.96\sqrt{0.0065} = 0.1241 \pm 0.1574 = (-0.0333, 0.2815)$$

This interval again includes zero, and we would again fail to reject the null hypothesis of independence or homogeneity.

A function of ρ that has seen most application is the "coefficient of mean square contingency" proposed by Pearson (1904). It is defined as

$$P = \sqrt{\frac{\rho^2}{\rho^2 + 1}}$$

An estimate of P is given by

$$\hat{P} = \sqrt{\frac{X^2}{X^2 + N}}$$

We give some of these measures as reproduced from a modified SAS software output.

```
OPTIONS NODATE NONUMBER LS=77 PS=66;
DATA MEASURE;
DO STATUS=1 TO 2;
  DO SMOKE=1 TO 2;
INPUT COUNT @@; OUTPUT;
END; END;
DATALINES;
14 61 8 72
;
PROC PRINT; TITLE 'MEASURES OF ASSOCIATION';
PROC FREQ; WEIGHT COUNT; TABLES STATUS*SMOKE/ALL NOCOL NOPCT;
RUN;
```

```
                   Statistics for Table of STATUS by SMOKE

           Statistic                     DF        Value      Prob
           ------------------------------------------------------
           Chi-Square                     1       2.3873    0.1223
           Likelihood Ratio Chi-Square    1       2.4068    0.1208
           Continuity Adj. Chi-Square     1       1.7288    0.1886
           Mantel-Haenszel Chi-Square     1       2.3719    0.1235
           Phi Coefficient                        0.1241
           Contingency Coefficient                0.1232
           Cramer's V                             0.1241

                          MEASURES OF ASSOCIATION
                            The FREQ Procedure

                   Statistics for Table of STATUS by SMOKE

           Statistic                               Value       ASE
           ------------------------------------------------------
           Pearson Correlation                    0.1241    0.0784
           Spearman Correlation                   0.1241    0.0784

                 Estimates of the Relative Risk (Row1/Row2)

      Type of Study                   Value       95% Confidence Limits
      -----------------------------------------------------------------
      Case-Control (Odds Ratio)      2.0656        0.8124          5.2521
      Cohort (Col1 Risk)             1.8667        0.8308          4.1941

              Estimates of the Common Relative Risk (Row1/Row2)

  Type of Study     Method                   Value      95% Confidence Limits
  ---------------------------------------------------------------------------
  Case-Control      Mantel-Haenszel         2.0656        0.8124          5.2521
    (Odds Ratio)    Logit                   2.0656        0.8124          5.2521
```

4.9.7 Choosing a Measure of Association for a 2 × 2 Table

Bishop et al. (1975) suggest that the choice of a measure of association is basically between Q and ρ, since both measures take the value 0 under H_0, the model of independence. Further, they always lie between -1 and $+1$ and have reasonable interpretations.

1. The measure Q takes the value $+1$ or -1 whenever any one of the cell probabilities in a 2×2 table is zero, whereas for $\rho = +1$ or -1, both entries on one of the diagonals must be zero.

2. In a given table, the marginal totals may constrain the cell entries in such a way that ρ cannot take the value of $+1$ or -1. That is, ρ is a margin sensitive measure. In fact, $\rho =\sim 1$ implies the margins are the same for both variables

    - The more different the margins, the lower is the upper bound for ρ. On the other hand, Q is not affected by row and column multiplications and can assume its full range of values irrespective of the distribution of the marginal totals.
    - Choice between Q and ρ depends to a great extent on whether we wish to use a *measure that is sensitive to margins* and whether one wishes to consider association complete when only one cell of the 2×2 table is zero (Q) rather than two cells zero (ρ).

4.10 Analyzing Several 2×2 Contingency Tables

We now consider in this section the extension to analyzing several twofold contingency tables arising from stratified studies. We are interested in combining the information in each of the several 2×2 tables without sacrificing the inherent associations in each of the several tables. The procedure for doing this is presented in the following sub sections.

4.10.1 Combining Several 2×2 Tables

Often, associations between two categorical variables are examined across two or more populations. The resulting data usually lead to several (say, h) 2×2 contingency tables. Table 4.19 relates to the effect of passive smoking on lung cancer. It summarizes results of case-control studies from three different countries among nonsmoking women married to smokers. These tables can sometime come from a single study that has been stratified by some factor (in this case, country) that might be a confounder. The goal is usually to be able to combine the tables in order to have a unified information across the tables.

We would like to combine the evidence from the three countries to make an overall statement about whether having lung cancer is independent of passive smoking. The conditional test for these data within each country can be obtained by computing Fisher's exact test separately for each subtable or obtain Pearson's X^2 for each table. Thus the value of X^2 is first computed for each country, and the results suggest that neither of them is significant, which suggests that lung cancer status is independent of passive smoking by wives whose husbands are smokers within each country. These results are given in the SAS software partial output below.

```
STATISTICS FOR TABLE OF SMOKE BY STATUS
```

Country	Spouse Smoked?	Cancer Status Cases	Cancer Status Control	Total	Cases Prop.	X^2
Japan	Yes	73	188	261	0.28	
	No	21	82	103	0.20	
	Totals	94	270	364	0.26	2.216
UK	Yes	19	38	57	0.33	
	No	5	16	21	0.24	
	Totals	24	54	78	0.31	0.653
USA	Yes	137	363	500	0.27	
	No	71	249	320	0.22	
	Totals	208	612	820	0.25	2.800
Totals	Yes	229	589	818	0.28	
	No	97	347	444	0.22	
	Total	326	936	1262	0.26	5.678

Table 4.19: Cancer status according to whether spouse smoked, from three countries

```
                     Analysis for JAPAN

Statistic                         DF       Value         Prob
-----------------------------------------------------------
Chi-Square                         1       2.216        0.137
Likelihood Ratio Chi-Square        1       2.288        0.130
Continuity Adj. Chi-Square         1       1.838        0.175
Sample Size = 364

                       Analysis for UK

Statistic                         DF       Value         Prob
-----------------------------------------------------------
Chi-Square                         1       0.653        0.419
Likelihood Ratio Chi-Square        1       0.674        0.412
Continuity Adj. Chi-Square         1       0.283        0.595
Sample Size = 78
                      Analysis for USA

Statistic                         DF       Value         Prob
-----------------------------------------------------------
Chi-Square                         1       2.800        0.094
Likelihood Ratio Chi-Square        1       2.833        0.092
Continuity Adj. Chi-Square         1       2.532        0.112
Sample Size = 820
```

For the countries, the estimated odds ratios are $\hat{\theta}_J = 1.5162$, $\hat{\theta}_{UK} = 1.600$, and $\hat{\theta}_{US} = 1.3236$, with corresponding 95% confidence intervals given respectively by $(0.8745, 2.6288), (0.5090, 5.0293)$, and $(0.9526, 1.8390)$. All the three intervals include 1, indicating independence. The three odds ratios above are essentially estimating the same population value, and the estimated values differ only because of sampling variability. The corresponding pvalues based on G^2 for the three countries are respectively $0.130, 0.412$, and 0.092. All the three are not significant at $\alpha = 0.05$.

In searching for an overall test of the above hypotheses, one may be tempted to collapse these tables across countries as displayed in Table 4.20, which gives the combined table collapsed across countries.

Analysis of the data in Table 4.20 gives $X^2 = 5.678$ on 1 d.f. (significant), which now suggests that lung cancer status is related to passive smoking among wives whose husbands smoked. The corresponding SAS software output is displayed below. Here, $\hat{\theta} = 1.3908$ and the corresponding confidence interval equals $(1.0598, 1.8258)$, which does not include 1, indicating once again that lung cancer is strongly associated with passive smoking among this group of women.

Spouse smoked?	Cancer status Cases	Control	Total
Yes	229	589	818
No	97	347	444
Total	326	936	1262

Table 4.20: Collapsed table over country

```
          Combined analysis collapsed over country
           STATISTICS FOR TABLE OF SMOKE BY STATUS

Statistic                       DF      Value         Prob
-----------------------------------------------------------
Chi-Square                       1      5.678        0.017
Likelihood Ratio Chi-Square      1      5.779        0.016
Continuity Adj. Chi-Square       1      5.362        0.021
Sample Size = 1262
```

This result obviously contradicts the results from the individual country analyses, where it was concluded that lung cancer is independent of smoking status of this group of women. This apparent contradiction (*change in direction*) with the earlier results is what is known as *Simpson's paradox.* This contradiction can immediately be attributed to the large differences in the overall number of individuals sampled from each country. While a total of 364 respondents were sampled from Japan, only 78 were sampled from the United Kingdom, and an even larger number (820) was sampled from the United States. There does not seem to be much differences in the case proportions for the three countries, being respectively Japan (0.258), United Kingdom (0.308), and United States (0.254). We will reanalyze these data later using a *log-linear* model approach.

Simpson's paradox can also be due to the magnitude of or strength of an association. Consider the following example, which was published in Pagano and Gauvreau (1993). The data relate to the study of the relationship between smoking and aortic stenosis, stratified by gender.

Gender	Aortic stenosis	Smoker Yes	No	Total
Males	Yes	37	25	62
	No	24	20	44
	Total	61	45	106
Females	Yes	14	29	43
	No	19	47	66
	Total	33	76	109
Combined	Yes	51	54	105
	No	43	67	110
	Totals	94	121	215

Table 4.21: Data on relationship between smoking and aortic stenosis: Stratified by gender

The SAS software partial output for both the individual tables (males and females) together with the combined table analysis is displayed below. In the above

SAS software output, we notice that we would fail to reject the null hypothesis of independence for either the individual gender tables or the combined table. In this case, there is no contradiction or directional changes between the individual tables and the combined table results. Notice, however, the magnitude of X^2 in each analysis (or the relative magnitudes of the pvalues). Clearly, the strength appears weaker for the combined data than for either males or females. In this case, Simpson's paradox manifests itself in terms of the strength of the association.

```
                     Analysis for MALES

            STATISTICS FOR TABLE OF AORTIC BY SMOKER

Statistic                      DF       Value          Prob
-------------------------------------------------------------
Chi-Square                      1       0.277         0.598
Likelihood Ratio Chi-Square     1       0.277         0.599
Sample Size = 106

                       Analysis for FEMALES

            STATISTICS FOR TABLE OF AORTIC BY SMOKER

Statistic                      DF       Value          Prob
-------------------------------------------------------------
Chi-Square                      1       0.175         0.675
Likelihood Ratio Chi-Square     1       0.175         0.676
Sample Size = 109

              Combined analysis collapsed over GENDER

            STATISTICS FOR TABLE OF AORTIC BY SMOKER

Statistic                      DF       Value          Prob
-------------------------------------------------------------
Chi-Square                      1       1.962         0.161
Likelihood Ratio Chi-Square     1       1.965         0.161
Sample Size = 215
```

In both examples that we have considered above, we observe that it would not be wise to collapse the data set in Table 4.19 over the factor variable *country* without serious distortion to the association between the two variables. Similarly, it would also not be wise to collapse the tables over the factor variable *gender* in the second example without distorting the strength of the association within the individual tables.

In general, we are interested in collecting information for each of several 2 × 2 tables across the levels of the subpopulations (which may be determined by various configurations of factor variables or covariates).

In many cases, the primary question involves the relationship between an independent variable (factor) that is either present (1) or absent (2) and a dependent (response) variable that is either present (1) or absent (2) in the presence of several covariates. This could give rise to frequency data that may be summarized as a set of 2 × 2 tables (see Table 4.22).

In these tables, $i = 1, 2, \cdots, h$ indexes the separate levels of the covariates set or the stratified subpopulations. Let $a_i, i = 1, 2, \cdots, h$ denote the number of subjects in the sample who are jointly classified as belonging to the i-th table, and the pivot cell for the i-th subtable.

Further, let $(a_i + b_i)$ and $(c_i + d_i)$ denote the row marginal totals and $(a_i + c_i)$ and $(b_i + d_i)$ the corresponding column marginal totals for subtable i.

	Response		
Factor	(1)	(2)	Total
(1)	a_1	b_1	a_1+b_1
(2)	c_1	d_1	c_1+d_1
Total	a_1+c_1	b_1+d_1	n_1

	Response		
Factor	(1)	(2)	Total
(1)	a_2	b_2	a_2+b_2
(2)	c_2	d_2	c_2+d_2
Total	a_2+c_2	b_2+d_2	n_2

⋮ ⋮

	Response		
Factor	(1)	(2)	Total
(1)	a_i	b_i	a_i+b_i
(2)	c_i	d_i	c_i+d_i
Total	a_i+c_i	b_i+d_i	n_i

	Response		
Factor	(1)	(2)	Total
(1)	a_h	b_h	a_h+b_h
(2)	c_h	d_h	c_h+d_h
Total	a_h+c_h	b_h+d_h	n_h

Table 4.22: Summary of the set of 2 × 2 Tables

In particular, let a_i be the "pivot" cell frequency of subjects in the i-th table who have both factor and response present. Under the assumption that the marginal totals are fixed, the overall null hypothesis of *no partial association* against the alternative hypothesis that on the average across the h subtables, there is a consistent relationship between the row and column variables is conducted by obtaining the Cochran-Mantel-Haenzel (CMH) test statistic X^2_{MH}, which is computed as follows:

For table i, for instance, $\mathbf{n}_i^{'} = (a_i, b_i, c_i, d_i)$ follows the hypergeometric distribution and therefore

$$P\{\mathbf{n}_i \mid H_0\} = \frac{(a_i+b_i)!(c_i+d_i)!(a_i+c_i)!(b_i+d_i)!}{n_i!a_i!b_i!c_i!d_i!}$$

From results in (4.5) and (4.7), it follows that the expected value for the pivot cell in the i-th subtable is given by:

$$\begin{aligned} E(a_i \mid H_0) = \hat{m}_i &= \frac{(a_i+b_i)(a_i+c_i)}{n_i} \\ \text{Var}(a_i \mid H_0) = V_{a_i} &= \frac{(a_i+b_i)(c_i+d_i)(a_i+c_i)(b_i+d_i)}{n_i^2(n_i-1)} \end{aligned}$$

where $n = \sum_i n_i$, $m = \sum_i m_i$, and $\hat{V} = \sum_i \hat{V}_i$ are the corresponding sums of the *observed frequencies, expected frequencies*, and *variance* of the pivot cell across the subtables. Then the Cochran-Mantel-Haenzel, (1959) statistic is obtained as

$$Q_{CMH} = \frac{(n-\hat{m})^2}{\hat{V}}$$

and it is distributed χ^2 with 1 d.f. for large n.

In the data in Table 4.19, we have:

$$
\begin{aligned}
n &= 73 + 19 + 137 && = 229 \\
E_1 &= \frac{261 \times 94}{364} && = 67.4011 \\
E_2 &= \frac{57 \times 24}{78} && = 17.5385 \\
E_3 &= \frac{208 \times 500}{820} && = 126.8293 \\
E &= \sum_{i=1}^{3} E_i && = 211.7689 \\
V_1 &= \frac{(261)(103)(94)(270)}{364^2(363)} && = 14.1860 \\
V_2 &= \frac{(57)(21)(24)(54)}{78^2(77)} && = 3.3115 \\
V_3 &= \frac{(500)(320)(208)(612)}{820^2(819)} && = 36.9848 \\
V &= \sum_{i=1}^{3} V_i && = 54.4823
\end{aligned}
$$

Then $X^2_{MH} = (229 - 211.7689)^2/54.482 = 5.45$, which is significant and clearly indicates cases of lung cancer are related to spousal smoking for these groups of women across the three countries. The following are the SAS® software statements (with a partial SAS software output) for implementing the CMH test.

```
DATA CMH;
INPUT COUNTRY$ SMOKE STATUS COUNT @@;
CARDS;
japan 1 1 77 japan 1 2 188 japan 2 1 21 japan 2 2 82
uk 1 1 19 uk 1 2 38 uk 2 1 5 uk 2 2 16
usa 1 1 137 usa 1 2 363 usa 2 1 71 usa 2 2 249
;
TITLE 'COCHRAN-MANTEL-HAENSZEL TEST';
PROC FREQ;WEIGHT COUNT;
TABLES COUNTRY*SMOKE*STATUS/NOPRINT CMH1;
RUN;
                    Summary Statistics for SMOKE by STATUS
                            Controlling for COUNTRY

          Cochran-Mantel-Haenszel Statistics (Based on Table Scores)

        Statistic    Alternative Hypothesis    DF        Value       Prob
        ---------------------------------------------------------------
            1        Nonzero Correlation        1       5.4497     0.0196

               Estimates of the Common Relative Risk (Row1/Row2)

  Type of Study      Method                      Value     95% Confidence Limits
  -----------------------------------------------------------------------------
  Case-Control       Mantel-Haenszel            1.3854       1.0536       1.8217
    (Odds Ratio)     Logit                      1.3839       1.0521       1.8203

  Cohort             Mantel-Haenszel            1.2779       1.0366       1.5753
    (Col1 Risk)      Logit                      1.2760       1.0351       1.5729

  Cohort             Mantel-Haenszel            0.9225       0.8642       0.9848
```

```
(Col2 Risk)      Logit                0.9223          0.8640          0.9845
                         Total Sample Size = 1262
```

The resulting Cochran-Mantel-Haenszel statistic computed from SAS software is 5.4497, which agrees with the value we computed above. The statistic is based on 1 d.f., giving a pvalue of 0.0196. This again confirms the conclusion that incidence of lung cancer among married women is associated with spousal smoking, after adjusting for country effects.

4.10.2 Estimating the Common Odds Ratio

While the Cochran-Mantel-Haenszel test provides the significance of the relationship between lung cancer status and spousal smoking across the subtables, it does not tell us the strength of this association. An estimator of the common odds ratio is given by:

$$\hat{\theta}_{MH} = \frac{\sum_{i=1}^{h} a_i d_i / n_i}{\sum_{i=1}^{h} b_i c_i / n_i}$$

In our example, the estimate of this common odds ratio in favor of being a lung cancer case with spousal smoking after controlling for the countries is:

$$\hat{\theta}_{MH} = \frac{(73 \times 82/364) + (19 \times 16/78) + (137 \times 249/820)}{(188 \times 21/364) + (38 \times 5/78) + (363 \times 71/820)} = 1.3854$$

SAS software gives the estimate of this common odds ratio as the case-control Cochran-Mantel-Haenszel odds ratio. Thus, the odds in favor of a married woman developing lung cancer are 1.3854 times higher for those whose spouses are smokers than for those whose spouses are nonsmokers. The expression for a confidence interval for this common odds ratio is a little complicated, but most statistical packages readily give these confidence bounds. The 95% confidence interval for this common odds ratio in our case is (1.0536, 1.8217). This interval does not include 1; therefore, there is dependence on lung cancer status and spousal smoking.

The above estimation of the common odds ratio assumes that the strength of association as measured by the odds ratios in each subtable is the same. This assumption is tested by the following homogeneity hypothesis:

$$H_0 : \theta_1 = \theta_2 = \cdots \theta_h$$

To test this hypothesis, the Breslow-Day test is often employed. SAS® software automatically gives this statistic, which should be compared to a standard χ^2 distribution with $(h - 1)$ degrees of freedom. In this example, the Breslow-Day test gives a value of 0.2381 on 2 d.f. and a pvalue of 0.8878, which indicates that we would fail to reject the null hypothesis of homogeneity of odds ratios across the subtables. The SAS software output of this test is given below. This result is part of the general output displayed earlier.

```
     Breslow-Day Test for
Homogeneity of the Odds Ratios
------------------------------
Chi-Square              0.2381
DF                           2
Pr > ChiSq              0.8878
```

The estimate of the common odds ratio is based on the assumption that the strength of the association is the same in each country. If this were not the case, then we would have believed that there is *interaction* or *effect modification* between country and spousal smoking. The factor variable (country) is often referred to as the *effect modifier.* Evidence of the homogeneity of the odds ratios across countries indicates that there is no significant effect modification in this case.

The logit odds ratio of 1.3839 displayed in the SAS software output above is the estimate obtained from the weighted regression approach, where

$$\log(\hat{\theta}) = \frac{\sum_i \omega_i \log(\hat{\theta}_i)}{\sum_i \omega_i} \tag{4.30}$$

where

$$\log(\hat{\theta}_i) = \log[(a_i d_i)/(b_i c_i)] \qquad \text{and}$$

$$\omega_i = \left\{ \frac{1}{a_i} + \frac{1}{b_i} + \frac{1}{c_i} + \frac{1}{d_i} \right\}^{-1}$$

For the data in Table 4.19,

$$\begin{aligned}
\log(\hat{\theta}_1) &= \log\left[\frac{73 \times 82}{21 \times 188}\right] && = \log(1.5162) = 0.4162 \\
\omega_1 &= \left[\frac{1}{73} + \frac{1}{188} + \frac{1}{21} + \frac{1}{82}\right]^{-1} && = (0.0788)^{-1} = 12.6852 \\
\log(\hat{\theta}_2) &= \log\left[\frac{19 \times 16}{38 \times 5}\right] && = \log(1.6) \quad = 0.4700 \\
\omega_2 &= \left[\frac{1}{19} + \frac{1}{38} + \frac{1}{5} + \frac{1}{16}\right]^{-1} && = (0.3414)^{-1} = 2.9287 \\
\log(\hat{\theta}_3) &= \log\left[\frac{137 \times 249}{363 \times 71}\right] && = \log(1.3236) = 0.2804 \\
\omega_3 &= \left[\frac{1}{137} + \frac{1}{363} + \frac{1}{71} + \frac{1}{249}\right]^{-1} && = (0.0282)^{-1} = 35.5181
\end{aligned}$$

From (4.30), therefore,

$$\begin{aligned}
\log(\hat{\theta}) &= \frac{(12.6852)(0.4162) + (0.4700)(2.9287) + (0.2804)(35.5181)}{(12.6852 + 2.9287 + 35.5181)} \\
&= 0.32495
\end{aligned}$$

Hence, $\hat{\theta} = e^{0.32495} = 1.38396$. This estimate agrees with the estimate of the logit odds ratio from SAS®. With this estimate, the test of homogeneity of odds ratios across the subpopulations is given by

$$X^2_{HOM} = \sum \omega_i \{\log(\hat{\theta}_i) - \log(\hat{\theta})\}^2 = 0.2377$$

Again, this result is very close to the Breslow-Day test statistic. The above statistic is also based on 2 degrees of freedom.

Most recently, von Eye and Indurkhya (2000) formulated representations of the Cochran-Mantel-Haenszel and Breslow-Day tests in terms of log-linear models, which are discussed in this text in Chapter 6.

4.11 Exercises

1. Generate all outcomes consistent with given marginal totals of the table in this exercise. Based on the definition for extreme or more extreme tables in (4.9a) to (4.9b), which of the tables so generated are in the tail of the observed table?

16	2	18
1	8	9
17	10	27

Conduct exact tests based on both X^2 and G^2 and compare your results with those obtained from probability ranking scheme.

2. Show that for a 2×2 table of cell counts $\{n_{ij}\}$, the odds ratio is invariant to:

   (a) interchange of rows with columns and

   (b) multiplication of cell counts within rows or within columns by a nonzero constant.

   (c) Show that the difference of proportions do not have these invariant properties.

3. The table for this exercise contains results of a study by Mendenhall et al. (1984) to compare *radiation* therapy with *surgery* in treating cancer of the larynx.

Treatment	Cancer controlled	Cancer not controlled
Surgery	21	2
Radiation	15	3

The distribution of the pivot cell n_{11} is also given by:

n_{11}	Probability
18	0.0449
19	0.2127
20	0.3616
21	0.2755
22	0.0939
23	0.0144

   (a) What theoretical sampling scheme is assumed in this analysis? Justify the range of the pivot cell n_{11}.

(b) Conduct exact tests to test the following two hypotheses separately:

$$H_0 : \theta = 1$$

against the alternatives

$$(i) \quad H_a : \theta > 1$$
$$(ii) \quad H_a : \theta \neq 1$$

where θ is the odds ratio for the table.

(c) Explain how you formed the pvalues in each case and draw your conclusions based on your analyses.

(d) Give the equivalent mid-pvalues for both hypotheses and compare your results here with that obtained in (b). Which is more conservative?

(e) Is it appropriate to conduct a large sample test in this study.?

4. For the 2×2 contingency table with counts n_{ij} following a multinomial distribution with probabilities $\pi_{11}, \pi_{12}, \pi_{21}, \pi_{22}$, obtain maximum likelihood estimates for testing the hypothesis $H_0 : \pi_{1+} = \pi_{+1}$ or equivalently, $\pi_{12} = \pi_{21}$, and use this to obtain an expression for Pearson's X^2.

5. The underlying probabilities for a 2×2 table under some sampling scheme are displayed in the following table.

Samples	Success	Failure	Total
1	π_{11}	π_{12}	1
2	π_{21}	π_{22}	1

(a) Identify the sampling scheme.

(b) Find the maximum likelihood estimators for π_{ij} given the homogeneity hypothesis $H_0 : \pi_{11} = \pi_{21}$.

6. Suppose in the last exercise the sampling scheme is multinomial; test the hypothesis that $\pi_{1+} = \pi_{+1}$ and show that in this case, the Pearson's X^2 statistic reduces to McNemar's test statistic for correlated binary data.

7. For the 2×2 contingency table, show that if n_{ij} are distributed as independent Poisson (λ_{ij}), the conditional distribution based on the fixed sample size n is the full multinomial.

8. The following data pertain to the presence of loss of memory for patients on hemodialysis at the University of Michigan Medical Center during two different time periods.

	Loss of memory		
Period	Present	Absent	Total
I	8	26	34
II	0	22	22
Total	8	48	56

The investigator is interested in testing whether the incidence of loss of memory has been significantly lowered in time period II. Test an appropriate hypothesis and briefly discuss the choice of model for this analysis.

9. If it is believed that treatment A is better than treatment B, list all possible outcomes that are extreme or more extreme than the observed table in the following fictitious table of data.

	Treatment		
Outcome	A	B	Total
Die	5	3	8
Live	9	15	24
Total	14	18	32

Conduct Fisher's exact test on these data. Calculate the corresponding mid-P probabilities and comment on your results.

10. The table in this exercise gives information for the percentages of applicants admitted to graduate programs in the six largest majors at the University of California, Berkeley, in the fall of 1973. There was no evidence to support the idea that men and women applicants were not equally well qualified.

	Men		Women	
Major	Number of applicants	Number admitted	Number of applicants	Number admitted
A	825	512	108	89
B	560	353	25	17
C	325	120	593	202
D	417	138	375	202
E	191	53	393	94
F	373	22	341	24

Does there appear any indication of biasedness in graduate admissions based on sex of the applicant?

Construct separate 2×2 tables for both men and women for each major and conduct individual tests for each subtables. Draw your conclusions.

11. The National Institute for Occupational Safety and Health has developed a case definition of carpal tunnel syndrome that incorporates three criteria: symptoms of median nerve involvement, a history of occupational risk factors, and the presence of physical exam findings. The sensitivity of this definition as a test for carpal tunnel syndrome is 0.67; its specificity is 0.58.

    (a) In a population in which the prevalence of carpal tunnel syndrome is estimated to be 15%, what is the predictive value of positive test results?
    (b) How does this probability change if the prevalence is only 5%?

12. The following data are taken from Pagano and Gauvreau (1993) and relate to the study to investigate the use of radionuclide ventriculography in detecting coronary disease.

Test	Disease		Total
	Present	Absent	
Positive	302	80	382
Negative	179	372	551
Total	481	452	933

(a) What is the sensitivity of radionuclide ventriculography in this study? What is its specificity?

(b) What is the predictive value negative of the test?

13. Show that Yule's Q falls between -1 and 1.
Give conditions under which $Q = -1$ or $Q = 1$.
Derive the relationship between Q and the odds ratio.

14. A study for examining the effectiveness of a drug product for the treatment of arthritis was conducted at four different centers. At each center, 90 patients were treated, 45 on the test drug and 45 on an identically appearing placebo. The data are reproduced in the following table (Ott, 1984).

Clinic	Treatment	Global Outcome			
		Worse	Same	Better	Completely well
1	Placebo	10	15	17	8
	Test drug	12	14	10	14
2	Placebo	6	20	22	2
	Test drug	4	15	10	21
3	Placebo	7	25	12	6
	Test drug	5	22	12	11
4	Placebo	2	14	20	14
	Test drug	1	12	15	22

(a) Suppose an investigator wishes to collapse the global outcome categories into improved and not improved. Comment on this.

(b) Conduct a Cochran-Mantel-Haenszel test on the collapsed data of part (a) using uniform scores. Draw conclusions.

15. A medical research team wished to evaluate a proposed screening test for Alzheimer's disease. The test was given to a random sample of 450 patients with Alzheimer's disease and an independent random sample of 500 patients without symptoms of the diseases. The two samples were drawn from populations of subjects who were 65 years of age or older where it is assumed that 11.3% of the US population aged 65 and over have Alzheimer's disease. The data from this study are presented as:

	Alzheimer's diagnosis?		
Test result	Yes (D)	No ($\bar{D}$)	Total
Positive (T)	436	5	441
Negative ($\bar{T}$)	14	495	509
Total	450	500	950

Obtain

(a) The sensitivity of the test.

(b) The specificity of the test.

(c) The predictive value positive of the test.

16. Suppose a retrospective study is conducted among men aged 50 – 54 in a specific county who died over a 1-month period. The investigators attempt to include approximately an equal number of men who died from CVD (the cases) and men who died from other causes (the controls). It is found that of 35 people who died from CVD (cardiovascular disease), 5 were on a high-salt diet before they died, whereas of 25 people who died from other causes, 2 were on such a diet. The result is displayed in the table below. Is there sufficient evidence to conclude that there is significant association between salt intake and cause of death?

	Type of diet		
Cause of death	High salt	Low salt	Total
CVD	5	30	35
Non-CVD	2	23	25
Total	7	53	60

(a) Generate all tables that are extreme or more extreme as the observed table.

(b) Compute the probabilities and conduct Fisher's exact test. Why was this test necessary?

(c) Compute the corresponding mid-P test and draw your conclusions.

(d) Conduct both the asymptotic tests and the exact tests. What is the equivalent exact test in this case?

17. Two drugs (A, B) are compared for the medical treatment of duodenal ulcer. For this purpose, patients are carefully matched on age, sex, and clinical condition. The treatment results based on 200 matched pairs show that for 89 matched pairs both treatments are effective; for 90 matched pairs, both treatments are ineffective; for 5 matched-pairs, drug A is effective, whereas drug B is ineffective; for 16 matched pairs drug B is effective, whereas drug A is ineffcetive.

(a) What test procedure can be used to assesss the results?

(b) Perform the test and report your result.

18. The following are data from two studies that investigated the risk factors for epithelial ovarian cancer (Pagano & Gauvreau, 1993).

Study I

Disease status	Term pregnancies		Total
	None	One or more	
Cancer	31	80	111
No cancer	93	379	472
Total	124	459	583

Study II

Disease status	Term pregnancies		Total
	None	One or more	
Cancer	39	149	188
No cancer	74	465	539
Total	113	614	727

(a) Estimate the odds ratio of developing ovarian cancer for women who have never had a term pregnancy versus women who have had one or more in the first study.

(b) Estimate the odds ratio of developing ovarian cancer for women who have never had a term pregnancy versus women who have had one or more in the second study.

(c) If possible, you would like to combine the evidence in these two strata to make an overall statement about the relationship between ovarian cancer and term pregnancies. What would happen if you were to simply sum the entries in the tables?

(d) Conduct a test of homogeneity. Is it appropriate to use the Cochran-Mantel-Haenszel method to combine the information in these two tables?

(e) Obtain the Cochran-Mantel-Haenszel estimate of the common odds ratio.

(f) Interpret the confidence interval estimate for the common odds ratio.

(g) Test the null hypothesis that there is no significant association between ovarian cancer and term pregnancies at the 0.01 level of significance.

19. Hansteen et al. (1982) reported the results of a clinical trial of propranolol on patients with mycordial infarction:

	Sudden death	No sudden death
Propranolol	11	267
Placebo	23	259

Conduct exact tests based on two-tsided and one-sided hypotheses. Calculate the corresponding mid-P probabilities and comment on your results.

Chapter 5

The General $I \times J$ Contingency Table

5.1 Introduction

Suppose a sample of N objects is jointly classified according to two different and independent classifications A and B with I and J classes, respectively. Let n_{ij} be the observed frequency in cell (i, j) with $i = 1, 2, \cdots, I$ and $j = 1, 2, \cdots, J$. The observed table can be displayed as in Table 5.1.

	B				
A	1	2	$\cdots$	J	Total
1	n_{11}	n_{12}	$\cdots$	n_{1J}	N_{1+}
2	n_{21}	n_{22}	$\cdots$	n_{2J}	N_{2+}
$\vdots$	$\vdots$	$\vdots$	$\vdots$	$\vdots$	$\vdots$
I	n_{I1}	n_{I2}	$\cdots$	n_{IJ}	N_{I+}
Total	N_{+1}	N_{+2}	$\cdots$	N_{+J}	N

Table 5.1: Observed $I \times J$ contingency table

The observed frequencies can be represented compactly by a vector

$$\mathbf{n}' = (\mathbf{n}_1', \mathbf{n}_2', \cdots, \mathbf{n}_I')$$

where

$$\mathbf{n}_i' = (n_{i1}, n_{i2}, \cdots, n_{iJ}), \quad i = 1, 2, \cdots, I$$

is the vector of observed frequencies from the i-th row of the table.

Similar to our discussion for the 2×2 contingency table, we shall again discuss the general $I \times J$ contingency table under the three sampling schemes A, B and C. We shall label the case when the row margin is fixed by $M_1 = \{N_{1+}, \cdots, N_{I+}\}$. Similarly, for the column totals $M_2 = \{N_{+1}, \cdots, N_{+J}\}$.

5.2 Multivariate Hypergeometric Distributions

5.2.1 The One-Sample Case

Consider a population of N subjects, consisting of N_1 of type 1, N_2 of type 2, $\cdots$, and N_J of type J, displayed in Table 5.2 such that $\sum_{j=1}^{J} N_j = N$.

	Types							
	1	2	$\cdots$	j	$\cdots$	$(J-1)$	J	Total
Population	N_1	N_2	$\cdots$	N_j	$\cdots$	$N_{(J-1)}$	N_J	N
Sample	n_1	n_2	$\cdots$	n_j	$\cdots$	$n_{(J-1)}$	n_J	n

Table 5.2: Distribution of subjects

If we take a simple random sample of size n from this population without replacement, then the joint distribution of the number n_j of type j within the sample is

$$P\{\mathbf{n}\} = \binom{N_1}{n_1}\binom{N_2}{n_2}\cdots\binom{N_J}{n_J}\Big/\binom{N}{n} = \prod_{j=1}^{J}\binom{N_j}{n_j}\Big/\binom{N}{n} \tag{5.1}$$

where $0 \leq n_j \leq n$ for $j = 1, 2, \cdots, J$ such that $\sum n_j = n$. This probability distribution is called the *multivariate hypergeometric distribution.* The proof of this can be found in appendix D.1.

The moments of the multivariate hypergeometric distribution are mostly readily obtained by computing factorial moments. It can be shown (see appendix D.2) that:

$$E\{n_j\} = \frac{nN_j}{N} \tag{5.2a}$$

$$\text{Var}\{n_j\} = \frac{n(N-n)N_j(N-N_j)}{N^2(N-1)} \tag{5.2b}$$

$$E\{n_j, n_{j'}\} = \frac{n(n-1)}{N(N-1)}N_jN_{j'} \tag{5.2c}$$

$$\text{Cov}\{n_j, n_{j'}\} = -\frac{n(N-n)N_jN_{j'}}{N^2(N-1)} \tag{5.2d}$$

5.2.2 Several-Samples Case

Consider an $I \times J$ table in which it is assumed that both marginal distributions M_1 and M_2 are fixed. In this context, the null hypothesis of interest is that the joint distribution of the I and J categories are randomly distributed. That is, the response variable B is randomly distributed with the categories of A. In other words, the data in the respective rows can be regarded as successive set of simple random samples of sizes $\{N_i+\}$ from a fixed population corresponding to the marginal distribution of B $\{N+j\}$.

On the basis of the null hypothesis, it can be shown that the vector $\mathbf{n}$ follows the product multivariate hypergeometric distribution given by the probability model,

$$P\{\underline{n}|H_0\} = \frac{\prod_{i=1}^{I} N_{i+}! \prod_{j=1}^{J} N_{+j}!}{N! \prod_{i=1}^{I}\prod_{j=1}^{J} n_{ij}!} \tag{5.3}$$

The proof of the results in (5.3) can be found in appendix D.3.

5.2.3 Moments

By expanding the results for the one-sample hypergeometric model in expressions (5.2a) and (5.2b), it follows that

$$E(n_{ij}|H_0) = \frac{N_{i+}N_{+j}}{N} \tag{5.4a}$$

$$\text{Var}(n_{ij}|H_0) = \frac{N_{i+}(N-N_{i+})N_{+j}(N-N_{+j})}{N^2(N-1)} \tag{5.4b}$$

However, the covariances are slightly different, depending on whether or not the two frequencies of interest share either a row or a column of the table.

For example, since marginal and conditional distributions of this probability model are also hypergeometric, it follows that if either $i = i'$ or $j = j'$, the covariance between n_{ij} and $n_{i'j'}$ can be obtained from the previous one-sample results in (5.2d). In particular, for same row but different columns observations,

$$\text{Cov}\{n_{ij}, n_{ij'}|H_o\} = -\frac{N_{i+}(N-N_{i+})N_{+j}N_{+j'}}{N^2(N-1)} \tag{5.5}$$

and similarly, for same column but different rows observations, we have

$$\text{Cov}\{n_{ij}, n_{i'j}\} = -\frac{N_{i+}N_{i'+}N_{+j}(N-N_{+j})}{N^2(N-1)} \tag{5.6}$$

The only remaining situation is to obtain the covariance of two frequencies that share neither the same row nor the same column, that is, $\text{Cov}(n_{ij}, n_{i'j'}|H_0)$. It can be shown that for this situation (see exercise 5.10) the covariance becomes:

$$\text{Cov}\{n_{ij}, n_{i'j'}|H_0\} = \frac{N_{i+}N_{i'+}N_{+j}N_{+j'}}{N^2(N-1)}, \quad \text{for } i \neq i', j \neq j' \tag{5.7}$$

In general, therefore, the covariance of any two frequencies n_{ij} and $n_{i'j'}$ can be written compactly as

$$\text{Cov}\{n_{ij}, n_{i'j'}|H_0\} = \frac{N_{i+}N_{+j}(S_{ii'}N - N_{i'+})(S_{jj'}N - N_{+j'})}{N^2(N-1)} \tag{5.8}$$

where

$$S_{ii'} = \begin{cases} 1 & \text{if } i = i' \\ 0 & \text{otherwise} \end{cases} \qquad S_{jj'} = \begin{cases} 1 & \text{if } j = j' \\ 0 & \text{otherwise} \end{cases}$$

It is not too difficult to see that the expression in (5.8) will generate those in (5.5), (5.6), and (5.7) for appropriate is and js. We also note here that each n_{ij} in Table 5.1 has marginal distributions that are themselves hypergeometric with parameters (N_{i+}, N_{+j}, N). That is, each has a mass univariate hypergeometric distribution.

5.2.4 Example 5.1

Stanley et al. (1996) examined the relative efficacy and side effects of morphine and pethidine (drugs commonly used for patient-controlled analgesia). The study was a prospective, randomized double-blind study where the subjects were 40 women between the ages of 20 and 60 years, undergoing total abdominal hysterectomy. The subjects were allocated randomly to recieve morphine or pethidine. At the end of the study, subjects described their appreciation of nausea and vomiting, pain, and

	Drug		
Pain	Pethidine	Morphine	Total
Unbearable/severe	2	2	4
Moderate	10	8	18
Slight/none	8	10	18
	20	20	40

Table 5.3: Joint distribution of drug and painful appreciation

satisfaction by means of a 3point verbal scale. The results for those who described their appreciation as painful is displayed below. It is obvious that both margins of this table were not fixed in the actual sampling scheme.
We are still, however, interested in the hypothetical question of whether or not the observed distribution of painful appreciation is randomly distributed with respect to the type of drugs administered. We shall assume that the marginals are fixed, and in this case, the multivariate hypergeometric probability model is appropriate for investigating this hypothesis.

However, unlike the simplicity of enumerating the distribution of the pivot cell n_{11} in the 2×2 table, the $I \times J$ table has $(I-1)(J-1)$ pivot cells whose distribution can be used to characterize the entire distribution of the vector $\mathbf{n}$. Specifically, once the values of $(I-1)(J-1)$ of the cell frequencies are fixed, the other $(I+J-1)$ elements can be determined immediately from the fixed margins M_1 and M_2. Thus, without loss of generality, the distribution of $\mathbf{n}$ in the example data set can be completely determined from the distribution of n_{11}, n_{21} as illustrated below.

	Drug		
Pain	Pethidine	Morphine	Total
Unbearable/severe	n_{11}	$4 - n_{11}$	4
Moderate	n_{21}	$18 - n_{21}$	18
Slight/none	$20 - n_{11} - n_{21}$	$n_{11} + n_{21} - 2$	18
	20	20	40

Table 5.4: Observed frequencies as functions of the pivot cells and fixed marginals subject to constraints $0 \leq n_{21} \leq 18$, and $n_{11} + n_{21} \leq 20$

As a result, n_{11} ranges from 0 to 4 and n_{12} ranges from 2 to 18. Under H_0, M_1 and M_2 are fixed, and the probabilities of each of the 82 possible tables that are consistent with M_1 and M_2 are listed below. The SAS software program used to generate this result is provided in appendix D.4.

Possible pairs of (n_{11}, n_{21}) and their
corresponding probabilities-generated using SAS

```
Obs    N11    N21       PROB          CUM
-----------------------------------------
  1     0      2     0.00000      0.00000
  2     0      3     0.00000      0.00000
  3     0      4     0.00000      0.00000
  4     0      5     0.00005      0.00005
  5     0      6     0.00041      0.00047
  6     0      7     0.00198      0.00244
  7     0      8     0.00589      0.00834
  8     0      9     0.01122      0.01956
  9     0     10     0.01389      0.03345
```

10	0	11	0.01122	0.04468
11	0	12	0.00589	0.05057
12	0	13	0.00198	0.05255
13	0	14	0.00041	0.05296
14	0	15	0.00005	0.05301
15	0	16	0.00000	0.05301
16	0	17	0.00000	0.05301
17	0	18	0.00000	0.05301
18	1	2	0.00000	0.05301
19	1	3	0.00000	0.05302
20	1	4	0.00007	0.05309
21	1	5	0.00076	0.05385
22	1	6	0.00462	0.05847
23	1	7	0.01714	0.07561
24	1	8	0.04041	0.11602
25	1	9	0.06174	0.17775
26	1	10	0.06174	0.23949
27	1	11	0.04041	0.27990
28	1	12	0.01714	0.29704
29	1	13	0.00462	0.30166
30	1	14	0.00076	0.30242
31	1	15	0.00007	0.30249
32	1	16	0.00000	0.30249
33	1	17	0.00000	0.30249
34	1	18	0.00000	0.30249
35	2	2	0.00000	0.30250
36	2	3	0.00003	0.30252
37	2	4	0.00041	0.30293
38	2	5	0.00320	0.30613
39	2	6	0.01500	0.32113
40	2	7	0.04408	0.36521
41	2	8	0.08334	0.44855
42	2	9	0.10289	0.55145
...	...	...	...	...
43	2	10	0.08334	0.63479
...	...	...	...	...
44	2	11	0.04408	0.67887
45	2	12	0.01500	0.69387
46	2	13	0.00320	0.69707
47	2	14	0.00041	0.69748
48	2	15	0.00003	0.69750
49	2	16	0.00000	0.69751
50	2	17	0.00000	0.69751
51	2	18	0.00000	0.69751
52	3	2	0.00000	0.69751
53	3	3	0.00007	0.69758
54	3	4	0.00076	0.69834
55	3	5	0.00462	0.70296
56	3	6	0.01714	0.72010
57	3	7	0.04041	0.76051
58	3	8	0.06174	0.82225
59	3	9	0.06174	0.88398
60	3	10	0.04041	0.92439
61	3	11	0.01714	0.94153
62	3	12	0.00462	0.94615
...	...	...	...	...
78	4	12	0.00041	0.99995
79	4	13	0.00005	1.00000
80	4	14	0.00000	1.00000
81	4	15	0.00000	1.00000
82	4	16	0.00000	1.00000

The observed table has $\{(n_{11}, n_{21}) = (2, 10)\}$, which is the 43rd configuration in the above list, with corresponding hypergeometric observed probability 0.08334 computed from (5.3). Observations numbered from 43 to 82 therefore are those configurations with extreme or more extreme probabilities (to the right of $(2, 10)$) than the observed table, that is, tables for which $P(n_{11}, n_{21}) \leq 0.08334$, the observed table probability. Similarly, observations numbered from 1 to 40 therefore are those configurations with extreme or more extreme probabilities to the left of $(2, 10)$] than the observed probability.

5.2.5 Fisher's Exact Test for $(I \times J)$ Tables

The choice of a rejection region of size exactly equal to α has the same difficulties as discussed previously for the 2×2 case. On the other hand, one can use the probabilities to compute the pvalue for the test of H_0 against the two-sided alternative that the joint distribution is not randomly distributed for a specific observed table. The principle behind Fisher's exact test is valid for tables of any size and the exact test procedure involves:

(i) Calculate null probabilities for each table that are consistent with $\mathbf{M}_1$ and $\mathbf{M}_2$ using expression in (5.3).

(ii) Compute the sum of the null probabilities of those tables that are as extreme or more extreme than the observed table. Here an extreme table is defined as any other table consistent with $\mathbf{M}_1$ and $\mathbf{M}_2$ having pivot cell frequencies (u_1, u_2) such that

$$P(u_1, u_2) \leq P(n_{11}, n_{21})$$

Note that for tables in which $(I - 1)(J - 1) > 1$, the notion of a primary and secondary tail probabilities is not well defined for a general alternative hypothesis since the tables cannot be ordered explicitly.

For our data in Table 5.3, the two-tailed pvalue is obtained as the sum of all the probabilities less than or equal to $P(2, 10) = 0.08334$. This total pvalue is $T(2, 10) = 0.89711$.

The above exact test was an extension of Fisher's exact test for the 2×2 table to the general $I \times J$ table. The test was due to Freeman and Halton (1951) and is well known as the *exact conditional test.* We can implement Fisher's exact test using SAS software again in this case because the sample size is not too large. Because about 33% of the cells in this example have expected values that are small, the exact test is most appropriate in this example. The ranked probabilities for all possible outcomes are presented in appendix D.5. The Fisher's exact test is implemented with the following program and partial output.

```
data tab52;
input trt $ pain $ count @@;
if pain='sever' then resp=3;
else if pain='mod' then resp=2;
else resp=1;
datalines;
peth sever 2 peth mod 10 peth none 8
morp sever 2 morp mod 8 morp none 10
;
proc print; run;
proc freq data=tab52 order=data; weight count;
tables trt*resp/exact; run;
                         Fisher's Exact Test
              ----------------------------------
              Table Probability (P)       0.0833
              Pr <= P                     0.8971
                          Sample Size = 40
```

The results obtained from SAS® agreed with those obtained from the exact test which uses probability ranking. Similar results can be obtained when the ranking criteria are the Pearson's X^2 and the likelihood ratio test statistic G^2 (appendices D.6 and D.7, respectively). We present the results of these in the table below where the first column denotes the criterion being used for ranking.

Test criterion	EXACT test	$P(\chi_2^2 \geq T_0^2)$
Probability	0.89711	
X^2	0.89711	0.801
G^2	0.89711	0.800

In the preceeding table, T_0^2 represents the observed value of the corresponding test criterion. In this example, corresponding observed values of X^2 and G^2 are respectively 0.44444 and 0.44536. Because of the near symmetry of the data in Table 5.3 (column marginals are equal and two of the row marginals are also equal), the number of distinct values of X^2 and G^2 are 28 each respectively (see appendices D.6 and D.7). Some of the 82 configurations yield the same value of the test statistics. In this example, the exact test probabilities computed from either ranked probabilities, X^2 or G^2 criteria yield the same exact pvalue. This is not often the case, but, as explained earlier, the symmetry of our table makes the distribution of the test statistics more discrete.

For larger tables, exact enumeration might not be possible because of the sheer number of possible configurations that would be consistent with the fixed margins. In such cases, the simulation approach implemented in the statistical package *StatXact* will be most appropriate.

The exact probabilities obtained for X^2 and G^2 are based on the ranking procedure suggested by Radlow and Alf (1975), that is: Carry out exact procedure as in (i) above. However, the pvalue is obtained by obtaining the sum of the null probabilities of those tables which have their X^2 values greater or equal to the observed X^2 value (X_0^2). The results from SAS software as shown earlier agree with the results obtained above.

5.2.6 The Mid-P Test

A recent alternative to the exact test is the mid-P test earlier discussed in Chapter 4. The mid-P as defined by Lancaster (1961), and Barnard (1989) is given by

$$\text{mid-P} = Pr\{T(u) > T(n_0)|H_0\} + \frac{1}{2}Pr\{T(u) = T(n_0)|H_0\}$$

where T(u) is the chosen test criterion, which in our case is the Pearson's X^2. For our data, the two-tailed mid-P value will be given as

$$0.73042 + 0.5\,(0.08334 + 0.08334) = 0.81376$$

In all the four cases (ranked probability, X^2, G^2, mid-P), we would fail to reject H_0 on the basis of the data given; that is, pain appreciation is randomly distributed with respect to the type of drug administered. In other words, the type of pain experienced does not depend on the type of drug administered.

5.3 Large Sample Test

As noted in Chapter 4, the main difficulty in carrying out the exact conditional tests is the sheer number of calculations involved. One must generate all $I \times J$ arrays of nonnegative integers in the set

$$S = \{n'_{ij}, \quad \text{for all} \quad i, j\}$$

having the same marginal totals $\mathbf{M}_1, \mathbf{M}_2$ as the given table. The conditional probability in (5.3) as well as the test statistic X^2 must be computed for each such table in S. The number of tables in S, which is denoted by $|S|$, increase very rapidly as a function of the sample size N, I, and J. The $|S|$ arrays generated are used to compute values of a GOF such as X^2.

For relatively large table dimension, $|S|$ is too large for the practical implementation of the exact tests even when asymptotic approximations would be crude. Discussion of the problem of enumerating the number of tables consistent with the marginal totals is provided by Good (1976), Agresti and Wackerly (1977), Gail and Mantel (1977), and Agresti et al. (1979). To illustrate this, Klotz and Teng (1977) gave $|S|$ as 12,798,781 for a 4×4 table data with sample size $N = 56$.

Earlier algorithms for exact enumerations of an $I \times J$ table are March (1972) algorithm 434, which enumerates all tables and conducts exact conditional tests. Baker (1977) algorithm AS 112, is applicable to two or one fixed margins or just a fixed sample size. More recent algorithms on the same subject are those by Mehta, C.R et al. (1983, 1990). The statistical software **StatXact** uses both the exact and simulation procedures to conduct the conditional tests for large sample sizes. Senchaudhuri et al. (1993) has just proposed a "smart" method of control variates *or Monte Carlo rescue* for estimating pvalues for the exact test in an $I \times J$ table.

Agresti (1992) has given a very recent review for two- and three-dimensional exact methods, and some of the important issues in exact inference for higher dimensional tables are discussed in Kreiner (1992). An exact test of significance for the 2^3 contingency table is provided (with the relevant FORTRAN algorithm) by Zelterman et al. (1993).

One solution to the exact test computational problem is to employ the results concerning the convergence of the multivariate hypergeometic distribution to the normal distribution in large samples.

Let $\mathbf{m}' = (m_{11}, m_{12}, \cdots, m_{IJ})$ be the vector of expected values of the $n'_{ij}s$ under H_0, that is,

$$\hat{m}_{ij} = E\{n_{ij}|H_0\} = \frac{N_{i+}N_{+j}}{N}$$

and let

$$\mathbf{A} = \left[\mathbf{I}_{(J-1)}, \mathbf{O}_{(J-1)}\right] \otimes \left[\mathbf{I}_{(I-1)}, \mathbf{O}_{(I-1)}\right]$$

be the matrix of appropriately positioned 1's and 0's such that the matrix product $\mathbf{An}$ eliminates the last column and last row of the frequencies in an $I \times J$ table. Here $\mathbf{I}_U$ denotes a $U \times U$ identity matrix, $\mathbf{O}_U$ denotes a $(U \times 1)$ vector of 0's and $\otimes$ denotes the Kr̃onecker product multiplication. Thus $\mathbf{A}$ is an $[(I-1)(J-1) \times IJ]$ matrix and if we let

$$\mathbf{G} = \mathbf{A}(\mathbf{n} - \mathbf{m})$$

be the vector of $(I-1)(J-1)$ differences of the observed and expected values under H_0, then the variances and covariances of $\mathbf{G}$ can be obtained from (5.7).

We observe that the matrix $\mathbf{A}$ is such that $\mathbf{A} = [\mathbf{I_2}, \mathbf{0_{(2,4)}}]$. Here $\mathbf{I_2}$ is the 2×2 identity matrix (null d.f.) and $\mathbf{0_{(2,4)}}$ is the 2×4 matrix of zeros (see Koch et al., 1982, and Koch and Bhapkar, 1982, for details).

Then for large N, such that $m_{ij} > 1$ for $i = 1, 2, \cdots, I$ and $j = 1, 2, \cdots, J$, it can be shown that

$$\mathbf{G} \sim AMN(\mathbf{0}, \text{Var}\{\mathbf{G}|H_0\})$$

where AMN denotes asymptotically multivariate normal. Under H_0 therefore, it follows that

$$Q = \mathbf{G}' \left[\text{Var}\{\mathbf{G}|H_0\}\right]^{-1} \mathbf{G}$$

is an $(I-1)(J-1)$ degree of freedom statistic for testing H_0, which follows the χ^2 distribution asymptotically. Moreover, it can be shown by matrix equivalence that

$$\begin{aligned} Q &= \frac{N-1}{N} \sum_{i=1}^{I} \sum_{j=1}^{J} (n_{ij} - \hat{m}_{ij})^2 / \hat{m}_{ij} \\ &= \frac{N-1}{N} X^2 \end{aligned}$$

Thus for large samples, the Pearson's X^2 criterion is essentially equivalent to the large sample test derived from the multivariate hypergeometric model.

5.3.1 Example 5.2

For the data in Table 5.3, we have

$$\mathbf{A} = \begin{pmatrix} 1 & 0 & 0 & 0 & 0 & 0 \\ 0 & 1 & 0 & 0 & 0 & 0 \end{pmatrix}$$

That is, $\mathbf{A}$ is a 2×6 matrix and hence $\mathbf{G}$ is given by:

$$\mathbf{G} = \begin{pmatrix} 1 & 0 & 0 & 0 & 0 & 0 \\ 0 & 1 & 0 & 0 & 0 & 0 \end{pmatrix} [\mathbf{n} - \mathbf{m}]$$

where $\mathbf{n}' = (n_{11}, n_{12}, n_{21}, n_{22}, n_{31}, n_{32})$ and $\begin{pmatrix} n_{11} \\ n_{21} \end{pmatrix}$ = pivot cells and $\mathbf{G}$ reduces to

$$\mathbf{G} = \begin{pmatrix} 2 \\ 10 \end{pmatrix} - \begin{pmatrix} 2.0 \\ 9.0 \end{pmatrix} = \begin{pmatrix} 0 \\ 1 \end{pmatrix}$$

Furthermore, the covariance structure of $\mathbf{G}$ under H_0 is from (5.8) computed as

$$\begin{aligned} \text{Var}\{\mathbf{G}|H_0\} &= \frac{1}{(40^2)(39)} \begin{pmatrix} (4)(36)(20)(20) & -(4)(18)(20)(20) \\ -(4)(18)(20)(20) & (18)(22)(20)(20) \end{pmatrix} \\ &= \begin{pmatrix} 0.9231 & -0.4615 \\ -0.4615 & 2.5385 \end{pmatrix} \end{aligned}$$

Consequently,

$$\left[\text{Var}\{\mathbf{G}|H_0\}\right]^{-1} = \begin{pmatrix} 1.1916 & 0.2166 \\ 0.2166 & 0.4333 \end{pmatrix}$$

From this, we have

$$Q = \mathbf{G}' \left[\text{Var}\{\mathbf{G}|H_0\}\right]^{-1} \mathbf{G} = \begin{pmatrix} 0 & 1 \end{pmatrix} \begin{pmatrix} 1.1916 & 0.2166 \\ 0.2166 & 0.4333 \end{pmatrix} \begin{pmatrix} 0 \\ 1 \end{pmatrix} = 0.4333$$

Alternatively, we can use the fact that under H_0, the expected values are given by

$$\hat{m}_{ij} = \frac{N_{i+}N_{+j}}{N}$$

to compute the Pearson's X^2 criterion as:

$$\begin{aligned}
X^2 &= \sum_{i=1}^{3}\sum_{j=1}^{2}(n_{ij} - \hat{m}_{ij})^2/\hat{m}_{ij} \\
&= \frac{(2-2)^2}{2} + \frac{(2-2)^2}{2} + \frac{(10-9)^2}{9} \\
&+ \frac{(8-9)^2}{9} + \frac{(8-9)^2}{9} + \frac{(10-9)^2}{9} \\
&= 0.4444
\end{aligned}$$

We note from the above results that

$$Q = \frac{39}{40}(0.4444) = 0.4333$$

which agrees with our earlier result.

The large sample test is implemented in SAS software as:

```
set tab52;
proc freq data=tab52 order=data;
weight count; tables trt*resp; run;
```

```
                STATISTICS FOR TABLE OF A BY B
-------------------------------------------------------
Chi-Square                          2     0.444       0.801
Likelihood Ratio Chi-Square         2     0.445       0.800
Mantel-Haenszel Chi-Square          1     0.228       0.633
Fisher's Exact Test (2-Tail)                          0.897
Sample Size = 40
WARNING: 33% of the cells have expected counts less
         than 5. Chi-Square may not be a valid test.
```

The warning in the SAS software ouput above indicates that almost one-third of the cells have small expected values, and hence the usual χ^2 large sample approximation may not be valid here. In such a situation, it is often desirable to use Fisher's exact test.

5.3.2 Example 5.3

The data below relate to the distribution of 59 female patients with rheumatoid arthritis who participated in a randomized clinical trial, (Koch et al., 1982). Table 5.5, is a two-way table obtained by collapsing over the covariate variable age.

Treatment	Patient response status					Total
	Excellent	Good	Moderate	Fair	Poor	
	1	2	3	4	5	n_{i+}
Active	5	11	5	1	5	27
Placebo	2	4	7	7	12	32
Total (n_{+j})	7	15	12	8	17	59

Table 5.5: Distributions of patient responses by treatment

The model of independence here will be based on $1 \times 4 = 4$ degrees of freedom. Consequently, we would expect $\mathbf{A}$ to be a 4×10 matrix such that $\mathbf{A} = [\mathbf{I_4}, \mathbf{0_{4,6}}]$. Here $\mathbf{I_4}$ is the 4×4 identity matrix (null d.f.) and $\mathbf{0_{4,6}}$ is the 4×6 matrix of zeros. Thus the matrix $\mathbf{G}$ that forms the observed and expected frequencies differences is of the form:

$$\mathbf{G} = \begin{pmatrix} 1 & 0 & 0 & 0 & 0 & 0 & 0 & 0 & 0 & 0 \\ 0 & 1 & 0 & 0 & 0 & 0 & 0 & 0 & 0 & 0 \\ 0 & 0 & 1 & 0 & 0 & 0 & 0 & 0 & 0 & 0 \\ 0 & 0 & 0 & 1 & 0 & 0 & 0 & 0 & 0 & 0 \end{pmatrix} [\mathbf{n} - \hat{\mathbf{m}}]$$

where $\mathbf{n}' = (n_{11}, n_{12}, \cdots, n_{25})$ and the expected values are those relating to the pivot cells n_{11}, n_{12}, n_{13}, and n_{14}. That is, $\hat{\mathbf{m}}' = (3.2034, 6.8644, 5.4915, 3.6610)$. The variance-covariance of $\mathbf{G}$ would be a 4×4 matrix, and similar arguments as in the previous example lead to Q being computed as:

$$Q = \frac{N-1}{N} \sum_{i=1}^{2} \sum_{j=1}^{5} \frac{(n_{ij} - \hat{m}_{ij})^2}{\hat{m}_{ij}}$$

For these data, it can be shown that $Q = 11.73$ and hence $X^2 = 11.9322$. This value gives a pvalue of 0.0179, which indicates that the null hypothesis is not tenable in this case.

We notice for this example that the response variable is ordinal in nature, and we could therefore exploit this ordinality in remodeling the data. Following Koch and Bhapkar (1982), let $\boldsymbol{\nu}' = (\nu_1, \nu_2, \nu_3, \nu_4, \nu_5)'$ be a set of scores that conforms to the ordinal response variable. A possible set of scores is integer scores $(1, 2, 3, 4, 5)$ or scores whose total sums to zero, e.g., $(-2, -1, 0, 1, 2)$. Other possible scores are the mid-rank scores, which are employed in *StatXact*, for instance. With our set scores obtained, we then compute the test statistic as:

$$Q_{RS} = \frac{N-1}{NV(\eta)} \sum_{i=1}^{2} n_{i+}(\hat{\eta}_i - \bar{\eta})^2$$

where

$$\hat{\eta}_i = \sum_{j=1}^{5} \frac{\nu_j n_{ij}}{n_{i+}}; \quad \bar{\eta} = \sum_{j=1}^{5} \frac{\nu_j n_{+j}}{N}; \quad \text{and} \quad \mathrm{V}(\eta) = \frac{n_{+j}}{N} \sum_{i=1}^{2} (\nu_j - \bar{\eta})^2$$

In the above, $\hat{\eta}_i$ are the sample mean responses for the two treatment groups, and $\bar{\eta}$ and $V(\eta)$ are the finite population mean and variance of subjects, respectively.

For the data in Table 5.5, $\hat{\eta}_1 = 2.6296$, and $\hat{\eta}_2 = 3.7188$ with $\bar{\eta} = 3.2203$ and $V(\eta)$=1.9684, where for instance

$$\hat{\eta}_1 = \frac{1(5) + 2(11) + 3(5) + 4(1) + 5(5)}{27}; \quad \bar{\eta} = \frac{1(7) + 2(15) + 3(12) + 4(8) + 5(17)}{59}$$

From the above, we have $Q_{RS} = 8.6764$, and it is distributed χ^2 with 1 degree of freedom (Koch et al., 1982; Koch & Bhapkar, 1982). The corresponding pvalue is 0.0032, which provides a stronger evidence of rejection of the null than the Q test based on 4 d.f. Thus, as argued by Koch et al. (1982), the Q_{RS} is more powerful than the Q test, and further, it does not require the stringent condition that expected values must be large since the test refers to means scores, which are linear combinations of the n_{ij} rather than the n_{ij} themselves. The Q_{RS} can be implemented in SAS software and is produced from the **CMH** and **CMH2** options in SAS software. However, the test statistic is given in the SAS software output by the "Row Mean Score Differ" line. The above result is implemented in SAS software (with partial output) as follows:

```
data kh;
input trt $ response $ count @@;
datalines;
active ex 5 active good 11 active mod 5 active fair 1
active poor 5 pla ex 2 pla good 4 pla mod 7 pla fair 7
pla poor 12
;
proc print; run;
proc freq data=kh order=data;
weight count;
tables trt*response/cmh scores=table;
tables trt*response/cmh2 scores=ridit; run;
```

```
                      Summary Statistics for trt by response

           Cochran-Mantel-Haenszel Statistics (Based on Table Scores)

         Statistic    Alternative Hypothesis    DF         Value      Prob
         ---------------------------------------------------------------
             1        Nonzero Correlation        1        8.6751    0.0032
             2        Row Mean Scores Differ     1        8.6751    0.0032
             3        General Association        4       11.7278    0.0195

                      Summary Statistics for trt by response

           Cochran-Mantel-Haenszel Statistics (Based on Ridit Scores)

         Statistic    Alternative Hypothesis    DF         Value      Prob
         ---------------------------------------------------------------
             1        Nonzero Correlation        1        8.7284    0.0031
             2        Row Mean Scores Differ     1        8.7284    0.0031
```

The SAS software results agree closely with the results obtained earlier. The test statistic for the general association we observed is based on the Q test. In either case, the null hypothesis is not tenable and would strongly be rejected. If we recognize that the response categories of the data in Table 5.3, are ordinal (severe, moderate, none), a trend model applied to this data giving severe, moderate, and none scores of $(3, 2, 1)$, respectively, gives the value of the test statistic $Q_{RS} = 0.228$ on 1 d.f. with a pvalue of 0.633, again indicating that there is no reason to believe that there is a trend in the degree of painful appreciation expressed in relation to the type of drug being administered.

5.4 Product Multinomial Probability Model

For data arising from stratified simple random sampling, only the one marginal distribution M_1 is assumed to be fixed, and again in this case, the underlying probability structure for the observed frequencies may be summarized as shown in Table 5.6.

The above probability structure can be written compactly in vector notation as $\mathbf{\Pi}' = (\pi'_1, \boldsymbol{\pi}'_2, \cdots, \boldsymbol{\pi}'_{\mathbf{I}})$ where

$$\mathbf{\Pi}'_i = (\boldsymbol{\pi}_{i1}, \boldsymbol{\pi}_{i2}, \cdots, \boldsymbol{\pi}_{iJ}), \quad \text{for} \quad i = 1, 2, \cdots, I.$$

Here the π_{ij} are unknown parameters such that $0 < \pi_{ij} < 1$ and they satisfy the constraints $\sum_j \pi_{ij} = 1$ for $i = 1, 2, .., I$

Sample	1	2	$\cdots$	J	Total
1	π_{11}	π_{12}	$\cdots$	π_{1J}	1
2	π_{21}	π_{22}	$\cdots$	π_{2J}	1
$\vdots$	$\vdots$	$\vdots$	$\vdots$	$\vdots$	$\vdots$
I	π_{I1}	π_{I2}	$\cdots$	π_{IJ}	1

Table 5.6: Probability structure with only M_1 fixed

The corresponding observed table is displayed in Table 5.7. The observed frequencies can also be represented compactly by a vector

$$\mathbf{n}' = (\mathbf{n}_1', \mathbf{n}_2', \cdots, \mathbf{n}_I'), \quad \text{where} \quad \mathbf{n}_i' = (n_{i1}', n_{i2}', \cdots, n_{iJ}')$$

and the latter is the vector of observed frequencies from the i-th row of the table.

Under this sampling scheme, $\mathbf{n}_i$ follows the multinomial distribution with parameters N_{i+} and π_i for $i = 1, 2, \cdots, I$. Since these are independent samples, it follows that the vector $\mathbf{n}$ follows the multivariate product multinomial probability model

$$\begin{aligned} P\{\mathbf{n}|M_1, \Pi\} &= \prod_{i=1}^{J} \binom{N_{i+}}{n_{i1} n_{i2} \cdots n_{iJ}} \pi_{i1}^{n_{i1}} \pi_{i2}^{n_{i2}} \cdots \pi_{ir}^{n_{iJ}} \\ &= \prod_{i=1}^{I} N_{i+}! \prod_{j=1}^{J} \pi_{ij}^{n_{ij}} / n_{ij}! \end{aligned} \tag{5.9}$$

with the constraints $\sum_j n_{ij} = N_{i+}$ and $\sum_j \pi_{ij} = 1$ for $i = 1, 2, \cdots, I$.

Sample	1	2	$\cdots$	J	Total
1	n_{11}	n_{12}	$\cdots$	n_{1J}	N_{1+}
2	n_{21}	n_{22}	$\cdots$	n_{2J}	N_{2+}
$\vdots$	$\vdots$	$\vdots$	$\vdots$	$\vdots$	$\vdots$
I	n_{I1}	n_{I2}	$\cdots$	n_{IJ}	N_{i+}
Total	N_{+1}	N_{+2}	$\cdots$	N_{+J}	N

Table 5.7: Observed table for the underlying probability Table 5.6

5.4.1 Homogeneity Hypothesis

In the framework of (5.9), the usual hypothesis involves the comparison of the distribution of say the column variable (B) from the I subgroups in the sense of homogeneity for specific categories. In particular, let the common parameter for the j-th level of B be denoted as π_{*j} for $j = 1, 2, \cdots, J$. Thus the null hypothesis of row homogeneity can be stated formally as

$$H_0 : \pi_{ij} = \pi_{*j}, \quad \text{for} \quad \begin{matrix} i = 1, 2, \cdots, I \\ j = 1, 2, \cdots, J \end{matrix} \tag{5.10}$$

Thus under H_0 in (5.10), the probability model (5.9) simplifies to

$$
\begin{aligned}
P\{\mathbf{n}|M_1, \mathbf{\Pi}\} &= \prod_{i=1}^{I} N_{i+}! \prod_{j=1}^{J} \pi_{*j}^{n_{ij}} / n_{ij}! \\
&= \left(\prod_{i=1}^{I} N_{i+}!\right)\left(\prod_{j=1}^{J} \pi_{*j}^{N_{+j}}\right) / \left(\prod_{i=1}^{I}\prod_{j=1}^{J} n_{ij}!\right)
\end{aligned} \tag{5.11}
$$

The number of degrees of freedom is computed as the total number of cells minus the number of independent linear constraints on the observations minus the number of independent parameters to be estimated from the data. That is,

Degrees of freedom = number of cells−number of Constraints−number of parameters

Specifically, for testing hypothesis (5.10), we have

$$
\text{d.f.} = IJ - I - (J-1) = IJ - I - J + 1 = (I-1)(J-1)
$$

since $(J-1)$ parameters are being estimated and there are I constraints.

5.4.2 Parameter Estimates

For each row indexed by $i = 1, 2, \cdots, I$, let the sample proportions be denoted by

$$
p_{ij} = \frac{n_{ij}}{N_{i+}}, \quad \text{for} \quad j = 1, 2, \cdots, J
$$

Then from previous multinomial results in Chapter 2, we have

$$
\begin{aligned}
E\{p_{ij}\} &= \pi_{ij} \\
\text{Var}\{p_{ij}\} &= \frac{\pi_{ij}(1-\pi_{ij})}{N_{i+}} \quad \text{and} \\
\text{Cov}\{p_{ij}, p_{ij'}\} &= -\left(\frac{\pi_{ij}\pi_{ij'}}{N_{i+}}\right)
\end{aligned}
$$

The variance-covariance matrix therefore can be written as:

$$
\begin{aligned}
V_{\mathbf{P_i}} &= \frac{1}{N_{i+}} \begin{pmatrix} p_{i1}(1-p_{i1}), & -p_{i1}p_{i2}, & \cdots, & -p_{i1}p_{iJ} \\ -p_{i2}p_{i1} & p_{i2}(1-p_{i2}), & \cdots, & -p_{i2}p_{iJ} \\ \vdots & \vdots & \cdots & \vdots \\ -p_{iJ}p_{i1}, & -p_{iJ}p_{i2}, & \cdots, & p_{iJ}(1-p_{iJ}) \end{pmatrix} \\
&= \frac{1}{N_{i+}} [\mathbf{D_{P_i}} - \mathbf{P_i P_i'}]
\end{aligned}
$$

where

$$
\mathbf{P}' = (\mathbf{p_1'}, \mathbf{p_2'}, \cdots, \mathbf{p_I'}); \text{with} \quad \mathbf{p_i} = (\mathbf{p_{i1}}, \mathbf{p_{i2}}, \cdots, \mathbf{p_{iJ}})
$$

Now under H_0, it can be shown that an unbiased MLE for the $\{\pi_{*j}\}$ can be obtained as

$$
p_{*j} = \frac{N_{+j}}{N}, \quad \text{for} \quad j = 1, 2, \cdots, J
$$

Therefore, for each row indexed by $i = 1, 2, \cdots, I$, the expected frequencies under H_0 are given by

$$
\hat{m}_{ij} = E\{n_{ij}|H_0\} = N_{i+}E\{p_{ij}|H_0\} = \frac{N_{i+}N_{+j}}{N}
$$

As a result, under H_0,

$$E\{(p_{ij} - p_{*j})|H_0\} = 0, \quad \text{for } i = 1, 2, \cdots, I \text{ and } j = 1, 2, \cdots, J \tag{5.12}$$

But since $\sum_j \pi_{ij} = 1$ for $i = 1, 2, \cdots, I$, without loss of generality, the $(I-1)(J-1)$ linearly independent differences in (9) have expected value 0. With the above results, Pearson's test statistic

$$X^2 = \sum_{i=1}^{I} \sum_{j=1}^{J} (n_{ij} - m_{ij})^2 / m_{ij}$$

can be computed by using the MLE cell estimates under H_0, and the statistic follows a χ^2 distribution approximately with $(I-1)(J-1)$ degrees of freedom for sufficiently large N.

5.4.3 Example 5.4

The example below is adapted from Lyman Ott (1984). Independent random samples of 83, 60, 56, and 62 faculty members of a state university system from four system universities were polled and by which of the three collective bargaining agents (union 101, union 102, union 103) was preferred. The resulting data rae displayed in Table 5.8.

	Bargaining agent			
University	101	102	103	Total
1	42	29	12	83
2	31	23	6	60
3	26	28	2	56
4	8	17	37	62
Total	107	97	57	261

Table 5.8: Observed table for the cross-classification

Interest centers on whether there is evidence to indicate a difference in the distribution of preference across the four state universities. For universities 1 and 2 the majority preference was for unit 101, while majority preference for universities 3 and 4 were from units 102 and 103 respectively.

Analysis

Let n_{ij} denote the frequency of bargaining agent $j = 1, 2, 3$ for the university $i = 1, 2, 3, 4$. Here $N_{i+} = \{83, 60, 56, 62\}$, for $i = 1, 2, 3, 4$, is fixed by design. Hence, $E(n_{ij}) = \hat{m}_{ij} = N_{i+}\pi_{ij}$ where $\sum_j \pi_{ij} = 1$ and N_{i+} is the total faculty in university i. That is, we let π_{ij} correspond to the probability of a bargaining agent being classified as of type j in university i. The hypothesis of homogeneity is:

$$H_0 : \pi_{ij} = \pi_{*j}, \quad j = 1, 2, 3$$

Parameter Estimates

Since we have shown that $p_{*j} = \dfrac{N_{+j}}{N}$, we therefore have,

$$p_{*1} = \frac{107}{261} = 0.40996; \qquad p_{*2} = \frac{97}{261} = 0.37165; \quad \text{and}$$
$$p_{*3} = \frac{57}{261} = 0.21839$$

The expected values for instance for the column "101" are respectively,

$$83(0.40996) = 34.0267 \qquad 60(0.40996) = 24.5976$$
$$56(0.40996) = 22.9578 \qquad 62(0.40996) = 25.4175$$

And similarly for the remaining two columns. These computed expected values, together with the observed values are displayed in Table 5.9.

	Bargaining agent			
University	101	102	103	Total
1	42	29	12	83
	34.0267	30.8470	18.1263	83.00
2	31	23	6	60
	24.5976	22.2990	13.1034	60.00
3	26	28	2	56
	22.9578	20.8124	12.2298	56.00
4	8	17	37	62
	25.4175	23.0423	13.5402	62.00
Total	107	97	57	261

Table 5.9: Observed and expected values for the data in Table 5.8

Hence, Pearson's X^2 is computed as

$$\frac{(42 - 34.0267)^2}{34.0267} + \cdots + \frac{(37 - 13.5402)^2}{13.5402} = 75.197$$

The corresponding likelihood ratio test statistic is

$$G^2 = 2\sum_{i=1}^{4}\sum_{j=1}^{3} n_{ij} \log\left(\frac{n_{ij}}{m_{ij}}\right) = 71.991$$

In both situations the pvalue = 0.0000 which indicates that we would have to reject the null hypothesis of no differences in the distribution of preference across the four universities. That is, we must conclude that bargaining agent was not uniform across the universities. The SAS software implementation for this data is presented in the next section.

Alternatively, we could compare the X^2 value of 75.197 with the standard χ^2 distribution with $(4-1)(3-1) = 6$ degrees of freedom.

Having established the fact that the null hypothesis is untenable, we would next wish to locate those cells that are not consistent with the null with a view to isolating them. We shall endeavor to answer specific questions with regard to the data above later in this chapter.

5.5 The Full Multinomial Probability Model

If we assume that neither of the marginal distributions M_1 or M_2 are fixed, then the underlying probability structure for the observed frequencies can be summarized as in Table 5.10. Data giving rise to this structure are usually derived from a simple random sample from a bivariate distribution.

Sample	1	2	$\cdots$	J	Total
1	π_{11}	π_{12}	$\cdots$	π_{1J}	π_{1+}
2	π_{21}	π_{22}	$\cdots$	π_{2J}	π_{2+}
$\vdots$	$\vdots$	$\vdots$	$\vdots$	$\vdots$	$\vdots$
I	π_{I1}	π_{I2}	$\cdots$	π_{IJ}	π_{I+}
Total	π_{+1}	π_{+2}	$\cdots$	π_{+J}	1

Table 5.10: Observed probability structure under this scheme

Here, the π_{ij} are unknown parameters such that $0 < \pi_{ij} < 1$ and they satisfy the constraints $\sum_i \sum_j \pi_{ij} = 1$ for $i = 1, 2, .., I$. The corresponding observed table of frequencies is displayed in Table 5.11.

Sample	1	2	$\cdots$	J	Total
1	n_{11}	n_{12}	$\cdots$	n_{1J}	N_{1+}
2	n_{21}	n_{22}	$\cdots$	n_{2J}	N_{2+}
$\vdots$	$\vdots$	$\vdots$	$\vdots$	$\vdots$	$\vdots$
I	n_{I1}	n_{I2}	$\cdots$	n_{IJ}	$N_{I.}$
Total	N_{+1}	N_{+2}	$\cdots$	N_{+J}	N

Table 5.11: Observed frequency table for the probability table in 5.10

The observed frequencies can be represented compactly by a vector

$$\mathbf{n}^{'} = n_{11}, n_{12}, \cdots, n_{IJ}$$

and $\mathbf{n}^{'}$ follow the multinomial distribution with parameters N and $\mathbf{\Pi}$, which can be written as

$$\begin{aligned} P\{\mathbf{n}|N, \mathbf{\Pi}\} &= \binom{N}{n} \pi_{11}^{n_{11}} \pi_{12}^{n_{12}} \cdots \pi_{1J}^{n_{1J}} \pi_{21}^{n_{21}} \cdots \pi_{IJ}^{n_{IJ}} \\ &= N! \prod_{i=1}^{I} \prod_{j=1}^{J} \pi_{ij}^{n_{ij}} / n_{ij}! \end{aligned} \tag{5.13}$$

with the constraints $\sum_i \sum_j n_{ij} = N$ and $\sum_i \sum_j \pi_{ij} = 1$.

5.5.1 Independence Hypothesis

The null hypothesis of independence is of primary interest here. This hypothesis can be stated formally as

$$H_0 : \pi_{ij} = \pi_{i+}\pi_{+j}, \quad \text{for } i = 1, 2, \cdots, I \text{ and } j = 1, 2, \cdots, J \tag{5.14}$$

Under H_0 as in (5.14), the joint probabilities can be obtained directly as products of the corresponding marginal probabilities. As a result,

$$\alpha_{ij} = \frac{\pi_{ij}\pi_{IJ}}{\pi_{Ij}\pi_{iJ}} = 1; \quad \text{for} \quad i = 1, 2, \cdots, (I-1);\ j = 1, 2, \cdots, (J-1)$$

The null hypothesis in (5.14) can equivalently be stated as

$$H_0 : \log(\alpha_{ij}) = 0; \quad \text{for} \quad i = 1, 2, \cdots, (I-1);\ j = 1, 2, \cdots, (J-1)$$

Under H_0, the probability model in (5.13) simplifies to

$$\begin{aligned} P\{\mathbf{n}|N, \mathbf{\Pi}; H_0\} &= N! \prod_{i=1}^{I} \prod_{j=1}^{J} (\pi_{i+}\pi_{+j})^{n_{ij}} / n_{ij}! \\ &= N! \left(\prod_{i=1}^{I} \pi_{i+}^{N_{i+}}\right) \left(\prod_{j=1}^{J} \pi_{+j}^{N_{+j}}\right) / \prod_{i=1}^{I} \prod_{j=1}^{J} n_{ij}! \end{aligned} \tag{5.15}$$

and the number of degrees of freedom is given by:

$$\text{d.f.} = IJ - 1 - (I + J - 2) = IJ - I - J + 1 = (I-1)(J-1)$$

Since there is one constraint and $(I-1) + (J-1)$ parameters to be estimated from the data.

5.5.2 Parameter Estimates

For each row indexed by $i = 1, 2, \cdots, I$, let the overall sample proportion be

$$p_{ij} = \frac{n_{ij}}{N}; \quad \text{for} \quad i = 1, 2, \cdots, (I-1);\ j = 1, 2, \cdots, (J-1)$$

Then from earlier results on the multinomial in Chapter 2, we again have

$$E\{p_{ij}\} = \pi_{ij}; \quad \text{Var}\{p_{ij}\} = \frac{\pi_{ij}(1-\pi_{ij})}{N}; \quad \text{Cov}\{p_{ij}, p_{ij'}\} = -\left(\frac{\pi_{ij}\pi_{ij'}}{N}\right)$$

Under H_0, it can be shown that the MLE (but not unbiased) for the π_{ij} can be obtained (using similar arguments) as in chapter 3 as:

$$\hat{\pi}_{ij} = \hat{\pi}_{i+}\hat{\pi}_{+j} = \frac{N_{i+}N_{+j}}{N} \tag{5.16}$$

so that the MLE of the n_{ij} can be obtained as

$$\hat{m}_{ij} = N\hat{\pi}_{ij} = \frac{N_{i+}N_{+j}}{N}$$

It should be noted here that the hypothesis of independence in either of the three forms $\{\pi_{ij}, \alpha_{ij}, \log(\alpha_{ij})\}$ is linear only on the log scale (i.e., log-linear) and a straight forward linear test similar to the previous quadratic form statistics is not possible here. We also note that the m_{ij} above are not unbiased. This hypothesis can, for the general $I \times J$ table be tested by a straight forward extension of the B^2 statistic (Lindley, 1964) discussed in chapter 4 for the 2×2 table. The formulation uses linear contrasts on the log scale.

In general, the hypothesis of independence is quite readily handled in the log-linear model formulation. Finally, we note that the test statistics for H_0 are identical to those for H_0 in the product-multinomial as the $\hat{m}_{ij}$ are the same in both models when using either the Pearson's criterion X^2 or the likelihood ratio statistic G^2.

5.5.3 Example 5.5

Again consider the example in Table 5.8, if the null hypotheses of interest in (5.14) can be written as:

H_0 : Bargaining agent preference is independent of university

H_a : Bargianing agent preference is associated with university

We then have under H_0 the following expected values:

$$\hat{m}_{11} = \frac{83 \times 107}{261} = 34.0268$$

$$\hat{m}_{12} = \frac{83 \times 97}{261} = 30.8467$$

$$\vdots = \vdots = \vdots$$

$$\hat{m}_{43} = \frac{62 \times 57}{261} = 13.5403$$

The above computed expected values are exactly the same as those computed earlier in Table 5.9 and hence lead to the same X^2 =75.197 as before. Once again, our result indicates that we would have to reject the hypothesis of independence. That is, the preference for a bargaining agent depends on the university the faculty belongs to.

The SAS software program using PROC FREQ to implement the above described model is presented below. The corresponding modified SAS software output is presented in appendix D.8. The summary statistics from the model fitting are also presented below.

```
data example; input unv agent count @@; datalines;
1 1 42 1 2 29 1 3 12 2 1 31 2 2 23 2 3 6
3 1 26 3 2 28 3 3 2 4 1 8 4 2 17 4 3 37
; proc freq order=data; weight count;
tables unv*agent/chisq expected cellchi2 nocol norow nopercdent; run;
```

```
                     Statistics for Table of unv by agent

             Statistic                     DF        Value       Prob
             ------------------------------------------------------
             Chi-Square                     6      75.1974     <.0001
             Likelihood Ratio Chi-Square    6      71.9911     <.0001
```

The first line in the SAS software output in appendix D.8 gives the observed frequencies, the second line gives the corresponding expected values $\hat{m}_{ij}$ while the third line gives the individual cell contributions to the overall X^2 of 75.1974.

5.6 Residual Analysis

Having established that there is a need to reject the null hypothesis of independence or homogeneity or randomness in either of the sampling schemes III, II, or I, our next concern is to locate the source or sources of lack of independence. The simplest way to do this is to examine one of the followings:

(i) The standardized residuals z_{ij} is defined (Haberman, 1978) as

$$z_{ij} = \frac{(n_{ij} - m_{ij})}{\sqrt{m_{ij}}}$$

Cells		H_{01}		H_{02}	
(i,j)	n_{ij}	$\hat{m}_{ij}$	z_{ij}	$\hat{m}_{ij}$	z_{ij}
11	42	34.0	1.4	39.16	0.45
12	29	30.8	-0.3	35.50	-1.09
13	12	18.1	-1.4	8.34	1.27
21	31	24.6	1.3	28.31	0.51
22	23	22.3	0.1	25.66	-0.53
23	6	13.1	-2.0	6.03	-0.01
31	26	23.0	0.6	26.42	-0.08
32	28	20.8	1.6	23.95	0.83
33	2	12.2	-2.9	5.63	-1.53
41	8	25.4	-3.5	13.11	-1.41
42	17	23.0	-1.3	11.89	1.48
43	37	13.5	6.4	37.00	0.00
X^2		75.20		10.756	
G^2		71.99		11.45	
df		6		5	

Table 5.12: Results of fit based on H_{01} and H_{02}

We note that $\sum_{ij} z_{ij}^2 = X^2$, the Pearson's test statistic. The distribution of z_{ij} when the model holds is asymptotically distributed normal with mean 0 and variance 1. A standardized residual will be said to exhibit lack of fit if $|z_{ij}| > 2.0$.

(ii) The adjusted residuals r_{ij} (Haberman, 1973) is the value of the standardized residuals z_{ij}, divided by its standard error. For the test of independence in a two-way contingency table, it simplifies to

$$r_{ij} = \frac{n_{ij} - \hat{m}_{ij}}{\sqrt{\hat{m}_{ij}(1 - p_{i+})(1 - p_{+j})}}$$

where as before $\hat{m}_{ij} = Np_{i+}p_{+j}$. Table 5.12 gives the values of z_{ij} for the data in Table 5.8 under the models of independence H_{01} and quasi-independence H_{02}. By quasi-independence, we mean that the table exhibits independence with repect to a reduced or imcomplete table (see definition later).

An examination of the values of the standardized residuals in Table 5.12 under model H_{01} shows that cells $(2,3), (3,3), (4,1)$, and $(4,3)$ all have $|z_{ij}| \geq 2$. The value for cell (4,3) is particularly high. Positive z_{ij} indicated that there are more respondents observed in that cell than the hypothesis is expecting. Similarly, a negative z_{ij} indicates that the observed number of respondents for that cell is less than what is expected under the model of interest. Thus for cell $(4,3)$, there are far more faculty respondents in that category than is expected under the model of independence. The four identified cells above have been referred to by various authors as the "rogue or aberrant" cells (Upton, 1982).

Corresponding values for the r_{ij} can similarly be obtained from the expression for r_{ij} above. For instance for cell $(4,3)$ we have

$$p_{4+} = 62/261 \quad = 0.23755$$
$$p_{+3} = 57/261 \quad = 0.21839$$

hence,

$$r_{4,3} = \frac{(37 - 13.5)}{\sqrt{13.5(1 - .2376)(1 - .2184)}} = 8.28.$$

PROC GENMOD in SAS software generates both z_{ij} and r_{ij} with the keywords **Reschi** and **Stdreschi**, respectively, in the output statement.

We now discuss two ways of looking at the above data more critically with a view to explaining why the data does not uphold our hypothesis of independence. The first method is to partition the G^2 (rather than X^2) from the 4×3 table into some smaller components. The other method is to fit the model of quasi-independence.

5.7 Partitioning of the G^2 Statistic

We now wish to decompose the G^2 (the likelihood ratio test) statistic, which was based on 6 d.f. in our analysis above. Specifically, we wish to sub-divide the 6 d.f. such that one degree of freedom relates in particular to cell $(4, 3)$. Since there are more than one cell with $|z_{ij}| \geq 2$, we usually start to partition first the cell with the highest $|z_{ij}|$, which in this case happens to be cell $(4, 3)$. Decomposition of G^2 requires some experience and alternative decomposition to the one adopted for this particular problem may lead to different conclusions. Maxwell (1961) and Iverson (1979) give detailed treatments of the technique, and we give below some rules that must be satisfied.

1. If there are g degrees of freedom for the original table, then there can be no more than g subtables formed. In this case there can not be more than 6 such subtables.
2. Each observed cell frequencies in the original table must appear as a cell frequency in one and only one subtable.
3. Each marginal total of the original table must appear as a marginal total of one and only one subtable.
4. Subtable cell frequencies not appearing in the original table must appear as marginal totals in a different subtable. Marginal totals not appearing in the original table must appear as either cells or grand totals.

Based on our observations above, Tables 5.13 to 5.16 give a decomposition of the observed G^2 for the original table into components that satisfy rules 1 to 4 above. The values of the goodness-of-fit statistic G^2 for each of the subtables in 5.13 to 5.16 are displayed and summarized in Table 5.17. We note that the components sum exactly to 71.9910 which is the G^2 for the original 4×3 table.
The decomposition works best for the G^2 statistic because it can be readily decomposed. This does not work well with X^2. To see the decomposing ability of G^2 (here into 4 components), suppose we expand G^2 as follows:

University	Bargaining agent 101 & 102	103	Total
1-3	179	20	199
4	25	37	62
Total	204	57	261

Table 5.13: Dichotomized university and agent, isolating cell (4,3)

University	Bargaining agent 101	102	Total
1	42	29	71
2	31	23	54
3	26	28	54
Total	99	80	179

Table 5.14: Units 101 and 102, universities 1 – 3

University	Bargaining agent 101	102	Total
1-3	99	80	179
4	8	17	25
Total	107	97	204

Table 5.15: Bargaining units by dichotomized university

University	Bargaining agent 101 & 102	103	Total
1	71	12	83
2	54	6	60
3	54	2	56
Total	179	20	199

Table 5.16: Bargaining units by universities (1,2,3)

$$\begin{aligned}
G^2 &= 2\sum_{i=1}^{I}\sum_{j=1}^{J} n_{ij}\log\left(\frac{n_{ij}}{\hat{m}_{ij}}\right)\\
&= 2\sum_{i=1}^{I}\sum_{j=1}^{J} n_{ij}\log\left[n_{ij}\frac{N}{N_{i+}N_{+j}}\right]\\
&= 2\left[\sum_{i=1}^{I}\sum_{j=1}^{J} n_{ij}\log(n_{ij}) - \sum_{i=1}^{I} N_{i+}\log(N_{i+})\right]\\
&- 2\left[\sum_{j=1}^{J} N_{+j}\log(N_{+j}) - N\ln(N)\right]
\end{aligned}$$

The results in Table 5.17 indicate that the model of independence is quite tenable for Tables 5.14 and 5.16 and for 5.15 (though not as strong as the latter two at $\alpha = 0.01$) but certainly not tenable for Table 5.13. It is evident therefore that only

Table	df	G^2	pvalue
4.13a	1	60.5440	0.0000
4.13b	2	1.6378	0.4411
4.13c	1	4.8441	0.0277
4.13d	2	4.9660	0.0835
Total	6	71.9910	0.0000

Table 5.17: Results of the decompositions (5.13 – 5.16)

the (101 & 102, 103) versus (universities 1 – 3 combined) table is significant. We can thus recombine the remaining tables to simplify our final summary. Tables 5.14 and 5.16 are now recombined to produce Table 5.18.

	Bargaining agent			
University	101	102	103	Total
1	42	29	12	83
2	31	23	6	60
3	26	28	2	56
Total	99	80	20	199

Table 5.18: Bargaining units by universities (1 – 3)

The final decomposition goodness-of-fit statistic G^2 values for these subtables are as displayed in Table 5.19.

Table	df	G^2	pvalue
11a	1	60.5440	0.0000
11c	1	4.8441	0.0277
11e	4	6.6038	0.1584
Total	6	71.9910	0.0000

Table 5.19: Results of the final decompositions

Our conclusions in light of the above analyses are :

(i) Table 5.18 shows that preference distribution is homogeneous between universities 1 to 3.

(ii) At $\alpha = 0.01$, Table 5.15 shows that preference for bargaining units 101 and 102 was homogeneous between universities (1 – 3) combined and university 4.

(iii) Table 5.13 shows that preference for a bargaining unit is independent of the faculty's university with the exception that if a faculty member belongs to university 4, then he or she is much more likely than would otherwise have been expected to show preference for bargaining unit 103 (and vice versa).

5.8 The Quasi-Independence Model

In our original analysis, we are interested in either the hypothesis of independence or the hypothesis of homogeneity. We showed that when either hypotheses holds, the expected values and the values of the test statistics are equivalent for both situations. Under the model of independence, we recall that

$$H_{01} : \pi_{ij} = \pi_{i+}\pi_{+j} \tag{5.17}$$

From the previous analysis, we recognize that this model does not fit the data. A natural extension of this hypothesis therefore is to see whether the responses (preference) are independent for most but not all the cells. That is, suppose we exclude cell $(4,3)$ from the model, would the table now exhibit independence on this reduced (or incomplete) table? The latter hypothesis would thus be termed the model of *quasi-independence* and would be given by

$$H_{02} : \pi_{ij} = \pi_{i+}\pi_{+j} \qquad \text{for } (i,j) \neq (4,3) \tag{5.18}$$

Hypothesis H_{02} will be termed the hypothesis of quasi-independence.

5.8.1 Computing Expected Values for Incomplete Tables

When rows and columns correspond to two variables, it is sometimes impossible to observe particular category combinations. If this logical impossibility occurs in cell (i,j), we say that we have a *structural zero* and $\pi_{ij} = m_{ij} = 0$, and we refer to array of nonzero cells as an incomplete table. For example, if cell $(4,3)$ were removed from our example data above, then the resulting Table 5.20 will be described as an incomplete table, where for brevity, we have coded "101" as 1, "102" as 2, and "103" as 3, respectively.

	Unit			
University	1	2	3	Total
1	42	29	12	83
2	31	23	6	60
3	26	28	2	56
4	8	17	-	25
Total	107	97	20	224

Table 5.20: Incomplete Table with cell (4,3) deleted

We notice that in this incomplete table, the marginal totals have been correspondingly reduced by the amount of the excluded cell count (in this case, 37). In order to calculate the expected values under the new hypothesis of quasi-independence in the complete table, Bishop et al. (1975) had enunciated several procedures. In particular their rule 3 for block-triangular tables can be employed to calculate expected for this case. Other situations of incomplete tables of exist and for these other cases, calculating the expected frequencies can be very time-consuming and it would be reasonable to employ any of the well-known computer algorithms for such situations. For now we consider the case in which the pattern is of type 3, the block-trianngular tables.

5.8.2 Block-Triangular Tables

We say that an incomplete table is in *block-triangular* form if after suitable permutation of rows and columns, $\delta_{ij} = 0$ implies $\delta_{kl} = 0$ for all $k \geq i$ and $l \geq j$ where

$\delta_{ij} = 1$ for $(i,j) \in S$, the set of cells which do not contain structural zeros, and $\delta_{ij} = 0$ for $(i,j) \notin S$.

Examples

We give three examples of tables of this form for 4×4, (i) and (ii), and 5×4 (iii) tables.

m_{11}	m_{12}	m_{13}	m_{14}	
m_{21}	m_{22}	m_{23}	-	(i)
m_{31}	m_{32}	-	-	
m_{41}	-	-	-	

-	-	m_{13}	m_{14}	
-	-	m_{23}	m_{24}	(ii)
m_{31}	m_{32}	m_{33}	m_{34}	
m_{41}	m_{42}	m_{43}	m_{44}	

m_{11}	m_{12}	m_{13}	m_{14}	
-	m_{22}	m_{23}	m_{24}	
-	m_{32}	m_{33}	m_{34}	(iii)
-	-	m_{43}	m_{44}	
-	-	m_{53}	m_{54}	

We call tables like (i) - (iii) block-triangular, because the nonstructural zero cells, after suitable permutation of rows and columns, form a right-angled triangle with a block of cells lying along the hypotenuse of the triangle.

Returning to our data, we recognize that the incomplete table in the previous section with cell $(4,3)$ presumed to be a structural zero satisfies the concept of block-triangular tables. Hence the expected values are computed as follows. First, we compute expected values of the adjacent cells to the structural zero cell, that is, cell $(4,2)$. Thus

$$\hat{m}_{42} = \frac{97 \times 25}{(224 - 20)} = 11.887$$

$$\hat{m}_{33} = \frac{20 \times 56}{(224 - 25)} = 5.628$$

Because of the marginal constraints, we thus have

$$\hat{m}_{41} = 25 - (11.887) = 13.113$$

Next, the marginal totals are adjusted to reflect these estimates, we now have the reduced table with the corresponding reduced marginal totals (brought forward from previous calculations).

	1	2	3	Total
1	42	29	12	83
2	31	23	6	60
3	26	28	-	50.372
New Total	93.887	85.113	14.372	193.372

Again

$$\hat{m}_{32} = \frac{50.372 \times 85.113}{(193.372 - 14.372)} = 23.952$$
$$\hat{m}_{23} = \frac{60 \times 14.372}{(193.372 - 50.372)} = 6.030$$

Because of the marginal constraints again, we thus have

$$\hat{m}_{31} = 50.372 - 23.952 = 26.420 \quad \text{and}$$
$$\hat{m}_{13} = 14.372 - 6.030 \quad = 8.342$$

And finally, we have the reduced 2×2 table after adjusting for the marginal totals

	1	2	Total
1	42	29	74.658
2	31	23	53.970
Total	67.467	61.161	128.628

Hence,

$$\hat{m}_{11} = \frac{74.658 \times 67.467}{128.628} = 39.159$$

Again, because of the new marginal constraints, we now have,

$$\hat{m}_{12} = 74.658 - 39.159 = 35.499$$
$$\hat{m}_{21} = 67.467 - 39.159 = 28.308 \quad \text{and}$$
$$\hat{m}_{22} = 53.970 - 28.308 = 25.662$$

These expected values are displayed earlier in Table 5.12 and has the computed G^2 =11.45 and is based on 5 d.f. In general, the number of degrees of freedom is given by

$$\text{d.f.} = (I - 1)(J - 1) - \text{number of excluded cells}$$

and in our case we have $(6 - 1) = 5$ d.f. The corresponding pvalue is 0.043, which indicated that we would fail to reject the hypothesis of quasi-independence at $\alpha =$ 0.01.

The above procedure for obtaining cell estimates for incomplete tables can be applied to any table with a block-triangular structure. For other structures, the extensive treatment provided in Bishop et al. (1975) will be found to cover most of all possible occuring structures. Of course, if we should have several cells with structural zeros, the cells estimation may not be straightforward as in the preceeding example, and iterative procedures would have to be employed for this situation (see chapter 6). In any case, any of the standard statistical package will handle this situation.

Other forms of hypotheses relating to incomplete tables will be discussed later in chapter 6.

Because of the extensive calculations involved above for obtaining the expected frequencies, there are standard algorithms for computing these values under the model of quasi-independence. We give the SAS software program for fitting the quasi-independence model using PROC GENMOD.

```
DATA EXAMPLE;
DO UNV=1 TO 4;
 DO AGENT=1 TO 3;
INPUT COUNT @@;
IF UNV EQ 4 AND AGENT EQ 3 THEN WT=0;
ELSE WT=1; OUTPUT;
END;
END;
DATALINES;
42 29 12 31 23 6 26 28 2 8 17 37
;
*** fits the model of independence***;
PROC GENMOD;
CLASS UNV AGENT;
MODEL COUNT=UNV AGENT/DIST=POI LINK=LOG; RUN;
*** fits the model of quasi-independence***;
PROC GENMOD DATA=EXAMPLE;
CLASS UNV AGENT WT;
MODEL COUNT=UNV AGENT WT/DIST=POI LINK=LOG;
RUN;
```

```
                        INDEPENDENCE MODEL
               Criteria For Assessing Goodness Of Fit

Criterion                       DF          Value          Value/DF
--------------------------------------------------------------------
Deviance                         6        71.9911           11.9985
Pearson Chi-Square               6        75.1973           12.5329

UNV    AGENT        COUNT         Pred       Reschi     Streschi
--------------------------------------------------------------------
 1       1            42       34.0268       1.3669       2.1547
 1       2            29       30.8467      -0.3325      -0.5079
 1       3            12       18.1265      -1.4390      -1.9709
 2       1            31       24.5977       1.2909       1.9150
 2       2            23       22.2988       0.1485       0.2134
 2       3             6       13.1035      -1.9624      -2.5293
 3       1            26       22.9579       0.6349       0.9326
 3       2            28       20.8123       1.5756       2.2427
 3       3             2       12.2299      -2.9252      -3.7334
 4       1             8       25.4176      -3.4548      -5.1508
 4       2            17       23.0422      -1.2587      -1.8185
 4       3            37       13.5403       6.3754       8.2586

                      QUASI-INDEPENDENCE MODEL
               Criteria For Assessing Goodness Of Fit

Criterion                       DF          Value          Value/DF
--------------------------------------------------------------------
Deviance                         5        11.4471            2.2894
Pearson Chi-Square               5        10.7552            2.1510

UNV    AGENT        COUNT         Pred       Reschi     Streschi
--------------------------------------------------------------------
 1       1            42       39.1590       0.4540       0.7877
 1       2            29       35.4993      -1.0908      -1.8170
 1       3            12        8.3417       1.2666       1.7492
 2       1            31       28.3077       0.5060       0.8139
 2       2            23       25.6621      -0.5255      -0.8117
 2       3             6        6.0302      -0.0123      -0.0155
 3       1            26       26.4205      -0.0818      -0.1300
 3       2            28       23.9513       0.8273       1.2626
 3       3             2        5.6281      -1.5293      -1.9022
 4       1             8       13.1127      -1.4119      -2.1859
 4       2            17       11.8873       1.4829       2.1859
 4       3            37       37.0000       0.0000        .
```

The expected values obtained from SAS software are displayed in Table 5.12 under hypothesis H_{02}. We shall elaborate more on the use of the above SAS software codes after our discussion of log-linear models in chapter 6. We shall also revisit the analysis of incomplete tables by discussing current approaches for anlyzing such data in the next chapter.

The corresponding SAS software program from using PROC CATMOD is also displayed below:

```
data new;
set example;
if unv eq 4 and agent eq 3 then count=0;
proc catmod; weight count;
model unv*agent= _response_/ml nogls freq pred=freq
                    noparm noresponse;
loglin unv agent; run;
```

5.9 Problems with Small Expected Frequencies

As discussed in chapter 4, when the sample size in a contingency table is small, the resulting expected values may become correspondingly small and consequently, the asymptotic χ^2 approximation may not be valid. Specifically, several authors have considered the χ^2 approximation to Pearson's X^2 test statistic under the hypothesis of independence. Hommel (1978), Larntz (1978, Koehler (1986), Lawal and Upton (1984), Lawal (1989b), and Lawal and Upton (1990) are but a few of them.

The approximation becomes suspect because as Lawal and Upton observed, for most cases,

$$Pr\{X^2 > \chi^2_d(\alpha)\} < \alpha$$

Lawal and Upton therefore sought a value k such that

$$Pr\{X^2 > k\chi^2_d(\alpha)\} = \alpha$$

Their results suggest that k varies with α, and they recommended that for any $I \times J$ contingency table containing N observations, the Pearson's X^2 should be calculated as usual with expected values based on the independence model, and its value compared with

$$\left(1 - \frac{3}{2N}\right)\chi^2_d(0.01) \quad \text{at the 1\% level, and}$$

$$\left\{1 - \frac{1}{N}(1 - d^{-1/2})\right\}\chi^2_d(0.05) \quad \text{at the 5\% level}$$

where $\chi^2_d(\alpha)$ is the $100\alpha\%$ critical value a χ^2 distribution with $d = (I-1)(J-1)$ degrees of freedom.
The following restrictions apply on the use of these procedures:

(i) The dimension of the table should satisfy

$$I, J > 2$$

(ii) The average cell frequency should be greater than 0.5:

$$2N > \text{IJ}$$

(iii) The smallest cell expected value, m, should satisfy the inequality

$$\hat{m} > sd^{-3/2}$$

where s is the number of cells having expected frequencies less than 3.

5.9.1 Example 5.6

The following data relate to a survey conducted by a video rental store of its customers. The two responses of most interest to the store were customer's frequency of renting (coded 1 for lowest frequency and 4 for highest frequency) and customers' rating of the adequacy of the store's selection (coded 1 for poorest rating and 4 for highest rating).

	Adequacy				
Freq.	1	2	3	4	Total
1	1 (4.38)	4 (11.94)	37 (35.04)	44 (34.64)	86
2	2 (2.41)	6 (9.31)	30 (27.30)	29 (26.99)	67
3	3 (2.04)	8 (5.56)	16 (16.30)	13 (16.11)	40
4	5 (1.17)	12 (3.19)	5 (9.37)	1 (9.26)	23
Total	11	30	88	87	216

Values in parentheses are the expected frequencies under the model of independence. We note the following:

(a) $X^2 = 61.04$ and $G^2 = 53.35$.

(b) The average observed frequency is 13.5.

(c) There are two expected values (1.17 and 2.04) that are less than 3.0. Hence by rule (*iii*) above, the smallest expected value allowed is $2(9)^{-3/2} = 0.074$. The above two expected values satisfy this condition. We should compare the computed X^2 with say at $\alpha = 0.05$ to

$$\left(1 - \frac{1}{216}(1 - \frac{1}{3})\right) \chi_9^2(.05) = 0.9969 \times 16.92 = 16.87$$

Based on the above we would therefore reject the hypothesis that the two classificatory variables are independent. A StatXact test based on 10,000 simulations gives a pvalue of 0.0000, indicating once again that the hypothesis would have to be rejected. A quasi-independence model may be needed after suitable residual analysis and diagnostics.

5.10 Association Measures in Two-Way Tables

Discussions of the global odds and local global odds ratios are presented in appendix D.9. In this section, we discuss *concordance* and *discordance* in the context of an $I \times J$ contingency table. An example is also presented.

5.10.1 Concordance and Discordance

For two variables A and B in a two-way table, we say that a pair of observations is *concordant* if the subject that ranks higher on variable A also ranks higher on variable B. Similarly, a pair is said to be *discordant* if the subject that ranks higher on A ranks lower on B. Formulae for calculating the number of concordant pairs (C) and number of discordant pairs (D) are given by Agresti (1984) as:

$$C = \sum_{i<k} \sum_{j<s} n_{ij} n_{ks}; \qquad D = \sum_{i<k} \sum_{j>s} n_{ij} n_{ks} \tag{5.19}$$

where for (C), the first summation is over all pairs of rows $i < k$ and the second summation is over all pairs of columns $j < s$.

Below is a computational procedure for obtaining C and D for a two-way table having I rows and J columns. The concordant pairs can be computed as follows:

$$C = n_{ij} \left(\sum_{u=i+1}^{I} \sum_{v=j+1}^{J} n_{uv} \right); \qquad D = n_{ij} \left(\sum_{u=i+1}^{I} \sum_{v=1}^{j-1} n_{uv} \right)$$

where $i = 1, 2, \cdots, (I-1)$, $j = 1, 2, \cdots, (J-1)$.

Example 5.7

The data in Table 5.21 are taken from Agresti (1984) and relate to a comparison of four (ordinal) different operations for treating duodenal ulcer patients.

	Dumping severity			
Operation	None	Slight	Moderate	Total
A	61	28	7	96
B	68	23	13	104
C	58	40	12	110
D	53	38	16	107
Total	240	129	48	417

Table 5.21: Data on duodenal ulcer patients

For the data in Table 5.21, we have

$$\begin{aligned} C &= n_{11}(n_{22} + n_{23} + n_{32} + n_{33} + n_{42} + n_{44}) \\ &\quad + n_{12}(n_{23} + n_{33} + n_{43}) + n_{21}(n_{32} + n_{33} + n_{42} + n_{43}) \\ &\quad + n_{22}(n_{33} + n_{43}) + n_{31}(n_{42} + n_{43}) + n_{32}(n_{43}) \\ D &= n_{12}(n_{21} + n_{31} + n_{41}) \\ &\quad + n_{13}(n_{21} + n_{22} + n_{31} + n_{32} + n_{41} + n_{42}) + n_{22}(n_{31} + n_{41}) \\ &\quad + n_{23}(n_{31} + n_{32} + n_{41} + n_{42}) + n_{32}(n_{41}) + n_{33}(n_{41} + n_{42}) \end{aligned}$$

and $M = \binom{I}{2}J$, where if $I > J$ *the larger of the two dimensions*, then M gives the number of product terms in the expressions for C and D for the general values of I and J. In this example, $M = 18$.

Similarly, we define, T_A to be the total number of pairs of observations for which $i = i'$ and it is computed as

$$T_A = \frac{1}{2} \sum_i n_{i+}(n_{i+} - 1)$$

T_B is defined as the total number of observations for which $j = j'$ and it is also computed as:

$$T_B = \frac{1}{2} \sum_j n_{+j}(n_{+j} - 1)$$

Also T_{AB} equals the total number of observations for which $i = i'$ and $j = j'$ and is also computed as:

$$T_{AB} = \frac{1}{2} \sum_i \sum_j n_{ij}(n_{ij} - 1)$$

We note here that

$$\binom{n}{2} = C + D + T_A + T_B - T_{AB}$$

We now describe three measures of association that are all based on the notion of concordant and discordant pairs for an $I \times J$ ordered table, having variables A and B respectively, and where we shall assume that category 1 is higher than category 2 and so forth.

5.10.2 Goodman and Kruskal's γ

The gamma measure proposed by Goodman and Kruskal (1954) is defined as:

$$\hat{\gamma} = \frac{C - D}{C + D} \tag{5.20}$$

The range of gamma is $-1 \leq \gamma \leq 1$ and under independence, $\gamma = 0$. However, $\gamma = 0$ does not necessarily imply that A and B are independent. For the data in Table 5.21, C= 21434, D= 15194, and hence,

$$\hat{\gamma} = \frac{21434 - 15194}{21434 + 15194} = 0.170$$

5.10.3 Somers's d

Somers (1962) proposed a measure which is a variation of γ, which is said to be more appropriate when one of the variables, say B, is dependent on the other variable. This statistic, defined as d_{BA}, is given by:

$$d_{BA} = \frac{(C - D)}{C + D + T_B} \tag{5.21}$$

The statistic is interpreted in Upton (1978)

> as the difference between the probabilities of like and unlike orders for two observations chosen at random from the population, conditional on their not having tied ranks for variable A.

Again for the data in Table 5.21, we have $T_B = 38064$ and hence $d_{BA} = 0.084$.

The corresponding statistic that assumes that A is the dependent variable is $d_{AB} = 0.102$, since $T_A = 21582$. Also, $T_{AB} = 9538$.

5.10.4 Kendall's τ

This statistic is defined as:

$$\tau = \frac{2(C - D)}{\sqrt{(C + D + T_A)(C + D + T_B)}} \tag{5.22}$$

Similarly, for the data in Table 5.21, we have $\tau = 0.189$.

It has been advocated that the gamma statistic be used because of its ease of computation, interpretation, and more importantly if the two variables are of equal interest. However, the gamma tends to depend on the number of categories and the way these categories are defined. Consequently, Somers's d_{BA} is particularly useful if one of the variables is dependent and is also particularly useful for the general $2 \times J$ tables in which the column variable B is an ordinal response variable. We can implement the various measures in SAS software as follows:

```
DATA MEASR;
DO A=1 TO 4;
  DO B=1 TO 3;
INPUT COUNT @@; OUTPUT; END;  END;
DATALINES;
61 28 7 68.....16
;
PROC FREQ; WEIGHT COUNT; TABLES A*B /MEASURES;
TITLE 'ASSOCIATION MEASURES'; RUN;
                         Association measures
                    Statistics for Table of a by b

         Statistic                                Value       ASE
         ------------------------------------------------------
         Gamma                                   0.1704    0.0647
         Kendall's Tau-b                         0.1108    0.0423
         Stuart's Tau-c                          0.1077    0.0412
         Somers' D C|R                           0.0958    0.0366
         Somers' D R|C                           0.1282    0.0490
         Pearson Correlation                     0.1222    0.0478
         Spearman Correlation                    0.1263    0.0482
         Lambda Asymmetric C|R                   0.0000    0.0000
         Lambda Asymmetric R|C                   0.0456    0.0395
         Lambda Symmetric                        0.0289    0.0253
         Uncertainty Coefficient C|R             0.0140    0.0083
         Uncertainty Coefficient R|C             0.0094    0.0056
         Uncertainty Coefficient Symmetric       0.0113    0.0067
                          Sample Size = 417
```

The special case of the $2 \times r$ ordered scored responses is presented in appendix D.10. Examples relating to this are also presented in the appendix.

5.11 Exercises

1. The data for this exercise relate the relationship between people's attitude on the government's role in guaranteeing jobs and their votes in state and local elections (Reynolds, 1977).

Voting in 1956 State and Local Elections by Attitudes on Government-Guaranteed Jobs

Vote	Attitude		
	Liberal	Moderate	Conservative
Democratic	312	34	115
Even	159	24	110
Republican	210	32	137

Fit a model of independence to the above data and conduct an analysis of residuals. Based on your residual analysis, partion the table into suitable components and again test for independence. Draw your conclusions.

2. Refer again to the data above, anf fit an appropriate model of quasi-independence.

3. For the data on voting attitudes, obtain Kruskal's γ, Kendall's τ, and Somers's d measures of association. Interpret these measures.

4. A survey of drivers was obtained to compare the proportions who use seat-belts regularly for various age categories. These data are displayed in the accompanying table.

Age Group	Regularity of seatbelt usage			
	Always	Regularly	Sometimes	Never
16-20	1	10	70	19
21-25	4	8	80	8
26-30	8	10	77	5
> 30	15	30	49	6

Analyze the data (including residual analysis) and draw conclusions.

5. The data in the next table show a cross-classification of level of smoking and myocardial infarction for a sample of young women (Agresti, 1989; Shapiro et al., 1979). Conduct a test of independence for the data, and comment on your results.

Patients	Smoking Level (Cigarettes/Day)		
	0	1-24	> 25
Control	25	25	12
Myocardial	0	1	3

6. The data below refer to 264 marriages in Surinam (Speckman, 1965). Here husbands and wives are categorized in terms of four religious groups (C=Christian, M=Moslems, S=Sanatin Dharm, A=Arya Samaj). S and A are two Hindustan religious groups.

Marriage in Surinam (Speckmann, 1965)

	Wives				
Husbands	C	M	S	A	Total
C	17	1	4	3	25
M	1	66	4	2	73
S	5	4	96	14	119
A	4	2	18	23	47
Total	27	73	122	42	264

(a) Fit the independence model.

(b) Conduct a residual analysis.

(c) Can you fit a quasi-independence model to these data?

7. The data in this example relate to the classification of 2000 sixth-grade children by school performance and weight category.

	Underweight	Normal	Overweight	Obese
Poor at school	36	160	65	50
Satisfactory at school	180	840	300	185
Above average	34	100	35	15

Fit the independence model to the data and perform residual analysis.

8. For the Surinam marriage data, obtain Kendall's, Somers's and Goodman and Kruskal's measures of association and interpret them. Relate these to your results above.

Chapter 6

Log-Linear Models for Contingency Tables

6.1 Introduction

The concept of log-linear analysis in contingency tables is analogous to the concept of analysis of variance (ANOVA) for the continously distributed factor-response variables. While response observations are assumed to be continuous with underlying normal distributions in ANOVA, the log-linear analysis assumes that the response observations are counts having Poisson distributions.

Basically, what we have done in the preceding chapters is to analyze simple (that is, two-factor or factor-response) two-way contingency tables where emphasis is on whether the classificatory variables are homogeneous or independent. In a way, the preceeding methods discussed in chapter 5 cannot be readily extended to situations where there are more than two underlying variables. Consequently, in this chapter, we develop a new methodology that will enable us to study various interactions for multiway contingency tables.

As discussed in chapter 4, the odds ratio is invariant under interchange of rows and columns. This property of the odds ratio makes it very attractive for use, especially when the dependent variable (i.e., the response) is not obvious. The invariance property ensures that the odds ratio can be utilized for studies of independence. The log-linear parameterization, which models association or dependence in terms of the odds ratios, therefore makes it suitable for the analysis of multiway contingency tables.

In this chapter, the odds ratio will be used to model multiple-response structure contingency tables. The log-linear model (Goodman, 1970, 1971a, 1972a,b; Bishop et al., 1975; Haberman, 1978) formulations are similar to analysis of variance (ANOVA) techniques, except that the models are applied to the natural logarithms of the multinomial probabilities or expected frequencies.

We begin by examining the log-linear model for a 2×2 table and then extend the formulation to a general $I \times J$ table and finally to a general multiway contingency table.

6.2 The 2×2 Table

If for a 2×2 table we assume a multinomial sampling scheme, then only the sample size n will be fixed and the observed frequency n_{ij} follows a multinomial distribution with parameters π_{ij} and n. Consider the following example on treatment of angina. In a study to evaluate the effectiveness of the drug Timolol (Samuels & Witmer, 1999) in preventing angina attacks, patients were randomly allocated to receive a daily dosage of either Timolol or placebo for 28 weeks. The numbers of patients who became free of angina attacks are displayed in Table 6.1.

	Response		
Treatment	Free	Not Free	Total
Timolol	44	116	160
Placebo	19	128	147
Total	63	244	307

Table 6.1: Response to angina treatment

Let T and R represent the treatment and response variables respectively. Let the joint probability that an observation falls in category i of variable T and category j of variable R be π_{ij}. That is, $P(\mathrm{T} = i, \mathrm{R} = j) = \pi_{ij} > 0$ for $i, j = 1, 2$. Let treatment T be indexed by $i = 1, 2$ for Timolol and placebo, respectively. Similarly, let the response variable R be indexed by $j = 1, 2$ for free and not free, respectively.

Under the multinomial sampling scheme, the expected frequencies is

$$\hat{m}_{ij} = n\pi_{ij}$$

where n is the sample size, and the natural logarithm of the expected values $\ln(\hat{m}_{ij})$, denoted henceforth as l_{ij}, is:

$$l_{ij} = \ln(n) + \ln(\pi_{ij})$$

Since the first term on the right-hand side of the above expression is constant for a fixed n, the model can be reexpressed in terms of either the underlying probabilities or the expected frequencies. The latter is often used because then maximum likelihood estimators are in terms of the observed frequencies n_{ij}.

The log-linear formulation for the 2×2 table in Table 6.1 in terms of l_{ij} is:

$$l_{ij} = \mu + \lambda_i^T + \lambda_j^R + \lambda_{ij}^{TR} \tag{6.1}$$

where the terms on the right-hand side of the above model are the parameters and correspond in order to the overall mean, the main effect of T at level i, the main effect of R at level j, and the interaction effects of T and R at level combination (i, j), respectively. For instance, the main effect for the Timolol treatment is the difference between the average $\bar{l}_{1+}$ and the overall average $\bar{l}_{++}$. In general, we define an estimate of the main effect of factor T at the ith level as:

$$\lambda_i^T = \bar{l}_{i+} - \bar{l}_{++}$$

Similarly, the estimate of the main effect of the jth level of factor R is defined as:

$$\lambda_j^R = \bar{l}_{+j} - \bar{l}_{++}$$

Model (6.1) above has too many parameters. We notice that there are at most four values of l_{ij}, but there are nine model parameters: $\mu, \lambda_1^T, \lambda_2^T, \lambda_1^R, \lambda_2^R, \lambda_{11}^{TR}, \lambda_{12}^{TR}$, λ_{21}^{TR}, and λ_{22}^{TR}. Since there are more parameters than underlying probabilities, this model is *overparameterized.* The above implies that we have four equations but nine unknowns, which would result in an infinite number of solutions.

To overcome this, we impose some *constraints* or *identifiability* conditions on the parameters of model (6.1). We describe below three forms of constraints usually imposed to solve this problem.

(i) Sum-to-zero constraints on the parameters. Here the parameters are constrained to sum to zero either row-wise or column-wise for main and interaction effects. We shall illustrate this in the next section. PROC CATMOD in SAS® employs this form.

(ii) Only the parameter of the last category of each variable and corresponding interaction terms are set to zero. This is the default approach employed by PROC GENMOD in SAS®.

(iii) Similar to **(ii)** except that the parameter of the first category of each variable and corresponding interaction terms are set to zero.[1]

We shall establish the correspondence of the first two approaches later in this chapter. We will however start with the sum-to-zero constraint approach, which is the most popular. In this case, the relevant constraint conditions for the log-linear formulation of the data in Table 6.1, viz. equation (6.1) are:

$$\sum_i \lambda_i^T = \sum_j \lambda_j^R = \sum_i \lambda_{ij}^{TR} = \sum_j \lambda_{ij}^{TR} = 0$$

The above conditions imply that

$$\lambda_1^T + \lambda_2^T = 0; \qquad \lambda_1^R + \lambda_2^R = 0; \qquad \lambda_{11}^{TR} + \lambda_{12}^{TR} = 0;$$
$$\lambda_{11}^{TR} + \lambda_{21}^{TR} = 0; \qquad \lambda_{12}^{TR} + \lambda_{22}^{TR} = 0$$

the sixth constraint $\lambda_{21}^{TR} + \lambda_{22}^{TR} = 0$ is not necessary because this constraint is the sum of the last three constraints involving the interaction terms.

With the above conditions, the number of parameters to be estimated is now equal to the number of equations minus the number of constraints, which in this case becomes $9 - 5 = 4$. Model with as many parameters as the number of cells in the table is called a *saturated model.* For example, the model in (6.1) with the above conditions imposed is a saturated model.

Solutions to these parameters lead to the following expressions for the parameter estimates:

$$\hat{\mu} = \frac{l_{++}}{4} \tag{6.2a}$$

$$\hat{\lambda}_i^T = \frac{l_{i+}}{2} - \frac{l_{++}}{4} \tag{6.2b}$$

$$\hat{\lambda}_j^R = \frac{l_{+j}}{2} - \frac{l_{++}}{4} \tag{6.2c}$$

$$\begin{aligned} \hat{\lambda}_{ij}^{TR} &= l_{ij} - \frac{l_{i+}}{2} - \frac{l_{+j}}{2} + \frac{l_{++}}{4} \\ &= l_{ij} - \bar{l}_{i+} - \bar{l}_{+j} - \bar{l}_{++} \end{aligned} \tag{6.2d}$$

[1]This is the default approach in GLIM. Parameterization in both (*ii*) and (*iii*) can be implemented in both GENMOD and GLIM.

where $l_{++} = \sum_{ij} \ln(n_{ij})$, $\quad l_{i+} = \sum_{j} \ln(n_{ij})$, and $l_{+j} = \sum_{i} \ln(n_{ij})$. Similarly, $\bar{l}_{i+}$, $\bar{l}_{+j}$, and $\bar{l}_{++}$ are respectively the averages of the row, column, and overall sums of the logarithms of the observations.

For the data in Table 6.1, we have in Table 6.2, the observed log of the frequency counts as well as the relevant means of the sums of the log-counts.

Treatment	Response 1	Response 2	Sum l_{i+}	Avg. $\bar{l}_{i+}$
1	$\ln(44) = 3.7842$	$\ln(116) = 4.7536$	8.5378	4.2689
2	$\ln(19) = 2.9444$	$\ln(128) = 4.8520$	7.7964	3.8982
Sum (l_{+j})	6.7286	9.6056	16.3342	-
Avg. ($\bar{l}_{+j}$)	3.3643	4.8028	-	4.0836

Table 6.2: Log of the observed counts $l_{ij} = \ln(n_{ij})$

In Table 6.2, for example, $8.5378 = 3.7842 + 4.7536$, $9.6056 = 4.7536 + 4.8520$, and $\bar{l}_{2+} = 7.7964/2 = 3.8982$.

From Table 6.2, and using equations (6.2a) to (6.2d), we have

$$\hat{\mu} = \frac{\ln(44) + \ln(116) + \ln(19) + \ln(128)}{4} = 4.0836$$

$$\hat{\lambda}_1^T = \frac{\ln(44) + \ln(116)}{2} - \hat{\mu} = 4.2689 - 4.0836 = 0.1853$$

$$\hat{\lambda}_1^R = \frac{\ln(44) + \ln(19)}{2} - \hat{\mu} = 3.3643 - 4.0836 = -0.7193$$

and

$$\hat{\lambda}_{11}^{TR} = \ln(44) - 4.2689 - 3.3643 + 4.0836 = 0.2346$$

Alternatively, we can obtain the latter as

$$\hat{\lambda}_{11}^{TR} = \ln(44) - \hat{\lambda}_1^T - \hat{\lambda}_1^R - \hat{\mu} = 3.7842 - 0.1853 - (-0.7193) - 4.0836 = 0.2346$$

We can similarly obtain $\hat{\lambda}_{ij}^{TR}$ by making use of the expressions in (6.2d) and substituting the relevant values. With the above estimates and utilizing the zero sum identifiability constraints, we have

$$\hat{\lambda}_2^T = -0.1853, \qquad \hat{\lambda}_2^R = 0.7193$$

and

$$\hat{\lambda}_{12}^{TR} = -\hat{\lambda}_{11}^{TR} = -0.2346 = \hat{\lambda}_{21}^{TR} = -\hat{\lambda}_{22}^{TR}$$

On the other hand, if we substitute the expected counts for l_{ij}, we obtain for $i = 1$.

$$\begin{aligned}\tilde{\lambda}_1^T &= \frac{\ln(\hat{m}_{11}) + \ln(\hat{m}_{12})}{2} - \frac{\ln(\hat{m}_{11}) + \ln(\hat{m}_{12}) + \ln(\hat{m}_{21}) + \ln(\hat{m}_{22})}{4} \\ &= \frac{\ln(\hat{m}_{11}) - \ln(\hat{m}_{21}) + \ln(\hat{m}_{12}) - \ln(\hat{m}_{22})}{4} = \frac{1}{4}\ln\left(\frac{\hat{m}_{11} \times \hat{m}_{12}}{\hat{m}_{21} \times \hat{m}_{22}}\right)\end{aligned}$$

Similarly for the classificatory variable R at $j = 1$,

$$\tilde{\lambda}_1^R = \frac{1}{4}\ln\left(\frac{\hat{m}_{11} \times \hat{m}_{21}}{\hat{m}_{12} \times \hat{m}_{22}}\right)$$

and the ML estimates (with the $\hat{m}$ terms replaced by the n_{ij}) are given for our example as

$$\hat{\lambda}_1^T = \frac{1}{4}\ln\left(\frac{n_{11} \times n_{12}}{n_{21} \times n_{22}}\right) = \frac{1}{4}\ln\left(\frac{44 \times 116}{19 \times 128}\right) = \frac{(0.74131)}{4} = 0.1853$$

$$\hat{\lambda}_1^R = \frac{1}{4}\ln\left(\frac{n_{11} \times n_{21}}{n_{12} \times n_{22}}\right) = \frac{1}{4}\ln\left(\frac{44 \times 19}{116 \times 128}\right) = \frac{(-2.87699)}{4} = -0.7192$$

The log-odds ratio for the data in Table 6.1 is computed as

$$\ln \hat{\theta} = \ln\left(\frac{44 \times 128}{19 \times 116}\right) = 0.9382 = 4(\hat{\lambda}_{11}^{TR})$$

The above result illustrates that the interaction term "parameter" estimate of the log-linear model is a function of the corresponding log odds ratio.

6.2.1 Estimates From other Constraints

Under PROC GENMOD constraints, we would have the following corresponding solutions.

$$\begin{aligned}
\breve{\mu} &= \ln(n_{22}) &&= l_{22} &&= 4.8520 \\
\breve{\lambda}_1^T &= l_{12} - l_{22} &&= 4.7536 - 4.8520 &&= -0.0984 \\
\breve{\lambda}_1^R &= l_{21} - l_{22} &&= 2.9444 - 4.8520 &&= -1.9076 \\
\breve{\lambda}_{11}^{TR} &= l_{11} - l_{12} - l_{21} + l_{22} &&= 0.9382
\end{aligned}$$

with

$$\breve{\lambda}_2^T = 0.0, \quad \breve{\lambda}_2^R = 0.0; \quad \text{and} \quad \breve{\lambda}_{12}^{TR} = \breve{\lambda}_{21}^{TR} = \breve{\lambda}_{22}^{TR} = 0.0$$

We shall give a general formula for estimating parameters for the general case later in this chapter.

6.2.2 Standard Errors of Parameter Estimates

For the sum to zero constraints approach, the expression for the estimator of the interaction term λ_{ij}^{TR} is:

$$\hat{\lambda}_{ij}^{TR} = l_{ij} - \frac{l_{i+}}{2} - \frac{l_{+j}}{2} + \frac{l_{++}}{4}$$

where the l's are as defined in the previous section. Letting $h_{ij} = \ln(n_{ij})$ be the logs of the observed frequencies, then we have in particular,

$$\begin{aligned}
\hat{\lambda}_{11}^{TR} &= h_{11} - \frac{h_{1+}}{2} - \frac{h_{+1}}{2} + \frac{h_{++}}{4} \\
&= h_{11} - \frac{h_{11} + h_{12}}{2} - \frac{h_{11} + h_{21}}{2} + \frac{h_{11} + h_{12} + h_{21} + h_{22}}{4} \\
&= \frac{1}{4}(h_{11} + h_{22} - h_{12} - h_{21})
\end{aligned}$$

As shown in chapter 4 using the delta method, the estimate of the asymptotic variance of $\ln(n_{ij}) = h_{ij}$ is

$$\widehat{Var}(h_{ij}) = \frac{1}{n_{ij}}$$

and in general, any estimator of the λ parameters can be written as a linear combination (a contrast) of the log of the observed cells as

$$\hat{\lambda} = \sum_i \sum_j a_{ij} h_{ij}$$

where the $\{a_{ij}\}$ satisfy $\sum_{ij} a_{ij} = 0$. Consequently, an estimate of the asymptotic variance is given by

$$\widehat{Var}(\hat{\lambda}) = \sum_i \sum_j \frac{a_{ij}^2}{n_{ij}}$$

The asymptotic standard error (a.s.e) for $\hat{\lambda}$ therefore equals:

$$\text{a.s.e.}(\hat{\lambda}) = \sqrt{\sum_i \sum_j \frac{a_{ij}^2}{n_{ij}}}$$

For any particular saturated model, the estimated asymptotic parameter variances need not all be the same. It generally depends on the number of categories for each variables. Thus to "put the λ's on equal footing" (Upton, 1982), it is necessary to standardize them so that the standardized value Z has variance 1.

$$Z(\hat{\lambda}) = \frac{\hat{\lambda}}{\sqrt{\widehat{Var}(\hat{\lambda})}} \tag{6.3}$$

Similar to the standardized residuals discussed in Chapter 5, a λ parameter will be considered important if $|Z(\hat{\lambda})| \geq 2.0$ (the upper 5% point of a standard normal) or $[Z(\hat{\lambda})]^2$ can be compared with the upper tail of a χ^2 distribution with 1 degree of freedom.

In the example above, $\{a_{ij}\} = \pm\frac{1}{4}$ for each of the λ's, and estimates of the asymptotic variances are given by

$$\frac{1}{16}\left(\frac{1}{44} + \frac{1}{116} + \frac{1}{19} + \frac{1}{128}\right) = 0.00574$$

The asymptotic standard error equals $\sqrt{0.00574} = 0.0758$. The Z's for the effects and interactions are $Z(T) = 0.1853/0.0758 = 2.445$, $Z(R) = -0.7193/0.0758 = -9.489$, and $Z(TG) = 0.2346/0.0758 = 3.095$. We notice immediately that all the effects have $|Z(.)| > 2$. We shall explore this result further, later in the chapter.

A $100(1-\alpha)\%$ asymptotic confidence interval may also be obtained using a standard normal approximation. For example, for the T-effect, a 95% confidence interval equals

$$0.1853 \pm 1.96(0.0758) = (0.0367, 0.3339)$$

Similar results may be obtained for the other effects and interaction terms.
Under the GENMOD constraints, the estimated asymptotic standard errors are:

$$\text{ase}[\hat{\lambda}_1^T] = \left(\frac{1}{n_{12}} + \frac{1}{n_{22}}\right)^{\frac{1}{2}} = 0.1282$$

$$\text{ase}[\hat{\lambda}_1^R] = \left(\frac{1}{n_{21}} + \frac{1}{n_{22}}\right)^{\frac{1}{2}} = 0.2459$$

$$\text{ase}[\hat{\lambda}_1^{TR}] = \left(\frac{1}{n_{11}} + \frac{1}{n_{12}} + \frac{1}{n_{21}} + \frac{1}{n_{22}}\right)^{\frac{1}{2}} = 0.3030$$

We present below the SAS software program and relevant output from PROC CATMOD for implementing the saturated model for the data in Table 6.1. The parameter estimates and their asymptotic standard error estimates agree with the ones computed from the previous sections.

```
data tab61;
/read in row, column variable names
 along with the corresponding count
the $ sign indicates we are reading the levels in
alphanumeric format.
/
input t $ r $ count @@;
datalines;
timo free 44 timo not 116 pcebo free 19 pcebo not 128
;
proc catmod order=data; weight count;
model t*r=_response_/ml;
loglin t|r; run;

OUTPUT
------
The CATMOD Procedure

Response            T*R        Response Levels     4
Weight Variable     COUNT      Populations         1
Data Set            TAB61      Total Frequency   307
Frequency Missing   0          Observations        4

    Maximum Likelihood Analysis of Variance

Source                DF   Chi-Square    Pr > ChiSq
--------------------------------------------------
T                      1         5.99        0.0144
R                      1        90.17        <.0001
T*R                    1         9.59        0.0020

Likelihood Ratio       0          .             .

                Analysis of Maximum Likelihood Estimates

                                      Standard        Chi-
Effect         Parameter   Estimate      Error      Square    Pr > ChiSq
-----------------------------------------------------------------------
T                      1     0.1853     0.0757        5.99        0.0144
R                      2    -0.7192     0.0757       90.17        <.0001
T*R                    3     0.2345     0.0757        9.59        0.0020
```

We observe here that the estimates produced by PROC CATMOD for effect R and interaction term TR above are different in the fourth decimal place from those earlier obtained. These differences are due to rounding errors.

The following are the corresponding SAS software program and relevant output for implementing the saturated model using PROC GENMOD.

```
set tab61;
proc genmod order=data; class t r;
model count=t|r/dist=poi link=log type3; run;

PARTIAL OUTPUT

The GENMOD Procedure
       Model Information

Data Set                 WORK.TAB61
Distribution                Poisson
Link Function                   Log
Dependent Variable            COUNT
Observations Used                 4
```

```
     Class Level Information
Class       Levels     Values

T                2     TIMO PCEBO
R                2     FREE NOT
           Criteria For Assessing Goodness Of Fit

Criterion                   DF             Value          Value/DF
--------------------------------------------------------------------
Deviance                     0            0.0000              .
Pearson Chi-Square           0            0.0000              .

                                 Analysis Of Parameter Estimates

                                         Standard       Wald 95%            Chi-
Parameter                  DF  Estimate     Error   Confidence Limits    Square  Pr > ChiSq
--------------------------------------------------------------------------------------------
Intercept                   1    4.8520    0.0884    4.6788    5.0253   3013.40      <.0001
T          TIMO             1   -0.0984    0.1282   -0.3497    0.1528      0.59      0.4425
T          PCEBO            0    0.0000    0.0000    0.0000    0.0000       .           .
R          FREE             1   -1.9076    0.2459   -2.3895   -1.4257     60.20      <.0001
R          NOT              0    0.0000    0.0000    0.0000    0.0000       .           .
T*R        TIMO     FREE    1    0.9382    0.3030    0.3444    1.5320      9.59      0.0020
T*R        TIMO     NOT     0    0.0000    0.0000    0.0000    0.0000       .           .
T*R        PCEBO    FREE    0    0.0000    0.0000    0.0000    0.0000       .           .
T*R        PCEBO    NOT     0    0.0000    0.0000    0.0000    0.0000       .           .
Scale                       0    1.0000    0.0000    1.0000    1.0000

      LR Statistics For Type 3 Analysis

                              Chi-
Source            DF        Square     Pr > ChiSq
-------------------------------------------------
T                  1          6.32         0.0119
R                  1        119.73         <.0001
T*R                1         10.24         0.0014
```

The saturated model is most useful in initial analysis to determine which effects or interaction terms are sufficiently important for a proper understanding of the data. In this example, our results from the sum-to-zero constraints indicate that all the three effects are important for future model consideration, while the results from the last-equal-zero constraint indicate that only the R effect and the interaction (TR) effect are important. We examine these further in the next section. Note that, the model degrees of freedom is zero because this is a saturated model.

6.2.3 Independence and Other Models for the 2×2 Table

The model in (6.1) with the identifiablility constraints imposed contained as many parameters as there are cells. Consequently, the model always fits perfectly. However, if we wish to find simpler models that may also explain the variations in our data, we would like to test models other than the saturated one. One of these possible models is the model based on the hypothesis of independence. For this case, we would have to set the interaction term $\lambda_{ij}^{TR} = 0$, such that (6.1) now becomes

$$l_{ij} = \mu + \lambda_i^T + \lambda_j^R \tag{6.4}$$

Under the null hypothesis of independence H_0, we recall that

$$\hat{m}_{ij} = \frac{n_{i+}n_{+j}}{n}$$

The relevant likelihood ratio test statistic G^2 can then be calculated and tests made. Other possible models are:

$$l_{ij} = \mu + \lambda_i^T \tag{6.5a}$$
$$l_{ij} = \mu + \lambda_j^R \tag{6.5b}$$
$$l_{ij} = \mu \tag{6.5c}$$

In (6.5a), we postulate that the R categories are equally probable. This implies that T and R are independent. Similarly, we could postulate that the categories of T are equally probable, leading to model (6.5b). We may finally postulate that all the categories are equally probable, which leads to the model in (6.5c). Table 6.3 lists the expected values under these models, their respective G^2 values, and the corresponding degrees of freedom. The models in (6.4), (6.5a), (6.5b), and (6.5c) are respectively designated as {T,R}, {T}, {R}, and {μ}. Table 6.3 gives the results of implementing these models on the data in Table 6.1.

		Models				
		(6.1)	(8.3)	(6.5a)	(6.5b)	(6.5c)
Cells	n_{ij}		{T,R}	{T}	{R}	{μ}
1 1	44	44.00	32.83	80.00	31.45	76.75
1 2	116	116.00	127.17	80.00	122.00	76.75
2 1	19	19.00	30.17	73.5	31.45	76.75
2 2	128	128.00	116.83	73.5	122.00	76.75
G^2		0.0	10.24	124.20	10.79	124.75
df		0	1	2	2	3

Table 6.3: Results when models are applied to data in Table 6.1

The models in (6.4), (6.5a) - (6.5c) are respectively implemented in SAS software with the following statements and partial SAS software outputs from each implementation to two decimal places only (see the format statement).

```
set tab61; proc genmod; make 'obstats' out=aa;
class t r; model count=t|r/dist=poi link=log obstats; run;
proc print data=aa noobs; var  t r count pred; format pred 8.2; run;

                t         r               count          Pred
                -----------------------------------------
                timo      free               44         44.00
                timo      not               116        116.00
                pcebo     free               19         19.00
                pcebo     not               128        128.00
***model 4***;
proc genmod order=data; class t r;
model count=t r/dist=poi obstats; run;

                t         r               count          Pred
                -----------------------------------------
                timo      free               44         32.83
                timo      not               116        127.17
                pcebo     free               19         30.17
                pcebo     not               128        116.83
***fit model 5a***;
proc genmod order=data; class t r;
model count=t/dist=poi; run;

                t         r               count          Pred
                -----------------------------------------
                timo      free               44         80.00
                timo      not               116         80.00
                pcebo     free               19         73.50
                pcebo     not               128         73.50
```

```
***fit model 5b***;
proc genmod order=data; class t r;
model count=r/dist=poi; run;
                 t          r              count          Pred
                 ----------------------------------------------
                 timo       free              44         31.50
                 timo       not              116        122.00
                 pcebo      free              19         31.50
                 pcebo      not              128        122.00

***fit model 5c***;
proc genmod order=data;
model count=/dist=poi; run;
                 t          r              count          Pred
                 ----------------------------------------------
                 timo       free              44         76.75
                 timo       not              116         76.75
                 pcebo      free              19         76.75
                 pcebo      not              128         76.75
```

6.3 Log-Linear Models for $I \times J$ Contingency Tables

We now consider an extension of the above formulation to the general twoway $I \times J$ tables. If we denote the row and column classificatory variables by A and B, respectively, then let A be indexed by $i = 1, 2, \cdots, I$ and let B be similarly indexed by $j = 1, 2, \cdots, J$. For either multinomial or product-multinomial sampling schemes, the log-linear formulation for the saturated model in an $I \times J$ table is given by

$$l_{ij} = \mu + \lambda_i^A + \lambda_j^B + \lambda_{ij}^{AB}, \qquad i = 1, \cdots, I; j = 1, \cdots, J \tag{6.6}$$

and where $l_{ij} = \ln(n_{ij})$. We again impose the sum-to-zero identifiability conditions

$$\sum_i \lambda_i^A = \sum_j \lambda_j^B = \sum_i \lambda_{ij}^{AB} = \sum_j \lambda_{ij}^{AB} = 0$$

The model now has as many parameters (IJ) as the total number of cells in the table, namely, $1 + (I-1) + (J-1) + (I-1)(J-1) = IJ$.

Then the parameter estimates are given by the following expressions:

$$\hat{\mu} = \frac{l_{++}}{IJ} \tag{6.7a}$$

$$\hat{\lambda}_i^A = \frac{l_{i+}}{J} - \frac{l_{++}}{IJ} \tag{6.7b}$$

$$\hat{\lambda}_j^B = \frac{l_{+j}}{I} - \frac{l_{++}}{IJ} \tag{6.7c}$$

$$\begin{aligned} \hat{\lambda}_{ij}^{AB} &= l_{ij} - \frac{l_{i+}}{J} - \frac{l_{+j}}{I} + \frac{l_{++}}{IJ} \\ &= l_{ij} - (\hat{\mu} + \hat{\lambda}_i^A + \hat{\lambda}_j^B) \end{aligned} \tag{6.7d}$$

where $l_{++} = \sum_{ij} \ln(n_{ij}); l_{i+} = \sum_j \ln(n_{ij})$ and $l_{+j} = \sum_i \ln(n_{ij})$.

If any of the observed counts n_{ij} equals zero, the standard practice is to use instead

$$l_{ij} = \ln(n_{ij} + 0.5)$$

That is, we add 0.5 to each observed value n_{ij}. This is necessary for the implementation of the saturated model in PROC CATMOD. The alternative to adding 0.5 to our data values when some have zero frequencies is to fit an unsaturated or reduced model to the data. For reduced models, the expected values are only functions of the marginal totals rather than the individual cell counts. With PROC GENMOD in SAS®, adding 0.5 to the observed data before fitting the saturated model is not necessary.

6.3.1 Example

The data in Table 6.4 (Hildebrand & Ott, 1991) relate to the popularity of three alternative flexible time-scheduling plans among clerical workers in four different offices. A random sample of 216 workers yields the following counts:

	Office				
Favored Plan	1	2	3	4	Total
1	15	32	18	5	70
2	8	29	23	18	78
3	1	20	25	22	68
Total	24	81	66	45	216

Table 6.4: Popularity of scheduling times among clerical workers

The corresponding natural logarithmm of the observed counts are displayed in Table 6.5. The results are given to three decimal places for brevity and clarity.

	Office					
Favored Plan	1	2	3	4	l_{i+}	$\bar{l}_{i+}$
1	2.708	3.466	2.890	1.609	10.673	2.668
2	2.079	3.367	3.136	2.890	11.472	2.868
3	0.000	2.996	3.219	3.091	9.306	2.327
l_{+j}	4.787	9.829	9.245	7.590	31.451	-
$\bar{l}_{+j}$	1.596	3.276	3.082	2.530	-	2.621

Table 6.5: Log of the observed counts $l_{ij} = \ln(n_{ij})$

In this example, $I = 3$ and $J = 4$; hence there are $IJ = 12$ cells in the table. Thus,

$$\hat{\lambda}_1^A = \bar{l}_{1+} - \bar{l}_{++} = 2.668 - 2.621 = 0.047$$

Similarly,

$$\hat{\lambda}_2^A = 2.868 - 2.621 = 0.247$$

and

$$\hat{\lambda}_3^A = -(0.047 + 0.247) \quad = -0.294$$

The latter result is obtained as a result of the sum-to-zero identifiability constraints.

Similarly, for the offices (B),

$$\hat{\lambda}_1^B = 1.596 - 2.621 \quad = -1.025$$

$$\hat{\lambda}_2^B = 3.276 - 2.621 \quad = 0.655$$

$$\hat{\lambda}_3^B = 3.082 - 2.621 \quad = 0.461$$

Again, because of the sum-to-zero constraints,

$$\hat{\lambda}_4^B = -(-1.025 + 0.655 + 0.461) = -0.091$$

For the interaction terms,

$$\hat{\lambda}_{11}^{AB} = l_{11} - \bar{l}_{1+} - \bar{l}_{+1} + \bar{l}_{++} = 2.708 - 2.668 - 1.596 + 2.621 = 1.065$$

Similarly,

$$\hat{\lambda}_{12}^{AB} = 3.466 - 2.668 - 3.276 + 2.621 = 0.143$$

These maximum likelihood (ML) estimates of the λ's are given in Table 6.6.

	Office				
Favored Plan	1	2	3	4	$\hat{\lambda}_i^A$
1	1.065	0.143	-0.239	-0.969	0.047
2	0.236	-0.156	-0.193	0.113	0.247
3	-1.301	0.013	0.432	0.856	-0.294
$\hat{\lambda}_j^B$	-1.025	0.655	0.461	-0.091	2.621

Table 6.6: ML Estimates of the λ_{ij}^{AB} parameters

Note that the parameter estimate $\hat{\lambda}_{31}^{AB} = -1.301$ is relatively small and $\hat{\lambda}_{11}^{AB} = 1.065$ relatively large. These correspond to a deficit among office type 1 respondents favoring plan 3 and an excess among office type 1 respondents favoring plan 1 respectively. We see that the results from a saturated log-linear model formulation allow us to characterize explicitly the variations among the table categories. Below is the corresponding SAS software output and program from PROC CATMOD. The results in Table 6.6 agree with the SAS software output under the sum-to-zero constraints only to the second place decimal with the SAS software results because of rounding errors. For example, $\hat{\lambda}_{12}^{AB} = 0.143$, but in SAS software output, it is 0.1421.

```
data tab64;
***we define labels here but will not be used
in the output ***;
label a ='plan'
      b ='office';
do a=1 to 3;
 do b=1 to 4;
input count @@;
output;
  end;
     end;
datalines;
15 32 18 5 8 29 23 18 1 20 25 22
;
proc catmod;
weight count;
model a*b=_response_/ml;
loglin a|b;
run;

The CATMOD Procedure

Response             a*b       Response Levels    12
Weight Variable      count     Populations         1
Data Set             TAB64     Total Frequency   216
Frequency Missing    0         Observations       12
```

Maximum Likelihood Analysis of Variance

Source	DF	Chi-Square	Pr > ChiSq
a	2	3.65	0.1611
b	3	27.85	<.0001
a*b	6	21.02	0.0018
Likelihood Ratio	0	.	.

Analysis of Maximum Likelihood Estimates

Effect	Parameter	Estimate	Standard Error	Chi-Square	Pr > ChiSq
a	1	0.0474	0.1396	0.12	0.7341
	2	0.2472	0.1323	3.49	0.0618
b	3	-1.0252	0.2797	13.43	0.0002
	4	0.6553	0.1362	23.14	<.0001
	5	0.4606	0.1409	10.69	0.0011
a*b	6	1.0648	0.3110	11.72	0.0006
	7	0.1421	0.1764	0.65	0.4207
	8	-0.2386	0.1910	1.56	0.2115
	9	0.2364	0.3232	0.54	0.4644
	10	-0.1561	0.1723	0.82	0.3649
	11	-0.1932	0.1802	1.15	0.2836

In the output above, the first column gives the effects (remember that a is the favored plan variable). Column 2 gives the parameter estimates. For instance, $\hat{\lambda}_1^A = 0.047$, while $\hat{\lambda}_2^A = 0.2472$. These are labeled as parameter 1 and 2, respectively, in column 2. Hence by the sum-to-zero constraint, $\hat{\lambda}_3^A$ would be -0.294 as displayed in Table 6.6. Similarly, the B main effects $\hat{\lambda}_1^B$, $\hat{\lambda}_2^B$, $\hat{\lambda}_3^B$ are designated as parameters 3, 4, and 5, respectively, in column 2. Corresponding $\hat{\lambda}_4^B$ can also be obtained from the sum-to-zero constraint. For the interaction term AB, the parameter estimates are given by the a*b effects. For instance, $\hat{\lambda}_{11}^{AB}, \hat{\lambda}_{12}^{AB}, \hat{\lambda}_{13}^{AB}, \hat{\lambda}_{21}^{AB}, \hat{\lambda}_{22}^{AB}, \hat{\lambda}_{23}^{AB}$ are labeled as 6, 7, 8, 9, 10, and 11, respectively. The others are again obtained from the use of the sum-to-zero constraints. Column 4 gives the estimated asymptotic standard errors for each of the parameter estimates presented. Column 5 gives the corresponding χ^2 statistics for testing the null hypothesis of the form $H_0 : \hat{\lambda} = 0$. For the $\hat{\lambda}_{11}^{AB}$ for instance, it is computed as $(1.0648/0.3110)^2 = 11.72$. The corresponding pvalue is given in the last column as 0.0006.

6.3.2 Asymptotic Standard Errors of Parameter Estimates

The asymptotic standard errors (a.s.e.) for the parameter estimates (under the sum-to-zero constraints) are not as simple to obtain in the $I \times J$ tables as for the 2×2 table. However, the principles remain basically the same.

To find the asymptotic standard error corresponding to the parameter estimate $\hat{\lambda}_1^A$ for instance, we observe that

$$\begin{aligned}\hat{\lambda}_1^A &= \frac{l_{11} + l_{12} + l_{13} + l_{14}}{4} - \frac{l_{11} + \cdots + l_{34}}{12} \\ &= \frac{(2l_{11} + 2l_{12} + 2l_{13} + 2l_{14}) - (l_{21} + l_{22} + \cdots + l_{34})}{12}\end{aligned}$$

hence,

$$\begin{aligned}\widehat{Var}(\hat{\lambda}_1^A) &= \frac{1}{144}\left[\frac{4}{n_{11}} + \frac{4}{n_{12}} + \frac{4}{n_{13}} + \frac{4}{n_{14}} + \frac{1}{n_{21}} + \frac{1}{n_{22}} + \cdots + \frac{1}{n_{34}}\right] \\ &= \frac{1}{144}\left[\frac{4}{15} + \frac{4}{32} + \frac{4}{18} + \frac{4}{5} + \frac{1}{8} + \frac{1}{29} + \cdots + \frac{1}{22}\right] \\ &= 0.0195\end{aligned}$$

The estimated a.s.e. therefore equals $\sqrt{0.0195} = 0.1396$ and the standardized value equals

$$\frac{\hat{\lambda}_1^A}{0.1396} = \frac{0.047}{0.1396} = 0.337$$

Similarly for the B's,

$$\begin{aligned}\hat{\lambda}_1^B &= \frac{l_{11} + l_{21} + l_{31}}{3} - \frac{l_{11} + \cdots + l_{34}}{12} \\ &= \frac{(3l_{11} + 3l_{21} + 3l_{31}) - (l_{12} + \cdots + l_{22} + \cdots + l_{32} + \cdots + l_{34})}{12}. \\ \widehat{Var}(\hat{\lambda}_1^B) &= \frac{1}{144}\left[\frac{9}{n_{11}} + \frac{9}{n_{21}} + \frac{9}{n_{31}} + \frac{1}{n_{12}} + \frac{1}{n_{13}} + \cdots + \frac{1}{n_{34}}\right] \\ &= \frac{1}{144}\left[\frac{9}{15} + \frac{9}{8} + \frac{9}{1} + \frac{1}{32} + \frac{1}{18} + \cdots + \frac{1}{22}\right] \\ &= 0.0783\end{aligned}$$

and the corresponding estimated asymptotic standard error equals $\sqrt{0.0783} = 0.2799$. Similar calculations yield the results for $\hat{\lambda}_2$.

For the interaction terms, the asymptotic standard error for $\hat{\lambda}_{11}^{AB}$, for instance, can be computed from the fact that:

$$\begin{aligned}\hat{\lambda}_{11}^{AB} &= l_{11} - \frac{l_{11} + l_{12} + l_{13} + l_{14}}{4} - \frac{l_{11} + l_{21} + l_{31}}{3} \\ &\quad - \frac{l_{11} + \cdots + l_{34}}{12} \\ &= \frac{6l_{11} - 2(l_{12} + l_{13} + l_{14}) - 3(l_{21} + l_{31})}{12} \\ &\quad + \frac{(l_{22} + l_{23} + \cdots + l_{34})}{12}\end{aligned}$$

Hence,

$$\begin{aligned}\widehat{Var}(\hat{\lambda}_{11}^{AB}) &= \frac{1}{144}\left[\frac{36}{n_{11}} + \frac{4}{n_{12}} + \frac{4}{n_{13}} + \frac{4}{n_{14}} + \frac{9}{n_{21}} + \frac{9}{n_{31}}\right] \\ &\quad + \frac{1}{144}\left[\frac{1}{n_{22}} + \frac{1}{n_{23}} + \cdots + \frac{1}{n_{34}}\right] \\ &= \frac{1}{144}\left[\frac{36}{15} + \frac{4}{32} + \frac{4}{18} + \frac{4}{5} + + \frac{9}{8} + \frac{9}{1}\right] \\ &\quad + \frac{1}{144}\left[\frac{1}{29} + \frac{1}{23} + \cdots + \frac{1}{22}\right] \\ &= 0.0968\end{aligned}$$

The estimated a.s.e. again equals $\sqrt{0.0968} = 0.3112$.

In general, for an $I \times J$ contingency table, the estimates of the asymptotic variances of the λ parameter estimates in a saturated model (under the sum-to-zero constraints) are given by:

$$\widehat{Var}(\hat{\lambda}_r^A) = \frac{1}{(IJ)^2}\left[\alpha^2 \sum_{j=1}^{J} \frac{1}{n_{rj}} + \sum_{i \neq r}^{I} \sum_{j=1}^{J} \frac{1}{n_{ij}}\right] \tag{6.8a}$$

$$\widehat{Var}(\hat{\lambda}_c^B) = \frac{1}{(IJ)^2}\left[\beta^2 \sum_{i=1}^{I} \frac{1}{n_{ic}} + \sum_{i=1}^{I} \sum_{j \neq c}^{J} \frac{1}{n_{ij}}\right] \tag{6.8b}$$

$$\widehat{Var}(\hat{\lambda}_{rc}^{AB}) = \frac{1}{(IJ)^2}\left[\frac{(\alpha\beta)^2}{n_{rc}} + \alpha^2 \sum_{j \neq c}^{J} \frac{1}{n_{rj}} + \beta^2 \sum_{i \neq r}^{I} \frac{1}{n_{ic}} + \sum_{i=c}^{I} \sum_{j \neq r}^{J} \frac{1}{n_{ij}}\right] \tag{6.8c}$$

where $\alpha = (I - 1)$, $\beta = (J - 1)$, $r = 1, \cdots, I$ and $c = 1, \cdots, J$.

	Office				
Favored Plan	1	2	3	4	Totals
1	0.0667	0.0313	0.0556	0.2000	0.3536
2	0.1250	0.0345	0.0435	0.0556	0.2586
3	1.0000	0.0500	0.0400	0.0455	1.1355
Totals	1.1917	0.1158	0.1391	0.3011	1.7477

Table 6.7: Table of reciprocals of cell counts

We give below examples of the use of these expressions. Suppose we wish to find the estimated asymptotic standard errors (a.s.e.) for λ_2^A, λ_3^B, and λ_{12}^{AB} for the data in Table 6.4, whose reciprocals are displayed in Table 6.7. From this table, we have $\alpha = 2$, $\beta = 3$, and $\alpha\beta = 6$. For λ_2^A, $r = 2$ and we have from (6.8a) and using the table of cell counts reciprocals in Table 6.7:

$$\text{a.s.e.} = \sqrt{\frac{4(0.2586) + (1.7477 - 0.2586)}{144}} = 0.1324$$

Similarly for λ_3^B, $c = 3$ and once again using (6.8b), we have

$$\text{a.s.e.} = \sqrt{\frac{9(0.1391) + (1.7477 - 0.1391)}{144}} = 0.1409$$

For the λ_{12}^{AB}, we have $r = 1, c = 2$, and using (6.8c), we also have

$$\text{a.s.e.} = \sqrt{\frac{36(0.0313) + 4(0.3536 - 0.0313) + 9(0.1158 - 0.0313) + 1.3096}{144}}$$
$$= 0.1765.$$

where

$$1.3096 = 1.7477 - 0.3536 - 0.1158 + 0.0313$$

Similar computations can be carried out to obtain other estimates of the asymptotic standard errors.

6.3.3 Estimates Based on SAS® PROC GENMOD

For the GLM constraints where the last category of each variable and corresponding interaction terms are set to zero, we have for the data in Table 6.4.

$$\begin{aligned}
\hat{\mu} &= \ln(n_{34}) &&= l_{34} &&= 3.0910 \\
\hat{\lambda}_1^A &= l_{14} - l_{34} &&= 1.6094 - 3.0910 &&= -1.4816 \\
\hat{\lambda}_2^A &= l_{24} - l_{34} &&= 2.8904 - 3.0910 &&= -0.2006 \\
\hat{\lambda}_3^A &= 0
\end{aligned}$$

Similarly,

$$\begin{aligned}
\hat{\lambda}_1^B &= l_{31} - l_{34} &&= 0.0 - 3.0910 &&= -3.0910 \\
\hat{\lambda}_2^B &= l_{32} - l_{34} &&= 2.9957 - 3.0910 &&= -0.0953 \\
\hat{\lambda}_3^B &= l_{33} - l_{34} &&= 3.2189 - 3.0910 &&= 0.1279 \\
\hat{\lambda}_4^B &= 0
\end{aligned}$$

For the interaction terms,

$$\begin{aligned}
\hat{\lambda}_{11}^{AB} &= l_{11} - l_{14} - l_{31} + l_{34} &&= 4.1897 \\
\hat{\lambda}_{12}^{AB} &= l_{12} - l_{14} - l_{32} + l_{34} &&= 1.9516 \\
\hat{\lambda}_{13}^{AB} &= l_{13} - l_{14} - l_{33} + l_{34} &&= 1.1531 \\
\hat{\lambda}_{12}^{AB} &= l_{12} - l_{14} - l_{32} + l_{34} &&= 1.9516 \\
\vdots &= \quad \vdots &&= \vdots \\
\hat{\lambda}_{23}^{AB} &= l_{23} - l_{24} - l_{33} + l_{34} &&= 0.1173
\end{aligned}$$

where $\hat{\lambda}_{14}^{AB} = \hat{\lambda}_{24}^{AB} = \hat{\lambda}_{34}^{AB} = 0$ and $\hat{\lambda}_{31}^{AB} = \hat{\lambda}_{32}^{AB} = \hat{\lambda}_{33}^{AB} = 0$. Below is the SAS software implementation using PROC GENMOD.

```
set tab64; proc genmod; class a b;
model count=a|b/dist=poi link=log type3; run;
The GENMOD Procedure
   Class Level Information
Class       Levels     Values

a                3     1 2 3
b                4     1 2 3 4
```

Analysis Of Parameter Estimates

Parameter			DF	Estimate	Standard Error	Wald 95% Confidence Limits		Chi-Square	Pr > ChiSq
Intercept			1	3.0910	0.2132	2.6732	3.5089	210.20	<.0001
a	1		1	-1.4816	0.4954	-2.4526	-0.5106	8.94	0.0028
a	2		1	-0.2007	0.3178	-0.8236	0.4222	0.40	0.5278
a	3		0	0.0000	0.0000	0.0000	0.0000	.	.
b	1		1	-3.0910	1.0225	-5.0951	-1.0870	9.14	0.0025
b	2		1	-0.0953	0.3090	-0.7009	0.5102	0.10	0.7577
b	3		1	0.1278	0.2923	-0.4451	0.7008	0.19	0.6619
b	4		0	0.0000	0.0000	0.0000	0.0000	.	.
a*b	1	1	1	4.1897	1.1455	1.9446	6.4348	13.38	0.0003
a*b	1	2	1	1.9516	0.5716	0.8313	3.0719	11.66	0.0006
a*b	1	3	1	1.1531	0.5840	0.0086	2.2976	3.90	0.0483
a*b	1	4	0	0.0000	0.0000	0.0000	0.0000	.	.
a*b	2	1	1	2.2801	1.1073	0.1099	4.4503	4.24	0.0395
a*b	2	2	1	0.5722	0.4307	-0.2719	1.4164	1.77	0.1840
a*b	2	3	1	0.1173	0.4295	-0.7246	0.9591	0.07	0.7848

```
            LR Statistics For Type 3 Analysis

                                Chi-
Source              DF        Square     Pr > ChiSq
----------------------------------------------------
a                    2          4.56         0.1024
b                    3         46.75         <.0001
a*b                  6         30.59         <.0001
```

The model statement in PROC GENMOD requests for the fit of a saturated model and the **type3** options requests for the partial tests of the effects as if the effect enters the model last. The interaction term is highly significant. We would expect from this analysis that the model of independence would not fit the data in Table 6.4. The type3 analysis presented is similar to the **Maximum Likelihood Analysis of Variance** presented in PROC CATMOD. Because of the identifiability constraints, we note that some of the parameters of a, b, and a*b effects are zero.

The output from PROC GENMOD above also gives in column 1 the parameter of interest. Immediately following the parameters are the parameter numbers, where 1 and 2 for A implies parameter estimates $\hat{\lambda}_1^A$ and $\hat{\lambda}_2^A$, respectively. We note that $\hat{\lambda}_3^A = 0$ in this case. The column headed DF gives the appropriated degrees of freedom for the parameter estimates. Similarly, the column headed standard error gives the asymptotic standard errors for each of the parameter estimates, while the Wald 95% C.I. gives the corresponding 95% confidence intervals for each of the parameter estimates. The chi-square column entries are obtained as explained earlier under PROC CATMOD and the last column gives the corresponding pvalues.

6.3.4 Estimating Asymptotic Standard Errors

The estimates of the asymptotic standard errors in PROC GENMOD are obtained for example as follows:

$$\text{a.s.e}(\hat{\lambda}_1^A) = \left(\frac{1}{n_{14}} + \frac{1}{n_{34}}\right)^{\frac{1}{2}} = 0.4955$$

$$\text{a.s.e}(\hat{\lambda}_2^B) = \left(\frac{1}{n_{32}} + \frac{1}{n_{34}}\right)^{\frac{1}{2}} = 0.3090$$

$$\text{a.s.e}(\hat{\lambda}_{22}^{AB}) = \left(\frac{1}{n_{22}} + \frac{1}{n_{24}} + \frac{1}{n_{32}} + \frac{1}{n_{34}}\right)^{\frac{1}{2}} = 0.4308$$

Again, these results agree with those produced by PROC GENMOD in SAS® displayed above.

6.3.5 Analysis of Parameter Estimates

We see that each of the λ parameter estimates can be written as a function of the log of the observed values. That is, if we let

$$\ell = \sum_{ij} a_{ij} \ln(n_{ij}); \quad \text{with} \quad \sum_{ij} a_{ij} = 0$$

then from previous results in chapter 5, the asymptotic estimated standard error of ℓ_{ij} is given by

$$\text{a.s.e.}(\ell) = \sqrt{\left(\sum_{ij} a_{ij}^2 n_{ij}^{-1}\right)}$$

and a test of any of the λ values being equal to zero can be conducted by the statistic

$$Z = \frac{\hat{\lambda}_{ij} - \lambda_{ij}}{\text{a.s.e}(\hat{\lambda}_{ij})} \sim N(0,1)$$

Under the null hypothesis, the test statistic for $H_0 : \lambda_{ij}^{AB} = 0$ versus $H_a : \lambda_{ij}^{AB} \neq 0$ is

$$Z = \frac{\hat{\lambda}_{ij}^{AB}}{\text{ase}(\hat{\lambda}_{ij}^{AB})}$$

The corresponding 95 % confidence interval for λ_{ij}^{AB} is given by

$$\hat{\lambda}_{ij}^{AB} \pm 1.96 \text{ ase}(\hat{\lambda}_{ij}^{AB})$$

For example, the test of the hypothesis that $\lambda_{11}^{AB} = 0$ gives a Z value of 3.42 with a corresponding 95% confidence interval

$$1.065 \pm 1.96(0.3112) = [0.455, 1.675]$$

Alternatively, Z^2 can be compared to a χ^2 distribution with one degree of freedom. SAS® PROC CATMOD and PROC GENMOD use this latter test.

6.3.6 Model of Independence

When the hypothesis of independence holds, the log-linear formulation reduces to

$$l_{ij} = \mu + \lambda_i^A + \lambda_j^B \tag{6.9}$$

From chapter 5, we have shown that irrespective of the sampling schemes

$$\hat{m}_{ij} = \frac{n_{i+}n_{+j}}{n} \quad \text{and} \tag{6.10a}$$

$$l_{ij} = -\ln(n) + \ln(n_{i+}) + \ln(n_{+j}) \tag{6.10b}$$

The formulation in (6.10a) can be seen to be equivalent to the formulation in (6.9) where $\mu = -\ln(n)$. That is, under the model of independence then, the formulation in (6.9) holds. On the other hand, given that the log-linear model (6.9) holds, can we show that $\hat{m}_{ij} = \dfrac{n_{i+}n_{+j}}{n}$?

Under the null hypothesis of independence, we can show that the above is always true and this is demonstrated below for the product-multinomial sampling scheme. If,

$$l_{ij} = \ln(m_{ij}) = \mu + \lambda_i^A + \lambda_j^B$$

holds, then,

$$\hat{m}_{ij} = e^{\mu + \lambda_i^A + \lambda_j^B} = \phi\phi_i\phi_j$$

where $\phi = e^{\mu}$; $\phi_i = e^{\lambda_i^A}$; and $\phi_j = e^{\lambda_j^B}$.

For the product-multinomial case, $\hat{m}_{ij} = n_{i+}\hat{\pi}_{*j}$, where $\hat{\pi}_{*j} = \frac{n_{+j}}{n}$, with $\sum_j \hat{\pi}_{*j} = 1$ and $\hat{m}_{i+} = n_{i+}$. Consequently,

$$\sum_j \hat{m}_{ij} = n_{i+} = \sum_j \phi\phi_i\phi_j$$

But

$$\hat{\pi}_{ij} = \hat{\pi}_{*j} = \frac{n_{+j}}{n} = \frac{\hat{m}_{ij}}{n_{i+}}$$

hence,

$$\hat{\pi}_{ij} = \frac{\phi\phi_i\phi_j}{N_{i+}} = \frac{\phi\phi_i\phi_j}{\sum_j \phi\phi_i\phi_j} = \frac{\phi_j}{\sum_j \phi_j}$$

since ϕ and ϕ_i are constants relative to the summation over j. The above is true for any i, so that

$$\frac{\phi_j}{\sum_j \phi_j} = \hat{\pi}_{*j}, \quad \text{for} \quad j = 1, 2, \cdots, J$$

In Table 6.8 are the results of fitting the models of independence {A,B}, {A}, and {B} to the data in Table 6.4, where model {A,B} implies the model containing the effects μ, A and B (this is the model of independence).

Models	df	G^2	p-value
{A,B}	6	30.586	< 0.001
{A}	9	67.428	< 0.001
{B}	8	31.355	< 0.001

Table 6.8: Results of fitting reduced models to the data in Table 6.4

Similarly, the model defined as {A} also implies the model containing μ and the effects of A and ditto for the model described by {B}. The degrees of freedom for models {A,B}, {A}, and {B} are computed respectively as follows:

$$\begin{aligned} \{A, B\} &: IJ - \{1 + (I-1) + (J-1)\} = (I-1)(J-1) \\ \{A\} &: IJ - \{1 + (I-1)\} = I(J-1) \quad \text{and} \\ \{B\} &: IJ - \{1 + (J-1)\} = J(I-1) \end{aligned}$$

The ML estimates under the various models are easily obtained. For instance, for the independence model {A,B}, we have

$$\hat{\lambda}_i^A = R_i^A - R_+$$
$$\hat{\lambda}_j^B = C_j^B - C_+$$

where R_i is the logarithm of the total for row i and C_j is similarly the logarithm of the total for column j. Further, $R_+ = \sum_i R_i$ and $C_+ = \sum_j C_i$. We leave the computations of these estimates and associated variances as an exercise to the reader.

We give below the SAS software codes required to fit the log-linear models presented in Table 6.8 to the data in Table 6.4.

```
set tab64;
***fit models using proc CATMOD***;
proc catmod; weight count;
model a*b=_response_/ml;
loglin a b;
run;
loglin a;
run;
loglin b;
run;
***fit models using proc GENMOD***;
```

```
proc genmod; class a b; make 'obstats' out=aa;
model count=a b/dist=poi link=log obstats; run;
proc print data=aa; var count pred xbeta resraw reschi streschi; run;
proc genmod; class a b; model count=a/dist=poi link=log; run;
proc genmod; class a b; model count=b/dist=poi link=log; run;
```

We give below partial log-linear model output from SAS software under the model of independence for the data in Table 6.4.

```
The CATMOD Procedure
     Maximum Likelihood Analysis of Variance

Source               DF   Chi-Square    Pr > ChiSq
--------------------------------------------------
a                     2         0.78        0.6782
b                     3        31.68        <.0001

Likelihood Ratio      6        30.59        <.0001

                Analysis of Maximum Likelihood Estimates
                                   Standard        Chi-
Effect         Parameter   Estimate      Error      Square    Pr > ChiSq
------------------------------------------------------------------------
a                      1    -0.0264     0.0970        0.07        0.7854
                       2     0.0818     0.0944        0.75        0.3864
b                      3    -0.7142     0.1629       19.21        <.0001
                       4     0.5022     0.1090       21.23        <.0001
                       5     0.2974     0.1153        6.66        0.0099

The GENMOD Procedure
            Criteria For Assessing Goodness Of Fit
Criterion                  DF            Value          Value/DF
----------------------------------------------------------------
Deviance                    6          30.5856            5.0976
Pearson Chi-Square          6          27.1350            4.5225
Log Likelihood                        427.1250

                     Analysis Of Parameter Estimates
                              Standard      Wald 95%            Chi-
Parameter     DF  Estimate       Error   Confidence Limits    Square  Pr > ChiSq
--------------------------------------------------------------------------------
Intercept      1    2.6509      0.1797    2.2987     3.0031   217.57      <.0001
a          1   1    0.0290      0.1703   -0.3047     0.3627     0.03      0.8648
a          2   1    0.1372      0.1659   -0.1880     0.4624     0.68      0.4083
a          3   0    0.0000      0.0000    0.0000     0.0000      .           .
b          1   1   -0.6286      0.2528   -1.1240    -0.1332     6.18      0.0129
b          2   1    0.5878      0.1859    0.2234     0.9522     9.99      0.0016
b          3   1    0.3830      0.1933    0.0041     0.7619     3.92      0.0476
b          4   0    0.0000      0.0000    0.0000     0.0000      .           .
Scale          0    1.0000      0.0000    1.0000     1.0000

                     Observation Statistics
Obs        count       Pred      Xbeta      Resraw      Reschi     Streschi
---------------------------------------------------------------------------
  1           15     7.7778     2.0513      7.2222      2.5897       3.3409
  2           32    26.2500     3.2677      5.7500      1.1223       1.7267
  3           18    21.3889     3.0629     -3.3889     -0.7328      -1.0695
  4            5    14.5833     2.6799     -9.5833     -2.5095      -3.4306
  5            8     8.6667     2.1595     -0.6667     -0.2265      -0.3005
  6           29    29.2500     3.3759     -0.2500     -0.0462      -0.0732
  7           23    23.8333     3.1711     -0.8333     -0.1707      -0.2563
  8           18    16.2500     2.7881      1.7500      0.4341       0.6104
  9            1     7.5556     2.0223     -6.5556     -2.3849      -3.0560
 10           20    25.5000     3.2387     -5.5000     -1.0892      -1.6644
 11           25    20.7778     3.0339      4.2222      0.9263       1.3428
 12           22    14.1667     2.6509      7.8333      2.0812       2.8258
```

The estimates of the parameters under the model of independence are parameterized and given in GENMOD as follows:

$$\hat{\lambda}_1^A - \hat{\lambda}_3^A = 0.0290; \quad \hat{\lambda}_2^A - \hat{\lambda}_3^A = 0.1372$$
$$\hat{\lambda}_1^B - \hat{\lambda}_4^B = -0.6286; \quad \hat{\lambda}_2^B - \hat{\lambda}_4^B = 0.5878$$
$$\hat{\lambda}_3^B - \hat{\lambda}_4^B = 0.3830$$

Thus to obtain the equivalent estimates based on PROC CATMOD (that is, under parameter sum-to-zero constraints), we note that by adding the two equations above involving the effects of A, we have

$$\hat{\lambda}_1^A + \hat{\lambda}_2^A - 2\hat{\lambda}_3^A = 0.0290 + 0.1372 \quad = 0.1662$$
$$-3\hat{\lambda}_3^A = 0.1662$$

The last expression is as a result of the constraint $\sum_{i=1}^{3} \hat{\lambda}_i^A = 0$. Hence,

$$\hat{\lambda}_3^A = -0.0554$$
$$\hat{\lambda}_1^A = 0.0290 + \hat{\lambda}_3^A \quad = -0.0264 \quad \text{and}$$
$$\hat{\lambda}_2^A = 0.1372 + \hat{\lambda}_3^A \quad = 0.0818$$

We can similarly use similar algebraic procedure to get the corresponding parameter estimates for the effects of B. These parameter estimates agree with those given by CATMOD. The above results therefore indicate the equivalence between the two parameterizations used by PROC CATMOD and PROC GENMOD in SAS®.

6.4 Interaction Analysis

The model of independence when applied to the data in Table 6.4 obviously does not fit the data. In order to determine which cells are responsible for the lack of fit of the independence model, we start by examining the standardized residuals z_{ij} or adjusted residuals r_{ij} under the model of independence. Either of these residuals designated in the SAS software output above as **reschi** and **Streschi**, respectively, indicated that cells $\{(1,1),(1,4),(3,1),(3,4)\}$ can be considered to be at variance with the null hypothesis of independence in view of their magnitudes being greater than 2.0.

We can test a model of quasi-independence by writing

$$l_{ij} = \begin{cases} \mu + \lambda_i^A + \lambda_j^B + \lambda_{ij}^{AB} & \text{for } (i,j) \in \text{S} \\ \mu + \lambda_i^A + \lambda_j^B & \text{for } (i,j) \notin \text{S} \end{cases}$$

where S is the set containing cells in which the interaction is significant. Thus the model of quasi-independence implies that the rows and columns are independent except in cells contained in S. We usually let the number of elements in S be as small as possible and generally $S < (I-1)(J-1)$.

For the model of quasi-independence, the expected values satisfy

$$\hat{m}_{ij} = \delta_{ij}\alpha_i\beta_j \tag{6.11}$$

where the α's and β's relate to the row and column variables, respectively, and $i = 1,2,..,I$, $j = 1,2,..,J$ with

$$\delta_{ij} = \begin{cases} 1 & \text{if } (i,j) \in \text{S} \\ 0 & \text{otherwise} \end{cases}$$

The MLE of the $\hat{m}_{ij}$ satisfy the marginal constraints

$$\hat{m}_{i+} = n_{i+} \quad i = 1, 2, .., I$$

$$\hat{m}_{+j} = n_{+j} \quad j = 1, 2, .., J$$

The Deming-Stephan iterative proportional fitting technique can then be used to obtain expected values under the model of quasi-independence. Basically, we start the iteration by setting at the 0th step

$$\hat{m}_{ij} = \delta_{ij} \tag{6.12}$$

for all (i, j). Then the ν-th cycle of the iteration has for all (i, j):

$$\hat{m}_{ij}^{(2\nu-1)} = \frac{\hat{m}_{ij}^{(2\nu-2)} n_{i+}}{\sum_k \hat{m}_{ik}^{(2\nu-2)}} \quad \text{and} \tag{6.13a}$$

$$\hat{m}_{ij}^{(2\nu)} = \frac{\hat{m}_{ij}^{(2\nu-1)} n_{+j}}{\sum_k \hat{m}_{ki}^{(2\nu-1)}} \tag{6.13b}$$

The iteration is continued for $\nu = 1, 2, \cdots$ until we achieve the desired accuracy. This procedure is demonstrated in the example below for the data in Table 6.4.

Example

For the data in Table 6.4, suppose we wish to fit the model of quasi-independence to the data with cells (1,4) and (3,1) removed (*these two cells have the largest absolute values of Z*). We would then have the reduced table, Table 6.9 where, observations in cells (1,4) and (3,1) have been deleted and these cells are accordingly being treated as structural zeros.

Favored Plan	**Office**				Total
	1	2	3	4	
1	15	32	18	-	65
2	8	29	23	18	78
3	-	20	25	22	67
Total	23	81	66	40	210

Table 6.9: Observed values in the reduced table

Here, the reduced marginal totals are:

$$n_{i+} = \{65, 78, 67\} \quad \text{and,} \quad n_{+j} = \{23, 81, 66, 40\}$$

and the initial estimates of the MLE are given by:

$$\hat{m}_{ij}^{(0)} = \begin{pmatrix} 1 & 1 & 1 & 0 \\ 1 & 1 & 1 & 1 \\ 0 & 1 & 1 & 1 \end{pmatrix}$$

For the first cycle of iteration, we start by first utilizing (6.13a), which matches the row n_{i+} marginal totals. That is,

$$\hat{m}_{ij}^{(1)} = \left[\begin{array}{cccc|c} 21.667 & 21.667 & 21.667 & - & 65 \\ 19.500 & 19.500 & 19.500 & 19.500 & 78 \\ - & 22.333 & 22.333 & 22.333 & 67 \\ \hline 41.167 & 63.500 & 63.500 & 41.167 & 210 \end{array}\right]$$

Now we can complete the first cycle of iteration by also matching the column marginal totals using (6.13b) with the reduced n_{+j} above. We notice that these marginal totals have been altered as a result of the first half of the cycle. The first cycle is completed with the new expected values below:

$$\hat{m}_{ij}^{(2)} = \left[\begin{array}{cccc|c} 12.105 & 27.638 & 22.520 & - & 62.263 \\ 10.895 & 24.874 & 20.268 & 18.645 & 74.682 \\ - & 28.488 & 23.213 & 21.355 & 73.055 \\ \hline 24.016 & 80.959 & 65.966 & 39.058 & 210.000 \end{array}\right]$$

The above completes the first iteration cycle, where

$$12.105 = \frac{21.667 \times 23}{41.167} \quad \text{and} \quad 24.874 = \frac{19.5 \times 81}{63.500} \quad \text{etc.}$$

The iteration continues in this way until we have convergence. By the seventh cycle we have convergence and the corresponding MLE are

$$\hat{m}_{ij}^{(14)} = \left[\begin{array}{cccc|c} 12.157 & 29.118 & 23.726 & - & 65.000 \\ 10.843 & 25.971 & 21.162 & 20.023 & 78.000 \\ - & 25.911 & 21.113 & 19.977 & 67.000 \\ \hline 23.000 & 81.000 & 66.000 & 40.000 & 210.000 \end{array}\right]$$

Below are the estimated expected values for the various models of quasi-independence considered. The model whose expected values have just been obtained is model 3 in the next table.

		Models				
		1	2	3	4	5
Cells	n_{ij}	$\hat{m}_{ij}$	$\hat{m}_{ij}$	$\hat{m}_{ij}$	$\hat{m}_{ij}$	$\hat{m}_{ij}$
1 1	15	7.780	10.878	12.157	10.878	12.292
1 2	32	26.25	24.942	29.118	25.541	30.796
1 3	18	21.390	20.323	23.726	19.392	21.912
1 4	5	14.58	13.887*	-	14.189*	-
2 1	8	8.67	12.122	10.843	12.122	10.708
2 2	29	29.25	27.792	25.971	28.459	26.828
2 3	23	23.830	22.646	21.162	21.608	19.088
2 4	18	16.25	15.440	20.023	15.811	21.375
3 1	1	7.56	-	-	-	-
3 2	20	25.50	28.266	25.911	27.000	23.375
3 3	25	20.75	23.031	21.113	-	-
3 4	22	14.17	15.703	19.977	15.000	18.625
G^2		30.585	18.178	6.269	17.786	4.676
df		6	5	4	4	3

Asterisk indicates cells with significant $\mid Z \mid$ values, and the n_{ij} are the observed cell counts. Models (1) to (5) are described below:

Model 1: This is the model of independence.

Model 2: This is the model of quasi-independence with cell (3,1) set to structural zero.

Model 3: This is the model of quasi-independence with cells (1,4) and (3,1) set to structural zeros.

Model 4: This is the model with cells (3,1) and (3,3) set to structural zeros based on our residual analysis earlier.

Model 5: This model sets cells (3,1), (3,2) and (1,4) to structural zeros because model 4 still has cell (1,4) with a significant standardized residual.

The basic approach to fitting the model of quasi-independence is to set cells having significant standardized residuals to zero in stages, beginning with the cell with the largest absolute value of Z in the model of independence. In our case, this was cell (3,1). Removing this cell and fitting a quasi-independence model (model 2) that does not fit the data because cell (1,4) still gives a significant $\mid Z \mid$ value. This cell is further removed, resulting in model 3, which now gives a fitted G^2 value of 6.269 on $(6 - 1 - 1) = 4$ degrees of freedom. This model fits the data well.

If we are interested in estimating the $\{\alpha\}$ and $\{\beta\}$ parameters that will ultimately lead to the estimation of the $\hat{m}_{ij}$ terms then, the following procedure, due to Goodman (1964, 1968), which is also presented in Bishop et al. (1975), is discussed.

The maximum likelihood equations above can be written as

$$\hat{\alpha}_i \sum_{j=1}^{J} \delta_{ij}\hat{\beta}_j = n_{i+}, \quad i = 1, 2, \cdots, I$$

$$\hat{\beta}_j \sum_{i=1}^{I} \delta_{ij}\hat{\alpha}_i = n_{+j} \quad j = 1, 2, \cdots, J$$

The above can be written succinctly as

$$\hat{\alpha}_i = \frac{n_{i+}}{\sum_{j=1}^{J} \delta_{ij}\hat{\beta}_j} \qquad i = 1, 2, \cdots, I$$

$$\hat{\beta}_j = \frac{n_{+j}}{\sum_{i=1}^{I} \delta_{ij}\hat{\alpha}_i} \qquad j = 1, 2, \cdots, J$$

The latter representation suggests an iterative procedure for estimating the $\{\hat{\alpha}_i\}$ and $\{\hat{\beta}_i\}$.

Thus if we begin by setting

$$\beta_j^0 = 1 \quad \text{for} \quad j = 1, 2, \cdots, J$$

and then continue at the ν-th cycle of the iteration $(\nu \geq 1)$ by setting

$$\hat{\alpha}_i^{(\nu)} = \frac{n_{i+}}{\sum_{j=1}^{J} \delta_{ij}\hat{\beta}_j^{(\nu-1)}} \quad \text{for} \quad i = 1, 2, \cdots, I$$

and

$$\hat{\beta}_j^{(\nu)} = \frac{n_{+i}}{\sum_{i=1}^{I} \delta_{ij}\hat{\alpha}_i^{(\nu)}} \quad \text{for} \quad j = 1, 2, \cdots, J$$

after the ν cycle, the estimates of the expected values are given by

$$\hat{m}_{ij}^{(2\nu)} = \delta_{ij}\alpha_i^{(\nu)}\beta_j^{(\nu)}$$

The model of quasi-independence can be implemented in SAS software as follows: In the first SAS software program and output below, we set cell (1,4) to a structural zero and fit a model of quasi-independence. This is accomplished in SAS software by creating a dummy variable **wt**. The results of this fit from the residual analyses indicate that while G^2 has been reduced considerably from 30.585 to 16.992 on 5 d.f., cells (1,1),(1,3), and (3,1) still exhibit significant residuals. This model corresponds to model 2 in the table above. We next set the cell with the largest absolute standardized residual to a structural zero and refit the model of quasi-independence. The next cell removed is cell (3,1), and the resulting model is model 3 above.

```
set tab64;
if a=1 and b=4 then wt=0;
else wt=1;
proc genmod;
make 'obstats' out=aa;
class a b wt;
model count=a b wt/dist=poi link=log obstats
          type3;
run;
proc print data=aa;
var count pred reschi streschi;
format pred reschi streschi 7.4;
run;
```

```
          Criteria For Assessing Goodness Of Fit

Criterion                 DF          Value       Value/DF
-----------------------------------------------------------
Deviance                   5        16.9922         3.3984
Scaled Deviance            5        16.9922         3.3984
Pearson Chi-Square         5        14.6320         2.9264
Scaled Pearson X2          5        14.6320         2.9264
Log Likelihood                     433.9217

Obs        count       Pred     Reschi    Streschi
--------------------------------------------------
  1           15     9.1228     1.9458     2.6656
  2           32    30.7895     0.2182     0.3819
  3           18    25.0877    -1.4151    -2.2936
  4            5     5.0000     0.0000     0.0000
  5            8     7.9481     0.0184     0.0239
  6           29    26.8248     0.4200     0.6539
  7           23    21.8572     0.2444     0.3604
  8           18    21.3699    -0.7290    -1.2536
  9            1     6.9291    -2.2524    -2.8333
 10           20    23.3857    -0.7001    -1.0511
 11           25    19.0550     1.3619     1.9392
 12           22    18.6301     0.7807     1.2536
```

We set cell (3,1) to structural zero and refitting, by creating again a dummy variable wt where

$$\text{wt} = \begin{cases} 1 & \text{for cell (1,4)} \\ 2 & \text{for cell (3,1)} \\ 0 & \text{elsewhere} \end{cases}$$

This is again implemented in SAS software as:

```
set tab64;
if a=1 and b=4 then wt=0;
else if a=3 and b=1 then wt=2;
else wt=1;
........
proc genmod; class a b wt;
model count=a b wt/dist=poi link=log obstats
          type3; run;
```

Criteria For Assessing Goodness Of Fit

Criterion	DF	Value	Value/DF
Deviance	4	6.2697	1.5674
Scaled Deviance	4	6.2697	1.5674
Pearson Chi-Square	4	6.0639	1.5160
Scaled Pearson X2	4	6.0639	1.5160
Log Likelihood		439.2830	

Obs	count	Pred	Reschi	Streschi
1	15	12.1568	0.8155	1.3075
2	32	29.1177	0.5341	0.9203
3	18	23.7255	-1.1755	-1.8748
4	5	5.0000	0.0000	.
5	8	10.8432	-0.8634	-1.3075
6	29	25.9715	0.5943	0.9093
7	23	21.1619	0.3996	0.5805
8	18	20.0234	-0.4522	-0.7581
9	1	1.0000	0.0000	0.0000
10	20	25.9108	-1.1612	-1.8622
11	25	21.1125	0.8461	1.2715
12	22	19.9766	0.4527	0.7581

The basic approach to fitting the model of quasi-independence is to set cells having significant standardized residuals to zero in stages, beginning with the cell with the largest absolute value of Z in the model of independence. In our case, this was cell (1,4). Removing this cell and fitting a quasi-independence model (model 2) that does not fit the data because cells (1,1), (1,3) and (3,1) still have significant $\mid Z \mid$ values. Cell (3,1) is then further removed, resulting in model 3, which now gives a fitted G^2 value of 6.269 on $(6 - 1 - 1) = 4$ degrees of freedom. This model now fits the data well. None of the standardized residuals is significant for this model.

Care must be taken in implementing the model of quasi-independence. However, if we had tried to simultaneously set to zero two or more cells initially exhibiting significant absolute Z values under the model of independence, model 4 would be an example of such a case. This model removes the two cells (3,1) and (3,3) simultaneously as they have the biggest absolute values of Z under independence. The resulting model, unfortunately, still does not fit, as we would still have to remove cell (1,4), which still exhibits significant Z. The resulting model from the removal of this cell gives a G^2 of 4.676 on 3 d.f., a good fit, but with a fewer degree of freedom than model 3. We shall consider parsimony of models later, but it seems preferable to adopt model 3 because it is based on 4 d.f. Since this model fits well, this implies that the four different offices are independent of the favored plans except for office 4 workers choosing favored plan 1 and office 1 workers choosing favored plan 3.

Alternatively, we could have fitted model 3 above, again in SAS software with the following program:

```
set tab64;
if a=1 and b=4 then count=e-20;
else if a=3 and b=1 then  count=e-20;
proc genmod;
make 'obstats' out=aa;
class a b;
model count=a b/dist=poi link=log obstats type3;
run;
```

```
         Criteria For Assessing Goodness Of Fit

Criterion                 DF           Value         Value/DF
-------------------------------------------------------------
Deviance                   4          6.2697           1.5674
Scaled Deviance            4          6.2697           1.5674
Pearson Chi-Square         4          6.0639           1.5160
Scaled Pearson X2          4          6.0639           1.5160
Log Likelihood                      437.2358

Obs         count       Pred      Resraw      Reschi    Streschi
-----------------------------------------------------------------
  1            15    12.1568      2.8432      0.8155      1.3075
  2            32    29.1177      2.8823      0.5341      0.9203
  3            18    23.7255     -5.7255     -1.1755     -1.8748
  4             .    22.4490           .           .           .
  5             8    10.8432     -2.8432     -0.8634     -1.3075
  6            29    25.9715      3.0285      0.5943      0.9093
  7            23    21.1619      1.8381      0.3996      0.5805
  8            18    20.0234     -2.0234     -0.4522     -0.7581
  9             .    10.8179           .           .           .
 10            20    25.9108     -5.9108     -1.1612     -1.8622
 11            25    21.1125      3.8875      0.8461      1.2715
 12            22    19.9766      2.0234      0.4527      0.7581
```

PROC CATMOD could also be used to fit the above models, but first we must specify that the counts or frequencies in these cells equal e^{-20} as in the case above. Yet another alternative for obtaining MLE under the quasi-independence model is presented in appendix E.1.

Other quasi-independence models will be discussed when we consider square tables having ordered classificatory variables in chapter 11.

6.5 Three-Way Contingency Tables

As an example of a three-way table, consider the data in Table 6.10, which relate to surveys carried out in 1975 (Y1) and 1976 (Y2) asking individuals whether they favored registration of guns (Aicken, 1983). The question was asked either in the form Q1 or in a form Q2 slanted against gun registration. The individual responses are (R1) *opposes* gun registration and (R2) *favors* gun registration. The data in Table 6.10 are the response R of whether individual favor gun registration for years 1975 and 1976 (Y) for two forms of mode of questionnaire (F). the example below.

		Question Form	
Response	Year	Q1	Q2
Opposes	1975	126	141
	1976	152	182
Favors	1975	319	290
	1976	463	403

Table 6.10: Gun registration Data

Our discussion of log-linear models for three-way contingency tables will use for illustration the 2^3 three-way table in Table 6.10, where the factor variables are Y and Q and the response variable is R, all having two categories. Let R, Y, and Q be indexed by $i = 1, 2$, $j = 1, 2$, and $k = 1, 2$ respectively.

The saturated log-linear formulation for this table is of the form:

$$l_{ijk} = \mu + \lambda_i^R + \lambda_j^Y + \lambda_k^Q + \lambda_{ij}^{RY} + \lambda_{ik}^{RQ} + \lambda_{jk}^{YQ} + \lambda_{ijk}^{RYQ} \tag{6.14}$$

subject to the parameters sum to zero constraints

$$\sum_i \lambda_i^R = \sum_j \lambda_j^Y = \sum_k \lambda_k^Q = 0 \tag{6.15a}$$

$$\sum_i \lambda_{ij}^{RY} = \sum_j \lambda_{ij}^{RY} = \sum_i \lambda_{ik}^{RQ} = 0 \tag{6.15b}$$

$$\sum_k \lambda_{ik}^{RQ} = \sum_j \lambda_{jk}^{YQ} = \sum_k \lambda_{jk}^{YQ} = 0 \tag{6.15c}$$

$$\sum_i \lambda_{ijk}^{RYQ} = \sum_j \lambda_{ijk}^{RYQ} = \sum_k \lambda_{ijk}^{RYQ} = 0 \tag{6.15d}$$

The above model contains as many parameters as the number of cells. If we let l_{ijk} denote the natural logarithm of the expected counts, then the estimates of the λ parameters are easily obtained by substituting the observed cell counts for the expected cell counts either in the previous expressions or by using standard statistical package such as SAS®. We give below expressions for the parameter estimates of the model in (6.14).

$$\begin{aligned}\hat{\mu} &= \frac{l_{+++}}{8} \\ &= \frac{\ln(\hat{m}_{111}\,\hat{m}_{112}\,\hat{m}_{121}\,\hat{m}_{122}\,\hat{m}_{211}\,\hat{m}_{212}\,\hat{m}_{221}\,\hat{m}_{222})}{8} \\ \hat{\lambda}_1^R &= \frac{l_{1++}}{4} - \frac{l_{+++}}{8} = \frac{1}{8}\ln\left(\frac{\hat{m}_{111}\,\hat{m}_{112}\,\hat{m}_{121}\,\hat{m}_{122}}{\hat{m}_{211}\,\hat{m}_{212}\,\hat{m}_{221}\,\hat{m}_{222}}\right)\end{aligned}$$

Similar expressions can be written for $\hat{\lambda}_1^Y$ and $\hat{\lambda}_1^Q$. For the first-order interaction terms such as $\hat{\lambda}_{11}^{RY}$, we have

$$\hat{\lambda}_{11}^{RY} = \frac{l_{11+}}{2} - \frac{l_{1++} + l_{+1+}}{4} + \frac{l_{+++}}{8} = \frac{1}{8}\ln\left(\frac{\hat{m}_{111}\,\hat{m}_{221}}{\hat{m}_{121}\,\hat{m}_{211}} \cdot \frac{\hat{m}_{112}\,\hat{m}_{222}}{\hat{m}_{122}\,\hat{m}_{212}}\right)$$

and similarly for $\hat{\lambda}_{11}^{RQ}$ and $\hat{\lambda}_{11}^{YQ}$. For the second-order or three-factor interaction term $\hat{\lambda}_{111}^{RYQ}$, we also have

$$\begin{aligned}\hat{\lambda}_{111}^{RYQ} &= l_{111} - \frac{l_{11+} + l_{1+1} + l_{+11}}{2} + \frac{l_{1++} + l_{+1+} + l_{++1}}{4} - \frac{l_{+++}}{8} \\ &= \frac{1}{8}\ln\left(\frac{\hat{m}_{111}\,\hat{m}_{221}}{\hat{m}_{121}\,\hat{m}_{211}} \cdot \frac{\hat{m}_{122}\,\hat{m}_{212}}{\hat{m}_{112}\,\hat{m}_{222}}\right)\end{aligned}$$

The maximum likelihood estimates of the parameters for the saturated model can be obtained from the above expressions by replacing the expected counts $\hat{m}_{ijk}$ by the observed counts n_{ijk}, we have

$$\hat{\mu} = \frac{\ln[126(141)(152)(182)(319)(290)(463)(403)]}{8} = 5.4481$$

$$\hat{\lambda}_1^R = \frac{1}{8}\ln\left(\frac{(126)(141)(152)(182)}{(319)(290)(463)(403)}\right) = -0.4449$$

$$\hat{\lambda}_1^Y = \frac{1}{8}\ln\left(\frac{126(141)(319)(290)}{152(182)(463)(403)}\right) = -0.1431$$

$$\hat{\lambda}_1^Q = \frac{1}{8}\ln\left(\frac{126(152)(319)(463)}{141(182)(290)(403)}\right) = -0.0073$$

$$\hat{\lambda}_{11}^{RY} = \frac{1}{8}\ln\left(\frac{126(463)(141)(403)}{152(182)(319)(290)}\right) = 0.0323$$

$$\hat{\lambda}_{11}^{RQ} = \frac{1}{8}\ln\left(\frac{126(290)(152)(403)}{141(319)(182)(463)}\right) = -0.0658$$

$$\hat{\lambda}_{11}^{YQ} = \frac{1}{8}\ln\left(\frac{126(182)(319)(403)}{141(152)(290)(463)}\right) = 0.0030$$

$$\hat{\lambda}_{111}^{RYQ} = \frac{1}{8}\ln\left(\frac{126(463)(182)(290)}{152(319)(141)(403)}\right) = 0.0139$$

The large sample standard errors for each of the estimated λ parameters is a straightforward generalization of our previous results. Since each effect is obtained as a difference of four sums (+ve) and four (-ve) sums, the asymptotic standard error becomes:

$$\text{a.s.e} = \frac{1}{8}\sqrt{\sum_i \sum_j \sum_k \left(\frac{1}{n_{ijk}}\right)} = 0.02447$$

An SAS software program and a revised output for fitting the saturated $2 \times 2 \times 2$ log-linear model to the data in Table 6.10 are displayed below.

```
data tab611;
do R=1,2;
  do Y=1,2;
    do Q=1,2;
input count @@; output;
end;
  end;
    end;
datalines;
126 141 152 182 319 290 463 403
;
run;
proc catmod; weight count;
 model r*y*q=_response_/ml;
loglin r|y|q;
run;
proc genmod; class r y q; model count=r|y|q/dist=poi type3;
run;
```

```
CATMOD Procedure
     Maximum Likelihood Analysis of Variance

Source                DF   Chi-Square    Pr > ChiSq
---------------------------------------------------
R                      1       330.45        <.0001
Y                      1        34.17        <.0001
R*Y                    1         1.75        0.1863
Q                      1         0.09        0.7651
R*Q                    1         7.24        0.0071
Y*Q                    1         0.02        0.9018
R*Y*Q                  1         0.32        0.5703
```

Analysis of Maximum Likelihood Estimates

Effect	Parameter	Estimate	Standard Error	Chi-Square	Pr > ChiSq
R	1	-0.4449	0.0245	330.45	<.0001
Y	2	-0.1431	0.0245	34.17	<.0001
R*Y	3	0.0323	0.0245	1.75	0.1863
Q	4	-0.00731	0.0245	0.09	0.7651
R*Q	5	-0.0658	0.0245	7.24	0.0071
Y*Q	6	0.00302	0.0245	0.02	0.9018
R*Y*Q	7	0.0139	0.0245	0.32	0.5703

The GENMOD Procedure

Analysis Of Parameter Estimates

Parameter				DF	Estimate	Standard Error	Chi-Square	Pr > ChiSq
Intercept				1	5.9989	0.0498	14502.9	<.0001
R	1			1	-0.7949	0.0893	79.23	<.0001
Y	1			1	-0.3291	0.0770	18.26	<.0001
R*Y	1	1		1	0.0738	0.1361	0.29	0.5875
Q	1			1	0.1388	0.0681	4.15	0.0416
R*Q	1	1		1	-0.3189	0.1293	6.08	0.0136
Y*Q	1	1		1	-0.0435	0.1059	0.17	0.6815
R*Y*Q	1	1	1	1	0.1111	0.1958	0.32	0.5703
Scale				0	1.0000	0.0000		

LR Statistics For Type 3 Analysis

Source	DF	Chi-Square	Pr > ChiSq
R	1	363.33	<.0001
Y	1	34.45	<.0001
R*Y	1	1.74	0.1867
Q	1	0.09	0.7650
R*Q	1	7.26	0.0071
Y*Q	1	0.02	0.9018
R*Y*Q	1	0.32	0.5703

Considering the maximum likelihood analysis of variance given by PROC CATMOD or the type 3 analysis from PROC GENMOD for the saturated model, only the interaction term λ^{RQ} is significantly different from zero with a pvalue of 0.0071. The third-factor interaction term λ^{RYQ} as well as the other second-factor or first-order interaction terms λ^{RY}, and λ^{YQ} are clearly not significantly different from zero. Hence, we may conclude (based on the non-significance of the three factor interaction term) that the association between any two of the variables {R,Y,Q} does not depend on the level of the third variable. For example, the association between R and Q does not depend on which year the survey was conducted. The preceeding conclusions imply that we would fail to reject the hypothesis of no three-factor interaction in the data. We summarize our findings below:

(a) That the three-factor interaction is zero is formally stated as $H_0 : \lambda_{ijk}^{RYQ} = 0$ and we fail to reject this hypothesis.

(b) That the association between year (Y) and response variable (R) is not significant and may be stated formally as the hypothesis $H_0 : \lambda^{RY} = 0$, which we also fail to reject.

(c) That there is a significant negative association between the form of question (Q) and the response variable (R). Here, the hypothesis $H_0 : \lambda^{RQ} = 0$ will be rejected and we would conclude that the response variable R and the form of question asked (Q) are dependent, after adjusting for the years of the survey.

(d) The association between years (Y) and form of question (Q) is not significant and we may again state that the hypothesis $H_0 : \lambda^{YQ} = 0$ and we would fail to reject this hypothesis.

Since only the first-order interaction term λ^{RQ} is significant, the estimated adjusted (or conditional) log-odds ratio equals (under the parameters sum to zero constraint - CATMOD)

$$\hat{\lambda}_{11}^{RQ} + \hat{\lambda}_{22}^{RQ} - (\hat{\lambda}_{12}^{RQ} + \hat{\lambda}_{21}^{RQ}) = 4(-0.0658) = -0.2632$$

since $\hat{\lambda}_{11}^{RQ} = -\hat{\lambda}_{12}^{RQ} = -\hat{\lambda}_{21}^{RQ} = \hat{\lambda}_{22}^{RQ}$. That is, the adjusted odds ratio is estimated to be $e^{(4\hat{\lambda}_{11}^{RQ})} = e^{(-0.2632)} = 0.769$. The estimated asymptotic standard error for this odds ratio is $2(.0245) = 0.0490$. This gives a 95% confidence interval for population conditional odds ratio of $-0.2632 \pm 1.96(.0490)$ or $(-0.3592, -0.1672)$ giving an odds ratio confidence interval (.70,.85). That is, we are 95% confident that the estimated odds of respondents who were asked the form of question Q1 are between 0.70 and 0.85 less likely to oppose gun registration than individuals who were asked the form of question Q2 after adjusting for the years of survey.

We can fit simpler models to the above data instead of the saturated model that we applied in the preceeding section. The next simpler model for the above data is the model that assumes that there is no third-factor interaction term. That is, the model that sets $\lambda^{RYQ} = 0$. This model {RY,RQ,YQ} has the log-linear model formulation:

$$l_{ijk} = \mu + \lambda_i^R + \lambda_j^Y + \lambda_k^Q + \lambda_{ij}^{RY} + \lambda_{ik}^{RQ} + \lambda_{jk}^{YQ}$$

with the relevant constraints. The model maintains homogeneous odds ratios between any two variables at each level of the third variable. The model has been described as the *homogeneous association* model. The model when fitted to our data has a $G^2 = 0.3222$ on 1 degree of freedom. A more simpler model is model {YQ,RQ}. Model {YQ,RQ} similarly has a $G^2 = 2.0154$ on 2 degrees of freedom. The difference between these two models leads to a difference of 1.6932 in G^2 with a corresponding 1 degree of freedom. Obviously, this indicates that the inclusion of the additional parameter λ^{RY} in the model is not significant. We shall explore other reduced models for the above data in a later section in this chapter. Both models as expected fit the data well. In particular, model {YQ,RQ} states that R and Y are conditionally independent given the levels of Q. It does appear as if there is no time effect association with the response G in respect of the above data. We give below the GENMOD output for both models together with predicted values and appropriate residuals.

```
MODEL {YQ,RY,RQ}
================
set tab611;
proc genmod;
make 'obstats' out=bb;
class r y q;
model count=r|y|q@2/dist=poi type3 obstats;
run;
proc print data=bb noobs;
var count pred Xbeta Resraw Reschi Streschi;
format pred Xbeta Resraw Reschi Streschi 8.4;
run;

The GENMOD Procedure
            Criteria For Assessing Goodness Of Fit
```

```
Criterion                   DF          Value       Value/DF
------------------------------------------------------------
Deviance                     1         0.3222         0.3222
Scaled Deviance              1         0.3222         0.3222
Pearson Chi-Square           1         0.3223         0.3223
Scaled Pearson X2            1         0.3223         0.3223

                         Analysis Of Parameter Estimates
                                    Standard     Wald 95%             Chi-
Parameter         DF  Estimate       Error   Confidence Limits     Square  Pr > ChiSq
--------------------------------------------------------------------------------------
Intercept          1    6.0061      0.0480    5.9120     6.1002   15647.6      <.0001
R          1       1   -0.8182      0.0797   -0.9744    -0.6619    105.28      <.0001
Y          1       1   -0.3463      0.0709   -0.4852    -0.2073     23.86      <.0001
R*Y        1  1    1    0.1274      0.0978   -0.0643     0.3192      1.70      0.1927
Q          1       1    0.1253      0.0638    0.0002     0.2505      3.85      0.0496
R*Q        1  1    1   -0.2705      0.0971   -0.4608    -0.0803      7.77      0.0053
Y*Q        1  1    1   -0.0109      0.0891   -0.1856     0.1637      0.02      0.9023
Scale              0    1.0000      0.0000    1.0000     1.0000

     LR Statistics For Type 3 Analysis

                             Chi-
Source             DF      Square    Pr > ChiSq
-----------------------------------------------
R                   1      363.23        <.0001
Y                   1       35.09        <.0001
R*Y                 1        1.69        0.1932
Q                   1        0.10        0.7532
R*Q                 1        7.79        0.0053
Y*Q                 1        0.02        0.9023


                        Observation Statistics

    count       Pred       Xbeta      Resraw      Reschi    Streschi
    ----------------------------------------------------------------
      126    123.101      4.8130      2.8992      0.2613      0.5677
      141    143.899      4.9691     -2.8992     -0.2417     -0.5677
      152    154.899      5.0428     -2.8992     -0.2329     -0.5677
      182    179.101      5.1879      2.8992      0.2166      0.5677
      319    321.899      5.7742     -2.8992     -0.1616     -0.5677
      290    287.101      5.6598      2.8992      0.1711      0.5677
      463    460.101      6.1314      2.8992      0.1352      0.5677
      403    405.899      6.0061     -2.8992     -0.1439     -0.5677
```

```
MODEL {YQ,RQ}
=============
set tab611;
proc catmod; weight count; model r*y*q=_response_/ml;
loglin r|q y|q; run;

proc genmod; make 'obstats' out=dd;
class r y q; model count=r|q y|q/dist=poi type3 obstats; run;
proc print data=dd noobs;
var count pred Xbeta Resraw Reschi Streschi;
format pred Xbeta Resraw Reschi Streschi 8.4; run;
```

```
CATMOD OUTPUT
==============
     Maximum Likelihood Analysis of Variance

Source               DF   Chi-Square    Pr > ChiSq
--------------------------------------------------
Y                     1        50.07        <.0001
Q                     1         0.11        0.7439
Y*Q                   1         0.04        0.8392
R                     1       343.22        <.0001
R*Q                   1         7.79        0.0052

Likelihood Ratio      2         2.02        0.3651
```

```
                 Analysis of Maximum Likelihood Estimates
                                        Standard        Chi-
Effect          Parameter    Estimate      Error      Square    Pr > ChiSq
--------------------------------------------------------------------------
Y                       1     -0.1573     0.0222       50.07        <.0001
Q                       2    -0.00801     0.0245        0.11        0.7439
Y*Q                     3    -0.00451     0.0222        0.04        0.8392
R                       4     -0.4494     0.0243      343.22        <.0001
R*Q                     5     -0.0677     0.0243        7.79        0.005

GENMOD OUTPUT
=============
            Criteria For Assessing Goodness Of Fit
Criterion                   DF           Value           Value/DF
--------------------------------------------------------------
Deviance                     2          2.0154             1.0077
Scaled Deviance              2          2.0154             1.0077
Pearson Chi-Square           2          2.0228             1.0114
Scaled Pearson X2            2          2.0228             1.0114
Log Likelihood                       9683.5985

                         Analysis Of Parameter Estimates
                                 Standard        Chi-
Parameter         DF  Estimate      Error      Square  Pr > ChiSq
-----------------------------------------------------------------
Intercept          1    5.9890     0.0466    16543.30      <.0001
Y          1       1   -0.3055     0.0635       23.16      <.0001
Q          1       1    0.1284     0.0643        3.99      0.0456
Y*Q        1   1   1   -0.0180     0.0889        0.04      0.8392
R          1       1   -0.7634     0.0674      128.39      <.0001
R*Q        1   1   1   -0.2709     0.0970        7.79      0.0052
Scale              0    1.0000     0.0000

     LR Statistics For Type 3 Analysis

                           Chi-
Source            DF     Square    Pr > ChiSq
---------------------------------------------
Y                  1      50.69        <.0001
Q                  1       0.11        0.7439
Y*Q                1       0.04        0.8392
R                  1     378.84        <.0001
R*Q                1       7.81        0.0052

                              Observation Statistics

    count        Pred       Xbeta      Resraw      Reschi    Streschi
    -----------------------------------------------------------------
      126     116.708      4.7597      9.2921      0.8601      1.3147
      141     137.021      4.9201      3.9793      0.3400      0.5425
      152     161.293      5.0832     -9.2928     -0.7317     -1.3148
      182     185.979      5.2256     -3.9793     -0.2918     -0.5425
      319     328.293      5.7939     -9.2925     -0.5129     -1.3148
      290     293.979      5.6835     -3.9793     -0.2321     -0.5425
      463     453.708      6.1175      9.2925      0.4363      1.3148
      403     399.021      5.9890      3.9793      0.1992      0.5425
```

In the Table 6.11 are displayed the expected counts under the homogeneous association model.

	Y1=1975		Y2=1976	
R	Q1	Q2	Q1	Q2
R1=1	123.10	143.90	154.90	179.10
R2=2	321.90	287.10	460.10	405.90

Table 6.11: Expected cell counts under the model of no three-factor interaction

Since we are usually interested in the ratio of the odds of an Y1 when the response was R1 relative to the odds of an Y1 when the response was R2, if there is no three-factor interaction, then we would expect the log odds ratio not to depend on the level of k of variable Q. That is,

$$\ln\left(\frac{\hat{m}_{11k}\hat{m}_{22k}}{\hat{m}_{12k}\hat{m}_{21k}}\right)$$

should be constant (Birch, 1963) for $k = 1, 2$. In our example, from expected counts in Table 6.11,

$$\ln\left(\frac{123.10 \times 460.10}{154.90 \times 321.90}\right) = 0.1274 = \ln\left(\frac{143.90 \times 405.90}{179.10 \times 287.10}\right)$$

The log-odds of 0.1274 equals the parameter estimate for λ^{RY} under the last equals zero GENMOD constraint. Similarly, it can be shown that the log-odds for the R-Q subtables at each of the two levels of Y would be equal to -0.2705, the estimate of λ^{RQ}.

6.5.1 General $I \times J \times K$ Contingency Tables

The above log-linear analysis for the 2^3 contingency table can readily be extended to the general $I \times J \times K$ three-way contingency tables, with variables A, B, and C that are indexed by $i = 1, 2, \cdots, I$, $j = 1, 2, \cdots, J$, and $k = 1, 2, \cdots, K$, respectively. In this case, the saturated log-linear model formulation for such a table is again of the form:

$$l_{ijk} = \mu + \lambda_i^A + \lambda_j^B + \lambda_k^C + \lambda_{ij}^{AB} + \lambda_{ik}^{AC} + \lambda_{jk}^{BC} + \lambda_{ijk}^{ABC}$$

subject to the constraints similar to those in (6.15). The l_{ijk} are the natural logarithms of the observed counts n_{ijk}. The saturated model above will succinctly be described as the model {ABC}.

The ML estimates in this case are given below for a few of the terms by:

$$\begin{aligned}
\hat{\mu} &= \frac{l_{+++}}{IJK} \\
\hat{\lambda}_i^A &= \frac{l_{i++}}{JK} - \frac{l_{+++}}{IJK} \\
\hat{\lambda}_{ij}^{AB} &= \frac{l_{ij+}}{K} - \frac{l_{i++}}{JK} - \frac{l_{+j+}}{IK} + \frac{l_{+++}}{IJK} \\
\hat{\lambda}_{ijk}^{ABC} &= l_{ijk} - \frac{l_{ij+}}{K} - \frac{l_{i+k}}{J} - \frac{l_{+jk}}{I} + \frac{l_{i++}}{JK} + \frac{l_{+j+}}{IK} + \frac{l_{++k}}{IJ} - \frac{l_{+++}}{IJK}
\end{aligned}$$

Similar expressions can be written for the other parameter estimates ($\hat{\lambda}_j^B, \hat{\lambda}_k^C, \hat{\lambda}_{ik}^{AC}$) and $\hat{\lambda}_{jk}^{BC}$. Simpler models are formed by deleting terms from the saturated model (ABC), and the reduced model formed is referred to as an unsaturated model or reduced model.

6.5.2 Hierarchy Principle

For the general $I \times J \times K$ three-way table, the saturated model, which can be written succinctly as {ABC}, implies that the λ parameters associated with the following are included in the model: μ, A, B, C, AB, AC, BC, and ABC. Similarly, the model written {AB} has included the parameters μ, A, B, and the AB interaction term, while the model of no three-factor interaction {AB,AC,BC} has μ, A, B, C, AB, AC, and BC parameters in the model and has the formulation:

$$l_{ijk} = \mu + \lambda_i^A + \lambda_j^B + \lambda_k^C + \lambda_{ij}^{AB} + \lambda_{ik}^{AC} + \lambda_{jk}^{BC}$$

Thus if an interaction, say {AB}, is included in a model, then, implicitly, the lower order terms A, B are also by the hierarchical principle included in the model, along with, of course, that interaction term. Conversely, if the interaction term $AC = 0$, then by the hierarchy principle, this implies that the higher order interaction ABC must also be zero. Further, we say that model {AB} is *nested* in model {AB,AC,BC} since the parameters of the former are a subset of the parameters of the latter model. We can take advantage of this in comparing nested log-linear models. We shall develop this further later in the chapter.

6.6 Sufficient Statistics for Log-Linear Models

The sufficient statistics are the configurations of sums that correspond to the effect terms of the loglinear model. To derive these statistics, we would need to relate the log- linear model of interest to the likelihood function.

Consider the saturated model {ABC} for the three-dimensional contingency table where A, B and C are indexed by $i = 1, 2, \cdots, I$, $j = 1, 2, \cdots, J$, and $k = 1, 2, \cdots, K$, respectively. The model has the formulation

$$\ln(\hat{m}_{ijk}) = \mu + \lambda_i^A + \lambda_j^B + \lambda_k^C + \lambda_{ij}^{AB} + \lambda_{ik}^{AC} + \lambda_{jk}^{BC} + \lambda_{ijk}^{ABC} \tag{6.16}$$

where l_{ijk} is the natural logarithm of the expected counts. The usual identifiability constraints are assumed imposed to ensure that the number of parameters are equal to the number of cells in the table.

The log-likelihood (ℓ) under the multinomial sampling scheme is again presented here as:

$$\ell = \ln\left\{\frac{n!}{\prod n_{ijk}!}\right\} + \sum_{ijk} n_{ijk} \ln(\hat{m}_{ijk}) - n \ln(n) \tag{6.17}$$

The first and last terms are constants for any set of $\hat{m}_{ijk}$, and we are only interested in the remaining term, known as the *kernel* of the function. That is, the kernel of the multinomial is

$$\sum_{ijk} n_{ijk} \ln(\hat{m}_{ijk}) \tag{6.18}$$

Similarly, for either the product-multinomial or Poisson sampling schemes, the kernels are given by the same expression as above. For example, for the Poisson distribution, we have the log of the likelihood being

$$\ell = \sum_{ijk} n_{ijk} \ln(\hat{m}_{ijk}) - \sum_{ijk} \hat{m}_{ijk} - \sum_{ijk} \ln(n_{ijk}!) \tag{6.19}$$

Again as before the kernel is given by the first term in the expression above since the last two terms are constants, with the second term on the RHS being equal to the sample size n.

Substituting for $\ln(\hat{m}_{ijk})$ in both (6.16) and (6.18) and applying the summation sign, we have

$$\begin{aligned}\sum_{ijk} n_{ijk} \ln(\hat{m}_{ijk}) &= n\mu + \sum_i n_{i++}\lambda_i^A + \sum_j n_{+j+}\lambda_j^B + \sum_k n_{++k}\lambda_k^C \\ &+ \sum_{ij} n_{ij+}\lambda_{ij}^{AB} + \sum_{ik} n_{i+k}\lambda_{ik}^{AC} \\ &+ \sum_{jk} n_{+jk}\lambda_{jk}^{BC} + \sum_{ijk} n_{ijk}\lambda_{ijk}^{ABC}\end{aligned} \tag{6.20}$$

Both the multinomial and Poisson distributions belong to the class of exponential probability density functions as explained in chapter 2. For this class of models, it is well known that the sufficient statistics are the coefficients of the parameters. In the above formulation of the kernel of the saturated model, the result is general, but for unsaturated models, terms will drop out as appropriate and the terms that remain will give coefficients that will be the sufficient statistics.

We consider sufficient statistics for some of these unsaturated models. We shall illustrate with the general three-way ABC contingency table.

6.6.1 The No Three-Factor Effect

For brevity, $\lambda_{ijk}^{ABC} = 0$, will be written succinctly as ABC $= 0$ for all i, j, k. If we therefore set ABC $= 0$, in this case, the reduced model is {AB,AC,BC} and so the last term in (6.20) disappears, and the term N and configurations with numbers $n_{i++}, n_{+j+}, n_{++k}, n_{ij+}, n_{i+k}$, and, n_{+jk} are the sufficient statistics for this model. The last three configurations yield the others and form the complete minimal set (since $n_{i++} = \sum_j n_{ij+}$ etc.). Using the more succinct notation of Bishop et al. (1975), we would say that the minimal sufficient statistics are C_{12}, C_{13}, and C_{23}, which are succinct forms of n_{ij+},n_{i+k}, and n_{+jk} respectively.

6.6.2 One-Factor Effect Term

If we further remove one additional term (say AC) by putting AC $=$ ABC $= 0$, the natural logarithm of the kernel of the likelihood function becomes

$$\begin{aligned}\sum_{ijk} n_{ijk} \ln(\hat{m}_{ijk}) &= n\mu + \sum_i n_{i++}\lambda_i^A + \sum_j n_{+j+}\lambda_j^B + \sum_k n_{++k}\lambda_k^C \\ &+ \sum_{ij} n_{ij+}\lambda_{ij}^{AB} + \sum_{jk} n_{+jk}\lambda_{jk}^{BC}\end{aligned}$$

and the sufficient statistics are C_{12} and C_{23}, that is, n_{ij+} and n_{+jk}. We note that by setting AC $= 0$, we are implying by the hierarchy principle that ABC is also zero.

We give in Table 6.12 the sufficient statistics for the complete models in a three-dimensional table.

There are three kinds of model type (2), namely, models with absent terms (AC, ABC), (AB, ABC), and (BC, ABC). Similarly, there are three kinds of model type (3), again with absent terms (AB, AC, ABC), (AB, BC, ABC), and (AC, BC, ABC).

Model Type	Absent term(s)	Sufficient Configurations	df
1	ABC	C_{12}, C_{13}, C_{23}	(I-1)(J-1)(K-1)
2	AC,ABC	C_{12}, C_{23}	(I-1)J (K-1)
3	AB,AC,ABC	C_{23}, C_1	(I-1)(JK-1)
4	AB,AC,BC,ABC	C_1, C_2, C_3	IJK-(I+J+K)+2

Table 6.12: Complete models in three dimensions

6.7 Maximum Likelihood Estimates, MLE

Again for the three-dimensional case, once a set of sufficient statistics has been identified, it follows for example for the model {AB,AC,BC} that the sufficient statistics are C_{12}, C_{13}, and C_{23} with corresponding cell marginal counts n_{ij+}, n_{i+k}, and n_{+jk} respectively. Then the following results are due to Birch (1963).

(1) The minimal sufficient statistics are the ML estimates of the corresponding marginal distributions. That is, for log-linear models, the likelihood equations match the sufficient statistics to their expected values. Thus the MLE of $\hat{m}_{ij+}, \hat{m}_{i+k}$, and $\hat{m}_{+jk}$ must correspond to the observed values. That is,

$$\begin{aligned} \hat{m}_{ij+} &= n_{ij+} \\ \hat{m}_{i+k} &= n_{i+k} \end{aligned} \tag{6.21}$$

and

$$\hat{m}_{+jk} = n_{+jk} \tag{6.22}$$

(2) There is a unique set of elementary cell estimates that satisfies the conditions of the model and these marginal constraints. For example, for the hypothesis (ABC) = 0, we have

$$\frac{\hat{m}_{ijk}\hat{m}_{rsk}}{\hat{m}_{rjk}\hat{m}_{isk}} = \frac{\hat{m}_{ijt}\hat{m}_{rst}}{\hat{m}_{rjt}\hat{m}_{ist}} \qquad \text{for} \quad i \neq r, j \neq s, k \neq t \tag{6.23}$$

Further, there is only one set of estimates satisfying relations (6.21), (6.22), and (6.23), and this set maximizes the likelihood function.

6.7.1 Two-Dimensional Case

In the two-dimensional $I \times J$ contingency table case, if we set AB = 0, which is the model of independence, the sufficient configurations are C_1, C_2 with corresponding cell entries n_{i+} and n_{+j}, respectively. Thus the likelihood equations are $n_{i+} = \hat{m}_{i+}$ and $n_{+j} = \hat{m}_{+j}$, for all i and j. The expected values are

$$\hat{m}_{ij} = \frac{n_{i+}n_{+j}}{n}$$

and they satisfy these equations.

6.7.2 Three-Dimensional Tables

Again let us consider the three-dimensional case with AC = (ABC) = 0. The sufficient statistics are C_{12} and C_{23} and they have in common C_2. Thus, the ML estimates are given by:

$$\hat{m}_{ijk} = \frac{n_{ij+}n_{+jk}}{n_{+j+}}$$

We give below types of direct estimates possible for three-dimensional tables.

Model	Sufficient configurations	Direct estimates
{A,B,C}	C_1, C_2, C_3	$\hat{m}_{ijk} = \frac{n_{i++}n_{+j+}n_{++k}}{n^2}$
{AB,C}	C_{12}, C_3	$\hat{m}_{ijk} = \frac{n_{ij+}n_{++k}}{n}$
{AB,AC}	C_{12}, C_{13}	$\hat{m}_{ijk} = \frac{n_{ij+}n_{i+k}}{n_{i++}}$
{AB,AC,BC}	C_{12}, C_{13}, C_{23}	No direct estimates
{ABC }	No Restrictions	$\hat{m}_{ijk} = n_{ijk}$

6.7.3 Four-Dimensional Tables

For four factors A, B, C, and D indexed by $i = 1, 2, \cdots, I$, $j = 1, 2, \cdots, J$, $k = 1, 2, \cdots, K$, and $l = 1, 2, \cdots, L$, respectively, there are 4 main effects, 6 two-factor effects, 4 three-factor effects, and 1 four-factor effect. In the next table, we give the direct estimates for various configurations when certain effect or effects are removed from the saturated model {ABCD}, where 4F, 3F, and 2F mean four-factor, three-factor, and two-factor terms, respectively.

Set	Effects removed 4F	3F	2F	Sufficient configurations	Direct $\hat{m}_{ijkl}$
1	1	2	1	C_{123}, C_{124}	$\frac{n_{ijk+}n_{ij+l}}{n_{ij++}}$
2	1	3	2	C_{123}, C_{14}	$\frac{n_{ijk+}n_{i++l}}{n_{i+++}}$
3	1	3	3	C_{123}, C_4	$\frac{n_{ijk+}n_{+++l}}{n}$
4	1	4	3	C_{12}, C_{13}, C_{14}	$\frac{n_{ij++}n_{i+k+}n_{i++l}}{n^2_{i+++}}$
				C_{12}, C_{13}, C_{24}	$\frac{n_{ij++}n_{i+k+}n_{+j+l}}{n_{i+++}n_{+j++}}$
5	1	4	4	C_{12}, C_{34}	$\frac{n_{ij++}n_{++jl}}{n}$
				C_{12}, C_{13}, C_4	$\frac{n_{ij++}n_{i+k+}n_{+++l}}{nn_{i+++}}$
6	1	4	5	C_{12}, C_3, C_4	$\frac{n_{ij++}n_{++k+}n_{+++l}}{n^2}$
7	1	4	6	C_1, C_2, C_3, C_4	$\frac{n_{i+++}n_{+j++}n_{++k+}n_{+++l}}{n^3}$

The general form of the direct estimates is predictable. For instance, the form of the ML estimates for set 1 is suggestive. The sufficient configurations C_{123} and

C_{124} have C_{12} in common, and hence, the denominator n_{ij++} in the expression. We give below the general rules for obtaining these estimates.

(i) The numerator has entries from each sufficient configuration.

(ii) The denominator entries from redundant configurations caused by "overlapping," terms in powers of N, appear either in the numerator or in the denominator to ensure the right order of magnitude. The order of the denominator is always *one less* than that of the numerator.

Sometimes, no combination of the sufficient statistics yields a direct estimate. In such cases, we shall employ iterative procedures. For instance, the estimates for the ABCD = 0 needs the use of an iterative procedure. {ABC, ABD, ACD, BCD} will be described as the *generating class* for this model, that is, the highest order interaction terms in the model, whereas for the model in set 1 above, we would say that the *generating class* is ABC and ABD. Further, the model {ABCD} is *comprehensive* because it includes all main effects for each of the factors in the table. A noncomprehensive model therefore is one with at least a missing main factor effect.

6.7.4 Closed Loops

When overlapping configurations can be linked to each other to form a closed loop, then no direct estimates exists. As an example, the model with the ABC term = 0 in a three-way table has as its sufficient configurations C_{12}, C_{13}, and C_{23}, which as we can see from Figure 1 in Figure 6.1, forms a closed loop. Hence, no direct estimates exists. These figures are presented in the next page.

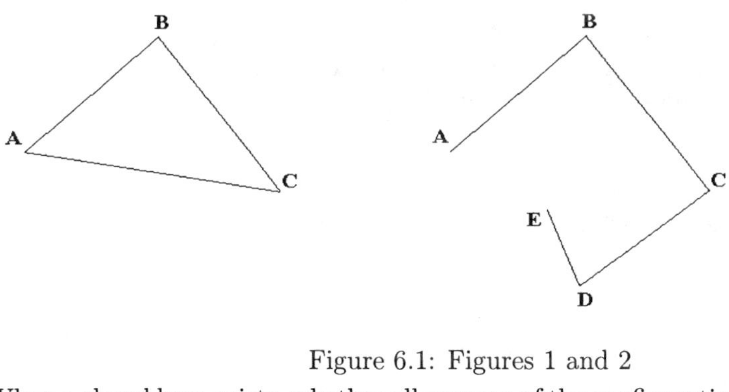

Figure 6.1: Figures 1 and 2

When a closed loop exists, whether all or some of the configurations are only involved in the loop, iterative methods are necessary to obtain ML estimates, as no direct estimates exist in this case.

As another example, in a five-dimensional table, direct estimates exist for the configurations, C_{12}, C_{23}, C_{34}, and C_{45}, that is, model {AB,BC,CD,DE}, because these do not form a closed loop (see Figure 2 in Figure 6.1). But we would be forced to use iterative procedures if we were to add a further two-factor term or configuration to the four above. For instance, adding C_{13}, that is, AC to the model gives a loop connecting C_{12}, C_{23}, and C_{13} (Figure 3). Figures 3 and 4 are again respectively presented in Figure 6.2.

However, we can obtain direct estimates from the configurations C_{123} and C_{345} even though these configurations have six two-factor effects because the configurations do have loops (Figure 4), but the generating classes {ABC} and {CDE} are also included in the model; that is, the model is decomposable (see next section). The model in Figure 3, {AB,BC,CD,DE,AC}, includes the generating class (ABC)

The above discussions bring us to the notions of decomposable and graphical log-linear models.

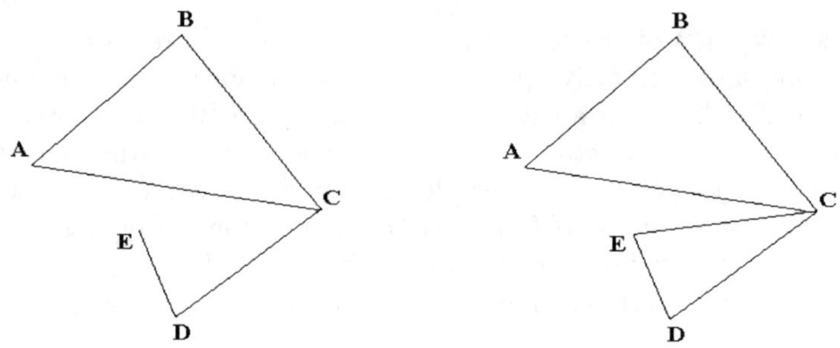

Figure 6.2: Figures 3 and 4

6.8 Decomposable and Graphical Log-Linear Models

A decomposable log-linear model (Goodman, 1970, 1971b; Haberman, 1978) is the class of models with closed form maximum likelihood estimates. They are very easy to interpret because they have very simple structures.

On the other hand, a graphical log-linear model (Christensen, 1990) is one that can be interpreted in terms of conditional independence arguments and can be formally defined as:

> A model is *graphical* if, whenever the model contains all two-factor terms generated by a higher order interaction, the model also contains the higher interaction term.

For a three-way table, the model {AB,AC,BC} is not graphical nor decomposable because by definition it should contain the generating three-factor interaction ABC. Also, model {AB,BC,AC,CD,DE} in a five-factor table is not decomposable because it does not have the three-factor generating term ABC (that forms the closed loop) in its model. If we were to add this term (i.e., ABC) to the model, then, the new model {ABC, CD, DE} would be decomposable and graphical. For the same five-way table, the model {ABC,CDE} is decomposable because it not only has the six two-factor terms but also their generating three-factor terms, even though the ABC and the CDE formed closed loops.

Bishop et al. (1975) have proposed the following rules for determining the existence of direct estimates (decomposability) for a general k-way log-linear model. They refer to these as *classifying rules.*

6.8.1 Rules for Detecting Existence of Direct Estimates

Example

Consider the configurations $C_{123}, C_{124}, C_{456}$ in a six-way contingency table. First:

1. Relabel variables that appear together (*here variables 1 and 2*) as $C_{1'3}, C_{1'4}, C_{456}$.

2. Delete common variables to all the three (*here none*).

3. Delete variables that appear once (*variables 3, 5, and 6*) $C_{1'}, C_{1'4}, C_4$.

4. Remove redundancies $C_{1'4}$.

The final configurations do not form a closed loop, hence direct estimates exist and the model is not only decomposable but is also graphical. This is not surprising as the model contains all three-factor generating terms that give rise to all the two-factor interaction terms in the model.

To obtain the corresponding maximum likelihood estimates of model, {ABC, ABD,DEF}, we note that the first two configurations C_{123} and C_{124} have C_{12} in common, so it yields an estimate of C_{1234}, that is,

$$\text{Estimates of Cells} \quad C_{1234} = \text{Cells of } \frac{C_{123}C_{124}}{C_{12}}.$$

Add the next configuration

$$\begin{aligned}\text{Estimates of Cells} \quad C_{123456} &= \text{Cells of } \frac{C_{1234}C_{456}}{C_4} \\ &= \text{Cells of } \frac{C_{123}C_{124}C_{456}}{C_{12}C_4}.\end{aligned}$$

The implications of the above rules are that:

1. If direct estimates exist, we can compute cell estimates in stages.

2. If direct estimates exist, then at least one of the two-factor terms is absent in the model.

3. We can always compute direct estimates by inspection if we can reduce to only two configurations.

For model {AB,AC,BC} in the three-way table, therefore, direct estimates do not exist because it has all the three two-factor terms in the model.

The main reasons why we always wish to know if direct estimates exist are:

- Difference(s) between two nested direct estimable models can be determined from the appropriate marginal tables rather than from the complete tables.

- Asymptotic variances of the estimated parameters can readily be obtained.

- Models can be interpreted in terms of hypotheses of independence, conditional independence, and equiprobability (Goodman, 1970).

6.8.2 Structural Breakdown of G^2

When we have a log-linear model where direct estimates exist, that is, with closed-form maximum likelihood estimates, we need not compute the expected values $\{\mathbf{m}\}$ to get G^2. For an $I \times J \times K$ three-way table, for example, consider the model given by setting AC = ABC = 0; that is, model {AB,AC}.

The sufficient statistics are C_{12}, C_{23} with

$$\hat{m}_{ijk} = \frac{C_{12}C_{23}}{C_2} = \frac{n_{ij+}n_{+jk}}{n_{+j+}}$$

and

$$\begin{aligned}
G^2 &= 2\sum_{ijk} n_{ijk} \ln(n_{ijk}n_{+j+}/n_{ij+}n_{+jk}) \\
&= 2\sum_{ijk} n_{ijk} \ln(n_{ijk}) + 2\sum_{ijk} n_{ijk} \ln(n_{+j+}) \\
&- 2\sum_{ijk} n_{ijk} \ln(n_{ij+}) - 2\sum_{ijk} n_{ijk} \ln(n_{+jk}) \\
&= 2\left[\sum_{ijk} n_{ijk} \ln(n_{ijk}) + \sum_{j} n_{+j+} \ln(n_{+j+})\right] \\
&- 2\left[\sum_{ij} n_{ij+} \ln(n_{ij+}) - \sum_{jk} n_{+jk} \ln(n_{+jk})\right] \\
&= 2\left[G_{123}(\mathbf{N}) + G_2(\mathbf{N}) - G_{12}(\mathbf{N}) - G_{23}(\mathbf{N})\right]
\end{aligned}$$

For a three-way contingency table, there are eight unsaturated hierarchical log-linear models, seven of which have direct estimates. We can compute G^2 for these seven models directly from the following quantities:

$$G_{123}(\mathbf{N}), G_{12}(\mathbf{N}), G_{13}(\mathbf{N}), G_{23}(\mathbf{N}), G_1(\mathbf{N}), G_2(\mathbf{N}), G_3(\mathbf{N}) \quad \text{and} \quad N\ln(N)$$

For any log-linear model with direct estimates therefore in any number of dimensions, the form of the structural breakdown of G^2 is always the same and of the form:

$$G^2 = 2\left[G(\mathbf{N}) - \sum_{*} G_{*}(\mathbf{N}) + \sum_{**} G_{**}(\mathbf{N})\right] \tag{6.24}$$

where $*$ indexes the minimal sufficient configurations associated with the given model and $**$ indexes the overlapping of the minimal sufficient statistics.

It must be stated here that since computer software is readily available for obtaining these estimates (whether direct or indirect), it is therefore not of serious importance to know of their direct MLE existence. It is, however, challenging to know how these estimates are derived.

Example

Consider fitting the model {RQ,YQ} to the data in Table 6.10. Then the sufficient statistics are n_{i+k} and n_{+jk}. In Tables 6.13 and 6.16 are the cell counts for this data when collapsed over Y and R, that is, n_{i+k}, and n_{+jk}, respectively, with

	Q(k)		
R(i)	1	2	Total
1	278	323	601
2	782	693	1475
Total	1060	1016	2076

Table 6.13: Data in Table 6.10 collapsed over Y

	Q(k)		
Y(j)	1	2	Total
1	445	431	876
2	615	585	1200
Total	1060	1016	2076

Table 6.14: Data in Table 6.10 collapsed over R

$$n_{++1} = 1060 \quad \text{and} \quad n_{++2} = 1016$$

Direct ML estimates are therefore given by:

$$\hat{m}_{ijk} = \frac{n_{i+k}n_{+jk}}{n_{++k}}$$

For instance, $\hat{m}_{111} = \dfrac{278(445)}{1060} = 116.708$. This value agrees with the expected values displayed from the previous SAS software output under this model. This model has a corresponding G^2 of 2.0154 on 2 d.f.

Now applying the structural breakdown of G^2 to obtain our result, we have:

$$\begin{aligned} G^2 = 2 &\left[\sum_{ijk} n_{ijk}\ln(n_{ijk}) + \sum_{k} n_{++k}\ln(n_{++k})\right] \\ - 2 &\left[\sum_{ik} n_{i+k}\ln(n_{i+k}) - \sum_{jk} n_{+jk}\ln(n_{+jk})\right] \end{aligned}$$

$$\begin{aligned} G^2 &= 2\{[126\ln(126) + 141\ln(141) + 152\ln(152) + 182\ln(182) \\ &+ 319\ln(319) + 290\ln(290) + 463\ln(463) + 403\ln(403)] \\ &+ [1060\ln(1060) + 1016\ln(1016)] \\ &- [278\ln(278) + 323\ln(323) + 782\ln(782) + 693\ln(693)] \\ &- [445\ln(445) + 431\ln(431) + 615\ln(615) + 585\ln(585)]\} \\ &= 2.0154 \end{aligned}$$

which agrees with the value of G^2 obtained via the computed expected values.

Consider, for example, a seven-dimensional table and a log-linear model with minimal sufficient configurations

$$C_{124}, C_{235}, C_{136}, C_{57}, C_{123}$$

Then we can write

$$G^2 = 2\left[G(\mathbf{N}) - G_{124}(\mathbf{N}) - G_{235}(\mathbf{N}) - G_{136}(\mathbf{N}) - G_{57}(\mathbf{N}) - G_{123}(\mathbf{N})\right] + 2\left[G_{12}(\mathbf{N}) + G_{23}(\mathbf{N}) + G_{13}(\mathbf{N}) + G_5(\mathbf{N})\right]$$

6.9 MLE via Iterative Procedures

Two standard methods of estimating the expected values in the general log-linear models are the *iterative proportional fitting* (IPF) algorithm due to Deming and Stephan (1940) and the *Newton-Raphson* NR algorithm, which fits a series of weighted regression analysis and is therefore some times referred to as the *iteratively reweighted least squares.* We describe below these two algorithms when applied to the data in Table 6.10, where the model of interest is the no three-factor interaction model that is, model {RY,RQ,YQ}, which does not have closed-form direct estimates.

6.9.1 Iterative Proportional Fitting Algorithm (IPF)

For all models where direct or indirect estimates do or do not exist, model {AB,AC, BC} in a three-way contingency table, for instance, does not have direct estimates. In this case, we can obtain maximum likelihood estimates for each of the elementary cells under such models by iterative fitting of the sufficient configurations. This method of successive proportional adjustments has the following properties.

- It always converges to the required unique set of MLE estimates.
- The estimates depend only on the sufficient configurations, and so no special provisions need be made for sporadic cells with no observations.
- Any set of starting values may be chosen that conforms to the model being fitted.
- If direct estimates exist, the procedure yields the exact estimates in one cycle.
- A stopping rule may be used that ensures accuracy to any desired degree (usually 0.001) in the elementary cell estimates, instead of a rule that only ensures accuracy in one or more summary statistics.

For example , consider again the $2 \times 2 \times 2$ gun registration data in Table 6.10 with variables R, Y, and Q. We give in the next table the observed cell counts for the data.

Cells	111	211	121	221	112	212	122	222
n_{ijk}	126	319	141	290	152	463	182	403

For model {RY,RQ,YQ}, therefore, the sufficient configurations (statistics) are C_{12}, C_{13}, and C_{23} respectively. That is, n_{ij+}, n_{i+k}, and n_{+jk}, respectively.

STARTING POINT:

To start the IPF procedure, we start by initially setting $\hat{m}_{ijk}^{(0)} = 1$ for every cell in the table. This correspondingly sets $\ln(\hat{m}_{ijk}^{(0)}) = 0$.

We then adjust the preliminary estimates to fit successively, C_{12}, C_{13}, and C_{23}.

FITTING C_{12}

Fitting C_{12} gives

$$\hat{m}_{ijk}^{(1)} = \hat{m}_{ijk}^{(0)} \frac{n_{ij+}}{\hat{m}_{ij+}^{(0)}}$$

FITTING C_{13}

We next fit the marginal C_{13}, giving

$$\hat{m}_{ijk}^{(2)} = \hat{m}_{ijk}^{(1)} \frac{n_{i+k}}{\hat{m}_{i+k}^{(1)}}$$

FITTING C_{23}

We now finally, complete the first cycle by fitting C_{23} to give

$$\hat{m}_{ijk}^{(3)} = \hat{m}_{ijk}^{(2)} \frac{n_{+jk}}{\hat{m}_{+jk}^{(2)}}$$

We will then repeat this three-step cycle until convergence to the desired accuracy is attained. A satisfactory stopping rule is to choose $\delta = 0.01$ or $\delta = 0.0001$ such that

$$\left| \hat{m}_{ijk}^{(3r)} - \hat{m}_{ijk}^{(3r-3)} \right| < \delta$$

The above procedure is implemented below for our data.

Cell	n_{ijk}	$\hat{m}_{ijk}^{(0)}$	$\hat{m}_{ijk}^{(1)}$	$\hat{m}_{ijk}^{(2)}$	$\hat{m}_{ijk}^{(3)}$	$\cdots$	$\hat{m}_{ijk}^{(12)}$
111	126	1	133.5	123.50	123.12	$\cdots$	123.10
211	319	1	304.5	322.87	321.88	$\cdots$	321.90
121	152	1	167.0	154.50	154.84	$\cdots$	154.90
221	463	1	433.0	459.13	460.16	$\cdots$	460.10
112	141	1	133.5	143.50	143.96	$\cdots$	143.90
212	290	1	304.5	286.13	287.04	$\cdots$	287.10
122	182	1	167.0	179.50	179.08	$\cdots$	179.10
222	403	1	433.0	406.87	405.92	$\cdots$	405.90

In the three tables below, we give the sufficient marginals for both the observed and estimated fitted values for the first cycle of iteration:

C_{12}	n_{ij+}	$\hat{m}_{ij+}^{(0)}$
11+	267	2
21+	609	2
12+	334	2
22+	866	2

C_{13}	n_{i+k}	$\hat{m}_{i+k}^{(1)}$
1+1	278	300.5
2+1	782	737.5
1+2	323	300.5
2+2	693	737.5

C_{23}	n_{+jk}	$\hat{m}_{+jk}^{(2)}$
+11	445	446.37
+21	615	613.63
+12	431	429.63
+22	585	586.37

where, for instance, the estimates for $\hat{m}_{111}$ and $\hat{m}_{222}$ for the three steps of the first cycle are (using the marginals in the above tables) given by:

$$\hat{m}_{111}^{(0)} = 1 \qquad \hat{m}_{222}^{(0)} = 1$$

$$\hat{m}_{111}^{(1)} = \frac{1 \times 267}{2} = 133.5$$

$$\hat{m}_{222}^{(1)} = \frac{1 \times 866}{2} = 433.60$$

$$\hat{m}_{111}^{(2)} = \frac{133.5 \times 278}{300.5} = 123.50$$

$$\hat{m}_{222}^{(2)} = \frac{433.0 \times 693}{737.5} = 406.873$$

$$\hat{m}_{111}^{(3)} = \frac{123.50 \times 445}{446.37} = 123.121$$

and

$$\hat{m}_{222}^{(3)} = \frac{406.87 \times 585}{586.37} = 405.922$$

We can show that the above computations are all equivalent. For example,

$$\hat{m}_{ijk}^{(2)} = \frac{\hat{m}_{ijk}^{(1)} n_{i+k}}{\hat{m}_{i+k}^{(1)}} = \frac{n_{ij+} n_{i+k}}{n_{i++}}$$

6.9.2 Successive Cycles of Iteration

At the end of the first cycle of iteration above, we obtained $\hat{m}_{ijk}^{(3)}$. These expected values are such that the constraints C_{23} are satisfied, That is, the marginals now add up to {445,615,431,585} but they no longer satisfy the constraints C_{12} and C_{13}; hence there is a need to repeat the process until the difference between the $\hat{m}_{ijk}$ in the $(3r)$ and $(3r-3)$ cycles satisfy our accuracy target δ. In the above example, we have convergence by $r = 4$, that is, at the 12th step or at the end of the fourth cycle of iteration with $\delta = 0.0$ as the convergence criterion. The iterative proportional algorithm can be implemented in SAS software by the use of the IPF algorithm in **PROC IML** in SAS®.

6.9.3 The Newton-Raphson Iterative Procedure

Any log-linear model can be reparameterized as a linear model. Consider, for example, the saturated model {AB} in a 2×2 table, where

$$l_{ij} = \mu + \lambda_i^A + \lambda_j^B + \lambda_{ij}^{AB}$$

where the usual constraints are imposed for identifiability and l_{ij} is the log of the expected counts.

The above model can be rewritten in terms of identifiable parameters as:

$$l_{ij} = \mu + \lambda_1^A Z_{1i} + \lambda_1^B Z_{2j} + \lambda_{11}^{AB} Z_{1i} Z_{2j}$$

where

$$Z_{1i} = \begin{cases} 1 & \text{if } i = 1 \\ -1 & \text{if } i = 2 \end{cases} \quad \text{and} \quad Z_{2j} = \begin{cases} 1 & \text{if } j = 1 \\ -1 & \text{if } j = 2 \end{cases}$$

Consequently, the models above can be written in the form:

$$\begin{bmatrix} l_{11} \\ l_{12} \\ l_{21} \\ l_{22} \end{bmatrix} = \begin{bmatrix} 1 & 1 & 1 & 1 \\ 1 & -1 & 1 & -1 \\ 1 & 1 & -1 & -1 \\ 1 & -1 & -1 & 1 \end{bmatrix} \begin{bmatrix} \mu \\ \lambda_1^A \\ \lambda_2^B \\ \lambda_{11}^{AB} \end{bmatrix}$$

where $l_{ij} = \ln(\hat{m}_{ij})$ have the familiar linear model formulation

$$\boldsymbol{\ell} = \mathbf{X}\,\boldsymbol{\lambda}$$

Similarly, for the saturated $2 \times 2 \times 2$ table, we have

$$l_{ijk} = \mu + \lambda_i^A + \lambda_j^B + \lambda_{ij}^{AB} + \lambda_k^C + \lambda_{ik}^{AC} + \lambda_{jk}^{BC} + \lambda_{ijk}^{ABC}$$

with the usual constraints again imposed for identifiability. The above can also be rewritten as:

$$\begin{aligned} l_{ijk} &= \mu + \lambda_1^A Z_{1i} + \lambda_1^B Z_{2j} + \lambda_{11}^{AB} Z_{1i}Z_{2j} + \lambda_1^C Z_{3k} + \lambda_{11}^{AC} Z_{1i}Z_{3k} \\ &+ \lambda_{11}^{BC} Z_{2j}Z_{3k} + \lambda_{111}^{ABC} Z_{1i}Z_{2j}Z_{3k} \end{aligned}$$

where

$$Z_{1i} = \begin{cases} 1 & \text{if } i = 1 \\ -1 & \text{if } i = 2 \end{cases}$$

the Z_{2j} and Z_{3k} are similarly defined. Again, the model reduces to

$$\begin{bmatrix} l_{111} \\ l_{112} \\ l_{121} \\ l_{122} \\ l_{211} \\ l_{212} \\ l_{221} \\ l_{222} \end{bmatrix} = \begin{bmatrix} 1 & 1 & 1 & 1 & 1 & 1 & 1 & 1 \\ 1 & 1 & 1 & 1 & -1 & -1 & -1 & -1 \\ 1 & 1 & -1 & -1 & 1 & 1 & -1 & -1 \\ 1 & 1 & -1 & -1 & -1 & -1 & 1 & 1 \\ 1 & -1 & 1 & -1 & 1 & -1 & 1 & -1 \\ 1 & -1 & 1 & -1 & -1 & 1 & -1 & 1 \\ 1 & -1 & -1 & 1 & 1 & -1 & -1 & 1 \\ 1 & -1 & -1 & 1 & -1 & 1 & 1 & -1 \end{bmatrix} \begin{bmatrix} \mu \\ \lambda_1^A \\ \lambda_1^B \\ \lambda_{11}^{AB} \\ \lambda_1^C \\ \lambda_{11}^{AC} \\ \lambda_{11}^{BC} \\ \lambda_{111}^{ABC} \end{bmatrix}$$

which can again be written as:

$$\boldsymbol{\ell} = \mathbf{X}\,\boldsymbol{\lambda}$$

For reduced models in the 2^2 and 2^3 tables, such models can similarly be written in the linear model format. Again, consider the models of independence and no three-factor interactions in the 2^2 and 2^3 tables, respectively, for example. The models have the following matrix formulations, respectively.

$$\begin{bmatrix} l_{11} \\ l_{12} \\ l_{21} \\ l_{22} \end{bmatrix} = \begin{bmatrix} 1 & 1 & 1 \\ 1 & -1 & 1 \\ 1 & 1 & -1 \\ 1 & -1 & -1 \end{bmatrix} \begin{bmatrix} \mu \\ \lambda_1^A \\ \lambda_2^B \end{bmatrix}$$

$$\begin{bmatrix} l_{111} \\ l_{112} \\ l_{121} \\ l_{122} \\ l_{211} \\ l_{212} \\ l_{221} \\ l_{222} \end{bmatrix} = \begin{bmatrix} 1 & 1 & 1 & 1 & 1 & 1 & 1 \\ 1 & 1 & 1 & 1 & -1 & -1 & -1 \\ 1 & 1 & -1 & -1 & 1 & 1 & -1 \\ 1 & 1 & -1 & -1 & -1 & -1 & 1 \\ 1 & -1 & 1 & -1 & 1 & -1 & 1 \\ 1 & -1 & 1 & -1 & -1 & 1 & -1 \\ 1 & -1 & -1 & 1 & 1 & -1 & -1 \\ 1 & -1 & -1 & 1 & -1 & 1 & 1 \end{bmatrix} \begin{bmatrix} \mu \\ \lambda_1^A \\ \lambda_1^B \\ \lambda_{11}^{AB} \\ \lambda_1^C \\ \lambda_{11}^{AC} \\ \lambda_{11}^{BC} \end{bmatrix}$$

In both cases, the last column in each of the X matrices (of constants) have been deleted. That is, the models can now be written as:

$$\boldsymbol{\ell} = \mathbf{X}^* \boldsymbol{\lambda}$$

where $\mathbf{X}^*$ is the reduced matrix of coefficients.

We see from the above representation of log-linear models that the standard least-squares approach (weighted) may be applied to estimate the parameters of interest and consequently, the expected values under a given model.

To demonstrate the implementation of Newton's algorithm, let us consider again, the data in Table 6.10 where we wish to again fit the model {RQ,RY,YQ} to the data. This model has RYQ = 0.

Let

$$c_{ijk} = \begin{bmatrix} 1 & -1 & -1 & 1 & -1 & 1 & 1 & -1 \end{bmatrix}$$

be the column containing the ABC (RYQ) interaction in the original matrix $\mathbf{X}$. Clearly, $\sum_{ijk} c_{ijk} = 0$. Also, let the initial estimates of the MLE be $\hat{m}_{ijk}^{(0)}$ and we would initially set these equal to the observed counts, that is, $\hat{m}_{ijk}^{(0)} = n_{ijk}$. To start the iteration, we also need to compute the followings:

$$\begin{aligned}
h^{(r)} &= \sum_{ijk} \left[\frac{c_{ijk}^2}{\hat{m}_{ijk}^{(0)}} \right] \\
&= \sum_{ijk} \frac{1}{\hat{m}_{ijk}^{(r)}} \\
\gamma^{(r)} &= \sum_{ijk} c_{ijk} l_{ijk}^{(r)} \\
f^{(r)} &= \frac{\gamma^{(r)}}{h^{(r)}} \\
\mu_{ijk}^{(r)} &= l_{ijk}^{(r)} - \frac{f^{(r)} c_{ijk}}{\hat{m}_{ijk}^{(r)}} \\
g^{(r)} &= \frac{N}{\sum_{ijk} \exp[\mu_{ijk}^{(r)}]} \\
\hat{m}_{ijk}^{(r)} &= g^{(r)} \exp[\mu_{ijk}^{(r)}] \qquad r = 0, 1, \cdots
\end{aligned}$$

where $l_{ijk}^{(0)} = \ln(n_{ijk})$ and N is the total sample size.

For our data, we have

$$\gamma^{(0)} = 0.11113 \quad h^{(0)} = 0.03833 \quad f^{(0)} = 2.89952$$

and using these we have $g^{(0)} = 0.99992$.

The results for the first iteration are given in the following table.

Cells	111	112	121	122	211	212	221	222
n_{ijk}	126	152	141	182	319	463	290	403
$\mu_{ijk}^{(0)}$	4.813	5.043	4.969	5.188	5.774	6.131	5.660	6.006
$\hat{m}_{ijk}^{(1)}$	123.12	154.92	143.92	179.11	321.89	460.07	287.09	405.88

To commence subsequent iterations, we would need to adjust the preceeding value of $l_{ijk}^{(r-1)}$ and repeat the above cycle of computations. The initial adjustment is of the form:

$$l_{ijk}^{(t)} = \ln(\hat{m}_{ijk}^{(t)}) + \frac{(n_{ijk} - \hat{m}_{ijk}^{(t)})}{\hat{m}_{ijk}^{(t)}} \qquad t = r+1, r+2, \cdots$$

For the second iteration, for example, we also have

$$\gamma^{(1)} = 0.11114 \quad h^{(1)} = 0.03834 \quad \text{and} \quad f^{(1)} = 2.89922$$

and using these again, $g^{(1)} = 1.000$. The results for the second iteration are given in the table below.

Cells	111	112	121	122	211	212	221	222
n_{ijk}	126	141	152	182	319	290	463	403
$\mu_{ijk}^{(1)}$	4.813	5.043	4.969	5.188	5.774	6.131	5.660	6.006
$\hat{m}_{ijk}^{(2)}$	123.10	154.90	143.90	179.10	321.90	460.10	287.10	405.90

Since $g^{(1)} = 1.000$, we have thus attained convergence and $m_{ijk}^{(2)}$ are the ML estimates based on the Newton-Raphson iterative procedure. We observe that these estimates agree to two decimal places with our previous estimates with the iterative proportional fitting algorithm.

However, if convergence has not yet been attained, the cycle of iterations would continue. In general,

$$g^{(t)} \longrightarrow 1.000 \quad f^{(t)} \longrightarrow f \quad \text{and} \quad m_{ijk}^{(t)} \longrightarrow \hat{m}_{ijk}$$

As pointed out in Haberman (1978), the G_0^2 test statistic under the above model is given by

$$G_0^2 = [f^{(0)}]^2 h^{(0)} = (2.89952)^2(0.03833) = 0.32222$$

This value of G_0^2 agrees with the G^2 computed for this model in the displayed SAS software output. The test statistic G_0^2 has been noted by Goodman (1964) to be equivalent to Woolf's (1955) test statistic for a no three-factor interaction in a $2 \times 2 \times 2$ contingency table.

Haberman (1978) has given a comprehensive approach to the Newton's iterative procedure. Most major statistical softwares adopt either the IPF algorithm or Newton's algorithm.

6.10 Interpretation of Parameters in Higher Tables

We consider in this section, the interpretations of parameters in log-linear models in three-way and four-way tables. Because of its simplicity relative to other higher dimensional tables, we will first consider the three-way tables in the next subsection.

6.10.1 The General Three-Way $I \times J \times K$ Tables

The two-factor effect AB represents the interaction between variables A and B averaged over variable C. We can therefore define two-factor effects as products of cross-product ratios. Let

$$\alpha^{(k)} = \frac{\hat{m}_{11k}\hat{m}_{22k}}{\hat{m}_{12k}\hat{m}_{21k}} \quad k = 1, 2$$

For the $2 \times 2 \times 2$, the above reduces to:

$$\hat{\lambda}_{ij}^{AB} = \frac{1}{8} \ln \left[\alpha^{(1)}\alpha^{(2)}\right]$$

The three-dimensional table has the three-factor effect ABC, which the two-dimensional model does not have. We derive ABC as the average value of (AB) across the tables and the particular value exhibited by k. Thus

$$\hat{\lambda}_{111}^{ABC} = \frac{1}{8} \ln \left[\frac{\alpha^{(1)}}{\alpha^{(2)}}\right] = \frac{1}{8} \left[\frac{\hat{m}_{111}\hat{m}_{221}\hat{m}_{122}\hat{m}_{212}}{\hat{m}_{121}\hat{m}_{211}\hat{m}_{112}\hat{m}_{222}}\right]$$

All the cells whose subscripts sum to *an odd number* appear in the numerator and all those whose subscripts sum to *an even number* appear in the denominator. The data in Table 6.10 are represented in the form below with computed relative risk estimates, where a.s.e. stands for asymptotic standard error.

Year	Quest.	R(i)			
Y(j)	Q(k)	1	2	Estimates	a.s.e
1975 (1)	Q 1	126	319	-0.9289	0.1052
	Q 2	141	290	-0.7211	0.1027
1976 (2)	Q 1	152	463	-1.1138	0.0935
	Q 2	182	403	-0.7949	0.0893

Table 6.15: Estimates of relative risk

Then, the relative risk of a response of *oppose gun registration* rather than *favor gun registration* is $\hat{m}_{1jk}/\hat{m}_{2jk}$, for persons in year j and form of question k. Hence

$$\hat{\tau}_{(12).jk}^{R.YQ} = \ln \left(\frac{n_{1jk}}{n_{2jk}}\right), \quad \begin{array}{l} i = 1, 2 \\ j = 1, 2 \end{array}$$

where $\hat{\tau}$ is the estimated log-odds and the estimated asymptotic standard error is given by:

$$\text{a.s.e.}(\hat{\tau}^{R.YQ}) = \left(\frac{1}{n_{1jk}} + \frac{1}{n_{2jk}}\right)^{\frac{1}{2}}$$

These estimates are given in the table above. We shall use these results in the next sections.

6.10.2 Three-Factor Effect Absent

Setting the ABC effect to zero enables us to describe a table with constant two-factor effects. That is, the model {AB,AC,BC}, which is given in log-linear formulation as

$$l_{ijk} = \mu + \lambda_i^A + \lambda_j^B + \lambda_k^C + \lambda_{ij}^{AB} + \lambda_{ik}^{AC} + \lambda_{jk}^{BC}$$

The model states that there is *"partial association"* between each pair of variables. The model has

$$\begin{aligned}\text{d.f.} &= IJK - [1 + (I-1) + (J-1) + (K-1) + (I-1)(J-1)] \\ &\quad - [(I-1)(K-1) + (J-1)(K-1)] \\ &= (I-1)(J-1)(K-1) \quad \text{degrees of freedom}\end{aligned}$$

The model assumes that every variable is directly associated with every other variable and that controlling one variable does not remove the association between the other two variables. In terms of odds ratios, the above implies that if we fix one factor variable, then the odds ratios relating the remaining two variables are constant for each categories of the fixed variable. For instance, if we hold variable C constant, then the odds ratio relating variables A and B at say fixed levels of C ($k = 1$ and $k = k$) are such that

$$\frac{\hat{m}_{111}\hat{m}_{ij1}}{\hat{m}_{i11}\hat{m}_{1j1}} = \frac{\hat{m}_{11k}\hat{m}_{ijk}}{\hat{m}_{i1k}\hat{m}_{1jk}}$$

corresponding to the hypothesis,

$$H_{01} : \frac{\pi_{111}\pi_{ij1}}{\pi_{i11}\pi_{1j1}} = \frac{\pi_{11k}\pi_{ijk}}{\pi_{i1k}\pi_{1jk}}$$

for $i = 1, 2, \cdots, (I-1)$, $j = 1, 2, \cdots, (J-1)$ and $k = 1, 2, \cdots, K$.

The above of course states that the model holds, if and only if the conditional log cross-products $\tau^{AB.C}_{(ii')(jj').k}$ are constants across variable C. That is,

$$\hat{\tau}^{AB(C)}_{(ii')(jj').k} = \ln\left(\frac{\hat{m}_{ijk}\hat{m}_{i'j'k}}{\hat{m}_{i'jk}\hat{m}_{ij'k}}\right) = \text{constant}, \quad k = 1, 2, .., I$$

The ML estimates for this model cannot be obtained directly, and iterative methods such as the IPF or Newton's algorithm are used to estimate the expected values. We have earlier used both algorithms to obtain the MLE for this model for the data in Table 6.10.

Under this model, the estimates of the partial log odds ratios are equal. That is,

$$\hat{\tau}^{AC(B)}_{(ii')(kk')} = \ln\left(\frac{\hat{m}_{ijk}\hat{m}_{i'jk'}}{\hat{m}_{i'jk}\hat{m}_{ijk'}}\right) = \ln\left(\frac{\hat{m}_{ij'k}\hat{m}_{i'j'k'}}{\hat{m}_{i'j'k}\hat{m}_{ij'k'}}\right)$$

$$\hat{\tau}^{AB(C)}_{(ii')(jj')} = \ln\left(\frac{\hat{m}_{ijk}\hat{m}_{i'j'k}}{\hat{m}_{i'jk}\hat{m}_{ij'k}}\right) = \ln\left(\frac{\hat{m}_{ijk'}\hat{m}_{i'j'k'}}{\hat{m}_{i'jk'}\hat{m}_{ij'k'}}\right)$$

$$\hat{\tau}^{BC(A)}_{(jj')(kk')} = \ln\left(\frac{\hat{m}_{ijk}\hat{m}_{ij'k'}}{\hat{m}_{ijk'}\hat{m}_{ij'k}}\right) = \ln\left(\frac{\hat{m}_{i'jk}\hat{m}_{i'j'k'}}{\hat{m}_{i'jk'}\hat{m}_{i'j'k'}}\right)$$

for $i \neq i'$, $j \neq j'$, and $k \neq k'$. The asymptotic variance for $\hat{\tau}^{AC(B)}_{(ii')(kk')}$ is given by:

$$s^2(\hat{\tau}^{AC(B)}_{(12)(12)}) = \left[\sum_j \left(\frac{1}{\hat{m}_{1j1}} + \frac{1}{\hat{m}_{1j2}} + \frac{1}{\hat{m}_{2j1}} + \frac{1}{\hat{m}_{2j2}}\right)\right]^{-1}$$

Similar expressions for the asymptotic variances of the other two parameter estimates can be easily formulated.

Model {RY, RQ, YQ} implies the following:

$$\begin{aligned}\hat{\tau}^{RQ(Y)}_{(12)(12)} &= \ln\left(\frac{\hat{m}_{111}\hat{m}_{212}}{\hat{m}_{112}\hat{m}_{211}}\right) = \ln\left(\frac{\hat{m}_{121}\hat{m}_{222}}{\hat{m}_{122}\hat{m}_{221}}\right) \\ &= -0.2705\end{aligned}$$

with estimated asymptotic variance computed as:

$$\frac{1}{\hat{m}_{111}} + \frac{1}{\hat{m}_{112}} + \frac{1}{\hat{m}_{212}} + \frac{1}{\hat{m}_{211}} = \frac{1}{123.10} + \frac{1}{460.10} + \frac{1}{321.90} + \frac{1}{154.90}$$
$$= 0.0199 \quad \text{for } j = 1$$

and

$$\frac{1}{\hat{m}_{121}} + \frac{1}{\hat{m}_{122}} + \frac{1}{\hat{m}_{221}} + \frac{1}{\hat{m}_{222}} = \frac{1}{287.10} + \frac{1}{179.10} + \frac{1}{143.90} + \frac{1}{405.90}$$
$$= 0.0185 \quad \text{for } j = 2$$

Hence,

$$\text{a.s.e.}(\hat{\tau}^{RQ(Y)}_{(12)(12)}) = \left(\frac{1}{0.0199} + \frac{1}{0.0185}\right)^{-\frac{1}{2}} = 0.0979$$

An approximate 95% confidence interval for $\tau^{RQ(Y)}_{(12)(12)}$ has bounds

$$-0.2705 - 1.96(0.0979) = -0.4624$$
$$-0.2705 + 1.96(0.0979) = -0.0786$$

These values agree with the results obtained from SAS® PROC GENMOD under model {RY,RQ,YQ}. Thus, given the year of the questionnaire, the odds that an individual will oppose gun registration rather than favor gun registration are estimated to be $\exp(-0.4624) = 0.63$ to $\exp(-0.0786) = 0.92$, from 0.63 to 0.92 times higher for those administered the question of form 1 (F1) than for those administered the question of form 2 (F2).

We can from the SAS software output obtain the parameter estimates and confidence intervals for the other two parameters. For instance, $\hat{\tau}^{RY(Q)}_{(12)(12)} = 0.1274$ with corresponding 95% confidence interval given as $(-0.0643, 0.3192)$. Thus given the type of question, the odds that an individual will oppose gun registration rather than favor gun registration are estimated to be $\exp(-0.0643) = 0.94$ to $\exp(0.3192) = 1.38$, from 0.94 to 1.38 times higher for those interviewed in 1975 than for those interviewed in 1976.

6.10.3 Three-Factor and One Two-factor Effect Absent

There are three versions of the model with the three-factor effect and one two-factor effect missing. The three generating classes have respectively

a ABC = BC = 0 implies model{AB,AC}

b ABC = AC = 0 implies model{AB,BC}

c ABC = AB = 0 implies model{AC,BC}

Selecting (a) above, that is, the model with BC (*and hence ABC*) absent, we have

$$l_{ijk} = \mu + \lambda^A_i + \lambda^B_j + \lambda^C_k + \lambda^{AB}_{ij} + \lambda^{AC}_{ik}$$

This model states that variables B and C are independent for every level of variable A, but each is associated with variable A. In other words, variables B and C are conditionally independent, given the level of variable A that is,

$$H_{02}: \quad \pi_{ijk} = \frac{\pi_{ij+}\pi_{i+k}}{\pi_{i++}}$$

where π_{ijk} is the underlying probability under multinomial sampling scheme. The model has direct MLEs given by

$$\hat{m}_{ijk} = \frac{n_{ij+}n_{i+k}}{n_{i++}}$$

with corresponding d.f. given by

$$\begin{aligned}\text{d.f.} &= IJK - [1 + (I-1) + (J-1)] \\ &- [(K-1) + (I-1)(J-1) + (I-1)(K-1)] \\ &= I(J-1)(K-1)\end{aligned}$$

The parameters λ_{ij}^{AB} and λ_{ik}^{AC} in the above log-linear formulation refer to the A-B and A-C partial associations, respectively. We may also note here that the conditional independence of B and C does not rule out the possibility that they may however be marginally dependent.

Under the conditional independence of B and C given A, written (Anderson, 1997) as $B \perp C|A$, that is, if model {AB,AC} is fitted to the data, then we would expect the estimates of the log cross-products $\tau_{(jj')(kk').i}^{BC.A} = 0$. That is,

$$\hat{\tau}_{(jj')(kk').i}^{BC.A} = \ln\left(\frac{\hat{m}_{ijk}\hat{m}_{ij'k'}}{\hat{m}_{ij'k}\hat{m}_{ijk'}}\right) = 0, \quad i = 1, 2, \cdots, I$$

Similar results and interpretations can also be obtained and made for models (b) and (c) above. Thus, if models {AB,BC} and {AC,BC} are fitted, then we would expect, respectively,

$$\hat{\tau}_{(ii')(kk').j}^{AC.B} = \ln\left(\frac{\hat{m}_{ijk}\hat{m}_{i'jk'}}{\hat{m}_{i'jk}\hat{m}_{ijk'}}\right) = 0$$

and

$$\hat{\tau}_{(ii')(jj').k}^{AB.C} = \ln\left(\frac{\hat{m}_{ijk}\hat{m}_{i'j'k}}{\hat{m}_{i'jk}\hat{m}_{ij'k}}\right) = 0$$

Again, when any one of these conditional independence models is applied, then the partial association coefficients are equal to their respective marginal association coefficients. That is,

$$\begin{aligned}\tau_{(jj')(kk').j}^{BC.A} &= \tau_{(jj')(kk')}^{BC} \\ \tau_{(ii')(jj').k}^{AB.C} &= \tau_{(ii')(jj')}^{AB} \\ \hat{\tau}_{(ii')(kk').j}^{AC.B} &= \hat{\tau}_{(ii')(kk')}^{AC}\end{aligned}$$

and the log-odds ratios (all zeros) reduce respectively to the following:

$$\begin{aligned}\ln\left(\frac{\hat{m}_{ijk}\hat{m}_{ij'k'}}{\hat{m}_{ij'k}\hat{m}_{ijk'}}\right) &= \ln\left(\frac{\hat{m}_{jk}\hat{m}_{j'k'}}{\hat{m}_{j'k}\hat{m}_{jk'}}\right) \\ \ln\left(\frac{\hat{m}_{ijk}\hat{m}_{i'j'k}}{\hat{m}_{i'jk}\hat{m}_{ij'k}}\right) &= \ln\left(\frac{\hat{m}_{ij}\hat{m}_{i'j'}}{\hat{m}_{i'j}\hat{m}_{ij'}}\right) \\ \ln\left(\frac{\hat{m}_{ijk}\hat{m}_{i'jk'}}{\hat{m}_{i'jk}\hat{m}_{ijk'}}\right) &= \ln\left(\frac{\hat{m}_{ik}\hat{m}_{i'k'}}{\hat{m}_{i'k}\hat{m}_{ik'}}\right)\end{aligned}$$

Only models {RY,YQ} and {RQ,YQ} are applied to the data in Table 6.10. The third model {RY, RQ} is not considered because that would mean fitting a model conditional on the response variable R. The results from the SAS software program below are presented below in Table 6.16.

```
set tab611;
proc genmod;
class r y q;
model count=r|y y|q/dist=poi;
run;
proc genmod;
class r y q;
model count=r|q y|q/dist=poi;
run;
```

Cell	n_{ijk}	**{RY,YQ}**	**{RQ,YQ}**
111	126	135.634	116.708
112	141	131.366	137.021
121	152	171.175	161.293
122	182	162.825	185.979
211	319	309.366	328.293
212	290	299.634	293.979
221	463	443.825	453.708
222	403	422.175	399.021
df		2	2
G^2		8.109	2.015
X^2		8.1059	2.0228

Table 6.16: Estimated cell counts $\hat{m}_{ijk}$ under the two models considered above

Model {RY, YQ} for instance has,

$$\hat{\tau}^{RQ.Y}_{(12)(12).j} = \ln\left(\frac{\hat{m}_{111}\hat{m}_{212}}{\hat{m}_{211}\hat{m}_{112}}\right) = \ln\left[\frac{(135.634)(299.634)}{(309.366)(131.366)}\right] = \ln(1.0000)$$

$$= \ln\left(\frac{\hat{m}_{121}\hat{m}_{222}}{\hat{m}_{221}\hat{m}_{122}}\right) = \ln\left[\frac{(171.175)(422.175)}{(443.825)(162.825)}\right] = 0$$

For the two-way table collapsed over the Y variable, that is, the R-Q subtable, the expected values under the model of marginal independence are $\hat{m}_{11} = 306.869$, $\hat{m}_{12} = 294.131$, $\hat{m}_{21} = 753.131$, $\hat{m}_{22} = 721.869$. Consequently,

$$\ln\left(\frac{\hat{m}_{11}\hat{m}_{22}}{\hat{m}_{12}\hat{m}_{21}}\right) = 0.0$$

Similar results can be obtained for model {RQ, YQ}. Of the two models, only model {RQ, YQ} fits the data with a $G^2 = 2.0154$ on 2 d.f. (pvalue = 0.3651). Therefore, let us consider the variation involving individuals interviewed in year j (Y1 = 1975, Y2 = 1976) given that an individual is administered the form of questions (Q) at levels $k = 1$ or $k = 2$, the estimated log-odds (log of relative risk) is estimated by:

$$\hat{\tau}^{RY.Q}_{(12)(12).j} = \ln\left(\frac{\hat{m}_{11k}\hat{m}_{22k}}{\hat{m}_{12k}\hat{m}_{21k}}\right)$$

For the form of question Q1, ($k = 1$, in this case), we have

$$\begin{aligned}\hat{\tau}^{RY.Q}_{(12)(12).1} &= \ln\left(\frac{\hat{m}_{111}\hat{m}_{221}}{\hat{m}_{121}\hat{m}_{211}}\right)\\ &= \ln\left(\frac{\hat{m}_{111}}{\hat{m}_{211}}\right) - \ln\left(\frac{\hat{m}_{121}}{\hat{m}_{221}}\right)\\ &= -1.0342 - (-1.0342)\\ &= 0\end{aligned}$$

with corresponding estimated asymptotic standard error

$$\left(\frac{1}{126} + \frac{1}{319} + \frac{1}{152} + \frac{1}{463}\right)^{\frac{1}{2}} = 0.1407$$

Thus, for question form 1 (Q1), we have a standardized value of $(0.0/0.1407) = 0$, which is not much strong evidence of different relative risks between the form of questions. An approximate 95% confidence lower and upper bound are estimated as $-0.2709 \pm 1.96(0.0970) = -0.2709 \pm 0.1902 = [-0.46, -0.08]$.

6.10.4 Three-Factor and Two Two-Factor Effects Absent

Again, for the three variables A, B, and C, there are three versions of the model with the three-factor effect and two two-factor effects missing. These are respectively:

a ABC = AB = AC = 0 implies model {A,BC}

b ABC = AB = BC = 0 implies model {B,AC}

c ABC = AC = BC = 0 implies model {C,AB}

Selecting (a) again, we have the log-linear model formulation

$$l_{ijk} = \mu + \lambda_i^A + \lambda_j^B + \lambda_k^C + \lambda_{jk}^{BC}$$

Variable A is now *completely independent* of the other two variables, while variables B and C are associated. The model has the probability structure

$$H_{03}: \quad \pi_{ijk} = \pi_{i++}\pi_{+jk}$$

and expected cell values

$$\hat{m}_{ijk} = \frac{n_{i++}n_{+jk}}{N}$$

with corresponding

$$\begin{aligned}\text{d.f.} &= IJK - [1 + (I-1) + (J-1) + (K-1) + (J-1)(K-1)]\\ &= (I-1)(JK-1)\end{aligned}$$

We say that A is jointly independent of B and C written succinctly as $A \perp B, C$ and this implies that A and B are conditionally independent given C. In addition, A and C are also conditionally independent given B. Further, A is independent of B and C in the A-B and A-C tables, and thus the partial odds ratios in the B-C table at the level of A and its corresponding marginal odds ratio are equal. A further implication of the above is that we can collapse the three-way table over variable A

Cell	n_{ijk}	{RQ,Y}	{RY,Q}	{YQ,R}
111	126	117.31	136.33	128.83
112	141	136.29	130.67	124.77
121	152	160.69	170.54	178.04
122	182	186.71	163.46	169.36
211	319	329.98	310.95	316.17
212	290	292.42	298.05	306.23
221	463	452.02	442.18	436.96
222	403	400.58	423.82	415.64
d.f.		3	3	3
G^2		2.057	8.151	9.829

Table 6.17: Estimated cell counts $\hat{m}_{ijk}$ under the three models considered above

as it is independent of B and C without affecting any of the remaining parameters of the subscripted terms.

Similar results can also be obtained for models (b) and (c) above. As an example, we apply the three models to the gun registration data in Table 6.10 and the results are presented in Table 6.17.

The model {RQ,Y}, which fits the data, postulates that the year of survey (Y) is completely independent of the remaining two variables R and Q. This model has a G^2 value of 2.0566 on 3 d.f. with a corresponding pvalue of 0.561. That is, Y and R are conditionally independent given Q and that Y and Q are also conditionally independent given R. From models {RQ,YQ} in Table 6.16 and model {RQ,Y} in the table above, the contribution of the interaction term {YQ} has a $G^2 = 0.042$ on 1 d.f., which is clearly not significant. Hence model {RQ,Y} seems the most parsimonious model to the data in Table 6.10. This model also suggests that the form of the question (Q) is associated with gun registration response (R) and that this association is independent of the level of Y. We can put this in terms of log odds ratio as

$$\tau^{RQ.Y}_{(ii')(kk').j} = \ln\left(\frac{\hat{m}_{ijk}\hat{m}_{i'jk'}}{\hat{m}_{i'jk}\hat{m}_{ijk'}}\right)$$

That is, for $j = 1, 2$ the expression below should be a constant:

$$\ln\left(\frac{\hat{m}_{1j1}\hat{m}_{2j2}}{\hat{m}_{2j1}\hat{m}_{1j2}}\right)$$

For the data, we have for $j = 1$ and $j = 2$ the partial association log odds ratios as:

$$\ln\left(\frac{117.31 \times 292.42}{329.98 \times 136.29}\right) = -0.2709 = \ln\left(\frac{160.69 \times 400.58}{452.02 \times 186.71}\right)$$

The corresponding observed log-odds of the marginal RQ table is

$$\ln\left(\frac{278 \times 693}{323 \times 782}\right) = -0.2709$$

The parameter estimates for $\lambda^{RQ}, \lambda^{R}, \lambda^{Q}$ should be equal to those of the saturated model based on the 2×2 R-Q marginal table (collapsed over Y). That is, $\lambda^{RQ}, \lambda^{R}, \lambda^{Q}$ in the marginal table. We give the SAS software program and modified outputs when these two models are fitted to the data in Table 6.10 below. As expected, the parameter estimates are equal.

```
set tab611;proc genmod;
class r y q; model count=r|q y/dist=poi; run;

MODEL {RQ,Y}
============


Class       Levels    Values

R                2    1 2
Y                2    1 2
Q                2    1 2

          Criteria For Assessing Goodness Of Fit

Criterion                  DF          Value         Value/DF
-------------------------------------------------------------
Deviance                    3         2.0566           0.6855
Pearson Chi-Square          3         2.0620           0.6873

                      Analysis Of Parameter Estimates
                                Standard        Chi-
Parameter          DF  Estimate    Error      Square  Pr > ChiSq
-----------------------------------------------------------------
Intercept           1    5.9929   0.0424      5.9099      <.0001
R            1      1   -0.7634   0.0674     -0.8954      <.0001
Q            1      1    0.1208   0.0522      0.0186      0.0206
R*Q          1  1   1   -0.2709   0.0970     -0.4610      0.0052
Y            1      1   -0.3147   0.0444     -0.4018      <.0001

MODEL {RQ}-Saturated Model.
=========
proc genmod;
class r q;
model count=r|q/dist=poi;
run;
***fit independence model to collapsed table****;
proc freq; weight count;
tables r*q/chisq;
run;
                      Analysis Of Parameter Estimates
                                Standard        Chi-
Parameter          DF  Estimate    Error      Square  Pr > ChiSq
-----------------------------------------------------------------
Intercept           1    6.5410   0.0380    29650.10      <.0001
r            1      1   -0.7634   0.0674      128.39      <.0001
q            1      1    0.1208   0.0522        5.36      0.0206
r*q          1  1   1   -0.2709   0.0970        7.79      0.0052
```

The resulting collapsed 2×2 R-Q table has fitted G^2=7.8133 on 1 d.f. for the test of independence. Clearly, the response R of an individual strongly depends on the form of question Q regardless of the year of survey. Since both marginal {R,Q} and partial {RY,YQ} indicate strong dependence of variables R and Q (no contradiction: both did not hold), we can thus collapse the three-way table over variable Y without distorting the primary association between variables R and Q.

On the other hand, the partial and marginal fits of models {RQ,YQ}, {R,Y} give $G^2 = 2.015$ and $G^2 = 1.7193$ on 2 and 1 degrees of freedom, respectively. However, model {RY,Q} does not fit the data. Thus in this case, both the partial association and marginal models fit the data but the model of joint independence does not fit the data. Hence, collapsibility is not possible over variable Q without distorting the primary association between variables R and Y.

It is obvious from the above results that a condition that must hold for the log odds ratios of the partial and marginal associations to be equal is that the joint independence model of the form {AB,C} must necessarily hold true. In our case, only model {RQ,Y}, fits the data as displayed in Table 6.17, and it should therefore be possible to collapse over variable Y. We shall explore collapsibility conditions further in chapter 7.

6.10.5 Three-Factor and All Two-Factor Absent

This model represents *complete independence* of all the three variables and has the log-linear representation

$$l_{ijk} = \mu + \lambda_i^A + \lambda_j^B + \lambda_k^C$$

The model is written as {A,B,C} and has the probability structure

$$H_{04}: \quad \pi_{ijk} = \pi_{i++}\pi_{+j+}\pi_{++k}$$

and expected cell values

$$\hat{m}_{ijk} = \frac{n_{i++}n_{+j+}n_{++k}}{N^2}$$

and corresponding $IJK - [1 + (I-1) + (J-1) + (K-1)]$, that is, $(IJK - I - J - K + 2)$ degrees of freedom. The model is sometimes described as the model of *mutual* or *pairwise independence.* Tables satisfying this hypothesis have the important property that collapsing the table over any of the variables retains the independence of the other two variables. That is, any two pairs of variables are conditionally independent (given the third variable) and also marginally independent. This model when applied to the data in Table 6.10, that is, model {R, Y, Q}, has a $G^2 = 9.8699$ and is based on 4 d.f. with a pvalue of 0.043. This model therefore does not fit the data, and indicates that the three variables are not pairwise independent. The model can be written succinctly as $A \perp B \perp C$.

A general relationship between the various forms of independence described in this section can be succinctly put in the form:

$$(a) \Rightarrow (b) \Rightarrow \{c(i) \quad \text{or} \quad c(ii)\}$$

where

(a) means A, B, C mutually independent {A,B,C}: $A \perp B \perp C$

(b) means B is jointly independent of A and C, {B,AC}: $B \perp A, C$

(c{i}) means A and B marginally independent, {A,B}: $A \perp B$

(c{ii}) means A and B conditionally independent, {AC,BC}: $A \perp B|C$: AB missing.

Generally therefore, *marginal independence implies conditional independence (or vice versa).* For example, model

$$\{R,Y\} \Rightarrow \{RQ,YQ\} \quad \text{if both models fit data}$$
$$\{R,Q\} \Rightarrow \{RY,YQ\} \quad \text{hence both model do not fit data}$$

Further, the partial association between variables A and B given level k of C (for example) has the log odds ratios defined by:

$$\tau_{(ii')(jj').k}^{AB.C} = \left(\frac{\hat{m}_{ijk}\hat{m}_{i'j'k}}{\hat{m}_{i'jk}\hat{m}_{ij'k}}\right) = H, \quad k = 1, 2, \cdots, K$$

where H is a constant. If this partial association model holds, then H would be equal to zero. That is, H = 0 when the model {AC,BC} holds. As discussed above, there are three variants of this model.

Also, the marginal association between variables A and B (variable C being ignored) has its log odds ratios defined by:

$$\tau^{AB}_{(ii')(jj')} = \left(\frac{\hat{m}_{ij}\hat{m}_{i'j'}}{\hat{m}_{i'j}\hat{m}_{ij'}}\right) = W$$

If this model holds, then $W = 0$.

The basic relationship between partial association and marginal association measures is that $\tau^{AB.C}_{(ii')(jj').k} = \tau^{AB}_{(ii')(jj')}$ holds if either A and C is conditionally independent given B (that is, model {AB, BC} holds) or B and C are conditionally independent given A (that is, model {AB, AC} holds). For three-way tables, therefore, if models {AB, BC} and {AB, AC} fit the data set, then it will be possible to collapse over variable C without incurring Simpson's paradox. In our example, the models {RY, RQ} and {RQ,YQ} hold, and therefore it is possible to collapse over variable Y.

6.10.6 Interpretation of Parameters

For the data in Table 6.10, the most parsimonious model is model {RQ, Y} with $G^2 = 2.0566$ on 2 d.f. The ML parameter estimates from both GENMOD and CATMOD procedures for the $\hat{\lambda}^{RQ}_{ik}$ are

Parameter estimates for $\hat{\lambda}^{RQ}_{ik}$ under GENMOD

	Q (k)	
R(i)	1	2
1	-0.2709	0
2	0	0

Parameter estimates for $\hat{\lambda}^{RQ}_{ik}$ under CATMOD

	Q (k)	
R(i)	1	2
1	-0.0677	0.0677
2	0.0677	-0.0677

The interaction between gun registration response and form of question indicates that fewer respondents oppose gun registration when asked the form of question 1 than question 2. Thus the odds of an individual opposing gun registration is $e^{-(-0.2709)} = 1.31 = e^{-(4*-.0677)}$ higher among those asked question form 2 than among those asked question form 1, averaged over the years. Similarly, more of the respondents favor gun registration when asked the form of question 1 than when asked the form of question 2. The odds here also are 1.31. Parameter estimates for the main effect $\hat{\lambda}^{Y}_{j} = (-0.1574, 0.1574)$ indicate that there are more respondents in 1976 than in 1975 in the survey. Similarly, parameter estimates for $\hat{\lambda}^{R}_{i} = (-0.4494, 0.4494)$ again indicate that there are more respondents overall favoring gun registration in the survey.

6.10.7 Example: Three-Way Table

The following example from Aicken (1983) relates to a follow-up survey in which voters were first asked to express their political preference (P), and then, upon followup, their actual voting pattern (V). The voters were further classified according to their income level (I).

		V (k)		
P (i)	I (j)	V0	V1	Totals
P0	$< 10,000$	112	7	119
	10-15,000	83	13	96
	15-20,000	76	8	84
	$> 20,000$	86	11	97
Totals		357	39	396
P1	$< 10,000$	67	37	104
	10-15,000	75	45	120
	15-20,000	67	57	124
	$> 20,000$	67	63	130
Totals		276	202	478
P2	$< 10,000$	5	35	40
	10-15,000	3	28	31
	15-20,000	9	35	44
	$> 20,000$	14	71	85
Totals		31	169	200

Table 6.18: Observed counts for the political preference survey

For Table 6.18,

P = *political preference* with categories:
- P0 = Democrat
- P1 = Independent
- P2 = Republican

V = *actual vote*
- V0 = Democrat
- V1 = Republican

I = *income level* (in \$) per year.
- I0 = $< 10,000$
- I1 = $10,000 - 15,000$
- I2 = $15,000 - 20,000$
- I3 = $> 20,000$

The above table is therefore a $3 \times 4 \times 2$ three-way contingency table. If we fit first a saturated model to the data, then we would be in a position to examine each main effect and interaction individually from the type3 tests. The saturated model from SAS software using PROC GENMOD is displayed in the modified output below:

```
data tab616;
do p=1 to 3; do i=1 to 4;  do v=1 to 2;
input count @@; output; end; end; end;
datalines;
112 7 83 13 76 8 86 11 67 37 75 45 67 57 67 63
```

```
5 35 3 28 9 35 14 71
;
proc freq; weight count; tables p*(I,V) I*V/chisq; run;
proc genmod; class p i v;
model count=p|i|v/dist=poi link=log type3; run;
```

Analysis Of Parameter Estimates

Parameter				DF	Estimate	Standard Error	Chi-Square	Pr > ChiSq
Intercept				1	4.2627	0.1187	1290.10	<.0001
P	1			1	-1.8648	0.3240	33.12	<.0001
P	2			1	-0.1195	0.1731	0.48	0.4898
I	1			1	-0.7073	0.2065	11.73	0.0006
I	2			1	-0.9305	0.2232	17.39	<.0001
I	3			1	-0.7073	0.2065	11.73	0.0006
P*I	1	1		1	0.2553	0.5258	0.24	0.6272
P*I	1	2		1	1.0975	0.4665	5.53	0.0186
P*I	1	3		1	0.3889	0.5085	0.58	0.4444
P*I	2	1		1	0.1751	0.2925	0.36	0.5494
P*I	2	2		1	0.5940	0.2965	4.01	0.0451
P*I	2	3		1	0.6072	0.2758	4.85	0.0277
V	1			1	-1.6236	0.2924	30.83	<.0001
P*V	1	1		1	3.6801	0.4336	72.02	<.0001
P*V	2	1		1	1.6852	0.3410	24.42	<.0001
I*V	1	1		1	-0.3223	0.5604	0.33	0.5652
I*V	2	1		1	-0.6100	0.6742	0.82	0.3656
I*V	3	1		1	0.2655	0.4745	0.31	0.5758
P*I*V	1	1	1	1	1.0384	0.7539	1.90	0.1684
P*I*V	1	2	1	1	0.4074	0.8038	0.26	0.6123
P*I*V	1	3	1	1	-0.0707	0.6826	0.01	0.9175
P*I*V	2	1	1	1	0.8545	0.6220	1.89	0.1695
P*I*V	2	2	1	1	1.0592	0.7217	2.15	0.1422
P*I*V	2	3	1	1	-0.1654	0.5371	0.09	0.7581

LR Statistics For Type 3 Analysis

Source	DF	Chi-Square	Pr > ChiSq	
P	2	179.28	<.0001	
I	3	14.07	0.0028	
P*I	6	16.47	0.0114	**
V	1	6.13	0.0133	
P*V	2	329.65	<.0001	**
I*V	3	2.47	0.4812	
P*I*V	6	7.51	0.2763	

The **type3** statement in the above program instructs GENMOD to produce a maximum likelihood analysis of the parameters. Examination of these indicate that only the PI and PV interaction terms are significantly different from zero (**). The interaction term IV is obviously not important in the model. SAS software also gives the parameter estimates (under last equal zero constraints) under the saturated model (together with the standard errors). The estimates are read in the other of appearance of the variables in the model. Thus $\hat{\lambda}_1^P = -1.86482, \hat{\lambda}_2^P = -0.1195$, with $\hat{\lambda}_3^P = 0$. The next three parameter estimates relate to I (i.e., #4, 5, 6), while the parameter estimate numbered 7 (#7) relates to V, etc.

The number of parameters for the PIV interaction term equals $(3-1) \times (4-1) \times (2-1) = 6$.

An examination of the pvalues for these parameters indicates that none of the six terms in the three-factor interaction is significantly different from zero. Similarly, of the two-factor interaction terms, all three IV terms indicate nonsignificant pvalues.

However, both PI and PV two-factor interaction terms each have one parameter term significantly different from zero.

While results from the above table do suggest that possible candidates for consideration in the final model might be the model {PI, PV}, the saturated approach has some drawbacks, especially when there are many factor variables. One drawback arose because for multifactor situations, it is not always clear which effects are to be removed without affecting the others in the model. Removing an effect of course implies that that effect has been tested to be significantly not different from zero. Thus the problem with the saturated procedure has been to determine which effect or effects are to be tested for zero. We shall consider two procedures that have been advocated for testing each of the individual effects, namely, *tests of partial and marginal associations*, in the next section.

Another starting point for determining the most parsimonious model for our data is to start by fitting log-linear models first to all the r one-factor effects, then to all the two-factor effects, then to all the three-factor effects and so on and so forth, that is, a stepwise approach. In our example above this would mean that we fit the following models:

(a) {P, I, V}

(b) {PI, PV, VI}

(c) {PIV}

The results of fitting the above models are displayed in the following table:

Model	df	G^2	X^2	P-value
{P,I,V}	17	392.9177	379.5464	< 0.0001
{PI, PV, VI}	6	7.5096	7.4134	0.276
{PIV}	0	0	0	

We can conduct goodness-of-fit tests on each of the above models to determine the smallest model that fits our data. In our example above, the smallest model that fits the data is model (b). We may therefore consider eliminating terms from model (b). This procedure also has the drawback that the choice of an initial model is very crucial, as different initial models may give different results. Having identified model (b) as our possible starting point, the next question of course would be if indeed we would need all the three terms PI, PV, and VI in the model. We shall discuss these data further at a later section in this chapter.

6.11 Model Interpretations in Higher Tables

Consider a four-dimensional contingency table with factors A, B, C, and D indexed by i, j, k, l and with I, J, K, L Categories, respectively. We give below some possible models for the four-way table. Some of these models are described in Christensen (1990).

6.11.1 No Four-Factor Interaction

Possible models and their interpretations are the following:

1. {ABC,ABD} This model is interpreted as given A and B, factors C and D are conditionally independent, and is written as $C \perp D|A, B$. The model is graphic and decomposable.

2. {ABC,AD,BD} Given A and B, factors C and D are conditionally independent, written as $C \perp D|A, B$

3. {ABC,AD,BD,CD} Each pair of factors is conditionally dependent, but at each level of D, the association between A and B or between A and C or between B and C varies across the levels of the remaining factor (Agresti, 1990). Further, the partial association between D and another variable is the same at each combination of levels of the other two factors.

4. {ABC,AD} Given A, factor D is independent of factors B and C, written as $D \perp B, C|A$. The model is graphic and decomposable.

5. {ABC,D} Factor D is independent of factors A, B, and C, written as $D \perp A, B, C$. The model is graphic and decomposable.

6.11.2 No Four-, and Three-Factor Terms in the Models

Possible models are:

6. {AB,AC,AD,BC,BD,CD} For this model, each pair of factors is conditionally dependent, given the other two factors. Thus if A and B are conditionally independent at each combination levels of C and D, then $\lambda_{kl}^{CD} = 0$. In general, conditional independence between two factors at the several combination levels of the other two factors in a four-way table implies the absence of some two-factor interaction term.

7. {AB,AC,AD,BC} Given A, factor D is independent of factors B and C, written as $D \perp B, C|A$.

8. {AB,BC,CD,AD} Given B and D, A and C are independent, and given A and C, B and D are independent. The model can be written as $A \perp C|B, D$ and $B \perp D|A, C$ and the model is graphical.

9. {AB,AC,AD} Given A, factors B, C, and D are all independent, written as: $B \perp C \perp D|A$.

10. {AB,AC,BD} Given A, factor C is independent of factors B and D, written as $C \perp B, D|A$, and given B, factor D is independent of factors A and C, written as $D \perp A, C|B$.

11. {AB,CD} Factors A and B are independent of factors C and D (and vice-versa), written as $A \perp B|C, D$. The model is graphic and decomposable.

12. {AB,AC,D} Factor D is independent of factors A, B, and C. Given factor A, factor B is independent of factor C. The model is written succinctly as $D \perp A, B, C$ and $B \perp C|A$, respectively. The model is both graphic and decomposable.

13. {AB,C,D} Factor C is independent of factors A, B, and D, that is, $C \perp A, B, D$, and factor D is independent of factors A, B, and C written as $D \perp A, B, C$. The model is graphic and decomposable. An alternative interpretation provided for this model in Anderson (1997) is that factors A, B are independent of factors C, D, written as: $A, B \perp C, D$, and factor C is independent of factor D, again written as $C \perp D$.

6.11.3 No Four-, Three-, and Two-Factor Terms in the Models

The only possible model here is the model of mutual independence:

14. {A,B,C,D} All factors are independent of all other factors, written as $A \perp B \perp C \perp D$. The model is both graphic and decomposable.

Note: For models 2 and 7, these models imply their interpretations, but the interpretations do not imply the models.

Expressions for the maximum likelihood estimates for all of these models when they exist are given in the next table.

Model Number	MLE $\hat{m}_{ijkl}$
1.	$\frac{n_{ijk+}n_{ij+l}}{n_{ij++}}$
2.	No closed form (ABD missing)
3.	No closed form
4.	$\frac{n_{ijk+}n_{i++l}}{n_{i+++}}$
5.	$\frac{n_{ijk+}n_{+++l}}{N}$
6.	No closed form
7.	No closed form (ABC missing)
8.	No closed form
9.	$\frac{n_{ij++}n_{i+k+}n_{i++l}}{n^2_{i+++}}$
10.	$\frac{n_{ij++}n_{i+k+}n_{+j+l}}{n_{i+++}n_{+j++}}$
11.	$\frac{n_{ij++}n_{++kl}}{N}$
12.	$\frac{n_{ij++}n_{i+k+}n_{+++l}}{Nn_{i+++}}$
13.	$\frac{n_{ij++}n_{++k+}n_{+++l}}{N^2}$
14.	$\frac{n_{i+++}n_{+j++}n_{++k+}n_{+++l}}{N^3}$

The models numbered 3, 6, and 8 do not have direct ML estimates because they do not contain their generating term ABCD.

6.12 Tests of Marginal and Partial Associations

When we are faced with higher dimensional contingency tables with several factors (presumably more than three), we do not usually have a defined hypothesis of interest, and in this case we are usually interested in the parameters that need to be included in a model that will fit the data well. One possible solution to this would be to set up a table of all possible hierarchical log-linear models for the data. However, for tables involving four or more factor variables, this can be so numerous that one would be motivated to search for a solution that would reduce considerably the number of such possible hierarchical models.

Brown (1976) proposed such a screening procedure that enables us to make an initial assessment on the significance of the individual parameters in the saturated model. The procedure proposed by Brown is described as the *marginal and partial association tests.*

We are therefore interested in testing whether or not to include a particular parameter in the model. Our approach would be that, for both marginal and partial tests, we would fit a model containing the effect of interest and another not containing the effect of interest, and assessing its significance by calculating the relevant G^2 and the corresponding difference in the number of degrees of freedom.

6.12.1 Tests of Partial Association

To illustrate this, consider a situation where we have four ($s = 4$) factor variables A, B, C and D. Then the test for partial association is conducted by first fitting the largest model of third order ($s - 1$), which in this case would be the model {ABC,ABD,ACD,BCD}. Next, we drop, say, the λ^{ABC} or the ABC-term from the model. That is, we would now fit the reduced model {ABD,ACD,BCD}. The difference in G^2 and the corresponding d.f., enables us to test for the partial association of the ABC term. Similarly, a partial association test of the BD-term, say, is again obtained by the test of {AB,AC,AD,BC,BD,CD} against {AB,AC,AD,BC,CD}. This procedure is implemented in SAS software by the type3 likelihood ratio tests for each of these terms. First the model of pairwise independence is fitted and type3 G^2 for each term is obtained; next the model of partial association is fitted, and finally the saturated model is fitted. Below is the result of such analysis for the $3 \times 4 \times 2$ data in Table 6.18.

```
set tab616;
proc genmod; class p i v; model count=p i v/dist=poi type3; run;
proc genmod; class p i v; model count=p|i|v@2/dist=poi type3; run;
proc genmod; class p i v; model count=p|i|v/dist=poi type3; run;
```

```
      LR Statistics For Type 3 Analysis
                              Chi-
Source              DF      Square    Pr > ChiSq
------------------------------------------------

P                    2      123.37        <.0001
I                    3        9.61        0.0222
V                    1       60.64        <.000

P*I                  6       17.69        0.0071
P*V                  2      336.10        <.0001
I*V                  3        4.30        0.2304

P*I*V                6        7.51        0.2763
```

The results here indicate that important two-factor terms are PI and PV.

6.12.2 Tests for Marginal Association

The test for marginal association of A, B, and C is made by collapsing the table to the A,B,C-margin. That is, we collapse over any factor or factors not included in the particular term of interest (in this case, the ABC term). We would then test that $\lambda^{ABC} = 0$ in the marginal table thus formed, that is, the test of {AB,AC,BC} against the model {ABC}. Similarly, a test for marginal association ABCD is the test of {ABC,ABD,ACD,BCD} against {ABCD}. Again, the test of marginal association AD is the test of {A,D} against {AD}. The above tests are each equivalent to dropping (i) λ^{ABC} from model {ABC}, (ii) λ^{ABCD} from model {ABCD}, and (iii) λ^{AD} from model {AD}, respectively.

Let us illustrate marginal association implementation with the $3 \times 4 \times 2$ data in Table 6.18. The highest marginal association models we can fit here are the two-factor models since there are only three factors in the data. These marginal association models are accomplished in SAS® by PROC FREQ and we give below modified results from these implementations.

```
set tab616;
proc freq; weight count; tables p*(v i) v*i/chisq; run;

Statistics for Table of P by I

Statistic                     DF        Value      Prob
-------------------------------------------------------
Chi-Square                     6      32.3191    <.0001
Likelihood Ratio Chi-Square    6      31.3466    <.0001

Statistics for Table of P by V

Statistic                     DF        Value      Prob
-------------------------------------------------------
Chi-Square                     2     319.8592    <.0001
Likelihood Ratio Chi-Square    2     349.7572    <.0001

Statistics for Table of I by V

Statistic                     DF        Value      Prob
-------------------------------------------------------
Chi-Square                     3      17.9056    0.0005
Likelihood Ratio Chi-Square    3      17.9649    0.0004
```

Clearly, the marginal association (or interaction) (PV) has the highest G^2 value of 319.8592 on 2 degrees of freedom, reflecting the strong dependence in the P-V subtable displayed in the next table.

	V		
P	1	2	Total
1	357	39	396
2	276	202	478
3	31	169	200
Total	664	410	1074

All the marginal associations are highly significant based on marginal association analyses here. Marginal associations are prone to the risk of Simpson's paradox, and a proper procedure to combine the information from the marginal association analyses with those from the partial associations analyses would be needed. We can summarize these results in Table 6.19.

No	k	df	G^2	Hypothesis	pvalue	Decision
(i)	3	6	7.510	PIV = 0	0.2763	Fail to reject
(ii)	2	17	392.918	PI=PV=VI=0	< 0.0001	Reject
(iii)	1	23	586.545	P=V=I=0	< 0.0001	Reject

Table 6.19: Tests that k-factor and higher order effects are simultaneously zero

The first line, $k = 3$, gives the G^2 for the model without the three-factor interaction PIV. That is, the line tests the hypothesis that PIV = 0. From this result, there is no sufficient reason not to accept this hypothesis. The line with $k = 2$ indicates the model without the third- and second-order effects (because of the hierarchy principle). The pvalue for this hypothesis strongly suggest that this hypothesis is not tenable. The last line, $k = 1$, corresponds to a model that has no effects (that is, all effects are zero), except for the grand mean. Again, this model is not tenable. We note that the above tests are based on fitting all k-factor marginals to the relevant data. A model with first- and second-order effects would seem adequate to represent our data.

It is sometimes necessary to test whether all interactions of a given order are simultaneously zero. In the above analysis, interest centers on whether all effects greater than a certain order are zero; here, however, our interest centers on whether that particular order interaction is zero. Table 6.20 displays the relevant information obtained from Table 6.19 by taking the difference.

No	k	d.f.	G^2	pvalue
(i)	1	6	193.627	0.0000
(ii)	2	11	385.408	0.0000
(iii)	3	6	7.510	0.2763

Table 6.20: Tests that k-factor effects are simultaneously zero

The G^2 value of 193.627 in Table 6.20 is the difference between the G^2 values of a model with only the mean and first-order effects ($586.545 - 392.918 = 193.627$). This value is an indication of how the inclusion of the first-order effects have improved our model. The pvalue thus indicates a significant contribution. On the other hand, the table also informs us that the contribution of the third-order term PIV is not significant. Once again, this indicates that only first- and second-order terms need be in our model.

The partial tests earlier displayed also indicate that of all the main-effect and second-order terms that have been identified as a plausible model, the second-order term IV is found not to be significant. We may note here that the degrees of freedom for conducting both marginal and partial tests are the same. Thus, although the marginal odds ratios describe the association when the third variable is ignored (i.e., when collapsed over the third variable), the partial odds ratios on the other hand, describe the associations present when the third variable is partially controlled or fixed.

Christensen (1990) suggested the following four ways in choosing an initial model using Brown's tests:

(a) We include in the model all terms with significant marginal tests.

(b) We include all terms with significant partial tests.

(c) We include all terms for which either the marginal or partial association tests are significant.

(d) We include all terms in which both the partial and marginal association tests are significant

It is obvious from the above that model **(d)** will always give the smallest initial model, while **(c)** will always give the largest initial model. Model **(d)** can therefore be used to determine an initial model for a forward selection procedure, while **(c)** can similarly be used for a corresponding backward selection procedure. (These procedures will be fully described in chapter 7.) Of course, we can always use as an initial model a model that is intermediate between **(c)** and **(d)**.

When Brown's tests are applied to the data in our example above, we have both the IP and PV being very significant. Thus an initial model would be the model {IP,PV}, which gives a G^2 value of 11.8140 on 9 degrees of freedom. This model turns out to be the most parsimonious model for describing this data, and we would thus conclude that income and voting are conditionally independent given political preference. In other words, political preference determines to a great extent the actual voting behavior of the respondents. (This is the last model implemented in the SAS software statements above.)

We note here that, because of the sampling constraint, the effect IP must necessarily be in the explanatory model.

I would like to caution here that the results from marginal and partial tests may be completely different in terms of describing the associations that are present in a given k-way table and they could sometimes lead to conflicting results. As an illustration, for the three-way table in Table 6.18, while the partial tests indicate significant effects for (PI) and (PV), with effect (IV) not being significant, indicating that this effect is not significantly important, marginal test gives a pvalue of 0.0004, very unlikely not to be important. This is illustrated below.

```
The GENMOD Procedure

     LR Statistics For Type 3 Analysis

                              Chi-
Source              DF      Square     Pr > ChiSq
---------------------------------------------------
I*V                  3        4.30         0.2304*    partial assoc. test
I*V                  3     17.9649         0.0004*    marginal assoc. test
```

The above contradictions in both marginal and partial tests indicate that one must therefore be very careful in eliminating possible significant effects at the initial stage of log-linear model analysis.

6.12.3 Interpretation of Chosen Model

The selected model can be interpreted as

$$I \perp V | P$$

P(i)	V (k) 1	2
1	0.9687	-0.9687
2	0.0177	-0.0177
3	-0.9864	0.9864

Table 6.21: Parameter estimates for $\hat{\lambda}_{ik}^{PV}$ under CATMOD

Of particular importance to us are the parameter estimates of the λ_{ik}^{PV} interaction, which are shown in Table 6.21.
The parameters indicate that there are more Democrats voting Democratic while there is a corresponding significant number of Republicans voting for the Republican party. The independents are least likely to vote for either of the parties. Democrats are $e^{2(.9687+.9864)} = e^{3.9102} = 49.91$ times more likely to vote Democratic than a Republican. Similarly, Independents are $e^{2(.0177+.9864)} = e^{2.008} = 7.45$ times more likely to vote Democratic than are Republicans, while the odds are $e^{(3.9102-2.008)} = e^{1.9022} = 6.70$ times more likely for a Democrat to vote Democratic than Independents.

6.13 Collapsibility Conditions for Contingency Tables

We have seen from the previous section that both partial and marginal associations may not be the same and indeed for many situations they are usually not the same. In this section we give sufficient conditions under which both the partial and marginal associations can be the same. First, let us consider again the difference between marginal and conditional independence.

Given three variables A, B, and C, the following condition is necessary for both the partial and marginal odds-ratios to be the same for the A-B association (or A-C or B-C as the case may be):

$$\tau_{ij}^{AB} = \tau_{(ij).1} = \tau_{(ij).2} = \cdots = \tau_{(ij).K}$$

where τ defines a log odds-ratio, variable C has K categories, and $1 \leq i \leq (I-1), 1 \leq j \leq (J-1)$.

As demonstrated from our previous analysis of the gun registration data, the above collapsibility conditions will hold if either (a) or (b) or both of the conditions below hold:

(a) $\tau_{(ii')(kk').j}^{AC.B} = \ln\left(\frac{\hat{m}_{ijk}\hat{m}_{i'jk'}}{\hat{m}_{i'jk}\hat{m}_{ijk'}}\right) = 1$, for $1 \leq i \leq (I-1), 1 \leq j \leq J, 1 \leq k \leq (K-1)$

(b) $\tau_{(jj')(kk').i}^{BC.A} = \ln\left(\frac{\hat{m}_{ijk}\hat{m}_{ij'k'}}{\hat{m}_{ij'k}\hat{m}_{ijk'}}\right) = 1$, for $1 \leq i \leq I, 1 \leq j \leq (J-1), 1 \leq k \leq (K-1)$

(a) above implies that A and C, are conditionally independent given B and is described by model {AB,BC}. That is, either model {AB,C} holds or both models {AB,AC} and {AC,BC} hold.

Similarly, **(b)** also implies that B and C are conditionally independent given A and is also described by the model {AB,AC}.

For **(a)** and **(b)** to hold, then model {AB,C} must hold. That is, C must be jointly independent of A and B.

6.13.1 Example

Consider again the data on voting preference that wERE analyzed in Table 6.18. We suggested that the best parsimonious model was the model {IP,PV}, which has a $G^2 = 11.814$ on 9 d.f. - that is, the model that says that I is conditionally independent of V given P. The odds ratios will be obtained as

$$\tau^{IV.P}_{(jj')(kk').i} = \ln\left(\frac{\hat{m}_{ijk}\hat{m}_{ij'k'}}{\hat{m}_{ij'k}\hat{m}_{ijk'}}\right)$$

These odds ratios would be expected to be equal to 1.00 under this model (since the model imposes certain marginal constraints). However, model {IP, IV} does not fit the data with $G^2 = 343.6063$ on 8 d.f. This should be obvious since model {IV, P} does not fit the data set. It has a $G^2 = 374.9528$ and is based on 14 degrees of freedom. Consequently we would not therefore be able to collapse the above three-way table into, say, a two-way I-V table over variable P without distorting the underlying association between I and V.

It is worth mentioning here that because the factor variable income (I) is ordinal in nature, we may take advantage of this fact and fit a less restrictive IP interaction term, I(1)*P, the linear component of I and P interaction term. This model when applied to the data in Table 6.18 gives a $G^2 = 20.577$ on 13 d.f. This model fits the data, but on examining the adjusted residuals for this model, we notice that the cells (212, 241, 312) with observed counts 37, 67, and 35 have expected counts 48.67, 80.93, and 27.38 respectively. The corresponding adjusted residual values are $-2.26, -2.51$, and 2.32, respectively. Thus, we would not adopt this model in favor of our earlier model that was based on 9 d.f.

6.13.2 Another Example

Aicken (1983) presents the data in Table 6.22, which relate to a sample of women during the Great Depression. The factors are:

E = Amount of full-time employment $(i = 1, 2, 3)$
E0 = None
E1 = 1 - 4 years
$E2 =>$ 4 years

C = Class $(j = 1, 2)$
C0 = Middle class
C1 = Working class

D = Deprivation $(k = 1, 2)$
D0 = Not deprived
D1 = Deprived

Interest here centers on fitting the models described by the following hypotheses to the data in Table 6.22.

	C0		C1	
	D0	D1	D0	D1
E0	4	5	2	3
E1	7	8	5	13
E2	2	7	4	9

Table 6.22: Three-way table relating employment, class, and deprivation

Hypothesis	Formulation	Model
H_1	$\pi_{ijk} = \pi_{i++}\pi_{+j+}\pi_{++k}$	{E, C, D }
H_2	$\pi_{ijk} = \pi_{i++}\pi_{+jk}$	{E, CD}
H_3	$\pi_{ijk} = \pi_{+j+}\pi_{i+k}$	{C, ED }
H_4	$\pi_{ijk} = \pi_{++k}\pi_{ij+}$	{D, EC }
H_5	$\pi_{ijk} = \frac{\pi_{i+k}\pi_{+jk}}{\pi_{++k}}$	{ED, CD}
H_6	$\pi_{ijk} = \frac{\pi_{ij+}\pi_{+jk}}{\pi_{+j+}}$	{EC, CD}
H_7	$\pi_{ijk} = \frac{\pi_{ij+}\pi_{i+k}}{\pi_{i++}}$	{EC, ED}
H_8	$\frac{\pi_{111}\pi_{ij1}}{\pi_{i11}\pi_{1j1}} = \frac{\pi_{11k}\pi_{ijk}}{\pi_{i1k}\pi_{1jk}}$	{EC, ED, CD}

Table 6.23: Hypotheses of interest

For each of the models, the likelihood ratio test statistic G^2 is computed and the corresponding degrees of freedom are also obtained. The expected values for these models are displayed in Table 6.24.

From Table 6.24, we see that model H_1 holds, that is, variables E, C and D are pairwise independent. It is not surprising therefore that models $H_2 - H_7$ also hold true since they are subsets of model H_1. Thus E is independent of C, C is independent of D, and D is independent of E. This model is interpreted as:

$$\text{E} \perp \text{C} \perp \text{D}$$

We may, if we so wish, consider for instance the interaction (CD) to be a single factor variable with four categories and test the hypothesis that E and (CD) are independent. The result of such a test will not be contradictory to the above result. We may also, if we wish, collapse the three-way tables over any of the variables, and our results above will still stand. We consider two of these possibilities in Tables 6.25 and 6.26.

In Table 6.25, under the model of independence between the two classificatory variables E and (CD), the expected values are equal as expected to those obtained for model H_2. The resulting $G^2 = 3.924$, and it is based on $(3-1)(4-1) = 6$ degrees of freedom. The result also confirms that E and (CD) are independent.

Cells	n_{ijk}	H_1	H_2	H_3	H_4	H_5	H_6	H_7	H_8
1 1 1	4	2.33	2.64	2.87	3.13	3.25	3.55	3.86	4.11
1 1 2	5	4.37	4.06	3.83	5.87	3.56	5.45	5.14	4.89
1 2 1	2	2.54	2.23	3.13	1.74	2.75	1.53	2.14	1.89
1 2 2	3	4.76	5.07	4.17	3.26	4.44	3.47	2.86	3.11
2 1 1	7	5.49	6.22	5.74	5.22	6.50	5.91	5.45	6.08
2 1 2	8	10.29	9.57	10.04	9.78	9.33	9.09	9.55	8.92
2 2 1	5	5.99	5.26	6.26	6.26	5.50	5.50	6.55	5.92
2 2 2	13	11.23	11.96	10.96	11.74	11.67	12.50	11.45	12.08
3 1 1	2	3.66	4.14	2.87	3.13	3.25	3.55	2.45	2.81
3 1 2	7	6.86	6.38	7.65	5.87	7.11	5.45	6.55	6.19
3 2 1	4	3.99	3.51	3.13	4.52	2.75	3.97	3.55	3.19
3 2 2	9	7.49	7.97	8.35	8.48	8.89	9.03	9.45	9.81
	G^2	4.52	3.92	3.52	2.48	2.93	1.89	1.49	1.08
	df	7	6	5	5	4	4	3	2
	p-value	0.72	0.69	0.62	0.78	0.57	0.76	0.69	0.58

Table 6.24: Results from the eight hypotheses considered

	CD				
E	C0D0	C0D1	C1D0	C1D1	Total
E0	4	5	2	3	14
E1	7	8	5	13	33
E2	2	7	4	9	22
Total	13	20	11	25	69

Table 6.25: The 3×4 two-way subtable relating employment and class-deprivation

Now suppose the three-way table is collapsed over class (C), for instance; the resulting table would be that shown in Table 6.26.

	D		
	D0	D1	Total
E0	9	5	14
E1	15	18	33
E2	9	13	22
Total	33	36	69

Table 6.26: The 3×2 collapsed two-way subtable relating employment and deprivation

Here again, $G^2 = 2.0329$ for testing independence and is based on 2 d.f., which again indicates that the model of independence still holds. That is, E $\perp$ D. The same will be true if we had collapsed instead over E or C. Analyses of the resulting collapsed tables will be consistent with earlier conclusions of pair wise independence.

As another example where collapsibility is possible, consider again the $2 \times 2 \times 3$ contingency table relating cancer status (C) to spousal smoking (S) from three countries (Y) discussed in chapter 4. Next consider the eight hypotheses corresponding to the earlier hypotheses H_1 to H_8. These hypotheses in terms of terms included are: $H_1 : \{Y, S, C\}$, $H_2 : \{Y, SC\}$, $H_3 : \{S, YC\}$, $H_4 : \{C, YS\}$, $H_5 : \{YC, SC\}$, $H_6 : \{YS, SC\}$, $H_7 : \{YS, YC\}$, $H_8 : \{YS, YC, SC\}$. Table 6.27 gives the results of the analysis for the eight hypotheses $H_1 - H_8$ considered in the previous section.

Models	df	G^2	pvalue
H_1	7	22.3399	0.0022
H_2	6	16.5606	0.0110
H_3	5	21.2911	0.0007
H_4	5	6.8440	0.2325
H_5	4	15.5118	0.0037
H_6	4	1.0647	0.8998
H_7	3	5.7952	0.1220
H_{07}	2	0.2396	0.8871

Table 6.27: G^2 and pvalues for all the models considered

We observe from the above analysis that models H_4, H_6, H_7, and H_8 are tenable for these data. The best parsimonious model is the model based on H_6, that is, model {YS,SC}. This is the model that assumes that given any level of factor S (spousal smoking status), cancer status (C) is independent of country (Y). However, because of this result, if we collapse the tables over variable S, we would have a $G^2 = 1.0488$ on 2 d.f. for the test of independence between C and Y. There is no contradiction in this case and *Simpson's paradox* will not manifest in this example.

The results in the above table are generated from the following SAS software programs in PROC CATMOD or GENMOD for the eight hypotheses of interest.

```
data 618a;
do e=1 to 3; do c=1 to 2; do d=1 to 2;
input count @@; output; end; end; end;
datalines;
4 5 2 3 7 8 5 13 2 7 4 9
;
proc catmod; weight count;
model e*c*d=_response_/ml nogls freq pred=freq
                        noparm noresponse;
loglin e c d; run; loglin e c|d; run; loglin c d|e; run;
loglin d e|c; run; loglin e|d c|d; run; loglin e|c c|d; run;
loglin e|c e|d; run; loglin e|c|d @2; run;
***Corresponding GENMOD codes***;
proc genmod; class e c d; model count=e c d/dist=poi link=log; run;
-------------------------------------
proc genmod; class e c d;
model count=e|c|d @@2/dist=poi link=log; run;
```

The reason why we are able to collapse is because the collapsibility conditions discussed above are fully met in these data, because models {C,YS}, {YS,SC}, and {YS,YC} all fit the data.

6.14 Problems Associated with Fitting Log-Linear Models

Clogg and Eliason (1988) identified what they described as "some common problems in log-linear analysis." We discuss briefly below some insights into some of these problems, and readers interested in a more thorough coverage of the topic should consult the aforementioned article by Clogg and Eliason and other references therein.

6.14.1 Problems with Small Cell Counts

In many situations, the sample size n may be very small relative to the dimension M of the contingency tables. In such cases, the observed counts tend to be thinly or sparsely spread across the cells of the table. Sampling zeros are not uncommon in this situation, and sometimes there are usually counts that are very small (say, 1's, 2's, and 3's). The data in Table 6.28 give an example of such a situation. The table is reproduced from Agresti (1990) and the data relate to the effectiveness of immediately injected or $1\frac{1}{2}$-hour-delayed penicillin in protecting rabbits against lethal injection with β-hemolytic streptococci.

Penicillin level	Delay	Response: Cured	Response: Died
1/8	None	0	6
	$1\frac{1}{2}$	0	5
1/4	None	3	3
	$1\frac{1}{2}$	0	6
1/2	None	6	0
	$1\frac{1}{2}$	2	4
1	None	5	1
	$1\frac{1}{2}$	6	0
4	None	2	0
	$1\frac{1}{2}$	5	0
Mar. Prob.		0.537	0.463

Table 6.28: The $5 \times 2 \times 2$ data for this example

This is a $5 \times 2 \times 2$ contingency table, with $n = 54$ observations spread across the 20 cells, of which there are 7 sampling zeros, and a total of 16 counts less than or equal to 5. The theories on which the log-linear formulation is based assume large sample or asymptotic theory. The consequences of the violations of this theory are:

1 The goodness-of-fit test statistics may not possess the approximating χ^2 distribution with the relevant degrees of freedom. Thus tests of significance are therefore considerably hampered.

2 The sparseness of the data often results in what Clogg and Eliason (1988) called the *existence problems*, which means that estimates of the λ parameters originally postulated for the model cannot be calculated because one or more of them take the value of $\pm\infty$.

3 The statistic $(\hat{\lambda} - \lambda)/(ASE(\hat{\lambda}))$ for each of the parameters of the model (except the constant term or other terms required by marginal constraints) may not follow the standardized normal variate because the asymptotic standard errors are grossly inflated. Indeed, the sampling distribution of estimates of λ may be far from the normal and thus the standardized parameter estimates may be misleading.

One possible solution suggested by Goodman (1984) is that model fits should be accompanied by both the likelihood ratio statistic G^2 and the corresponding Pearson's statistic X^2 since both have the same asymptotic distribution when the sample size is large and different distributions when the sample size is small (Lawal, 1984). Goodman argues that when the two statistics lead to different conclusions, then the sample size is not large enough to make a good decision. However, when the two statistics lead to the same conclusion, then we are assured that the sample size may be adequate for the model of interest. Thus inspecting either G^2, X^2, or indeed $I(\lambda)$ may be sensible when we are confronted with sparse data.

Very attractive and famous in the analysis of contingency tables is the addition of 0.5 to each cell count when we are confronted with data with sampling zeros. This procedure is particularly recommended for fitting the saturated model; however, it is not uncommon to see analysts add 0.5 to each cell counts when analyzing unsaturated models. This approach may be misleading. We illustrate this with the above data.

Suppose we wish to fit the model {PD,PR} to the above data. Here, P refers to penicillin level, D to delay, and R to the response. Let these variables be indexed by $i = 1, \cdots 5; j = 1, 2; k = 1, 2$, respectively. ML estimates do not exist for this model (by Birch's theorem) because the three marginals n_{1+1}, n_{11+}, and n_{5+2} are zero and estimates exist only when all cell marginal counts are positive.

Parameter estimates (main effects only) for model {PD,PR}

Parameter	Poisson		+0.50		+Weighted values	
	Estimate	ASE	Estimate	ASE	Estimate	ASE
$\hat{\mu}$	-24.000	0.1.136	-0.406	1.027	-1.147	1.489
$\hat{\lambda}_1^P$	25.609	1.221	2.117	1.108	2.804	1.550
$\hat{\lambda}_2^P$	25.504	1.219	2.015	1.108	2.699	1.549
$\hat{\lambda}_3^P$	24.693	1.275	1.322	1.152	1.947	1.587
$\hat{\lambda}_4^P$	23.307	0.612	0.406	1.276	0.832	1.725
$\hat{\lambda}_1^D$	-0.916	0.837	-0.693	0.707	-0.789	0.763
$\hat{\lambda}_1^R$	25.609	1.045	2.079	1.061	2.790	1.514
G^2	14.2938		9.7138		11.2021	

Model {PD,PR} applied to the data gives a $G^2 = 14.2938$ on 5 d.f., with parameter estimates having standard errors of zeros. Further, PROC GENMOD warns with the following statement:

WARNING: Negative of Hessian not positive definite.

In this case therefore our parameter estimates are most unreliable. If we add 0.5 to each of the 20 observations in the data set and we refit this model, we have a summary of the SAS software output for main effects only in column 2 of the table above. The SAS software program below will implement the models in columns 1 and 2 of the above table, using PROC GENMOD.

```
data example;
do p=1 to 4; do d=1 to 2;  do r=1 to 2;
input count @@; output; end; end; end;
datalines;
0 6 0 5 3 3 0 6 6 0 2 4 5 1 6 0 2 0 5 0
;
proc genmod; class p d r; model count=p|d p|r/dist=poi; run;
data new; set example; count=count+0.5;
proc genmod order=data; class p d r;
model count= p|d p|r/dist=poi link=log type3; run;
```

In this case, we have what appear to be stable asymptotic standard errors. However, we have achieved this stability by the addition of 0.5 to each cell count, which in turn has inadvertently increased the sample size from 54 to 64, an almost 19% increase. This might lead to serious distortions in the interpretations of the underlying associations within the original data. To minimize this distortion, Clogg has proposed the approach that suggests that, we shrink the data toward the marginal distribution of the response variable, in this case R. From the table, the marginal distribution of R is (0.537,0.463). The model of interest {PD,PR} has 5 degrees of freedom, hence 5 observations must be distributed across the entire 20 cells. For cells with R (cured), the required constant is $0.537 \times (5/10) = 0.2685$, and similarly for cells with R (died), we add $0.463 \times (5/10) = 0.2315$.

The result of fitting model {PD,PR} to the augmented data is displayed in the SAS software output below and in column 3 of the table above. We notice that this model has a G^2 of 11.9791 on 5 d.f. The estimates of the parameters of the model and the corresponding ASE are very much in agreement with those of adding 0.5, and this time around, we have only added 9% of sample size to obtain our very consistent results. We would therefore suggest that we adopt the Clogg et al. (1990) approach to analyzing this type of data set.

```
data new1;
set example;
if r=1 then count=count+0.2685;
else count=count+0.2315;
proc genmod order=data;
class p d r; model count= p|d p|r/dist=poi link=log type3; run;
```

6.14.2 Log-Linear Analysis of Incomplete Contingency Tables

We consider in this section the analysis of incomplete multiway contingency tables where sampling zeros are observed. Table 6.29 is such an example and gives the number of pediatric contacts by subscriber's age, number of children, and age of youngest children (Freeman, 1987).

Age of Subscriber	Number of Children	Age of Youngest	Number of Pediatric Contacts						
			0	1	2	3	4-6	7-9	10+
< 25	1	< 5	4	8	2	5	10	8	19
	2+	< 5	3	2	2	1	8	8	25
25-44	1	< 5	3	2	4	7	14	7	26
	2+	< 5	17	7	9	18	37	41	172
	2+	5-9	17	22	12	19	33	9	46
	2+	10+	25	5	4	7	5	5	9
45-59	1	10+	63	5	5	5	4	3	3
	2+	5-9	5	5	7	3	8	1	4
	2+	10+	32	4	6	2	6	5	3

Table 6.29: Data for this example

In the above four-way contingency table, variable (D) "age of subscriber" has three levels (<25, 25-44, 45-59), variable (C) "number of children" has two levels (1, 2+), and variable (B) "age of youngest" also has three levels (<5, 5-9, 10+), while the number of contacts (variable A) has seven categories. Thus a complete contingency table would have observations in each of the $3 \times 2 \times 3 \times 7 = 126$ cells of the four-way table. However, the observed counts in the above table indicate that we do not have counts in all the 126 cells. Indeed, we have counts in only 63 of the cells. Thus the table is incomplete.

We next want to ask if the missing cell entries should be treated as *structural zeros*, in which we would wish to fit *quasi-log-linear models,* or as *sampling zeros.* In this example, the number of contacts is ordinal, and this is designated as variable A here. In the data, there are no subscribers <25 years in age who have the youngest child age 5 or more. Also, for the ages 25-44 and 45-59, the youngest child is 5 or more. That is, none has a young child <5 in these two groups. Hence, instead of having $3 \times 2 \times 3 = 18$ samples, we only have 9 samples in our data. The resulting table is therefore incomplete. However, these missing cells are not inherently zero. It is just that they were not observed in our sample. We are thus dealing here with sampling zeros, rather than structural zeros. We present in Table 6.30 the observed cell counts for the data in Table 6.29, including the sampling zeros. Table 6.30 is the complete table of the number of pediatric contacts by subscriber's age, number of children, and age of youngest children with the sampling zero cells included.

			Number of pediatric contacts						
Age of subscriber	Number of children	Age of youngest	0	1	2	3	4-6	7-9	10+
< 25	1	< 5	4	8	2	5	10	8	19
		5-9	0	0	0	0	0	0	0
		10+	0	0	0	0	0	0	0
	2+	< 5	3	2	2	1	8	8	25
		5-9	0	0	0	0	0	0	0
		10+	0	0	0	0	0	0	0
25-44	1	< 5	3	2	4	7	14	7	26
		5-9	0	0	0	0	0	0	0
		10+	0	0	0	0	0	0	0
	2+	< 5	17	7	9	18	37	41	172
		5-9	17	22	12	19	33	9	46
		10+	25	5	4	7	5	5	9
45-59	1	< 5	0	0	0	0	0	0	0
		5-9	0	0	0	0	0	0	0
		10+	63	5	5	5	4	3	3
	2+	< 5	0	0	0	0	0	0	0
		5-9	5	5	7	3	8	1	4
		10+	32	4	6	2	6	5	3

Table 6.30: The complete table for the data in Table 6.29

Our analysis of the data in Table 6.30 is complicated by the fact that there are many sampling zeros in the data, leading to nine of the marginal totals being zero. The log-linear model analysis is based on large sample statistical theory and this will no doubt be violated if care is not taken in modeling these data. Further, the number of degrees of freedom on which our various models will be based is seriously compromised by the presence of so many sampling zeros. Other problems associated with modeling such data as in Table 6.30 are that the relevant goodness-of-fit statistics, such as G^2, may not posses the desired null distribution even with the use of the correct degrees of freedom. Further, there is also the problem relating to the *existence* of parameters, as mentioned in the preceeding section.

The above and similar problems are considered in the analysis below of the data in Table 6.30. Any program that is based on Newton's iterative procedure will be able to handle the analysis of this type of data. The IPF based algorithms (software) usually give the wrong number of degrees of freedom when there are many sampling zeros resulting in one or more zero group or marginal totals.

The constructed weight (**wt**) variable in our SAS software program below instructs the program to treat the cells having zero values as sampling zeros and take this into account when calculating the relevant degrees of freedom for all the models that will be considered. Logit models that fit the marginal distribution of the set of explanatory variables B, C, and D are considered. In PROC GENMOD, this is specified by including the $B|C|D$ model in the model specification. In PROC CATMOD, this is accomplished by specifying POPULATION B C D. The logit models displayed in appendix E.2 whose results are displayed in Table 6.31, are fitted to the data in Tables 6.30 and 6.29.

Both PROC CATMOD and GENMOD give the same goodness-of-fit test statistics for all the models.

Number	Log-linear models	Logit models	G^2	X^2	d.f.	pvalue
(i)	BCD,A	μ	395.8815	422.7455	48	0.0000
(ii)	BCD, AB	B	58.4539	64.4891	36	0.0104
(iii)	BCD,AC	C	357.0956	363.0394	42	0.0000
(iv)	BCD,AD	D	172.8302	190.2218	36	0.0000
(v)	BCD, AB,AC,AD	B, C, D	9.7354	9.5321	18	0.9402
(vi)	BCD,AB,AD	B, D	25.0227	25.3667	24	0.4046
(vii)	BCD,AB,AC	B, C	33.3738	33.6263	30	0.3063

Table 6.31: Results of logit and equivalent log-linear models

Results in Table 6.31 indicate that the response pattern cannot be wholly explained in terms of B, C or D variables alone, as observed in the pvalues in models (*ii*), (*iii*), and (*iv*). The difference in G^2 between models (*v*) and (*vi*) provides a test of the hypothesis that the term AC is zero. In this case, we have $G^2 = 25.0227 - 9.7354 = 15.2873$ and it is based on 6 d.f. (pvalue = 0.0181). This clearly indicates that the interaction term AC cannot be considered zero. Similarly, the difference between models (*v*) and (*vii*) provides a test of the hypothesis that the interaction term AD is zero. Again, $G^2 = 33.3738 - 9.7354 = 23.6384$ and it is based on 12 d.f. (pvalue = 0.0228). This result also indicates that the term AD cannot be considered zero in the model. Of course, from models (i) and (ii), we can show that the AB interaction is very important for the model. Thus based on the above results, we can say that both the logit models {B,D} and {B,C} are adequate for an explanation of the model, but both models are missing either the AC or the AD terms; terms that have been shown to be important in the model. Hence, we propose the final logit model {B,C,D}, which is equivalent to the log-linear model {BCD,AB,AC,AD }, for a proper explanation of the data. This model fits the data well and the analysis of residuals indicates a very good fit to the data. That is, the number of pediatric contacts is dependent on the age of the subscriber, the number of children registered, and the age of the youngest child.

We also try fitting linear by linear association models to the data, all such models fail to fit the data.

Another Example

Andersen (1997) analyzed the following table of cross-classified data from the Danish Welfare Study. The sample is classified by variables

A : Strenuous work with categories yes, yes sometimes and no.

B : Type of employment with categories blue collar employee, white collar employee and employer.

C : Social group with four categories.

A: Strenuous work	**B**: Type of employment	**C**: Social group I-II	III	IV	V
Yes	Blue collar	0	0	64	182
	White collar	79	98	110	0
	Employer	38	126	19	0
Yes, sometimes	Blue collar	0	0	131	265
	White collar	156	166	292	0
	Employer	28	150	52	0
No	Blue collar	0	0	156	556
	White collar	136	166	382	0
	Employer	18	180	54	0

Table 6.32: Table of counts for this problem

Because the sampling design in this example has fixed the B-C marginal totals, the expected values $\hat{m}_{ijk}$ for any model that is employed therefore have to satisfy the condition that:

$$\hat{m}_{+jk} = n_{+jk}$$

Possible models satisfying these are: (i) {AB,AC,BC}, (ii) {AB,BC}, (iii) {AC, BC}, (iv) {ABC}, and (v) {A,BC}. We fit the log-linear models (i), (ii), and (iii) to the above data. These models correspond to the first three models of Table 5.3 in Andersen (1997). We decide not to implement the fourth model {AB,AC} in Andersen because it does not satisfy the condition imposed by the sampling design. The equivalent logit models for the three models above are respectively, (ia) {B,C}, (iia) {B}, and (iiia) {C}. The implementation of these models is carried out with the SAS software programs in appendix E.3 using PROC CATMOD and in appendix E.4 using GENMOD. We have also presented relevant partial outputs based on the logit model (ia). The model is based on 4 degrees of freedom with $G^2 = 13.5575$.

```
           Maximum Likelihood Analysis of Variance

  Source                  DF    Chi-Square    Pr > ChiSq
  ------------------------------------------------------
  Intercept                2        185.64        <.0001
  b                        4         14.98        0.0047
  c                        6         55.48        <.0001
  Likelihood Ratio         4         13.56        0.0088
```

```
                 Criteria For Assessing Goodness Of Fit
Criterion                    DF            Value            Value/DF
---------------------------------------------------------------------
Deviance                      4          13.5575              3.3894
Pearson Chi-Square            4          13.6843              3.4211
```

We have presented above a partial output from implementation of the log-linear model (*i*), together with the expected values under the model in appendices E.3 and E.4. The results of fitting the above logit models are presented in the table below.

Number	Log-linear models	Logit models	G^2	X^2	d.f
(i)	BC, AB, AC	B, C	13.5575	13.6843	4
(ii)	BC, AB	B	70.3204	69.3636	10
(iii)	BC, AC	C	28.4867	30.0944	8

The values of the goodness-of-fit test statistics obtained here are different from those obtained in Andersen (1997), although our degrees of freedom do agree. We have therefore presented the expected values obtained under model (i) with our results.

6.15 Weighted Data and Models for Rates

Most data in sociological studies usually arise from stratified sampling scheme in which the data of interest are weighted on a case-by-case basis. The solution to this is to fit a weighted log-linear model of the form:

$$\ln\left(\hat{\mathbf{m}}_r/\mathbf{z}_r\right) = \mathbf{X}\,\boldsymbol{\lambda}$$

where r refers to a particular cell in the table, $\mathbf{X}$ and $\boldsymbol{\lambda}$ are as defined previously, and $\mathbf{z}_r$ is referred to as the *start table* in programs based on the IPF algorithm (BMDP) or *exposure, cell weights* or *offset* in programs based on the Newton-Raphson method (SPSS, SAS® GENMOD). This approach will be fully explored in chapter 9.

For the analysis of rare or not so rare events, the log-rate models have been proposed, where for a three-way table in which the response variable C is dichotomous, we would have

$$\ln\ (\hat{m}_{ij1}/n_{ij+}) = \delta + \delta_i^A + \delta_j^B$$

where n_{ij+} is the "group totals," the marginal distribution of the joint variables composed of all predictors and the IJ responses pertain to rare events. The above model is equivalent to the weighted log-linear model discussed above, if we set the cell weight or start table to n_{ij+}. We shall explore this model further in chapter 9.

6.16 Problems with Interpretation

Most log-linear models produce too many parameters in the final model, which consequently makes full interpretations of these parameters almost impossible. The problem of interpretations in log-linear models has been echoed among others by Upton (1991), Kaufman and Schervish (1986), Amemiya (1981), and Alba (1987).

These authors have provided some insights into this problem and have also suggested some measures for reducing some of these models to simpler and interpretable models.

Concurrent with the problem of interpretations are the following: the problem of retrieving the odds ratios in the final model, interpreting higher-order interactions, and the choice of an appropriate scale since log-linear model parameters are expressed in logarithmic or exponential forms. Clogg et al. (1986) has proposed a *purging method* as a possible solution to the latter problem.

6.17 Exercises

1. In the saturated log-linear model for a four dimensional table, let $\lambda^{BD} = 0$ and let all of the corresponding higher order terms also be zero, e.g., $\lambda^{ABD} = 0$. Show whether the resulting model is decomposable or not and find a formula for $\hat{m}_{ijkl}$ without using the ITP algorithm.

2. Obtain estimates of μ, λ_i^A, and λ_j^B in a 3×2 table displaying independence. Hence, for the general $I \times J$ table, consider the log-linear model
$$\log(m_{ij}) = \mu + \lambda_i^A$$
Show that $\hat{m}_{ij} = \dfrac{n_{i+}}{J}$ and that its residual d.f. = I(J-1).

3. A $2 \times 2 \times 2$ table with observed probabilities $\{\pi_{ijk}\}$ satisfies $\pi_{i++} = \pi_{++k} = \pi_{++k} = \frac{1}{2}$ for all i, j, and k. Give examples of cell probabilities that satisfy the log-linear models

   (a) (A,B,C)
   (b) (AB,C)
   (c) (AB,BC)
   (d) (AB,AC,BC)
   (e) (ABC)

   For cases (a), (b), and (c) find the following:

   (i) minimal sufficient statistics,
   (ii) likelihood equations
   (iii) fitted values and
   (iv) residual d.f. for testing goodness of fit.

4. In a general three-way $I \times J \times K$ contingency table with observed counts n_{ijk}, give a decomposition of the likelihood ratio test statistic
$G^2 = 2\sum_{ijk} n_{ijk} \log\left(\dfrac{n_{ijk}}{\hat{m}_{ijk}}\right)$ for the model {AB,C}.

5. The table below is a national study on 15- to 16-year old adolescents and is described in Morgan and Teachman (1988). The event of interest is ever having sexual intercourse.

		Intercourse	
Race	Gender	Yes	No
White	Male	43	134
	Female	26	149
Black	Male	29	23
	Female	22	36

Calculate the conditional odds ratios between gender and intercourse and between race and intercourse. Interpret.

6. Explain what is meant by "no statistical interaction" in modeling response C and explanatory A and B in the following cases. Use graphs or tables to illustrate

   (a) All variables are continuous (multiple regression).

   (b) C and A are continuous, B is categorical (analysis of covariance).

   (c) C is continuous, A and B are categorical (two-way ANOVA).

   (d) All variables are categorical.

7. Suppose $\{\pi_{ij}\}$ satisfy the log-linear model of independence.

   (a) Show that $\lambda_a^B - \lambda_b^B = \ln(\pi_{+a}/\pi_{+b})$.

   (b) Show that $\{all\lambda_j^B = 0\}$ is equivalent to $\pi_{+j} = 1/J$ for all $j = 1, 2, \cdots, J$.

8. For a $2 \times 2 \times 2$ table, show that for the model (AB,AC,BC), $\lambda_{11}^{AC} = (1/4)\log\theta_{11}(k)$.

9. Suppose the log-linear model (AB, AC) holds, find an expression for m_{ijk} and hence calculate m_{ij+} and $\ln(m_{ij+})$. Show that when (AB,AC) holds, the log linear model for the A-B marginal totals has the same association parameters as $\{\lambda_{ij}^{AB}\}$ in model (AB,AC). Hence deduce the odds ratios are the same in the A-B marginal table as in the partial tables. Derive a similar result for model (AB,BC) and hence the collapsibility condition in this chapter.

10. The table below from DiFrancisco and Critelman (1984) relates to effects of nationality and education level on whether one follows politics regularly. Analyze these data using log-linear models. How do the countries differ with regard to following politics regularly?

Country	Education level	FP regularly Yes	No
USSR	Primary	94	84
	Secondary	318	120
	College	473	72
USA	Primary	227	112
	Secondary	371	71
	College	180	8
UK	Primary	356	144
	Secondary	256	76
	College	22	2
ITALY	Primary	166	256
	Secondary	142	103
	College	47	7
MEXICO	Primary	447	430
	Secondary	78	25
	College	22	2

11. The data below relate to a health survey in which a single sample of respondents was taken from a population of adults in a certain county. The survey included questions on health status and access to health care. The results from two of the questions are displayed in the table below. The first question was, "In general, would you say your health is *excellent, good, fair or poor*?" The second question was, "Are you able to afford the kind of medical care you should have?"

Health status	Afford medical care: Almost never	Not often	Often	Always	Total
Excellent	4	20	21	99	144
Good	12	43	59	195	309
Fair	11	21	15	58	105
Poor	8	9	8	17	42
Total	35	93	103	369	600

(a) Test the hypothesis that the answers to the health status question are independent of the answers to the access to health care question for adults in the selected county. Fit the appropriate log-linear $\alpha = 0.05$.

(b) Are there any cells contributing more than usual to the overall X^2?

(c) Fit an appropriate quasi-independence model and draw your conclusions.

12. The table below is reproduced from Agresti (1990) and refers to the effect of academic achievement on self-esteem among black and white college students.

		Black		White	
Gender	Cumul. GPA	High self-esteem	Low self-esteem	High self-esteem	Low self-esteem
Males	High	15	9	17	10
	Low	26	17	22	26
Females	High	13	22	22	32
	Low	24	23	3	17

Which log-linear model do you think would be most appropriate for the data? Interpret the data using your chosen model.

13. Repeat problem 12 above for model (ABC,AD,BD)

14. Consider the log-linear model (A,B,C) for a $2 \times 2 \times 2$ table.

    (a) Express the model in the form $\ln \mathbf{m} = \mathbf{X}\boldsymbol{\beta}$.

    (b) Show that the likelihood equations $\mathbf{X}'\mathbf{n} = \mathbf{X}'\hat{\mathbf{m}}$ equate $\{n_{ijk}\}$ and $\{\hat{m}_{ijk}\}$ in the one-dimensional margins.

15. Apply IPF to log-linear model (A,BC), and show the ML estimates are obtained within one cycle.

16. Write out the terms in the log-linear model with generating class

$$\{AB, AF, BC, DE, EF\}$$

Show that it is decomposable. Find the functional form of its maximum likelihood estimates.

17. The table below gives a three-factor table based on gender, socio-economic status, and opinion about legalized abortion. The status has two categories: low and not low. The table of counts is given below. Analyze these data.

		Opinion (k)		
Gender (i)	Status (j)	Support	Not support	Total
Female	Low	171	79	250
	Not low	138	112	250
	Total	309	191	500
Male	Low	152	148	300
	Not low	167	133	300
	Total	319	281	600

18. Find closed-form estimated expected counts under the following models:

    (a) {A,B,C}

    (b) {ABC}

    (c) {AB,AC}

    (d) {AB,BC}

    (e) {AB,C}

Interpret all the above models in the context of the discussion in this chapter.

19. Fit log-linear models to the data in exercise 6.17. Which is the most parsimonious model? Can we collapse the table over gender?

Chapter 7

Strategies for Log-Linear Model Selection

7.1 Introduction

We now consider some strategies for building the most parsimonious model for the general contingency table. For a k-way contingency table, the number of effects and interactions increases very rapidly as k increases. For instance, for $k = 3$, there are 7 effects and interactions, and for $k = 4$ this number increases to 15, while the number is further increased to 31 for $k = 5$. In general, the number of effects and interactions equals $(2^k - 1)$. Thus trying to fit all possible models, for instance, if k were 5, would be unwieldily and a more selective procedure would in this case be most desirable.

Some important notions (Santner & Duffy, 1989) for model selection would be appropriate at this point before we discuss the various procedures for model selection strategies.

1. *Parsimony*: The model should have as few parameters as possible while adequately explaining the data. That is, we balance between a model that has as enough parameters to explain the data, while at the same time, it is easy enough to interpret.

2. *Interpretability*: It would not be reasonable to fit, for instance, a model with interaction effects without the main effects being included in the model. This situation does not seem to pose problems for as long as we restrict our selves to hierarchical log-linear models.

3. *Significant effects*: The removal or addition of any effect or interaction must be either significant or not significant.

4. *Coherence*: If a model is rejected, then all submodels should also be rejected, and conversely, if a model is accepted, then all models containing it should be accepted. This result is due to Gabriel (1969).

5. *Factor variables*: If there are factor and response variables, then all possible models to be considered should include the highest interaction term among

the factor variables, and the selection process must focus on terms linking the response variables or the response with the factor variables. For example, if A, B, C are factor variables and D is a response variable, then it would be advisable to include the interaction effect ABC in the model, and subsequent models should explore the associations between D and any of the other six interaction and main effect terms for the factor variables. That is, the marginal total n_{ijk+} is fixed. It would not make much sense to determine whether there is piecewise or conditional independence between the factor variables.

The last consideration above is often referred to as the sampling condition consideration, because in the example above, the marginal totals n_{ijk+} are assumed fixed and corresponding ML cell estimates must satisfy these marginal constraints.

As discussed in chapter 6, the initial exploratory analysis of a k-way contingency table could be in the form of examining the significant Z values (or chi-squared values) when a saturated model is fitted. But as also noted, this method is not by all means without its drawbacks. We also consider, as an alternative to examining the individual effects in a saturated model, the tests based on marginal and partial associations. These usually provide a good starting point for model building.

We will consider two other important methods for the next phase of the model building process for a k-way table. These two methods are the *forward selection* and *backward selection* techniques discussed next.

7.1.1 The Forward Selection Procedure

The forward selection method belongs to the general class of *stepwise selection procedures*. For the stepwise procedure, rules will be developed for adding or removing terms from an initial model with the sole purpose of arriving at a more parsimonious final model.

For this procedure, terms are added to an initial or working smaller model. The procedure is sequential in that it adds to or deletes terms from a model one at a time. At each stage, the test statistic G^2 is computed both for the current and larger model in which additional term has been added. The corresponding G^2 values are obtained for both models and a pvalue is computed to test for the significance of the extra term added. For a predetermined α level, the term is rejected if the pvalue is less than or equal to α.

The procedure assumes that we have identified an initial model by either of the methods suggested in section 6.15. For a k-factor table, for instance, the forward selection procedure

(a) Adds the k factor term not already in the model that has the most significant pvalue.

(b) Continues with the process until further additions do not show any appreciable significance (based on the chosen α) in fit.

7.1.2 The Backward Selection Procedure

For the backward selection procedure, we begin with an initial large or complex model and terms are then sequentially deleted from the model in turn. Again, just as for the forward selection procedure, at each stage, we compute corresponding test statistic G^2 for both the current model and the reduced model. Deletion of a term is based on the nonsignificance of its corresponding pvalue when compared with a predetermined α. This process continues until further deletion would lead to significantly poorer fit to the data. Thus:

(a) Delete the k factor term with the least significant pvalue from the model.

(b) Continue with the process until further deletions show deviations of pvalues from the predetermined α level.

Care must be taken especially when using the backward selection procedure to ensure that we do not delete an effect that may cause the saturated model test to be rejected.

As discussed earlier, a term may be forced into the model by the sampling scheme (e.g., factors) or by the presence of a higher order interaction term. For example, consider the table with four variables A, B, C, and D. Suppose we fit the model {ABD,CD}. Then, when considering all two-factor effects to be eliminated, all of {AB,AD,BD} are forced into the model as a result of {ABD}. The only candidate for elimination would have to be the effect {CD}.

7.2 Example 7.1: The Stepwise Procedures

We present in Table 7.1 an example of the forward and backward selection procedures. The data are from Aicken (1983), and relate to the study of marijuana use among adults in a large city and a suburb. The variables are G = geographical location with G1 = San Francisco and G2 = Contra Costa. Family status F has levels F1: married with children and F2: unmarried or without children. Religion is R with R1 Protestant or Catholic and R2 Others. Response is M; with M1:used marijuana and M2:did not use marijuana. We have therefore an $2 \times 2 \times 2 \times 2$ contingency table.

		Geographical region			
		G1		G2	
M	R	F1	F2	F1	F2
1	1	52	3	23	12
	2	37	9	69	17
2	1	35	15	23	35
	2	130	67	109	136

Table 7.1: Marijuana use (M) by religion (R) by family status (F) by geographical region (G) for a sample of adults (Aicken, 1983)

We start by first fitting the all three-factor, the all two-factor models, the model of complete independence (that is, the all one-factor model), and the equiprobable model to the data. The results of these fits are displayed below together with the relevant SAS software codes for implementing them.

Model	df	G^2	pvalue
{MRG,MRF,MGF,RGF}	1	3.823	0.051
{MR,MG,MF,RG,RF,GF}	5	16.495	0.006
{M,R,G,F}	11	147.978	0.000
$\{\mu\}$	15	534.797	0.000

Only the all three-factor model fits the data relative to the saturated model since pvalue >0.05 in this case.

```
data chap71;
do M=1 to 2; do R=1 to 2; do G=1 to 2;  do F=1 to 2;
input count @@; output; end; end; end; end;
datalines;
52 3 23 12 37 9 69 17 35 15 23 35 130 67 109 136
;
proc genmod order=data; class M R G F;
model count=M|R|G|F@3/dist=poi; run;
proc genmod; model count=M|R|G|F@2/dist=poi; run;
proc genmod; model count=M R G F/dist=poi; run;
proc genmod; model count=/dist=poi; run;
```

7.2.1 The Forward Selection Procedure

Since only the all three-factor model fits the data, we would therefore start the forward selection procedure with the all two-factor model, {MR,MG,MF,RG,RF,GF}. We shall adopt here $\alpha = 0.05$ as the criterion for adding terms. That is, we add a term to a current model if it has the most significant pvalue.

Because the initial model is the model {MR,MG,MF,RG,RF,GF}, we next start by adding each of the three factor terms to this model in turn and check for the significance of their pvalues. For instance, to test for the MRG term, we compute the difference of G^2 under the models {MR,MG,MF,RG,RF,GF} and {MF,RF,GF,MRG} giving a value $(16.495 - 8.411) = 8.084$ on $(5 - 4) = 1$ d.f. These differences are also displayed below.

				Differences		
Model	Added term	d.f.	G^2	d.f.	G^2	pvalue
{MR,MG,MF,RG,RF,GF}	-	5	16.495			
{MF,RF,GF,MRG}	MRG	4	8.411	1	8.084	0.0045
{MG,RG,GF,MRF}	MRF	4	16.207	1	0.288	0.5915
{MR,RG,RF,MGF}	MGF	4	16.429	1	0.066	0.7973
{MR,MG,MF,RGF}	RGF	4	9.141	1	7.354	0.0067

From the above, the term MRG has the smallest pvalue and so the effect {MRG} must next be added to the initial model, giving a revised current model {MF,RF,GF, MRG}. The next step is now to proceed by adding another simple effect to the new model. To determine which effect to add next, we again start with the revised (or new or current) model {MF,RF,GF,MRG} and add each of the three-factor terms MRF, MGF, and RGF in turn. Once again, the fitted models and their corresponding differences are displayed in Table 7.2. We note here that adding for instance the term MRF to the new model {MF,RF,GF,MRG} changes the model to {GF,MRG,MRF} since the new term already incorporates the two-factor terms MF and RF into the model.

Model	Added term	d.f.	G^2	Differences d.f.	Differences G^2	pvalue
{MF,RF,GF,MRG}	-	4	8.411	-	-	
{GF,MRG,MRF}	MRF	3	8.411	1	0.000	1.0000
{RF,MRG,MGF}	MGF	3	8.351	1	0.060	0.8065
{MF,MRG,RGF}	RGF	3	4.340	1	4.071	0.0436

Table 7.2: Models and corresponding differences

Only the term RGF has the smallest pvalue that is less than our cut-off criterion $\alpha = 0.05$. Hence, the term RGF should now be included in the model, yielding the model {MF,MRG,RGF}

We are still interested in whether the inclusion of the remaining two three-factor terms MRF and MGF would be necessary in our model (given that MF, MRG, and RGF are already in the model). The above question could simply be answered by noting that the pvalues in the last display are all not significant, indicating that we need to stop at this point. Alternatively, we could proceed as before and fit additional terms MRF and MGF to the new current model {MF,MRG,RGF}, yielding the following display once again in Table 7.3.

Model	Added term	d.f.	G^2	Differences d.f.	Differences G^2	pvalue
{MF,MRG,RGF}	-	3	4.340	-	-	
{MRG,MGF,MRF}	MRF	2	4.329	1	0.011	0.9165
{MRG,RGF,MGF}	MGF	2	3.856	1	0.484	0.4860

Table 7.3: Models with additional terms

As expected, none of the two pvalues is significant at the cutoff point $\alpha = 0.05$. Consequently, we have arrived at the most parsimonious model {MF,MRG,RGF} for our data based on the forward selection procedure. This model has a $G^2 = 4.340$ (pvalue $= 0.2270$) and is based on 3 degrees of freedom. We will consider this final model later.

7.2.2 The Backward Selection Procedure

We next consider the backward selection procedure for our data. For this procedure, it is always desirable to start with the most complex model, which in this case would be the all three-factor model {MRG,MRF,MGF,RGF}. We will again use a cutoff point of $\alpha = 0.05$ as our criterion. At each stage of our selection, we delete the term for which the pvalue will be least significant (that is, highest value). These results are presented in Table 7.4.

At the first stage, deletion of the term MRF has the least acceptable pvalue and it will thus be deleted from the model to have a new reduced model {MRG,MGF, RGF}. At the second stage, we next delete in turn the terms MRG, MGF, and RGF. Only the term MGF gives the least nonsignificant pvalue and it is thus the next candidate for deletion, leaving us with a new reduced model {MRG,RGF}. At this point, none of the other two effects not yet deleted give a pvalue greater than $\alpha = 0.05$. Hence at this stage, it will not be desirable to delete any further

Stage	Model	Deleted term	d.f.	G^2	Differences d.f.	Differences G^2	pvalue
1	{MRG,MRF,MGF,RGF}	-	1	3.823	-	-	
	{MRF,MGF,RGF}	MRG	2	7.970	1	4.147	0.0417
	{MRG,MGF,RGF}	MRF	2	3.856	1	0.033	0.855*
	{MRG,MRF,RGF}	MGF	2	4.329	1	0.506	0.4768
	{MRG,MRF,MGF}	RGF	2	8.349	1	4.526	0.0334
2	{MRG,MGF,RGF}	-	2	3.856	-		
	{MGF,RGF}	MRG	3	8.528	1	4.672	0.0306
	{MF,MRG,RGF}	MGF	3	4.340	1	0.484	0.486*
	{RF,MRG,MGF}	RGF	3	8.351	1	4.495	0.0340
3	{MF,MRG,RGF}	-	3	4.340	-		
	{MG,MR,MF,RGF}	MRG	4	9.141	1	4.801	0.0285
	{MF,RF,GF,MRG}	RGF	4	8.411	1	4.071	0.0436
	{MRG,RGF}	MF	4	56.521	1	52.181	0.0000

Table 7.4: Models based on the backward selection procedure

three-factor term, but we could proceed to the third stage by noting that the five two-factor terms MR, MG, RG, RF, and GF are automatically included in the chosen current model. The only two-factor term not included is the term MF. Consequently, fitting the model and using the backward procedure to test for the deletion of the terms MRG, RGF, and MF. None of the terms can be deleted based on our cutoff point. Hence the final model is once again given by {MRG,RGF,MF} and is based on 3 degrees of freedom.

7.2.3 Selection Based on the Saturated Parameters

For a saturated model fitted to the above data set, we give below the results of the type3 analysis on each of the terms in the model as provided by SAS® PROC GENMOD.

```
set chap71;
proc genmod order=data; class M R G F;
model count=M|R|G|F/dist=poi type3; run;

     LR Statistics For Type 3 Analysis
                          Chi-
Source           DF     Square     Pr > ChiSq
---------------------------------------------
M                 1      91.22         <.0001
R                 1      75.57         <.0001
M*R               1      10.96         0.0009
G                 1       9.55         0.0020
M*G               1       0.87         0.3496
R*G               1       0.68         0.4109
M*R*G             1       0.37         0.5423
F                 1      74.05         <.0001
M*F               1      39.94         <.0001
R*F               1       0.52         0.4706
M*R*F             1       0.56         0.4524
G*F               1      24.36         <.0001
M*G*F             1       0.00         0.9464
R*G*F             1       7.93         0.0049
M*R*G*F           1       3.82         0.0506
```

Clearly, from the results above, the significant (those with pvalue < 0.05) first-order and second-order effects are (RGF), (MR), (MF), and (GF) at $\alpha = 0.05$. An initial possible model would therefore be {RGF,MR,MF}, since the term GF is already in the model as a result of RGF being in the model. This model when fitted to the data has $G^2 = 13.382$ on 5 d.f. Obviously, this model does not fit the data as it has a pvalue of 0.02.

The next obvious decision would be to introduce the two-factor term MG, which is the only two-factor term not yet in the model. This model, designated as *(ii)* in the table below, barely fits the data with a pvalue of 0.058 and shows that the inclusion of the MG term is significant. Having included all the six two-factor terms, the next possible models will be the inclusion of either the MRG, the MGF, or the MRF. These additional models are displayed in *(iii)* to *(v)* in Table 7.5. Using the model in *(ii)* as a baseline model, only the term MRG is significant and is worth being added to the model in *(ii)*. Once, again, the chosen model is the model {MF,MRG,RGF} by this procedure.

					Differences		
		Added					
No	Model	term	d.f.	G^2	d.f.	G^2	pvalue
(i)	{MF,MR,RGF}	-	5	13.382	-	-	
(ii)	{MR,MF,MG,RGF}	MG	4	9.141	1	4.241*	0.0395
(iii)	{MF,MRG,RGF}	MRG	3	4.340	1	4.801*	0.0284
(iv)	{MR,RGF,MGF}	MGF	3	8.528	1	0.6130	0.4337
(v)	{MG,RGF,MRF}	MRF	3	8.601	1	0.5400	0.4624

Table 7.5: Additional models having {MF,MR,RGF} as baseline

7.3 Selection Based on Marginal and Partial Associations

Brown (1976) has suggested marginal and partial associations tests for screening effects in multidimensional contingency tables. To implement the marginal association screening tests in SAS software, we give below the relevant GENMOD statements to fit the following marginal association models. The first GENMOD statement fits the equiprobable model, while the last fits the all three-factor model.

```
set chap71;
proc genmod order=data; class M R G F;
(i)   model count=/dist=poi type3; run;
proc genmod order=data; class M R G F;
(ii)  model count=M R G F/dist=poi type3;run;
proc genmod order=data; class M R G F;
(iii) model count=M| R |G| F@2/dist=poi type3;run;
proc genmod order=data; class M R G F;
(iv)  model count=M| R |G| F@3/dist=poi type3;run;
```

The results of these fits are once again summarized in Table 7.6 in the order they are fitted in the program above.

The marginal association test, for instance, that all one-factor effects (M,R,G,F) are simultaneously zero is conducted by the difference $G^2 = (534.797 - 147.978) =$

Model	d.f.	G^2	pvalue
$\{\mu\}$	15	534.797	0.000
{M,R,G,F}	11	147.978	0.000
{MR,MG,MF,RG,RF,GF}	5	16.495	0.006
{MRG,MRF,MGF,RGF}	1	3.823	0.051

Table 7.6: Marginal association models

386.819 on (15 − 11) = 4 degrees of freedom. These marginal tests are displayed in Table 7.7.

k	d.f.	G^2	pvalue	Decision
1	4	386.819	0.000	reject
2	6	131.483	0.000	reject
3	4	12.672	0.013	reject
4	1	3.823	0.051	fail to reject

Table 7.7: Marginal tests: Tests that k-way effects are zero

The first line $k = 1$ in the table tests the hypothesis that:

$$H_0 : \lambda_i^M = \lambda_j^R = \lambda_k^G = \lambda_l^G = 0 \tag{7.1}$$

against the alternatives that at least one of the above parameters is not equal to zero. The result indicates that the main effect terms are necessary in the model. In any case, we do not wish to eliminate zero-order terms (noncomprehensive models) in whatever model we finally adopt. The marginal associations for both the six two-factor and the four three-factor effects are also necessary in our eventual model, though the actual individual terms are not apparent from this result, but hopefully, the partial associations test would shed more light on these individual terms. The marginal four-factor term is not significant. We again would not wish to include this term in the final model in any case as it would represent the saturated model. The marginal associations tests thus tell us to look out for *some* two-factor and three-factor terms.

The partial association model is implemented in SAS® GENMOD with the same above statements by simply requesting **type3** likelihood ratio test statistics. Below is presented a modified output for the partial association test from PROC GENMOD. These results are obtained from model statements in (*ii*), (*iii*), and (*iv*) in the SAS software statements above. That is, the upper section of the table below gives Type 3 LR values for the main effects are obtained from model designated as (ii) in the SAS software program at the beginning of this section. Similarly, the middle Type 3 values for the firstorder interaction effects are produced from the model statement in (iii), while the bottom section was produced from the model statement in (iv).

The partial associations test below tell us that all terms whose pvalues are <0.05 are important. These are MRG, RGF, MR, MG, RG, MF, and GF, plus the main effects that are all embedded in the effects listed because of the hierarchy principle. Indeed, likewise for the effects MR, MG, and RG as they are contained in MRG. Similarly, GF is contained in RGF. Consequently, the effects to be fitted are {MRG,RGF,MF} as the basic model. Our earlier results indicate that this is indeed the final model selected by the previous procedures. We thus see that if Brown's marginal and partial tests are conducted right from the onset, it usually leads to a faster selection of the final model. PROC CATMOD can also be employed to obtain these partial association tests. We present in Table 7.8 the partial and marginal association tests for each effect in the model.

LR Statistics For Type 3 Analysis

Source	DF	PARTIAL ASSOC. TESTS Chi-Square	Pr > ChiSq	MARGINAL ASSOC. TESTS Chi-Square	Pr > ChiSq
M	1	143.89	<.0001		
R	1	191.16	<.0001		
G	1	7.49	0.0062		
F	1	44.28	<.0001		
M*R	1	33.09	<.0001	34.44	<.0001
M*G	1	3.68	0.0551	0.02	0.8822
R*G	1	6.98	0.0082	6.78	0.0092
M*F	1	54.97	<.0001	54.58	<.0001
R*F	1	0.10	0.7511	3.16	0.0754
G*F	1	35.44	<.0001	33.48	<.0001
M*R*G	1	4.15	0.0417	4.15	0.0417
M*R*F	1	0.03	0.8552	0.03	0.8552
M*G*F	1	0.51	0.4768	0.51	0.4768
R*G*F	1	4.53	0.0334	4.53	0.0334

Table 7.8: Partial and marginal association tests: Tests that each effect is zero

7.4 Aitkin's Selection Method

Aitkin (1979) proposed a method for testing the significance of effects in an all $(k-1)$ factor model against an all k-factor model. The method consists basically of fitting an all $k, (k-1), (k-2), (k-3), \cdots, 1$ models to the k-way data and noting both the corresponding G^2 and the relevant degrees of freedom.

For instance, for a four-way table, we would fit an all four-factor model (that is, the saturated model), an all three-factor, an all two-factor and finally an all main effect models. For each of these, G^2 and their corresponding degrees of freedom would be obtained.

We let G^2_{s-1} and G^2_s be the computed G^2 values for the all $(s-1)$-factor and s-factor models $(s \leq k)$, respectively. Also, let d_{s-1} and d_s be their corresponding degrees of freedom. Then by Aitken's method, we would reject the hypothesis of no s-factor effects if

$$\mathrm{G}^2_{s-1} - \mathrm{G}^2_s > \chi^2(1-\gamma_s, d_{s-1} - d_s) \tag{7.2}$$

where γ_s is chosen so that

$$1 - \gamma_s = (1-\alpha)^{\binom{k}{s}}, \qquad s = 2, 3, \cdots, k$$

The above test is a test for the adequacy of all $(s-1)$ factor model.

In the above, $\gamma_k = \alpha$ is the assigned level (usually 0.05 or 0.01), k is the number of factors in the table, and $s = 2, 3, \cdots, k$ is all the s-factor effects. Further, the $\gamma's$ are such that γ must lie within 0.25 and 0.5, where:

$$\gamma = 1 - \prod_{s=2}^{k}(1 - \gamma_s)$$

For example, consider the case in which $k = 4$ and $\gamma_4 = \alpha = 0.05$. Then

$$1 - \gamma_2 = (1 - 0.05)^6 = 0.7351 \quad \text{since} \binom{4}{2} = 6$$

$$1 - \gamma_3 = (1 - 0.05)^4 = 0.8145 \quad \text{since} \binom{4}{3} = 4$$

$$1 - \gamma_4 = (1 - 0.05)^1 = 0.9500 \quad \text{since} \binom{4}{4} = 1$$

Consequently,

$$\gamma_2 = (1 - 0.7351) = 0.2649$$
$$\gamma_3 = (1 - 0.8145) = 0.1855$$

and of course

$$\gamma_4 = 0.05$$

Hence,

$$\gamma = 1 - (0.7351)(0.8145)(1 - 0.05) = 1 - 0.5688 = 0.4312$$

Clearly, this value of $\gamma = 0.4321$ satisfies the condition that it lies between 0.25 and 0.5.

For our example in Table 7.1, Table 7.9 displays for the all s-factor $s = 1, 2, 3, 4$ the computed G^2 and their corresponding degrees of freedom.

s	Model	G_s^2	d_s
1	{M,R,G,F}	147.978	11
2	{MR,MG,MF,RG,RF,GF}	16.495	5
3	{MRG,MRF,MGF,RGF}	3.823	1
4	{MRGF}	0	0

Table 7.9: Computed G^2 for the s-factor models

Aitkin's method would on the above basis, to select the all three-factor effects model. As pointed out by Christensen (1990), Aitkin's method may be problematic and at best could only provide a starting point to which higher order interactions could be added. Comparisons of s- and $(s-1)$-factor models using Aitken's proecedure are dispalyed in Table 7.10.

s	Tests	$G_{s-1}^2 - G_s^2$	$\chi^2(\gamma_s, d_{s-1} - d_s)$
4	3 Vs 4	3.823	$\chi^2(0.95, 1) = 3.841$
3	2 Vs 3	12.672	$\chi^2(0.815, 4) = 6.178$
2	1 Vs 2	131.483	$\chi^2(0.735, 6) = 7.638$

Table 7.10: Aitkin's method results

We feel that the selection methods described above, when used effectively, will more than take care of most problems relating to log-linear model selection situations for any k-way contingency table.

7.5 Selection Criteria

As indicated at the beginning of this chapter, one of the basic characteristics of any selection procedure would have to be parsimony: that is, the marrying together of a model complex enough to explain the data, while at the same time it is easy enough to interpret. To this end, several criteria have been advocated for selecting the best parsimonious model in a log-linear formulation. Most of these have been regression type measures such as the R^2, adjusted R^2, and Mallow's C_P. Others are Akaike's information criterion (AIC). A good review of these criteria are given in Clayton et al. (1986).

Since it does not make much sense to have several selection criteria simultaneously, previous results have shown that the Akaike's information criterion AIC is by far the best of all the model selection criteria, and it will be adopted in this book.

7.5.1 Akaike's Information Criterion AIC

Akaike (1974) considered the expected value of the logarithm of the joint likelihood of a set of data and proposed that the model selected should be that which maximized this quantity. For log-linear models, maximizing this quantity is equivalent to choosing the model that minimizes (Sakamoto et al., 1986).

$$\text{AIC} = G^2 - 2\,\text{d}$$

where G^2 is the likelihood ratio test statistic computed for the model and d is its corresponding degrees of freedom.

Obviously, as we consider every sequence of simpler models (in terms of fewer parameters), both G^2 and d will increase. Akaike suggested that the preferred model would be the one with a minimum AIC value. However, as Raftery (1986) pointed out, Akaike's procedure does not take into account the increasing certainty of selecting a particular model as the sample size increases, but a Bayesian approach due to Schwarz (1978) does have this desirable property. The strength of this approach is that Schwarz was able to show that, for large enough sample size, the exact form of the prior distribution was irrelevant and that the correct model was certain to be selected. As Raftery shows, this approach is equivalent to selecting the model for which the quantity BIC (Bayesian information criterion) is a minimum, where BIC is given by

$$\text{BIC} = G^2 - \text{d}\,\ln(N)$$

where N is the total sample size. Unlike the AIC statistic, BIC takes account not only of the model complexity, but also of the sample size. Upton (1992) strongly advocated the use of the BIC criterion. We give in Table 7.11 the computation of these criteria for all models that seemingly fitted our data when our selection cutoff point was $\alpha = 0.05$ where $N = 772$ for these data.

Our results indicate that if the AIC criterion were employed as the sole selection criterion for these data, the best parsimonious model would be the model {MRG,RGF,MF}. This is the final model selected by all our selection procedures. However, if the BIC is employed, the best parsimonious model is now model number 5, that is, model {MRG,MF,RF,GF}.

The above results clearly demonstrate the fact that the selection criteria may sometimes result in the selection of different models. If the original cutoff point had been $\alpha = 0.01$ instead of 0.05, the former model would still have been preferred

No	Models	d.f.	G^2	AIC	BIC
1.	{MRG,MRF,MGF,RGF}	1	3.823	1.82	-2.83
2.	{MRG,MGF,MRF}	2	4.329	0.33	-8.97
3.	{MRG,MGF,RGF}	2	3.856	-0.14	-9.44
4.	{MRG,RGF,MF}	3	4.340	-1.66	-15.61
5.	{MRG,MF,RF,GF}	4	8.411	0.41	-18.18
6.	{RGF,MR,MG,MF}	4	9.141	1.14	-17.45

Table 7.11: Model selection based on AIC and BIC criteria

while the latter model would not have qualified for a good fit in the first instance. Thus the former model would have been more robust than the model chosen by the BIC were the selection cutoff point changed. In any case, the model selected by the BIC is very much like the former in that the interaction effects of interest MRG and MF are in both models. Really, the effects RF and GF do not tell us much in relation to M (the response variable). Thus the data can be interpreted in terms of the effects MRG and MF. We would in any case prefer the former model because of its robustness to cutoff point α.

One final check on the models is to examine how well the models fit our data by examining both the standardized and adjusted residuals. These are presented below for the two models.

```
MODEL: {MGR, RGF, MF}
=====================
set tab71;
proc genmod order=data; make 'obstats' out=aa; class M R G F;
model count=M|G|R R|G|F M|F/dist=poi link=log type3 obstats; run;

   count       Pred      Resraw      Reschi    Streschi
   ----------------------------------------------------
      52    50.3218      1.6782      0.2366      1.0377
       3     4.6782     -1.6782     -0.7759     -1.0377
      23    24.4340     -1.4340     -0.2901     -0.7312
      12    10.5660      1.4340      0.4412      0.7312
      37    40.2815     -3.2815     -0.5170     -1.7107
       9     5.7185      3.2815      1.3722      1.7107
      69    65.9627      3.0373      0.3740      1.2265
      17    20.0373     -3.0373     -0.6785     -1.2265
      35    36.6782     -1.6782     -0.2771     -1.0377
      15    13.3218      1.6782      0.4598      1.0377
      23    21.5660      1.4340      0.3088      0.7312
      35    36.4340     -1.4340     -0.2376     -0.7312
     130    126.718      3.2815      0.2915      1.7107
      67    70.2815     -3.2815     -0.3914     -1.7107
     109    112.037     -3.0373     -0.2870     -1.2265
     136    132.963      3.0373      0.2634      1.2265

MODEL: {RGF, MR, MF, MG}
======================
set tab71; proc genmod order=data; make 'obstats' out=bb; class M R G F;
model count=R|G|F M|R M|F M|G/dist=poi link=log type3 obstats; run;

   count       Pred      Resraw      Reschi    Streschi
   ----------------------------------------------------
      52    45.2572      6.7428      1.0023      2.4188
       3     3.7844     -0.7844     -0.4032     -0.4967
      23    27.9829     -4.9829     -0.9420     -1.9527
      12    12.9755     -0.9755     -0.2708     -0.4244
      37    45.5498     -8.5498     -1.2668     -2.7221
       9     6.4086      2.5914      1.0236      1.2901
      69    62.2101      6.7899      0.8609      2.2842
      17    17.8315     -0.8315     -0.1969     -0.3131
```

```
 35   41.7428   -6.7428   -1.0436   -2.4188
 15   14.2156    0.7844    0.2080    0.4967
 23   18.0171    4.9829    1.1739    1.9527
 35   34.0245    0.9755    0.1672    0.4244
130   121.450    8.5498    0.7758    2.7221
 67   69.5914   -2.5914   -0.3106   -1.2901
109   115.790   -6.7899   -0.6310   -2.2842
136   135.168    0.8315    0.0715    0.3131
```

It is evident from the results above that while both models have no significant **Reschi** value, their standardized values show that model {MGR,RGF,MF} is much better than the competing model. Consequently, this model will be chosen as the most parsimonious model for this data.

We may note here that $1 - \dfrac{\text{reschi}}{\text{streschi}}$ equals the leverage a_{ijkl} of the $ijkl$-th cell defined in Christensen (1990).

7.6 Interpretation of Final Model

Since the final chosen model is {MGR,RGF,MF}, Table 7.12 displays the observed and expected values under this model.

		Geographical region			
		G1		G2	
M	R	F1	F2	F1	F2
1	1	52	3	23	12
		(50.33)	(4.68)	(24.41)	(10.55)
	2	37	9	69	17
		(40.30)	(5.73)	(65.96)	(20.04)
2	1	35	15	23	35
		(36.69)	(13.32)	(21.59)	(36.44)
	2	130	67	109	136
		(126.68)	(70.28)	(112.04)	(132.96)

Table 7.12: Observed and expected counts for the marijuana data in Table 7.1

The model has a G^2 value of 4.340 on 3 d.f., with a pvalue of 0.227. Clearly, this model fits the data well. The model implies that given the response M_h, religion R_h and geographical location G_h of the respondents are conditionally independent of family status F. That is, religion and geographical location do not affect the relationship between family status and the response M. The equivalent log odds ratio of this relationship can be formulated as:

$$\tau_{(il)(i'l').jk}^{MF.RG} = \ln\left(\frac{m_{ijkl}m_{i'jkl'}}{m_{i'jkl}m_{ijkl'}}\right)$$

for $i, i', j, k, l, l' = 1, 2$. Note here that we are keeping the levels jk constant. There are four such combinations of j and k, namely, $(1,1), (1,2), (2,1)$, and $(2,2)$. For each of these, we have

j,k	i,i'	j,j'	$\hat{\tau}^{MF.RG}$	Result
1 1	1 2	1 2	$\ln\left(\dfrac{m_{1111}m_{2112}}{m_{2111}m_{1112}}\right)$	1.361
1 2	1 2	1 2	$\ln\left(\dfrac{m_{1121}m_{2122}}{m_{2121}m_{1122}}\right)$	1.361
2 1	1 2	1 2	$\ln\left(\dfrac{m_{1211}m_{2212}}{m_{2211}m_{1212}}\right)$	1.361
2 2	1 2	1 2	$\ln\left(\dfrac{m_{1221}m_{2222}}{m_{2221}m_{1222}}\right)$	1.361

Below are presented the log-parameter estimates produced by PROC GENMOD when this model is fitted to the data in Table 7.1. Clearly, the estimate of the parameter $\lambda^{MF} = 1.3627$, which corresponds to the values of $\hat{\tau}^{MF.RG}$ computed from expected values above.

```
                     Analysis Of Parameter Estimates
                                Standard         Chi-
Parameter          DF  Estimate    Error       Square  Pr > ChiSq
-----------------------------------------------------------------
Intercept           1    4.8901   0.0847      3333.23      <.0001
M         1         1   -1.8925   0.1929        96.26      <.0001
G         1         1   -0.6376   0.1420        20.15      <.0001
M*G       1  1      1   -0.6163   0.2139         8.30      0.0040
R         1         1   -1.2946   0.1753        54.52      <.0001
M*R       1  1      1    0.6546   0.2617         6.26      0.0124
R*G       1  1      1   -0.3685   0.3207         1.32      0.2505
M*R*G     1  1  1   1    0.8077   0.3701         4.76      0.0291
F         1         1   -0.1712   0.1216         1.98      0.1591
R*F       1  1      1   -0.3532   0.2487         2.02      0.1556
G*F       1  1      1    0.7607   0.1829        17.29      <.0001
R*G*F     1  1  1   1    0.7765   0.3900         3.96      0.0465
M*F       1  1      1    1.3627   0.2019        45.55      <.0001
```

That is, the model stipulates that there is a constant association as measured by the log odds ratio between the M and F in each of the four (R-G) subtables. That is, if we form the R-G subtables of expected counts, then the log odds ratio would be constant in each of these subtables: These can be further demonstrated clearly as follows:

Subtable G1R1			**Subtable G1R2**		
	M1	M2		M1	M2
F1	50.33	36.69	F1	40.30	126.68
F2	4.68	13.32	F2	5.73	70.28

Subtable G2R1			**Subtable G2R2**		
	M1	M2		M1	M2
F1	24.41	21.59	F1	65.96	112.04
F2	10.55	36.44	F2	20.04	132.96

Table 7.13: R-G Subtables of expected counts

In each of subtables (G1R1, G1R2, G2R1, G2R2), the estimates of the odds ratios are:

$$\hat{\tau}_{G1R1} = \frac{50.33 \times 13.32}{36.69 \times 4.68} = 1.361$$

$$\hat{\tau}_{G1R2} = \frac{40.30 \times 70.28}{126.68 \times 5.73} = 1.361$$

$$\hat{\tau}_{G2R1} = \frac{24.41 \times 36.44}{21.59 \times 10.55} = 1.361$$

$$\hat{\tau}_{G2R2} = \frac{65.96 \times 132.96}{112.04 \times 20.04} = 1.361$$

In each of the cases, the log of the odds ratio equals 1.361 and can be seen to be constant from one-subtable to another. Consequently, $\hat{\theta} = e^{1.3627} = 3.91$. In other words, for a given religion and geographical location, the odds that an individual would have smoked marijuana is 3.9 times higher for married respondents with children than those without children. Similarly, the odds that an individual would have responded as smoking marijuana is $\hat{\theta} = e^{0.6546} = 1.92$ higher for Protestant or Catholic than for those with religious affiliation. On the other hand, the odds are lower for those respondents resident in San Francisco than for those from Contra Costa, being about 54% of those from Contra Costa.

Again, since the MRG effect is very important, we present the observed counts table of religion, geographical location, and response (smoked marijuana) in Table 7.14.

		Yes	No	
G	R	M1	M2	Total
S-Francisco	Christians	55	50	105
	Others	46	197	243
C-Costa	Christians	35	58	93
	Others	86	245	331
Total		222	550	772

Table 7.14: Observed marginal table for MRG

We observe that there are more respondents in the study from Contra Costa than from San Francisco. Among the individuals from San Francisco, the proportions of those who reported that they had used and those that had not used marijuana are about the same among Catholic and Protestants. On the other hand, the proportion who responded that they have used marijuana is very much lower than those who reported they have never used marijuana among the non-Christians. This same pattern is exhibited among the respondents from Contra Costa.

	Smoked marijuana?		
Family status	Yes	No	Total
With children	181	297	478
Without children	41	253	294
Total	222	550	772

Table 7.15: Marginal table MF collapsed over R and G.

When the table is again collapsed over the variables R and G (religion and geographical location), we see that the proportion of individuals who responded yes to having used marijuana are much more higher for those that are married with children than those that are unmarried or without children. Clearly, the reporting of having used marijuana or not is highly associated with the family status of the individual.

7.7 Danish Welfare Study Data: An example

The data for our second example relate to the response to the question of whether there was a freezer in the household or not among subjects in the 1976 Danish Welfare Study (from Andersen, 1997). The data are presented in Table 7.16.

A :Sex	B:Age	C:Income	D:Sector	E: Freezer in the household Yes	No
		High	Private	152	39
			Public	82	18
	Old	Medium	Private	135	31
			Public	35	12
		Low	Private	89	45
Male			Public	20	9
		High	Private	259	46
			Public	101	26
	Young	Medium	Private	183	55
			Public	54	15
		Low	Private	108	54
			Public	22	13
		High	Private	82	17
			Public	85	16
	Old	Medium	Private	46	16
			Public	60	11
		Low	Private	29	29
Female			Public	40	18
		High	Private	160	23
			Public	152	28
	Young	Medium	Private	89	17
			Public	56	21
		Low	Private	57	41
			Public	34	28

Table 7.16: Danish Welfare Study cross-classified according to possession of a freezer

The data are a five-way $2 \times 2 \times 3 \times 2 \times 2$ contingency table, with variables: A sex, B age, C family taxable income, D employment sector, and E whether there is a freezer in the household. The age categories were:

$$\text{age} = \begin{cases} \text{Old} & \text{if} > 40 \\ \text{Young} & \text{if} \leq 40 \end{cases} \qquad \text{income} = \begin{cases} \text{Low} & \text{if} < 60{,}000 \text{ D.kr} \\ \text{Med.} & \text{if } 60,000 - 100,000 \text{ D.kr} \\ \text{High} & \text{if} > 100,000 \text{ D.kr} \end{cases}$$

We shall adopt the selection cutoff point of $\alpha = 0.05$ in this analysis.Results produced from PROC GENMOD are displayed in the table below for both the partial and marginal associations. The SAS software program for implementing the partial and marginals are: First, for the **partial results**, we need to execute GENMOD to fit the following:

```
data tab77;
```

```
do a=1 to 2; do b=1 to 2;  do c=1 to 3; do d=1 to 2; do e=1 to 2;
input count @@; output; end; end; end; end; end; datalines;
152 39 82 18 135 31 35 12 89 45 20 9 259 46 101 26 183 55 54 15 108 54 22
13 82 17 85 16 46 16 60 11 29 29 40 18 160 23 152 28 89 17 56 21 57 41 34 28
;
proc genmod; class a b c d e; model count=a b c d e/dist=poi type3; run;
model count=a|b|c|d|e@2/dist=poi type3;run;
model count=a|b|c|d|e@3/dist=poi type3; run;
model count=a|b|c|d|e@4/dist=poi type3; run;
proc freq; weight count; tables a b c d e a*(b c d e) b*(c d e) c*(d e) d*e/chisq; run;
```

The partials for zero-order, two-factor, three-factor, and four-factor effects are obtained sequentially from the above GENMOD statements, respectively. The two-factor marginal partials are obtained from the PROC FREQ statements. The G^2 here is what would be expected under the model of independence between the two variables involved. These marginal tests can also be obtained from the fourth GENMOD statement. The G^2 obtained for the two-factor interactions are exactly those obtained from the various tests of independence provided by PROC FREQ statements. As mentioned in chapter 6 care should be taken in relying solely on the marginal tests because of problems associated with collapsibility of contingency tables, namely, Simpson's paradox. Brown's partial association takes care of this problem as it only measures the contribution of each term as if it enters last in the model. In general, the partial and marginal G^2 values are often very close. These G^2 values are presented in Table 7.17.

		Partial assoc. tests		Marginal assoc. tests	
Source	d.f.	G^2	pvalue	G^2	pvalue
a	1	73.10	< .0001		
b	1	100.93	< .0001		
c	2	235.73	< .0001		
d	1	263.74	< .0001		
e	1	864.16	< .0001		
a*b	1	3.50	0.0615	2.09	0.1485
a*c	2	6.69	0.0353	8.28	0.0160
a*d	1	144.78	< .0001	144.86	< .0001
a*e	1	0.00	0.9817	0.03	0.8536
b*c	2	6.63	0.0364	6.13	0.0466
b*d	1	4.78	0.0288	2.44	0.1187
b*e	1	0.03	0.8631	0.41	0.5245
c*d	2	25.75	< .0001	26.11	< .0001
c*e	2	98.82	< .0001	98.98	< .0001
d*e	1	0.33	0.5683	0.07	0.7979
a*b*c	2	1.43	0.4882	1.80	0.4068
a*b*d	1	0.66	0.4160	0.83	0.3630
a*b*e	1	0.35	0.5524	0.04	0.8373
a*c*d	2	6.44	0.0400	6.35	0.0417
a*c*e	2	7.07	0.0291	6.83	0.0329
a*d*e	1	0.31	0.5795	0.23	0.6335
b*c*d	2	0.77	0.6809	0.37	0.8302
b*c*e	2	2.76	0.2520	2.38	0.3045
b*d*e	1	4.98	0.0256	3.57	0.0589
c*d*e	2	1.04	0.5950	0.26	0.8789
a*b*c*d	2	4.96	0.0839	3.82	0.1481
a*b*c*e	2	0.06	0.9696	0.16	0.9214
a*b*d*e	1	2.35	0.1250	1.68	0.1953
a*c*d*e	2	0.39	0.8228	0.43	0.8072
b*c*d*e	2	0.22	0.8976	0.31	0.8556

Table 7.17: Results of Association tests from PROC GENMOD.

We present the G^2 values for the equiprobable, zero-order, first-order, second-order, third-order, and saturated models for the data in Table 7.16 in Table 7.18.
The corresponding marginal tests are presented in Table 7.19:

k	df	G^2	p-value
0	47	1863.4400	0.0000
1	41	325.7866	0.0000
2	27	35.8556	0.1186
3	11	11.3376	0.4154
4	2	3.9186	0.1410
5	0	0.0000	-

Table 7.18: G^2 values under various models for the data in Table 7.16

k	d.f.	G^2	pvalue	Decision
1	6	1537.6534	0.0000	Reject
2	14	289.9310	0.0000	Reject
3	16	24.5180	0.0788	Fail to reject
4	9	7.4190	0.5936	Fail to reject
5	2	3.9186	0.1410	Fail to reject

Table 7.19: Marginal tests: Tests that k-way effects are zero

The marginal k-way tests indicate that four-way and three-way marginals are not necessary in the model. Because marginal tests are subject to Simpson's paradox, we examine below the partial association tests obtained from the SAS software output displayed in Table 7.17.

Results from the marginal and partial tests indicate that the following effects need to be included in the final model as they have significant partial and marginal effects. These are presented in order of their importance (pvalues): AD, CE, CD, AC, BC, BD, CE, BDE, ACE, ACD. We may note that the BDE interaction is only significant in the partial association test. A possible all two-factor initial model would therefore be the model with generating class {AD,CE,CD,AC,BC,BD}. However, we started our model building by the initial model {AD,CE,B} designated here as model (1). We present in Table 7.20 some possible hierarchical log-linear models for the data.

Number	Added term	d.f.	G^2	pvalue	Differences d.f.	Differences G^2
1	-	38	81.9450	0.0000	-	-
2	CD	36	55.8432	0.0185	2	26.1018
3	AC	34	49.1389	0.0449	2	6.7043
4	BC	32	43.0077	0.0926	2	6.1312
5	BD	31	39.7386	0.1351	1	3.2691
6	na	29	36.1553	0.1691	na	
7	ACD	27	29.8022	0.3231	2	6.3531
8	AB	26	27.8265	0.3670	1	1.9757
9	ABCD	18	16.5724	0.5527	na	na

Table 7.20: Possible hierarchical models

From Table 7.21, we see that model 1 has the two most significant of the 10 two-factor effects as the generating class. Closed-form expression for the expected values

Model	Generating Class
1	AD, CE, B
2	AD,CE,CD,B
3	AD,CE,CD,AC, B
4	AD,CE,CD,AC, BC
5	AD,CE,CD,AC, BC, BD
6	ACE, AD, CD, BC
7	ACE, ACD,BC
8	ACE, ACD, BC, AB
9	ABCD, ACE

Table 7.21: Generating classes for the models in Table 7.20

exists because the model is decomposable. The second model introduce the next important effect, namely, the CD. The resulting model (2) does not fit the data. We next introduce the AC term. The resulting model (3) is not decomposable since it does not have its three-factor generating class ACD. We next introduce the BC term, leading to model (4). The new model fits the data with a pvalue of 0.0926. We introduce the BD term in model (5). While this model fits, the contribution of the BD term is not significant. Hence, we remove the BD term from the model and then introduce the ACE term in model (6). This model fits the data with a pvalue of 0.1691. We next introduce the ACD term in the model, leading to model (7). The model fits the data well with a G^2 value of 29.8022 on 27 degrees of freedom (pvalue = 0.3231). The model is also decomposable as its has the generating class ACE and ACD in the model.

If we let A, B, C, D, and E be indexed by i, j, k, l, and p, respectively, then under model 7, we have the estimates as:

$$\hat{m}_{ijklp} = \frac{n_{i+kl+p}\, n_{i+k+p}\, n_{+jk++}}{n_{i+k++}\, n_{++k++}}$$

Each of models 4 to 7 fit the data based on their corresponding pvalues. Model 4 is thus the minimal model for this data. If a four-factor interaction term, representing the factor variables, is not included in our choice of model, then model (7) is the most parsimonious, simplest model for these data. Examination of its standardized and adjusted standardized residuals, however, indicates that four of the cells have their adjusted standardized residuals greater in absolute values than 2, although all its standardized residuals seem acceptable as none of them is significant.

Admitting the two-factor interaction term AB into model {ACD,ACE,BC} leads to model {ACD,ACE,BC,AB}. This is designated as model (8) above. The model fits even better from the pvalue, but again, three of the cells have significant adjusted standardized residuals, hence not acceptable. The model, however, lacks its generating term ABCD. If we include this term in the model, we now have a better model as all the adjusted standardized residuals are no longer significant. Hence, the final model is model {ABCD, ACE} with a $G^2 = 16.5724$ on 18 d.f. (pvalue = 0.5527), which is model (9) above.

7.7.1 Equivalent Logit Model

If we consider variable E as a response variable, then it would be necessary to start off by fitting logit models to the data. A logit model would necessarily include the four-factor interaction term ABCD in its log-linear equivalence. The saturated logit model for the data in Table 7.16 is implemnted in PROC LOGISTIC as follows together with the type 3 statistics.

```
set tab77; proc logistic; class a b c d; weight count;
model e=a|b|c|d/scale=none aggregate; run;
```

Type III Analysis of Effects

Effect	DF	Wald Chi-Square	Pr > ChiSq
a	1	0.0063	0.9366
b	1	0.0082	0.9278
a*b	1	0.0073	0.9320
c	2	78.3490	<.0001
a*c	2	5.3566	0.0687
b*c	2	1.3830	0.5008
a*b*c	2	0.0669	0.9671
d	1	0.0076	0.9306
a*d	1	0.7091	0.3997
b*d	1	5.3445	0.0208
a*b*d	1	3.1979	0.0737
c*d	2	1.1274	0.5691
a*c*d	2	0.4192	0.8109
b*c*d	2	0.1556	0.9252
a*b*c*d	2	3.9139	0.1413

Significant terms from the type 3 analysis above are C and BD with pvalues of <0.0001 and 0.0208, respectively. Hence our baseline model would be the logit model {C,BD} which is equivalent to the log-linear model {ABCD,CE,BDE}. Fitting this model, we have $G^2 = 19.0454$ on 18 d.f. (pvalue = 0.3890). The model fits very well and is implemented with the following SAS software program and corresponding partial output.

```
set tab77;
proc logistic;class a (ref=last) b (ref=last)  c (ref=last) d (ref=last)/param=ref;
weight count; model e=c b|d/scale=none aggregate=(a b c d); run;
```

Deviance and Pearson Goodness-of-Fit Statistics

Criterion	DF	Value	Value/DF	Pr > ChiSq
Deviance	18	19.0454	1.0581	0.3890
Pearson	18	19.0695	1.0594	0.3876

Analysis of Maximum Likelihood Estimates

Parameter		DF	Estimate	Standard Error	Chi-Square	Pr > ChiSq
Intercept		1	0.3768	0.1271	8.7887	0.0030
c	1	1	1.1054	0.1120	97.4328	<.0001
c	2	1	0.7884	0.1179	44.7073	<.0001
b	1	1	0.2432	0.1614	2.2720	0.1317
d	1	1	0.2223	0.1272	3.0575	0.0804
b*d	1 1	1	-0.3960	0.1987	3.9728	0.0462

Odds Ratio Estimates

Effect	Point Estimate	95% Wald Confidence Limits	
c 1 vs 3	3.020	2.425	3.762
c 2 vs 3	2.200	1.746	2.772

The class statement in the SAS software program above instructs SAS software to use the cell reference coding scheme as in PROC GENMOD. The parameter estimates with this coding therefore will be exactly the same as those that would be obtained using PROC GENMOD to fit the equivalent log-linear model. The logit model {C,BD} fits the data very well. We also consider a linear effect of variable C on the model. The model that assumes the linear effect of variable C has $G^2 = 24.530$ on 19 d.f. (pvalue = 0.1754) and a log-parameter estimate −0.5469.

7.7.2 Interpretation

C	Freezer Status	
k	1	2
1	1073	213
2	658	178
3	399	237

Table 7.22: Observed CE marginal table with C fixed at $k = 1, 2, 3$

From the parameter estimates, the odds are 3.02 times higher for an individual with a low income to respond to yes as against no for having a freeze in the household, and 2.20 times higher among the middle income group. The linear effect parameter also indicate that as the income increases, the odds reduce since the estimate is negative.

The BDE interaction being significant indicates that given the response E_p, age and employment sector are conditionally independent of sex and income. That is, age and employment do not affect the relationship between sex and the response variable.

The above indicates that age and freezer response given the employment sector do not depend on sex and income.

Let us form the three-way BDE table as in Table 7.23 and we will use this to obtain the estimated log of the odds ratio.

B	D	E	
		p	
j	l	1	2
1	1	533	177
2		856	236
1	2	322	84
2		419	131

Table 7.23: Observed BDE marginal table with D fixed at $l = 1, 2$

Estimated log odds ratios for $\tau^{BE.D}$ are −0.186 and 0.181 at levels of D = 1 and 2, respectively. Thus the odds of a yes response are about 17% lower among the over 40's than for the under 40's among the private-sector respondents, while the odds are about 20% higher between the two age groups among the public sector respondents.

Logit models based on the forward, backward and stepwise procedures from PROC LOGISTIC when applied to the data in Tables 7.1 and 7.16 will be discussed in later chapters.

7.8 Exercises

1. The data in Table 7.24 make uo a four-way $2 \times 2 \times 2 \times 2$ table of severity of drivers' injuries in auto accident data (Christensen, 1990).

Car	Driver ejected	Injury	Accident type: Collision	Accident type: Rollover
Small	No	Not severe	350	60
		Severe	150	112
	Yes	Not severe	26	19
		Severe	23	80
Standard	No	Not severe	1878	148
		Severe	1022	404
	Yes	Not severe	111	22
		Severe	161	265

Table 7.24: Auto Accident Data

   (a) Use the forward selection procedure to build a suitable model.

   (b) Use the backward elimination procedure to build a model.

   (c) For both situations above, identify the graphical models as described in 6.8 and hence give the MLE expressions for your models. Interpret your model. (You may also examine the odds and odds ratios for a proper interpretation.)

2. Refer to problem 1 above; use backward and forward selection models to build a model that treats type of car, driver ejection and accident type as explanatory variables.

3. The data for this problem, presented in Table 7.25 are from Stern et al. (1977) and relate to a survey to study the relationship between Pap testing and hysteectomies. Interest in the data centers on whether there is association between Pap testing and ethnicity.

   (a) Fit the saturated log-linear model to the data and identify the important interactions and hence fit the most parsimonious model based on this result.

   (b) Identify a log-linear model based on both the forward and backward elimination procedures.

   (c) Identify a log-linear model based on partial associations.

   (d) Which of the forward, backward, and partial associations procedure would you prefer and why? Based on this, select the final model and perform residual analysis. Draw your conclusions.

4. Refer to problem 6.10. Use backward, saturated, and partial association procedures to determine the most parsimonious model for the data.

Income	Age	Ethnicity White < 2	White 2+	Black < 2	Black 2+	Spanish < 2	Spanish 2+
Middle	35-44	22	3	37	6	36	4
	45-54	23	3	27	4	30	9
	55-64	23	5	13	1	16	4
	65+	21	8	6	4	3	4
Low	35-44	21	10	39	16	15	5
	45-54	14	11	44	16	5	15
	55-64	11	15	24	18	2	4
	65+	13	15	11	25	3	9

Table 7.25: Data relating ethnicity and pap testing (Freeman, 1987)

5. The data for this problem is from Andersen (1997) and relate to the results of an investigation of 1314 employees who had left their jobs during the second half of the year. These layoffs were cross classified by A length of employment prior to the lay-off, B cause of layoff, and C employment status. The data are presented in Table 7.26

A: Length of employment	B: Cause of layoff	C: Employment status Got a new job	Still unemployment
<1 month	Closure	8	10
	Replacement	40	24
1 - 3 months	Closure	35	42
	Replacement	85	42
3 months - 1 year	Closure	70	86
	Replacement	181	41
1 - 2 years	Closure	62	80
	Replacement	85	16
2 - 5 years	Closure	56	67
	Replacement	118	27
> 5 years	Closure	38	35
	Replacement	56	10

Table 7.26: Data for exercise problem 7.5

(a) Use the backward procedure to find the most parsimonious model.

(b) Use marginal and partial association procedures to determine the most parsimonious model for the data.

6. The data in Table 7.27 relate to the attitudes toward nontherapeutic abortions among white Christian subjects in the 1972 - 1974 General Social Surveys as reported by Haberman (1978). Find an appropriate model for the data and interpret your results.

Year	Religion	Education in years	Positive	Mixed	Negative
			Attitudes		
1972	N. Prot.	≤ 8	9	16	41
		9-12	85	52	105
		≥ 13	77	30	38
	S. Prot.	≤ 8	8	8	46
		9-12	35	29	54
		≥ 13	37	15	22
	Catholic	≤ 8	11	14	38
		9-12	47	35	115
		≥ 13	25	21	42
1973	N. Prot.	≤ 8	17	17	42
		9-12	102	38	84
		≥ 13	88	15	31
	S. Prot.	≤ 8	14	11	34
		9-12	61	30	59
		≥ 13	49	11	19
	Catholic	≤ 8	6	16	26
		9-12	60	29	108
		≥ 13	31	18	50
1974	N. Prot.	≤ 8	23	13	32
		9-12	106	50	88
		≥ 13	79	21	31
	S. Prot.	≤ 8	5	15	37
		9-12	38	39	54
		≥ 13	52	12	32
	Catholic	≤ 8	8	10	24
		9-12	65	39	89
		≥ 13	37	18	43

Table 7.27: Attitudes towards abortion, Haberman (1978)

Chapter 8

Models for Binary Responses

8.1 Introduction

We shall consider here a response variable having binary or dichotomous categories such as "yes or no," "alive or dead," "present or absent," "survive or does not survive," etc., with factor variable(s) denoted by **X**. The terms "success" and "failure" are usually used generically for these two categories. If the binary random variable is defined as:

$$Y = \begin{cases} 1 & \text{if outcome is a success} \\ 0 & \text{if outcome is a failure} \end{cases}$$

where $P(Y = 1) = \pi$ (the underlying probability of success) and $P(Y = 0) = 1 - \pi$, then with n such independent random variables $Y_1, Y_2, \ldots, Y_n$ with constant probability of success π, the random variable

$$R = \sum_{i=1}^{n} Y_i$$

has the binomial distribution $b(n, \pi)$:

$$P(R = r) = \binom{n}{r} \pi^r (1 - \pi)^{n-r} \qquad r = 0, 1, 2, \cdots, n \tag{8.1}$$

To motivate the discussion in this chapter, let us consider the following data in Table 8.1, relating to a 30-year coronary heart disease (CHD) mortality study of Black and White men aged 35 to 74 years in Charleston, SC, and Evans County, GA (Keil et al., 1995).

City or county	Black men		White men	
	Examined	Died	Examined	Died
Charleston	319	56	635	139
Evans	407	69	711	184

Table 8.1: Coronary disease mortality in the Charleston, SC and Evans County, GA, Heart Studies, 1960 to 1990

In the above example, the outcome Y is binary (died or not died), that is,

$$Y = \begin{cases} 1 & \text{if died} \\ 0 & \text{if alive} \end{cases}$$

and the explanatory variables are sites (Charleston or Evans county) and race (Black or White). Evidently, here, deaths are coded as successes. Interest centers on the relationship between the sites and race on coronary heart disease mortality. We would like to ask whether there is any difference (and to what degree) in the rates of mortality between the two sites and if such differences depend on the race of the individual. These and other similar questions will be considered in this chapter by fitting models that can adequately describe the relationships that are inherent in the data.

To develop the necessary tools for handling such data, let us first consider k independent random variables $R_1, R_2, \cdots, R_k$ (each having the binomial distribution described above) corresponding to the number of successes in k different subgroups. Let the probability of success $P(R_i = \pi_i)$ be constant within each subgroup and let n_i be the corresponding number of trials in each subgroup. We display this in Table 8.2:

	Subgroups					
Outcome	1	2	$\cdots$	i	$\cdots$	k
Success	R_1	R_2	$\cdots$	R_i	$\cdots$	R_k
Failure	$n_1 - R_1$	$n_2 - R_2$	$\cdots$	$n_i - R_i$	$\cdots$	$n_k - R_k$
Totals	n_1	n_2	$\cdots$	n_i	$\cdots$	n_k

Table 8.2: Outcome display for k subgroups

With the above setup, it follows that $R_i \sim b(n_i, \pi_i)$. That is,

$$P(R_i = r_i) = \binom{n_i}{r_i} \pi_i^{r_i}(1-\pi_i)^{n_i - r_i} \quad r_i = 0, 1, 2, \cdots, n_i \quad \text{and} \quad i = 1, 2, .., k$$

and the log-likelihood function is given by:

$$L(\boldsymbol{\pi}, \mathbf{r}) = \sum_{i=1}^{k} \left[r_i \ln\left(\frac{\pi_i}{1-\pi_i}\right) + n_i \ln(1-\pi_i) + \ln\binom{n_i}{r_i} \right] \tag{8.2}$$

With $R_i \sim b(n_i, \pi_i)$ under the assumption that π_i is constant within each subgroup, it follows that an estimate of π_i is $p_i = \dfrac{R_i}{n_i}$ and that

$$E(p_i) = \pi_i, \quad \text{and} \quad \text{Var}(p_i) = \frac{\pi_i(1-\pi_i)}{n_i}$$

8.2 Generalized Linear Model

Our goal is to be able to describe the observed proportion of success in each subgroup in terms of the explanatory variables $\mathbf{X}$. That is, we wish to model the probabilities as

$$g(\pi_i) = \mathbf{X}\,\boldsymbol{\beta} \tag{8.3}$$

where $\mathbf{X}$ is a vector of explanatory or factor variables, $\boldsymbol{\beta}$ is a vector of parameters and g is a link function as defined in section 6 of chapter 2. (We note here that we usually use dummy variables for factor levels and measured values for covariates.)

The well-known *general linear model* has

$$\pi(x_i) = \mathbf{x}_i' \boldsymbol{\beta}$$

One disadvantage of the linear model is that estimates of π_i may sometimes lie outside the interval [0,1]. To ensure that this does not happen, we use instead the cumulative distribution

$$F(x) = g^{-1}(\mathbf{x}_i'\boldsymbol{\beta}) = \int_{-\infty}^{x} f(y)dy$$

where $f(y)$ is the probability density function, and is sometimes referred to as the p.d.f. of the *tolerance distribution.* Tolerance distributions that have been well documented in the literature and are commonly used are the normal, the logistic, and the extreme value distributions. We discuss these distributions in the next subsections.

8.2.1 Dose-Response Models

Let us consider in particular a dose-response experiment and let π_i be the theoretical population survival rate using drug (X) at dosage level x_i (or usually log-dosage). Let p_i be the corresponding observed value of π_i and we will be interested in modeling π_i as a function of the dose levels x_i. The simplest form of this relationship is

$$g(\pi_i) = \beta_0$$

or

$$g(\pi_i) = \beta_0 + \beta_1 x_i$$

8.2.2 The Normal Tolerance Distribution

If the normal distribution is employed as the tolerance distribution, then we would have

$$\begin{aligned} F(x) &= \frac{1}{\sigma\sqrt{2\pi}} \int_{-\infty}^{x} e^{\left[-\frac{1}{2}\left(\frac{y-\mu}{\sigma}\right)^2\right]} dy \\ &= \Phi\left(\frac{x-\mu}{\sigma}\right) \end{aligned}$$

where Φ denotes the cumulative probability distribution for the standard normal $N(0, 1)$. Thus

$$g(\pi_i) = \Phi^{-1}(\pi_i) = \beta_0 + \beta_1 x_i$$

where $g(\pi_i)$ is the link function, $\beta_0 = \frac{-\mu}{\sigma}$ and $\beta_1 = \frac{1}{\sigma}$. The link function $\mathbf{g}$ is the inverse Cumulative Normal probability function Φ^{-1}. When the tolerance distribution is the normal, the relevant model is called the *probit model.* The probit of a probability π is defined for each dose level $x_i, i = 1, 2, \cdots, k$ to be a value s such that

$$\begin{aligned} \pi_i &= \frac{1}{\sqrt{2\pi}} \int_{-\infty}^{s} e^{-\frac{y^2}{2}} dy, \quad \text{where } s_i = \frac{X_i - \mu}{\sigma} \\ &= P(Z < s_i), \qquad \text{that is, } \pi_i = \Phi(s_i) \end{aligned}$$

The model is appropriately fitted by invoking the probit link in PROC GENMOD in SAS®. SAS® also has a PROC PROBIT that can be readily employed. Finney (1971) and Bliss (1935) have given extensive discussions on probit analysis and interested readers are encouraged to consult these texts. We display in Figure 8.1 a graph of the probit model for values of $0 \leq p \leq 1.0$.

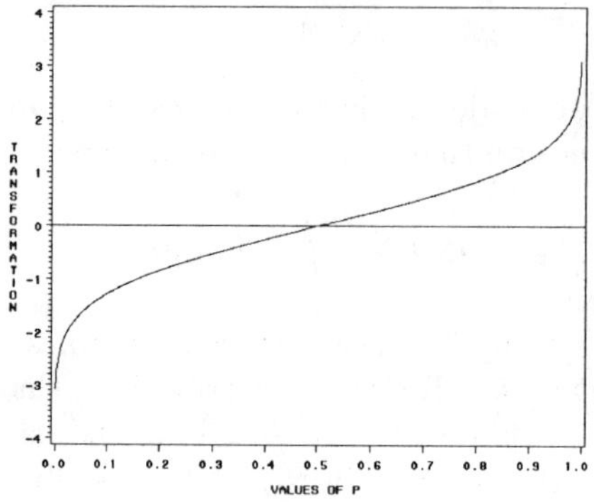

Figure 8.1: Probit Plot

8.2.3 The logistic distribution

The logistic tolerance distribution (Berkson, 1953, 1955) has density

$$f(y) = \frac{\beta_1 e^{\beta_0+\beta_1 y}}{[1+e^{\beta_0+\beta_1 y}]^2}$$

and

$$\pi(x) = \int_{-\infty}^{x} f(y)\,dy = \frac{\exp(\beta_0+\beta_1 x)}{1+\exp(\beta_0+\beta_1 x)}$$

An alternative definition of $F(x)$ is:

$$\pi(x_i) = \frac{1}{1+e^{-(\beta_0+\beta_1 x_i)}} \tag{8.4}$$

which leads to:

$$\begin{aligned} \ln\left[\frac{\pi(x_i)}{1-\pi(x_i)}\right] &= \beta_0+\beta_1 x_i \\ &= \lambda_i \end{aligned} \tag{8.5}$$

for $(i = 1, 2, \cdots, k)$, which is referred to as the *logistic function.* The corresponding link function is given by the logit:

$$g(\pi) = \ln\left[\frac{\pi(x_i)}{1-\pi(x_i)}\right]$$

Clearly from the above expression, the odds of response 1 (success) are

$$\begin{aligned} \frac{\pi_i(x)}{1-\pi_i(x)} &= e^{(\beta_0+\beta_1 x_i)} \\ &= e^{\beta_0} \times (e^{\beta_1})^{x_i} \end{aligned}$$

In general,

$$\boldsymbol{\lambda}_i = \mathbf{x}_i'\boldsymbol{\beta} \tag{8.6}$$

where $\mathbf{x}_i'\boldsymbol{\beta} = \beta_0 + \beta_1 X_1 + \beta_2 X_2 + \cdots + \beta_k X_k$.

Equation (8.6) is described as a *linear logistic regression* model because it is a regression type model if the explanatory variables are quantitative. Similarly, (8.6) will be an ANOVA-type model when the explanatory variables are categorical. In this case, we often refer to it as the *logit model.*

λ_i is sometimes referred to as the logit or (log-odds) of the probability of survival for the i-th dosage level. "Logit" is a contraction of the phrase "logarithmic unit" in analogy to Bliss's (1935) probit for "probability unit." The logit is a transformation that takes a number π between 0 and 1 and transforms it to $\ln\left(\frac{\pi}{1-\pi}\right)$. Similarly, the logistic transformation takes a number x on the real line and transforms it to $e^x/(1+e^x)$. We note here that the logistic and logit transformations are inverses of each other, with the former transformation giving π and the latter giving x.

The shapes of the function $f(y)$ and $\pi(x)$ are similar to those of the probit model except in the tails of the distributions (see Cox & Snell, 1989). We display in Figure 8.2 a plot of the logistic distribution for $0 \leq p \leq 1$. Any value of π in the range (0,1) is transformed into a value of the logit(π) in $(-\infty, \infty)$. Thus as $\pi \to 0$, logit(π) $\to -\infty$.

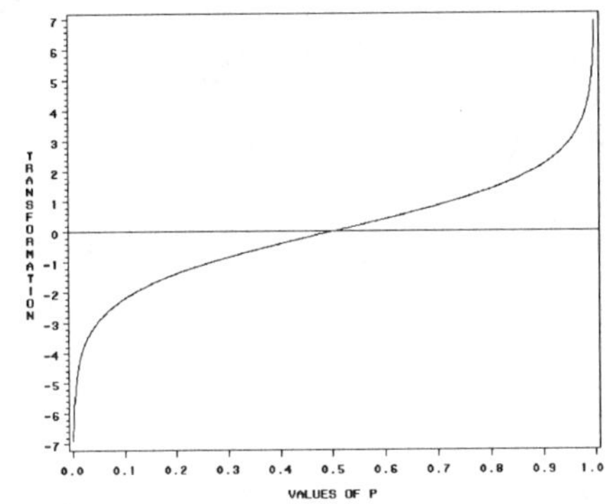

Figure 8.2: Logistic plot

In Figure 8.3 is the plot of the transformation of $0 \leq X \leq 80$ to $\pi(x)$.

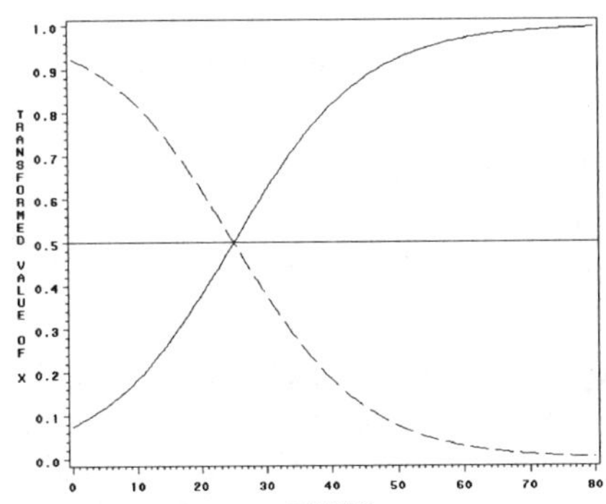

Figure 8.3: Transformation of X to $\pi(x)$

In Figure 8.3, $\pi(x)$ is defined either as in (8.5) or (8.6). Thus, as $x \to \infty, \pi(x) \downarrow 0$ when $\beta_1 < 0$ and $\pi(x) \uparrow 1$ when $\beta_1 > 0$. As $\beta_1 \to 0$, the curve flattens to a horizontal straight line and when the model holds true with $\beta = 0$, the dichotomous or binary response is independent of X.

$\pi(x)$ has $\partial\pi(x)/\partial x = \beta_1\pi(x)[1-\pi(x)]$. Thus the curve has its steepest slope at $\hat{x}$ value where $\pi(\hat{x}) = \frac{1}{2}$, which is $x = -\beta_0/\beta_1$. Logit(π) is a sigmoid function that is symmetric about $\pi = 0.5$. The graph is approximately linear between $\pi = 0.2$

and $\pi = 0.8$, but is definitely nonlinear outside this range.

For more details on the properties of $\pi(x)$ see Agresti (1990).

The probit (discussed earlier) and the logit links are similar, and it can be shown that the variances of the two are equal if the logistic β_1^2 is $\frac{\pi^2}{3}$ times the probit β_1^2, and the two models usually give similar fitted values, with parameter estimates for logit models being usually 1.6 to 1.8 times those of the probit model.

8.2.4 Extreme Value Distribution

Other models that have also been considered for the dose-response data include the *complementary log-log* and the *log-log* link models. The latter, sometimes referred to as the extreme value model, is characterized by:

$$f(y) = \beta_1 e^{[(\beta_0+\beta_1 y)-e^{(\beta_0+\beta_1 y)}]}$$

and

$$\pi(x) = 1 - \exp[-e^{(\beta_0+\beta_1 x)}]$$

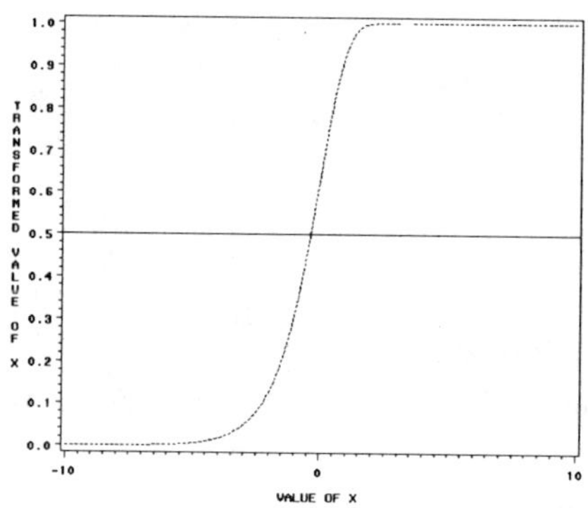

Figure 8.4: Complimentary log-log

Figure 8.4 displays the graph of

$$F(x) = 1 - \exp[-e^x]$$

against x for values of x ranging over $[-10, 10]$. A transformation of the form

$$\ln[-\ln(1-\pi)] = \beta_0 + \beta_1 x$$

transforms $\pi(x)$ above to the well-recognized linear model form. The link function, $\ln[-\ln(1-\pi)]$, is called the *complementary log-log function* or **CLL** for short. The model is usually preferred to either the probit or the logistic models for values of π near 0 or 1. In Figure 8.4 is a plot of the function that transforms a probability π in the range (0,1) to a value in $(-\infty, \infty)$. From the graph, we see that the function is not symmetric about $\pi = 0.5$ in this case, and as $\pi \to 1$, $\ln(\pi) \to \infty$ and for $\pi = 0.5$, $\ln(\pi) = 0$. The function is a special form of the *Gumbel distribution*, which has been found very useful in modeling the breaking strength of materials. In Figure 8.5, the plot of the link transformation against p uses values of p ranging from 0.001 to 0.999.

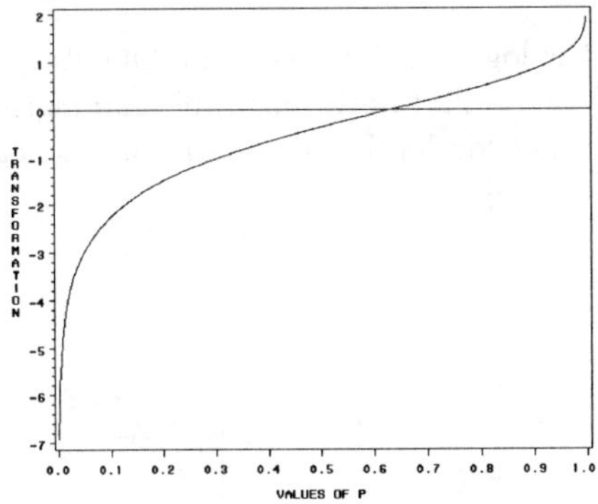

Figure 8.5: Complimentary log-log

For particular values of X (say $x_1 < x_2$), we have

$$\ln\left[1 - \ln\left\{1 - \pi(x_2)\right\}\right] - \ln\left[1 - \ln\left\{1 - \pi(x_1)\right\}\right] = \beta_1(x_2 - x_1)$$

so that

$$\frac{\ln\left[1 - \pi(x_2)\right]}{\ln\left[1 - \pi(x_1)\right]} = \exp[\beta_1(x_2 - x_1)], \quad \text{and}$$
$$1 - \pi(x_2) = \{1 - \pi(x_1)\}^{\exp[\beta_1(x_2 - x_1)]}$$

That is, the probability of surviving at x_2 equals the probability of surviving at x_1 raised to the power $\exp(\beta_1)$ higher for each unit increase in the distance $x_2 - x_1$.

The complimentary log-log link is not symmetric, but for small π, it is very close to the logit link since in this case $\ln(1 - \pi) \approx -\pi$, so that

$$\ln(-\ln(1 - \pi)) \approx \ln \pi, \quad \ln(\pi/(1 - \pi)) \approx \ln \pi + \pi$$

The complimentary log-log link model is most suitable in asymmetric situations when $\pi(x)$ increases from 0 fairly slowly, but approaches 1 quite rapidly as X increases.

On the other hand, if $\pi(x)$ departs from 1 slowly but approaches 0 sharply as X increases, the appropriate model is the log-log link, which is defined as:

$$g(\pi) = -\ln\left[-\ln\left(\pi(x)\right)\right] = \beta_0 + \beta_1 x$$

where $F(x) = \exp(-e^{-v}) = \exp\left\{-\exp[-(\beta_0 + \beta_1 x)]\right\}$. $F(x)$ is the CDF for the reversed extreme value distribution. We note here that if v has the extreme value distribution with CDF

$$F(x) = 1 - \exp(-e^{v})$$

then $r = -v$ has the reversed extreme value distribution with

$$F(r) = \exp(-e^{-r})$$

For $\pi(x)$ near 1, the log-log link is very close to the logit link.

The SAS® PROC GENMOD and PROC LOGISTIC can be used to fit some of the models described in the preceding sections by specifying the following link functions in PROC GENMOD and LOGISTIC.

a. Logistic model (*logit link function*): **LOGIT.**

b. Probit model (*inverse cumulative Normal link function* Φ^{-1}): **PROBIT.**

c. Extreme value model (*complimentary log log function*): **CLL**

Figure 8.6 gives the graphs of the logit, probit, and complimentary log-log against p ranging from 0.001 to 0.999. We note that both the probit and logistic are symmetric about $p = 0.5$. The complimentary log-log is asymmetric in this case.

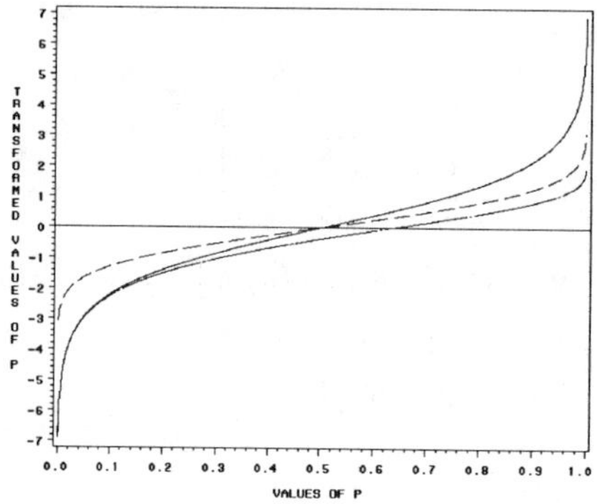

Figure 8.6: Plots of the three functions

8.3 Estimating the Parameters of a Logistic Regression Model

For the simple linear logistic regression, we have

$$\lambda_i = \ln\left(\frac{\pi_i}{1-\pi_i}\right) = \beta_0 + \beta_1 x_i \tag{8.7}$$

where λ_i is the logit of the probability of "survival" or "death" for the i-th level of X.

Thus the logistic model in (8.7) can be fitted by a linear model

$$E_A(\delta_i) = E_A\left\{\ln\left(\frac{p_i}{1-p_i}\right)\right\} = \beta_0 + \beta_1 x_i$$

where

$$\delta_i = \ln\left(\frac{p_i}{1-p_i}\right)$$

are the observed logits. Accordingly,

$$E(\delta_i) = \beta_0 + \beta_1 x_i$$

We can compare the above with the classical simple linear regression model $E(y_i) = \beta_0 + \beta_1 x_i$, where consistent estimates of the parameters β_0 and β_1 can be easily obtained by ordinary least squares (OLS) method. The OLS assumes that the y_i are homoscedastic, that is, the y_i has constant variance denoted by σ^2 and hence the OLS estimates are given by:

$$\hat{\boldsymbol{\beta}} = (X'X)^{-1}X'Y$$

where

$$\boldsymbol{\beta} = \begin{bmatrix} \beta_0 \\ \beta_1 \end{bmatrix}$$

However, in our case, the variance of δ_i is not constant, being dependent on p_i (the observed success probability) and n_i (the binomial denominator), the probability of survival when the i-th dosage level is applied and the number of subjects receiving the dosage, respectively. We encourage students to show, using the delta method, that the variance of δ_i is given by (appendix F.1)

$$\text{Var}(\delta_i) = \frac{1}{n_i p_i (1 - p_i)} \quad \text{for} \quad i = 1, 2, \cdots, k \tag{8.8}$$

Thus an estimate of $\text{Var}(\delta_i) = [n_i p_i (1 - p_i)]^{-1}$ is an element of a diagonal matrix.

If we assume independent sampling for each subpopulation $i = 1, 2, \cdots, k$, we have

$$\text{Cov}(\delta_i, \delta_j) = 0 \quad \text{for } i \neq j$$

If we therefore write the observed logits in vector form as

$$\boldsymbol{\delta}' = (\delta_1, \delta_2, \cdots, \delta_k)$$

and the *theoretical* logits similarly in vector form as:

$$\boldsymbol{\lambda}' = (\lambda_1, \lambda_2, \cdots, \lambda_k)$$

then the model in (8.7) can be written as:

$$\boldsymbol{\lambda} = E_A(\boldsymbol{\delta}) = \mathbf{X}\,\boldsymbol{\beta}$$

where $\mathbf{X}$ is an $k \times 2$ and $\boldsymbol{\beta} = 2 \times 1$ matrix. That is,

$$\mathbf{X} = \begin{bmatrix} 1 & x_1 \\ 1 & x_2 \\ \vdots & \vdots \\ 1 & x_k \end{bmatrix} \quad \text{and} \quad \boldsymbol{\beta} = \begin{bmatrix} \beta_0 \\ \beta_1 \end{bmatrix}$$

Furthermore, the unrestricted linearized covariance matrix of $\boldsymbol{\delta}$ in (8.6) is given by:

$$\mathbf{V}_{\boldsymbol{\delta}} = \begin{bmatrix} [n_1 p_1 (1-p_1)]^{-1} & 0 & 0 & \cdots & 0 \\ 0 & [n_2 p_2 (1-p_2)]^{-1} & 0 & \cdots & \vdots \\ 0 & 0 & \ddots & 0 & \vdots \\ \vdots & \vdots & 0 & \ddots & 0 \\ 0 & \cdots & \cdots & 0 & [n_k p_k (1-p_k)]^{-1} \end{bmatrix}$$

We see from the above that the variances of the δ's are not constant and hence the method of weighted least squares (WLS) will be employed to obtain estimates of our parameters. From WLS theory, we have:

$$\hat{\boldsymbol{\beta}} = (\mathbf{X}'\mathbf{V}_{\boldsymbol{\delta}}{}^{-1}\mathbf{X})^{-1}\mathbf{X}'\mathbf{V}_{\boldsymbol{\delta}}{}^{-1}\boldsymbol{\delta}$$

and

$$\text{Var}(\hat{\boldsymbol{\beta}}) = (\mathbf{X}'\mathbf{V}_{\boldsymbol{\delta}}{}^{-1}\mathbf{X})^{-1}\sigma^2$$

where $\mathbf{V}_{\boldsymbol{\delta}}{}^{-1}$ is the inverse of $\mathbf{V}_{\boldsymbol{\delta}}$ given above.

8.3.1 Example 8.1: A Bioassay Example

The data in Table 8.3 give the effect of different concentrations of nicotine sulphate in a 1% saponin solution on an insect *Drosophila melanogaster,* the fruit fly. We usually employ the logarithm of x_i, and in this case we chose to use $\log(x_i)$, that is, the log to base 10 of the doses. Table 8.4 gives the relevant initial calculations

Nicotine sulphate g/100 cc	Number killed	Number of insects
x_i	r_i	n_i
0.10	8	47
0.15	14	53
0.20	24	55
0.30	32	52
0.50	38	46
0.70	50	54
0.95	50	52

Table 8.3: Effect of different concentrations of nicotine sulphate on Drosophila melanogaster, (Hubert, 1992)

$\log_{10}(x_i)$	Observed proportions p_i	Observed logits δ_i	Expected proportions $\hat{\pi}_i$
-1.000	0.170	-1.5856	0.1448
-0.824	0.264	-1.0253	0.2864
-0.699	0.436	-0.2574	0.4253
-0.523	0.615	0.4684	0.6369
-0.301	0.826	1.5575	0.8387
-0.155	0.926	2.5268	0.9141
-0.022	0.962	3.2314	0.9532

Table 8.4: Results of analysis

for the data. While I do not for a moment think that the linear logistic moment should be fitted by what follows in this section, it is nevertheless incorporated here so that students can have a proper understanding of what is really going on from the use of statistical packages.

The linear logistic model can be fitted by writing

$$\begin{bmatrix} -1.5856 \\ -1.0253 \\ -0.2574 \\ 0.4684 \\ 1.5575 \\ 2.5268 \\ 3.2314 \end{bmatrix} = \begin{bmatrix} 1 & -1.000 \\ 1 & -0.824 \\ 1 & -0.699 \\ 1 & -0.523 \\ 1 & -0.301 \\ 1 & -0.155 \\ 1 & -0.022 \end{bmatrix} \begin{bmatrix} \beta_0 \\ \beta_1 \end{bmatrix}$$

that is,

$$\boldsymbol{\delta} = \mathbf{X}\,\boldsymbol{\beta}$$

and

$$\mathbf{V}_{\boldsymbol{\delta}}^{-1} = \text{diag}\{6.6317, 10.2981, 13.5247, 12.3123, 6.6113, 3.7003, 1.9009\}$$

which is a 7×7 diagonal matrix, where, for instance, 6.6317 is obtained as:

$$6.6317 = \left[\frac{1}{47(0.170)(1-0.170)}\right]^{-1} = n_1 p_1 (1 - p_1)$$

Hence,

$$(\mathbf{X}'\mathbf{V}_{\boldsymbol{\delta}}{}^{-1}\mathbf{X})^{-1} = \begin{bmatrix} 0.1189 & 0.1652 \\ 0.1652 & 0.2712 \end{bmatrix} \quad \text{and}$$

$$\mathbf{X}'\mathbf{V}_{\boldsymbol{\delta}}{}^{-1}\boldsymbol{\delta} = \begin{bmatrix} 7.5005 \\ 13.4499 \end{bmatrix}$$

$$\Rightarrow \begin{bmatrix} \hat{\beta}_0 \\ \hat{\beta}_1 \end{bmatrix} = \begin{bmatrix} 0.1189 & 0.1652 \\ 0.1652 & 0.2712 \end{bmatrix} \begin{bmatrix} 7.5005 \\ 13.4499 \end{bmatrix}$$

$$= \begin{bmatrix} 3.1137 \\ 4.8867 \end{bmatrix}$$

Hence, $\hat{\beta}_0 = 3.1137$, $\hat{\beta}_1 = 4.8867$, with corresponding variances $0.1189\sigma^2$ and $0.2712\sigma^2$, respectively. Thus for a unit change in $\log_{10}$ of nicotine sulphate (X), the estimated odds of the number of insects killed are multiplied by $\exp\{4.8867\} = 132.5156$.

The expected or predicted logits from the fitted model are:

$$[-1.7730, -0.9125, -0.3020, 0.5585, 1.6427, 2.3567, 3.0048]$$

with estimated proportions

$$\hat{\pi}_i = [0.145, 0.286, 0.425, 0.637, 0.838, 0.914, 0.953]$$

With $x_i = \log_{10} x_i$, the above are obtained from the estimated regression equation:

$$\ln\left(\frac{\hat{\pi}_i}{1-\hat{\pi}_i}\right) = \hat{\beta}_0 + \hat{\beta}_1 X_i = 3.1137 + 4.8867 x_i$$

Thus with $x_i = -1.00$, we have

$$\ln\left(\frac{\hat{\pi}_i}{1-\hat{\pi}_i}\right) = -1.7730 \Rightarrow \frac{\hat{\pi}_i}{1-\hat{\pi}_i} = 0.1698228$$

That is, $\hat{\pi}_i = \dfrac{0.1698228}{1+0.1698228} = \dfrac{0.1698228}{1.1698228} = 0.1452.$

The level of the dosage that would result in a 50% response by subjects in the population under study is an important parameter in dose-response models. A measure of the potency of the drug is the statistic LD50 *median lethal dose* (or ED50 for *effective median dose*, LC50 for *median lethal concentration*, and its corresponding EC50, *median effective concentration*). In this example, LD50 is the lethal dosage at which 50% of the subjects (insects) are expected to be killed, and in experiments where the response is not death, we refer to the ED50, median effective dose. Thus,

$$\ln\left(\frac{50}{100-50}\right) = \hat{\beta}_0 + \hat{\beta}_1 x_i \quad \Rightarrow \hat{x}_M = -\frac{\hat{\beta}_0}{\hat{\beta}_1}$$

That is, $\log_{10}(\text{LD50}) = -0.6372 \quad \Rightarrow \quad \text{LD50} = 10^{-0.6372} = 0.2306$ g/100 cc.

Similarly, an LD90 is given by 10^U, where $U = \dfrac{(2.1972 - \hat{\beta}_0)}{\hat{\beta}_1}$, which when computed gives a value of 0.6493 g/100 cc.

The χ^2_{logit} for the $k \times 2$ table is given by

$$\chi^2_{\text{logit}} = \sum_{i=1}^{k} (\delta_i - \bar{\delta}_\omega)^2$$

where if we define

$$u_i = n_{i+}p_i(1 - p_i) \quad \text{for} \quad i = 1, 2, \cdots, k$$

which are simply the inverses of the asymptotic variances of the observed logits δ_i, then

$$\bar{\delta}_\omega = \frac{\sum u_i \delta_i}{\sum u_i}$$

is the weighted mean of the observed logits.

For the data in Table 8.3, we have the following calculations:

i	n_i	p_i	$p_i(1-p_i)$	u_i	δ_i	$u_i\delta_i$	δ_i^2	$u_i\delta_i^2$
1	47	0.1702	0.1412	6.6383	-1.5841	-10.5159	2.5094	16.6584
2	53	0.2642	0.1944	10.3019	-1.0245	-10.5543	1.0496	10.8130
3	55	0.4364	0.2460	13.5273	-0.2559	-3.4621	0.0655	0.8861
4	52	0.6154	0.2367	12.3077	0.4700	5.7847	0.2209	2.7188
5	46	0.8261	0.1437	6.6087	1.5581	10.2973	2.4278	16.0447
6	54	0.9259	0.0686	3.7037	2.5257	9.3546	6.3793	23.6271
7	52	0.9615	0.0370	1.9231	3.2189	6.1901	10.3612	19.9253
Total				55.0106		7.0944		90.6733

Hence,

$$\bar{\delta}_\omega = \frac{\sum u_i \delta_i}{\sum u_i} = \frac{7.0944}{55.0106} = 0.1290$$

and as a result

$$\begin{aligned} \text{Total } \chi^2_{\text{logit}} &= \sum_{i=1}^{5} u_i \delta_i^2 - (\sum u_i)\bar{\delta}_\omega^2 \\ &= 90.6733 - (55.0106) * (0.1290)^2 \\ &= 90.6733 - 0.9149 \\ &= 89.7584 \end{aligned}$$

Comparison with the WLS Solution

The total sum of squares is given by:

$$Q_T = \boldsymbol{\delta}'\mathbf{V}_{\boldsymbol{\delta}}^{-1}\boldsymbol{\delta} - \bar{\boldsymbol{\delta}}'\mathbf{V}_{\boldsymbol{\delta}}^{-1}\bar{\boldsymbol{\delta}}$$

where $\bar{\delta}_i = \sum u_i \delta_i$ and $u_i = n_i p_i (1 - p_i)$.

Thus Q_T = 90.6733 - 0.9149 = 89.7584 on 6 degrees of freedom. The regression contribution is accordingly given by

$$\frac{\hat{\beta}_1^2}{\text{Var}\,(\hat{\beta}_1)} = \frac{(4.8867)^2}{0.2712} = 88.0525$$

These results are summarized in Table 8.5.

Source	d.f.	χ^2	pvalue
Regression	1	88.0525	0.0000
Lack of fit	5	1.7059	0.8882
Total	6	89.7584	

Table 8.5: Lack of fit test

This model fits the data well, as can be seen from the nonsignificant χ^2 lack of fit pvalue (= 0.8882), while the test of $\beta_1 = 0$ was significant with $\chi^2_{reg} = 88.0525$, indicating that the shape parameter should be kept in the model.

We give in the next table, the fits of the various models discussed in the preceding sections to the data in Table 8.3 using the statistical package SAS® (equivalent SPSS implementation can be found in the appendix). The MLE for these models are obtained by the iterative re-weighted least squares discussed in section 2.6 of chapter 2. For the logistic model for instance, the likelihood equations are derived using the Fisher's scoring algorithm discussed in the same section of chapter 2, which are derived from

$$\sum_i y_i x_{ij} = \sum_i \mu_i x_{ij}$$

where $\mu_i = n_i p_i$. The derivation of this is provided in appendix F.2.

We observe here that the ratio of the logistic parameter model estimates to those of the logit models is (3.1236/1.8255)=(4.8995/2.8749)=1.7. This ratio is within the expected range of 1.6 to 1.8.

Models	d.f.	G^2	pvalue	Parameters	
				$\hat{\beta}_0$	$\hat{\beta}$
Logistic	5	0.7336	0.9811	3.1236	4.8995
Probit	5	0.5437	0.9904	1.8255	2.8749
CLL	5	1.5578	0.9063	1.3888	2.9196
E-V	5	4.1421	0.5291	-2.7309	-3.5081

- CLL: complimentary log-log model
- E-V: Extreme value model

All the models provide adequate fits of the data, but both the probit and the logistic models provide better fits for the data, although the probit models seems better both in terms of G^2 values and the standardized residuals (not printed). All the models are of course based on 5 degrees of freedom.
The following are the corresponding SAS software statements for fitting the above models using PROCs LOGISTIC and GENMOD.

```
data tab83;
input x r n @@;
dose=log10(x);
surv=n-r;
datalines;
datalines;
0.10 8 47 0.15 14 53 0.20 24 55 0.30 32 52
0.50 38 46 0.70 50 54 0.95 50 52
;
*** (i) fit logistic model ***;
proc logistic data=tab83;
model r/n=dose/scale=none aggregate influence covb;
output out=aa p=phat stdxbeta=selp h=lev; run;
proc print data=aa; run;
*** (ii) Fits probit model with Proc Logistic ***;
proc logistic;
model r/n=dose/link=normit scale=none aggregate; run;
(iii)
*** (iii) Fits CLL model with Proc Logistic ***;
proc logistic;
model r/n=dose/link=cloglog scale=none aggregate; run;
*** (iv) Fits Extreme value model with Proc Logistic ***;
proc logistic;
model surv/n=dose/link=cloglog scale=none aggregate; run;
```

Equivalent program using PROC GENMOD is also provided below:

```
set tab83;
proc genmod;
model r/n=dose/dist=bin link=logit obstats; run;
proc genmod;
model r/n=dose/dist=bin link=probit obstats; run;
proc genmod;
model r/n=dose/dist=bin link=cll obstats; run;
proc genmod;
model surv/n=dose/dist=bin link=cll obstats; run;
```

In PROC LOGISTIC, **P**, **STDXBETA**, **PRED**, and **H** contain respectively the predicted values of the probabilities, the standard errors of the linear predictor, the predicted logits, and the leverages for the model. The AGGREGATE and influence statements in (i) allow the goodness-of-fit tests (deviance or G^2 and the Pearson's X^2) to be presented, while the influence statement allows the detection of influential observations. The INFLUENCE option may result in a large volume of output especially if there are many observations in the data. Below is the result from PROC LOGISTIC when the option aggregate, which must accompany the SCALE=NONE option in the logistic model statement, is invoked.

```
                     The LOGISTIC Procedure
                       Model Information

          Data Set                        WORK.TAB83
          Response Variable (Events)      r
          Response Variable (Trials)      n
          Number of Observations          7
          Model                           binary logit
          Optimization Technique          Fisher's scoring

         Deviance and Pearson Goodness-of-Fit Statistics

  Criterion        DF         Value      Value/DF     Pr > ChiSq
  ---------------------------------------------------------------
  Deviance          5        0.7336        0.1467         0.9811
  Pearson           5        0.7351        0.1470         0.9810

                 Number of unique profiles: 7
```

The parameter estimates from the logistic model using PROC LOGISTIC are displayed as:

```
              Analysis of Maximum Likelihood Estimates

                                  Standard          Wald
 Parameter    DF    Estimate        Error    Chi-Square    Pr > ChiSq
 ----------------------------------------------------------------------
 Intercept     1      3.1236       0.3349       86.9809        <.0001
 dose          1      4.8995       0.5098       92.3620        <.0001

                        Odds Ratio Estimates

                      Point           95% Wald
            Effect    Estimate       Confidence Limits
            ------------------------------------------
            dose       134.228       49.419     364.582

     Association of Predicted Probabilities and Observed Responses

          Percent Concordant      80.0    Somers' D     0.692
          Percent Discordant      10.8    Gamma         0.762
          Percent Tied             9.2    Tau-a         0.333
          Pairs                  30888    c             0.846
```

Estimated Covariance Matrix

Variable	Intercept	dose
Intercept	0.112172	0.15627
dose	0.15627	0.259907

The parameter estimates $\hat{\beta}_0 = 3.1236$ and $\hat{\beta}_1 = 4.8995$ are both significantly ($p <$ .0001) different from zero. The odds ratio for the intercept would be $e^{3.1236} = 22.73$. Thus at log (base 10) dosage level (dose = 0), it is almost 23 times most likely that the insect *Drosophila melanogaster* would die than not die. Note that this dose level is equivalent to 1.0 g/100 cc nicotine sulphate concentration. Similarly with each unit increase in $\log_{10}$ dosage level, the odds of insect dying increase by 134.23 times.

A test of the hypothesis concerning the parameters of the logistic model is carried out as follows:

To test whether $H_0 : \beta = 0$ against $H_a : \beta \neq 0$, the statistic is:

$$t^* = \frac{\hat{\beta}}{se(\hat{\beta})}$$

For binomial data, however, this is not usually distributed as Student's t distribution but as a normal distribution. Hence, when H_0 holds, t^* will be distributed approximately normal. In our case, $t^* = 9.6105$, which clearly indicates that the null hypothesis is not tenable in this case. This test is provided in SAS software output below plus the parameter estimates when the logistic model is fitted:

Testing Global Null Hypothesis: BETA=0

Test	Chi-Square	DF	Pr > ChiSq
Likelihood Ratio	145.2882	1	<.0001
Score	127.2173	1	<.0001
Wald	92.3620	1	<.0001

We observe that $9.6106^2 = 92.3636$, which should have been equal to the Wald's value of 92.3620 in the above test except for rounding error.

The estimated killing probability as a function of the log dosage is estimated as:

$$\hat{p} = \frac{\exp{(3.1236 + 4.8995\text{dose})}}{1 + \exp{(3.1236 + 4.8995\text{dose})}}$$

which for the first dosage becomes

$$\begin{aligned} &= \frac{\exp{[3.1236 + 4.8995\,(-1)]}}{1 + \exp{[3.1236 + 4.8995\,(-1)]}} \\ &= \frac{0.1693}{1.1693} \\ &= 0.1448 \end{aligned}$$

The expected number of deaths for this dosage level is $n_i * \hat{p} = 47 * 0.1448 = 6.8056$ or about 7 insects. These and other relevant parameters are displayed below.

Obs	r	n	yhat	phat	selp	lev	resid
1	8	47	6.8056	0.1448	0.2440	0.3465	0.4838
2	14	53	15.1792	0.2864	0.1763	0.3368	-0.3604
3	24	55	23.3915	0.4253	0.1439	0.2782	0.1657
4	32	52	33.1188	0.6369	0.1408	0.2382	-0.3206
5	38	46	38.5802	0.8387	0.2041	0.2591	-0.2305
6	50	54	49.3614	0.9141	0.2646	0.2968	0.3172
7	50	52	49.5664	0.9532	0.3246	0.2442	0.2926

To fit a probit model, specify the link function as NORMIT in the model statement (see ii) in PROC LOGISTIC. Similarly, we specify the CLOGLOG statement for the complimentary log-log model as in (iii). The default in SAS software is the LOGIT link. The above SAS software results can also be realized by the use of PROCEDURE CATMOD in SAS software. The final model based on the logistic regression is given by

$$\ln\left(\frac{\hat{\pi}_i}{1-\hat{\pi}_i}\right) = 3.1236 + 4.8995 \log_{10} \text{dose}_i, \quad i = 1, 2, \cdots, 7 \tag{8.9}$$

The logistic model can also be implemented with PROC CATMOD. We present below the SAS software program required to implement this together with a partial SAS software output, again for the data in Table 8.3.

```
DATA NEW; SET TAB83;
COUNT=R; Y=0; OUTPUT; COUNT=N-R; Y=1; OUTPUT; DROP N R;
PROC CATMOD DATA=NEW; WEIGHT COUNT; DIRECT DOSE;
MODEL Y=DOSE/FREQ ML NOGLS; RUN;
```

```
                        The CATMOD Procedure

               Maximum Likelihood Analysis of Variance

           Source                DF   Chi-Square    Pr > ChiSq
           --------------------------------------------------
           Intercept              1        86.98        <.0001
           dose                   1        92.36        <.0001

           Likelihood Ratio       5         0.73        0.9811

               Analysis of Maximum Likelihood Estimates

                                  Standard         Chi-
         Parameter    Estimate       Error       Square    Pr > ChiSq
         ------------------------------------------------------------
         Intercept      3.1236      0.3349        86.98        <.0001
         dose           4.8996      0.5098        92.36        <.0001
```

Results obtained from the application of PROC CATMOD agree with those obtained earlier for the logistic model.

8.3.2 Implementing a Probit Model

We also implement both the probit and logistic models for the data in Table 8.3 with PROC PROBIT in SAS®. In the program below to implement these models, notice that there is no need to formally transform the dose to log10 or log dose. PROC Probit does this automatically by specifying either LOG 10 or LOG in the PROC statement. The option LACKFIT conducts the usual lack-of-fit tests

(X^2 and G^2: this test assumes that the data has been sorted by the explanatory variable), while **INVERSECL** computes confidence limits for values of the first continuous explanatory variable, in this case, dose together with the response rates. To implement the logistic regression in this case, specify distribution (D = logistic) in the model statement as in the second model statement below.

```
DATA TAB83; INPUT DOSE R N @@; DATALINES;
0.10 8 47 0.15 14 53 0.20 24 55 0.30 32 52
0.50 38 46 0.70 50 54 0.95 50 52
;
PROC PROBIT DATA=TAB83 LOG10;  MODEL R/N=DOSE/LACKFIT INVERSECL;
 OUTPUT OUT=AA P=PROB; RUN;
PROC PROBIT DATA=TAB83 LOG10; MODEL R/N=DOSE/D=LOGISTIC LACKFIT INVERSECL;
 OUTPUT OUT=BB P=PROB2; RUN;
DATA NEW; MERGE AA BB; PROC PRINT DATA=NEW; RUN;
```

```
Probit Procedure
                        Goodness-of-Fit Tests

Statistic                                 Value        DF    Pr > ChiSq
---------------------------------------------------------------------
Pearson Chi-Square                       0.5451         5        0.9904
L.R.     Chi-Square                      0.5437         5        0.9904

                       Analysis of Parameter Estimates

                            Standard   95% Confidence      Chi-
Parameter    DF Estimate      Error        Limits        Square Pr > ChiSq
---------------------------------------------------------------------------
Intercept     1   1.8255     0.1753    1.4819    2.1690  108.47      <.0001
Log10(DOSE)   1   2.8749     0.2710    2.3438    3.4061  112.53      <.0001

Probit Model in Terms of Tolerance Distribution

        MU            SIGMA
 -0.634963      0.34783305

       Estimated Covariance Matrix
        for Tolerance Parameters
                        MU               SIGMA
MU                0.000766           -0.000173
SIGMA            -0.000173            0.001075

                 Probit Analysis on Log10(DOSE)
Probability            Log10(DOSE)          95% Fiducial Limits

       0.01                -1.44414       -1.64886          -1.30058
       0.02                -1.34932       -1.53355          -1.21955
       0.03                -1.28917       -1.46052          -1.16801
      .................................................
       0.45                -0.67867       -0.73963          -0.62476
       0.50                -0.63496       -0.69216          -0.58172
       0.55                -0.59125       -0.64591          -0.53746
      .................................................
       0.98                 0.07940       -0.03443           0.23977
       0.99                 0.17422        0.04691           0.35477

                   Probit Analysis on DOSE
Probability            DOSE         95% Fiducial Limits

       0.01          0.03596        0.02245          0.05005
       0.02          0.04474        0.02927          0.06032
       0.03          0.05138        0.03463          0.06792
      .........................................
       0.45          0.20957        0.18213          0.23727
       0.50          0.23176        0.20316          0.26199
       0.55          0.25630        0.22599          0.29009
      .........................................
       0.98          1.20060        0.92378          1.73688
       0.99          1.49354        1.11406          2.26344
```

Partial results for the probit analysis on log10(dose) and dose are presented above. The complete result are displayed in appendix F.3. The partial results above indicate that the tolerance distribution parameter mean has an estimate of -0.634963. The variance covariance matrix of the parameter estimates are also displayed. The LD50 for LOG10(Dose) is -0.63496, that is, the log10(dose) corresponding to a probability of 0.5. The LD50 in this case has a 95% confidence interval (C.I.) of $(-0.69216, -0.58172)$. Similarly, the LD50 for dose is 0.23176 with a 95% C.I. of $(0.20316, 0.26199)$.

The partial results below are those obtained again when PROC PROBIT is employed for implementing the logistic regression approach. Again, the results obtained here agree with those obtained earlier from the use of PROC LOGISTIC. The LD50 under this model is 0.23039 with a 95% C.I. of $(0.20194, 0.26096)$. Again the detailed results are displayed in appendix F.4.

```
                    Goodness-of-Fit Tests

Statistic                          Value       DF     Pr > ChiSq
-----------------------------------------------------------------
Pearson Chi-Square                0.7351        5         0.9810
L.R.     Chi-Square               0.7336        5         0.9811

                 Analysis of Parameter Estimates

                      Standard   95% Confidence      Chi-
Parameter   DF Estimate   Error        Limits        Square Pr > ChiSq
-----------------------------------------------------------------------
Intercept    1   3.1236  0.3349   2.4672   3.7800    86.98      <.0001
Log10(DOSE)  1   4.8996  0.5098   3.9003   5.8988    92.36      <.0001

Probit Model in Terms of Tolerance Distribution

       MU           SIGMA
 -0.637529    0.20409985

       Estimated Covariance Matrix
        for Tolerance Parameters

                     MU               SIGMA
MU             0.000773           -0.000080
SIGMA         -0.000080            0.000451

              Probit Analysis on Log10(DOSE)

Probability          Log10(DOSE)        95% Fiducial Limits

       0.01             -1.57539      -1.83216        -1.40317
       0.02             -1.43185      -1.65304        -1.28275
       ..................................................
       0.45             -0.67849      -0.73938        -0.62429
       0.50             -0.63753      -0.69479        -0.58342
       0.55             -0.59657      -0.65152        -0.54121
       ..................................................
       0.99              0.30033       0.13984         0.53908

              Probit Analysis on DOSE

Probability          DOSE          95% Fiducial Limits

       0.01       0.02658        0.01472        0.03952
       0.02       0.03700        0.02223        0.05215
       ..........................................
       0.45       0.20966        0.18223        0.23752
       0.50       0.23039        0.20194        0.26096
       0.55       0.25318        0.22309        0.28760
       ..........................................
       0.99       1.99680        1.37988        3.46001
```

Below are presented the predicted probabilities under both the probit and logistic models. As mentioned before, the probit model fits better.

```
Obs    DOSE     R     N     PROBIT     LOGISTIC
------------------------------------------------
 1     0.10     8    47    0.14698    0.14480
 2     0.15    14    53    0.29349    0.28635
 3     0.20    24    55    0.42700    0.42530
 4     0.30    32    52    0.62636    0.63685
 5     0.50    38    46    0.83148    0.83871
 6     0.70    50    54    0.91623    0.91409
 7     0.95    50    52    0.96092    0.95322
------------------------------------------------
```

In order to test for the adequacy of this model, we would need to examine the plots of residuals and conduct other diagnostics procedures.

8.3.3 Example 8.2: Beetle Mortality Data

While both the logistic and probit models seem to fit well the data in Table 8.3, there are other data sets in which both models would be found inadequate. We give below in Table 8.6 the beetle mortality data (Bliss, 1935) that relate to the numbers of insects dead after 5 hours of exposure to gaseous carbon disulphide at various concentrations. The data have also been analyzed by Dobson (1990) and Agresti (1990).

$\log_{10} X_i$			**Models**			
x_i	n	r	Logistic	Probit	CLL	E-V
1.6907	59	6	3.457	3.358	5.589	56.637 (53)
1.7242	60	13	9.842	10.722	11.281	47.447 (47)
1.7552	62	18	22.451	23.482	20.954	34.214 (44)
1.7842	56	28	33.898	33.815	30.369	19.568 (28)
1.8113	63	52	50.096	49.615	47.776	13.437 (11)
1.8369	59	53	53.291	53.319	54.143	7.618 (6)
1.8610	62	61	59.222	59.664	61.113	4.898 (1)
1.8839	60	60	58.743	59.228	59.947	2.941 (0)
d.f.			6	6	6	6
G^2			11.232	10.120	3.446	27.917
X^2			10.025	9.513	3.295	25.093

Table 8.6: Expected values under various models for the beetles mortality data (Bliss, 1935)

We present below extracts from the SAS software output for the CLL model from PROCs LOGISTIC and GENMOD, respectively.

```
DATA TAB86;
INPUT DOSE N R @@;
DATALINES;
1.6907 59 6 1.7242 60 13 1.7552 62 18
1.7842 56 28 1.8113 63 52 1.8369 59 53
1.8610 62 61 1.8839 60 60
;
PROC LOGISTIC DATA=TAB86 DESCENDING;
       MODEL R/N=DOSE/LINK=CLOGLOG PLCL PLRL LACKFIT  AGGREGATE
                        SCALE=DEVIANCE;
```

```
OUTPUT OUT=AA PREDICTED=PROBS;
RUN;
PROC PRINT DATA=AA; RUN;
proc gslide; run;
GOPTIONS CBACK=WHITE
         COLORS=(BLACK)
         vsize=6
         hsize=6;
PROC SORT DATA=AA;
BY dose; RUN;
SYMBOL2 I=spline VALUE=none HEIGHT=.75;
axis1 label=(angle=-90 rotate=90 'EXPECTED PROBS');
axis2 label=('LOG OF DOSAGE');
PROC GPLOT DATA=AA; PLOT PROBS*dose/vaxis=axis1 haxis=axis2 VREF=.5;
RUN;
```

Deviance and Pearson Goodness-of-Fit Statistics

Criterion	DF	Value	Value/DF	Pr > ChiSq
Deviance	6	3.4464	0.5744	0.7511
Pearson	6	3.2947	0.5491	0.7711

Number of unique profiles: 8

Analysis of Maximum Likelihood Estimates

Parameter	DF	Estimate	Standard Error	Chi-Square	Pr > ChiSq
Intercept	1	-39.5725	3.2403	149.1487	<.0001
DOSE	1	22.0412	1.7994	150.0498	<.0001

USING PROC GENMOD:

```
DATA GEN;
SET TAB86;
PROC GENMOD DATA=TAB86 DESCENDING; MODEL r/n=dose/link=cloglog;
RUN;
```

Criteria For Assessing Goodness Of Fit

Criterion	DF	Value	Value/DF
Deviance	6	3.4464	0.5744
Pearson Chi-Square	6	3.2947	0.5491

Analysis Of Parameter Estimates

Parameter	DF	Estimate	Standard Error	Chi-Square	Pr > ChiSq
Intercept	1	-39.5723	3.2290	150.19	<.0001
DOSE	1	22.0412	1.7931	151.10	<.0001

The results suggest that the complimentary log-log model with $G^2 = 3.446$ on 6 d.f. adequately fits the data better than either the logistic or the probit model. The extreme-value (E-V) model fits poorly. For the E-V model, the figures in parentheses are the observed number of insects that survived, that is, $(n-r)$ which the model tries to match. The final CLL and logistic models for these data are given as:

$$\ln\left[-\ln\left(1-\hat{\pi}_i\right)\right] = -39.5725 + 22.0412\, x_i$$

where $x_i = \log_{10}(\text{dose})$, and

$$\ln\left(\frac{\hat{\pi}}{1-\hat{\pi}}\right) = -60.7114 + 34.2669\, x_i$$

The plot of the estimated complimentary log-log model for the data in Table 8.6 is displayed in Figure 8.7.

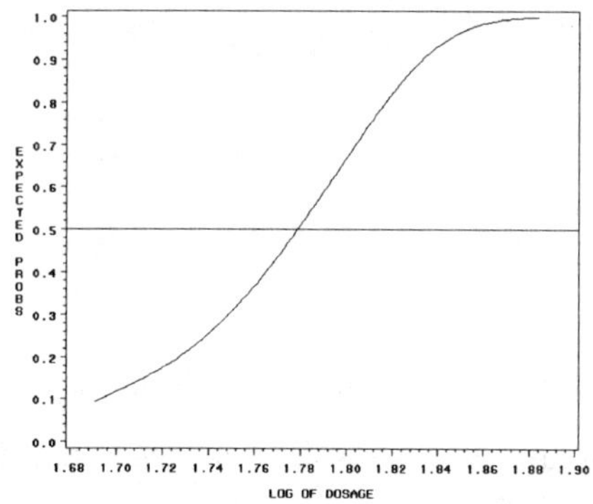

Figure 8.7: Fitted complimentary log-log model

8.3.4 Example 8.3: Status Data

The following example comes from Schork and Remington (2000). The data represent the outcome variable, HIV status (0 = no, 1 = yes), factor variables IV (intravenous) drug status (0 = no, 1 = yes), and number of sexual partners for 25 men selected from a homeless shelter.

ID	HIVSTAT	IVDRUG	SEXPART	ID	HIVSTAT	IVDRUG	SEXPART
1	0	0	4	14	0	0	5
2	0	1	4	15	1	1	9
3	1	1	3	16	1	0	19
4	0	0	2	17	0	0	7
5	0	0	7	18	1	1	10
6	1	0	12	19	0	0	5
7	1	1	8	20	1	1	8
8	0	0	1	21	0	0	14
9	1	0	9	22	0	1	8
10	0	0	5	23	1	0	14
11	0	0	6	24	1	1	9
12	0	1	4	25	1	1	17
13	0	1	2				

Table 8.7: Data for the HIV status example

Let the response variable be Y_i from individual i be defined as:

$$Y_i = \begin{cases} 1 & \text{if HIVSTAT is yes} \\ 0 & \text{otherwise} \end{cases}$$

Suppose the probability of a yes HIV status depends on the drug, sexpart, and the interaction between drug and sexpart, that is,

$$\begin{aligned} \pi_i &= \text{pr}[\text{Y}_i = 1|\text{drug}_i, \text{part}_i, \text{drug*part}_i] \\ &= \frac{\exp\left(\beta_0 + \beta_1\text{drug}_i + \beta_2\text{part}_i + \beta_3\text{drug*part}_i\right)}{1 + \exp\left(\beta_0 + \beta_1\text{drug}_i + \beta_2\text{part}_i + \beta_3\text{drug*part}_i\right)}, \end{aligned} \tag{8.10}$$

where π_i is the probability of the i-th individual having the HIV, *drug* represents the IV drug effect, *part* represents the effect of the number of sexual partners, and *drug*part* represents the interaction effect of IV drug and numbers of sexual partners. Also,

$$\text{drug}_i = \begin{cases} 1 & \text{if yes} \\ 0 & \text{if otherwise} \end{cases}$$

Model (8.10) therefore becomes:

$$\ln\left(\frac{\pi_i}{1-\pi_i}\right) = \boldsymbol{\beta} \tag{8.11a}$$

that is

$$\text{logit}_i = \beta_0 + \beta_1 \,\text{drug}_i + \beta_2 \,\text{part}_i + \beta_3 \,\text{drug} * \text{part}_i \tag{8.11b}$$

The above model is implemented in SAS software with the following SAS software program presented with a partial output.

```
data ex84;
input hiv drug sexpart @@;
datalines;
0 0 4 0 1 4 1 1 3 0 0 2 0 0 7 1 0 12 1 1 8 0 0 1 1 0 9 0 0 5
0 0 6 0 1 4 0 1 2 0 0 5 1 1 9 1 0 19 0 0 7 1 1 10 0 0 5 1 1 8
0 0 14 0 1 8 1 0 14 1 1 9 1 1 17
;
proc logistic data=ex84 descending;
model hiv=drug|sexpart/scale=none aggregate lackfit; run;
```

Deviance and Pearson Goodness-of-Fit Statistics

Criterion	DF	Value	Value/DF	Pr > ChiSq
Deviance	13	11.5979	0.8921	0.5609
Pearson	13	10.4732	0.8056	0.6549

Testing Global Null Hypothesis: BETA=0

Test	Chi-Square	DF	Pr > ChiSq
Likelihood Ratio	16.1069	3	0.0011
Score	12.2969	3	0.0064
Wald	7.5477	3	0.0563

Analysis of Maximum Likelihood Estimates

Parameter	DF	Estimate	Standard Error	Chi-Square	Pr > ChiSq
Intercept	1	-5.3304	2.5223	4.4659	0.0346
DRUG	1	2.5151	3.2281	0.6070	0.4359
SEXPART	1	0.4816	0.2417	3.9707	0.0463
DRUG*SEXPART	1	0.0342	0.3867	0.0078	0.9295

The descending statement in the PROC line above tells SAS software to model HIV=1 rather than HIV=0. That is, we wish to model those having the positive HIV status. The deviance or G^2 for this model is 11.5979 on 13 d.f. (p-value=0.5609). This model fits the data well. The "Testing Global Null Hypothesis" output tests the hypothesis

$$H_0 : \beta_1 = \beta_2 = \beta_3 = 0; \quad \text{versus} \quad H_1 : \text{at least one of the } \beta\text{s is} \neq 0$$

Three alternative tests are provided for this hypothesis utilizing the likelihood ratio, Fisher's scoring, and Wald based tests. The likelihood and score tests indicate that we would reject H_0, suggesting that at least one of the β parameters is not equal to zero.

Examination of the parameter estimates indicates that the interaction term is not significant and can be removed from the model, given that both drug and sexpart are already in the model. Hence, we next fit a reduced model

$$\text{logit}_i = \beta_0 + \beta_1 \text{ drug}_i + \beta_2 \text{ part}_i \tag{8.12}$$

The model in (8.12) is implemented in SAS software with the following SAS software program with again a partial output display.

```
set ex84; proc logistic data=ex84 descending;
model hiv=drug sexpart/scale=none aggregate lackfit rsq plcl plrl
                     waldcl waldrl;
                     units sexpart=1 2 3 drug=1;
output out=aa p=phat; run;
```

Deviance and Pearson Goodness-of-Fit Statistics

Criterion	DF	Value	Value/DF	Pr > ChiSq
Deviance	14	11.6057	0.8290	0.6379
Pearson	14	10.3988	0.7428	0.7325

Number of unique profiles: 17

Analysis of Maximum Likelihood Estimates

Parameter	DF	Estimate	Standard Error	Chi-Square	Pr > ChiSq
Intercept	1	-5.4649	2.0683	6.9816	0.0082
DRUG	1	2.7748	1.3881	3.9960	0.0456
SEXPART	1	0.4954	0.1901	6.7938	0.0091

Odds Ratio Estimates

Effect	Point Estimate	95% Wald Confidence Limits	
DRUG	16.035	1.056	243.551
SEXPART	1.641	1.131	2.382

Wald Confidence Interval for Adjusted Odds Ratios

Effect	Unit	Estimate	95% Confidence Limits	
DRUG	1.0000	16.035	1.056	243.551
SEXPART	1.0000	1.641	1.131	2.382
SEXPART	2.0000	2.693	1.279	5.674
SEXPART	3.0000	4.420	1.446	13.514

Hosmer and Lemeshow Goodness-of-Fit Test

Chi-Square	DF	Pr > ChiSq
4.7626	6	0.5746

The implementation of the model described in (8.12) gives a $G^2 = 11.6057$ on 14 d.f. The model fits the data very well. Parameter estimates of β_1 and β_2 are significant. The analysis show that the odds of an HIV status being positive is 16.035 times higher for IV drug users than those not on IV drugs when the effect of sexual partner is controlled. Similarly, the odds increase by 1.641 for a unit increase in the number of sexual partners. The odds increase by $2.693 = e^{2*0.4954} = 1.641^2$ and $4.420 = e^{3*0.4954} = 1.641^3$ for 2 units and 3 units increases in the number of sexual partners, respectively. The **UNITS** option in the model statement generates these additional results along with their Wald-based confidence intervals. The Hosmer and Lemeshow goodness-of-fit test indicate that our model is adequate.

We also obtain the expected probabilities ($\hat{\pi}_i$), of having a positive HIV status based on the estimated logistic model:

$$\ln\left(\frac{\hat{\pi}_i}{1-\hat{\pi}_i}\right) = -5.4649 + 2.7748\,\text{drug}_i + 0.4954\,\text{part}_i \tag{8.13}$$

These expected probabilities are presented below. They are generated in the program with the **OUTPUT** statement into a file named **aa**, the contents of which are printed below with a SAS software **PRINT** statement.

```
proc print data=aa noobs; run;
```

Obs	HIV	DRUG	SEXPART	_LEVEL_	PHAT
1	0	0	4	1	0.02979
2	0	1	4	1	0.32991
3	1	1	3	1	0.23077
4	0	0	2	1	0.01127
5	0	0	7	1	0.11950
6	1	0	12	1	0.61770
7	1	1	8	1	0.78125
8	0	0	1	1	0.00690
9	1	0	9	1	0.26769
10	0	0	5	1	0.04797
11	0	0	6	1	0.07638
12	0	1	4	1	0.32991
13	0	1	2	1	0.15455
14	0	0	5	1	0.04797
15	1	1	9	1	0.85425
16	1	0	19	1	0.98106
17	0	0	7	1	0.11950
18	1	1	10	1	0.90583
19	0	0	5	1	0.04797
20	1	1	8	1	0.78125
21	0	0	14	1	0.81314
22	0	1	8	1	0.78125
23	1	0	14	1	0.81314
24	1	1	9	1	0.85425
25	1	1	17	1	0.99677

The expected probabilities are very much consistent with the observed data. For instance, among those who had HIV positive in the sample data, their expected probabilities are quite high except for individuals 3 and 9. Similarly, individuals 21 and 22 have high expected probabilities but were HIV negative. We therefore decided to see to what degree do the observed HIV status agree with the predicted probabilities if we classify an individual to be HIV positive if his expected probability is greater or equal to 0.5, that is, if $\hat{\pi}_i \geq 0.5$. We must confess here that this value is somehow subjective. We present again below the result of McNemar's test (test of agreement) for this data based on this classification.

```
data new;
set aa;
predicts=(phat ge 0.5);
proc freq data=new; hiv*predicts/norow nocol nopercent agree;
run;
                    Statistics for Table of HIV by PREDICTS

                               McNemar's Test
                           -------------------
                           Statistic (S)    0.0000
                           DF                    1
                           Pr > S           1.0000
```

```
        Simple Kappa Coefficient
        ------------------------------
        Kappa                        0.6753
        ASE                          0.1487
        95% Lower Conf Limit         0.3840
        95% Upper Conf Limit         0.9667
               Sample Size = 25
```

Table of HIV by PREDICTS

HIV	PREDICTS 0	PREDICTS 1	Total
0	12	2	14
1	2	9	11
Total	14	11	25

The estimate of κ, the agreement statistic, and McNemar's test of agreement indicate that there is very strong agreement between the observed HIV status and the expected HIV status based on the model employed. The graph of the predicted probabilities versus the number of sexual partners for the two levels of IV drug are presented below with the accompanying SAS software statements.

```
set aa;
proc gslide;
run;
GOPTIONS CBACK=WHITE
         COLORS=(BLACK)
         vsize=6
         hsize=6;

PROC SORT DATA=AA;
BY sexpart;
RUN;
SYMBOL1 I=spline VALUE=+ HEIGHT=.75;
SYMBOL2 I=spline VALUE=none HEIGHT=.75;
axis1 label=(angle=-90 rotate=90 'EXPECTED PROBS');
axis2 label=('NO OF SEXUAL PARTNERS');
PROC GPLOT DATA=AA;
PLOT Phat*sexpart=drug/vaxis=axis1 haxis=axis2;
RUN;
```

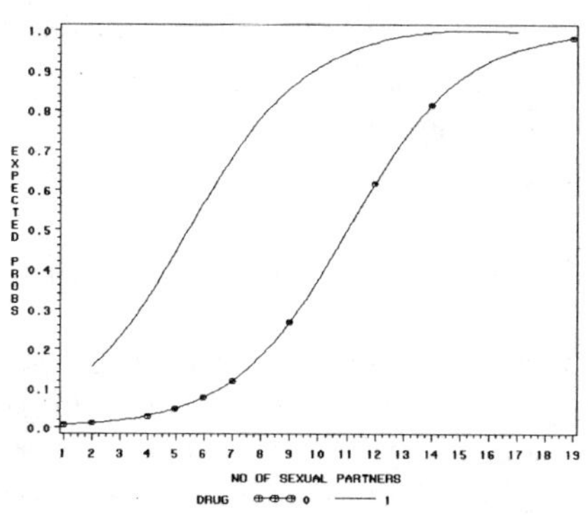

Figure 8.8: Predicted probabilities plot from model (8.13)

8.4 Example 8.4: Relative Potency in Bioassays

We give below an example relating to the fit of linear logistic model when we have two drugs administered to mice.

The following data are an example in which we are interested in the comparison of two or more groups after adjusting for some factor. In this case, the groups are two drugs A and B, while the factor is the dosage levels and the response is categorical (that is, binary). This of course leads to what we know as the quantal assay, in analogy to the covariance analysis where the response variable is continuous. In the example below, the acute toxicities of two drugs A and B were tested by intravenous injection into mice. Each drug was given at different doses to four groups of 20 mice, and deaths were recorded 5 minutes later.

Drug	Dose x_i	Coded level X_i	Number dead r_i	Proportion p_i	Logits δ_i
A	2	0	2	0.10	-2.20
	4	1	9	0.45	-0.20
	8	2	14	0.70	0.85
	16	3	19	0.95	2.94
B	0.3	0	1	0.05	-2.94
	0.6	1	6	0.30	-0.85
	1.2	2	14	0.70	0.85
	2.4	3	17	0.85	1.73

Table 8.8: Data for Example 8.4

It has been established (see Collett, 1991) that responses of subjects to drugs tend to vary proportionately to the log dosage rather than to dose levels themselves, and hence an arbitrary log-scale can be established.

In the example above, there are four dosage levels each for drugs A and B respectively. Of most interest to us is to compare the relative potency of the drugs after adjusting for the various dosage levels.

To answer the above question, first let y_{ij} denote the number of deaths in the j-th dose for the i-th drug and n_{ij} the corresponding sample size ($n_{ij} = 20$ for all i and j).

Let P (death | i-th drug, j-th dose) $= \pi_i > 0$. Then a linear logistic model of the following form fitted to the data is considered for the two drugs (that is, we fit two separate logistic regression lines, one for each drug):

$$\text{logit}\,(\pi_{ij}) = \beta_{0i} + \beta_{1i}\text{dose}_{ij}$$

where $i = 1, 2$; $j = 1, 2, 3, 4$; and $\text{dose}_{ij} = \ln \text{dose}$ for the i-th drug and j-th level of dose.

We are also interested to know if the dosage effect is the same for the two groups and, given that this is the case, is there a significant difference between the two drugs? To answer this, let us consider fitting a model with drugs as a factor variable with two levels (A and B) and the dosage levels as the covariate. Because the dosage levels are not to the same scale, it would be necessary to transform the dosage levels by either using the natural or common logarithmic transformations to ensure uniformity from one drug to the other. We chose to use the common logarithmic transformation, that is, $\ln(\text{dose}_j)$.

Let us first examine how PROC LOGISTIC and PROC GENMOD code factor variables that are invoked with the CLASS statement in both procedures. If we let drug be the drug dummy variable, then in PROC LOGISTIC, drug takes the values:

$$\text{drug} = \begin{cases} 1 & \text{if drug A} \\ -1 & \text{if drug B} \end{cases}$$

Drug A is coded 1 since alphabetically, it comes before B while in PROC GENMOD, drug is coded as:

$$\text{drug} = \begin{cases} 1 & \text{if drug A} \\ 0 & \text{if drug B} \end{cases}$$

Again, drug A is coded as 1 because of its alphabetical order compared with drug B. We can achive the same coding scheme from PROC LOGISTIC by specifying: **class drug (ref=last)/param=ref;** in the class statement.

All the models implemented for this example are shown in the statements below:

```
data tab88;
input drug $ x r n @@;
dose=log10(x);
datalines;
A 2 2 20 A 4 9 20 A 8 14 20 A 16 19 20
B 0.3 1 20 B 0.6 6 20 B 1.2 14 20 B 2.4 17 20
;
proc logistic order=data;
class drug;
(i)    model r/n=drug/scale=none aggregate; run;
(ii)   model r/n=drug|dose/scale=none aggregate; run;
(iii)  model r/n=drug dose/scale=none aggregate; run;
(iv)   model r/n=dose drug*dose/scale=none aggregate; run;
(v)    model r/n=dose/scale=none aggregate; run;
*ditto for proc genmod;
```

(a) The model in (i) is:

$$\text{logit}(\pi_i) = \beta_0 + \beta_1\,\text{drug}_i, \quad i = 1, 2$$

and tests the hypothesis that the drug differences are significantly small. For this model, we have $G^2 = 74.184$ on 6 d.f. Obviously, this model is untenable and thus the effects of the drugs are significantly different.

```
set data;
model r/n=drug/scale=none aggregate=(dose); run;

        Deviance and Pearson Goodness-of-Fit Statistics

Criterion          DF          Value      Value/DF      Pr > ChiSq
------------------------------------------------------------------
Deviance            6        74.1835       12.3639          <.0001
Pearson             6        64.1999       10.7000          <.0001
```

(b) The second fit statement fits two separate regression lines of the form

$$\text{logit}(\pi_{ij}) = \beta_0 + \beta_{1i}\text{drug}_i + \beta_{2j}\text{dose}_j + \beta_{3ij}(\text{drug} * \text{dose})_{jk} \tag{8.14}$$

where

- drug_i
  - is the effect of the i-th drug

- dose_j
  - is the effect of the j-th dosage
- $(\text{drug} * \text{dose})_{jk}$
  - is the interaction between drugs and dose levels

The model has a $G^2 = 1.6398$ on 4 d.f. That is, the model is adequate. The SAS software program and partial outputs are displayed for both the PROC LOGISTIC and GENMOD.

```
set tab88;
proc logistic order=data;
class drug; model r/n=drug|dose/scale=none aggregate; run;
proc genmod; class drug; model r/n=drug|dose/dist=b link=logit; run;
```

Deviance and Pearson Goodness-of-Fit Statistics

Criterion	DF	Value	Value/DF	Pr > ChiSq
Deviance	4	1.6398	0.4100	0.8016
Pearson	4	1.6412	0.4103	0.8014

Analysis of Maximum Likelihood Estimates

Parameter		DF	Estimate	Standard Error	Chi-Square	Pr > ChiSq
Intercept		1	-1.6678	0.4346	14.7279	0.0001
drug	a	1	-1.8623	0.4346	18.3629	<.0001
dose		1	5.0969	0.7608	44.8766	<.0001
dose*drug	a	1	0.0411	0.7608	0.0029	0.9569

GENMOD:

Criteria For Assessing Goodness Of Fit

Criterion	DF	Value	Value/DF
Deviance	4	1.6398	0.4100
Pearson Chi-Square	4	1.6412	0.4103

Analysis Of Parameter Estimates

Parameter		DF	Estimate	Standard Error	Wald 95% Confidence Limits		Chi-Square	Pr > ChiSq
Intercept		1	0.1945	0.2928	-0.3794	0.7684	0.44	0.5066
drug	a	1	-3.7246	0.8692	-5.4282	-2.0211	18.36	<.0001
dose		1	5.0558	1.0667	2.9650	7.1465	22.46	<.0001
dose*drug	a	1	0.0822	1.5217	-2.9002	3.0647	0.00	0.9569

The estimated overall regression models under PROC LOGISTIC and GENMOD are respectively.

$$\begin{aligned} \text{logit}(\hat{\pi}_{ij}) &= -1.6678 - 1.8623\text{drug}_i + 5.0969\text{dose}_j \\ &\quad + 0.0411(\text{drug} * \text{dose})_{ij} \end{aligned} \tag{8.15}$$

$$\begin{aligned} \text{logit}(\hat{\pi}_{ij}) &= 0.1945 - 3.7246\text{drug}_i + 5.0558\text{dose}_j \\ &\quad + 0.0822(\text{drug} * \text{dose})_{ij} \end{aligned} \tag{8.16}$$

where drug takes the values $(1, -1)$ and $(1, 0)$ for drug A and B, respectively, in models (8.15) and (8.16). These are the default coding schemes in PROC LOGISTIC and GENMOD, respectively.

Notice that the parameter estimate for drug in PROC LOGISTIC is half that from PROC GENMOD. This is also true for the interaction term as well as their corresponding standard errors.

Specifically, if we adopt the PROC logistic output, this model reduces to the two separate estimated regression lines

$$\text{logit}(\pi_j) = \begin{cases} -3.5301 + 5.138\text{dose}_j & \text{For Drug A} \quad \text{drug} = 1 \\ 0.1945 + 5.056\text{dose}_j & \text{For Drug B} \quad \text{drug} = -1 \end{cases}$$

Although this model fits the data very well, further examination of the parameters, shows that the interaction term β_3 is not significant with a pvalue of (0.9569). Hence this term can be removed from the model, leading to parallel response models.

(c) The third model fits parallel regression lines with a common slope but with different intercepts. That is,

$$\text{logit}(\pi_{ij}) = \beta_0 + \beta_{1i}\text{drug}_i + \beta_{2j}\text{dose}_j$$

This model gives a $G^2 = 1.6427$ on 5 degrees of freedom. Again the SAS software program for implementing this and a partial output from PROC LOGISTIC is presented below:

```
set data;
proc logistic order=data;
class drug; model r/n=drug dose/scale=none aggregate; run;

        Deviance and Pearson Goodness-of-Fit Statistics

Criterion         DF          Value      Value/DF      Pr > ChiSq
-----------------------------------------------------------------
Deviance           5         1.6427        0.3285          0.8960
Pearson            5         1.6496        0.3299          0.8952

               Analysis of Maximum Likelihood Estimates

                                      Standard
Parameter        DF     Estimate         Error     Chi-Square    Pr > ChiSq
---------------------------------------------------------------------------
Intercept         1      -1.6522        0.3232        26.1349        <.0001
drug        a     1      -1.8486        0.3520        27.5840        <.0001
dose              1       5.0964        0.7607        44.8826        <.0001
```

This model fits our data well and indicates that there is a common dose effect for the two drugs, which is another way of stating the parallelism hypothesis. The model has the estimated response function,

$$\text{logit}(\hat{\pi}_{ij}) = -1.6522 - 1.8486\text{drug}_i + 5.0964\,\text{dose}_j$$

where drug and dose are as defined previously.

The "relative potency" of the two drugs from this analysis under PROC LOGISTIC (effect coding scheme) is obtained as

$$\text{Relative potency} = 10^{\frac{-2\hat{\beta}_1}{\hat{\beta}_2}} = 10^{0.7253} = 5.313$$

It is not too difficult to see that the individual estimated regression lines are:

$$\text{logit}(\hat{\pi}_j) = \begin{cases} -3.5008 + 5.0964\,\text{dose}_j & \text{If drug A} \\ 0.1964 + 5.0964\,\text{dose}_j & \text{If drug B} \end{cases}$$

The relative potency from these individual models can similarly be obtained as:

$$10^{\frac{-(-3.5008-0.1964)}{5.0964}} = 10^{3.6974/5.0964} = 5.316$$

If PROC GENMOD is employed, the relative potency will be computed from the overall estimated equation as simply $10^{\frac{-\hat{\beta}_1}{\hat{\beta}_2}}$. With PROC GENMOD, $\hat{\beta}_1 = -3.6972$ and $\hat{\beta}_2 = 5.0964$; hence, relative potency equals

$$10^{3.6972/5.0964} = 10^{0.7255} = 5.3150$$

(d) The fourth fit statement fits a model with different slopes and a common intercept to the data. That is, the model is given by:

$$\text{logit}(\pi_{ij}) = \beta_0 + \beta_{2j}\text{dose}_j + \beta_{3ij}(\text{drug} * \text{dose})_{ij}$$

The model has a $G^2 = 26.634$ on 5 d.f. The model is inadequate.

(e) The final fit statement (v) fits the covariate X ignoring the grouping to see if a single line fits the data. That is, the model is

$$\text{logit}(\pi_j) = \beta_0 + \beta_{2j}\text{dose}_j$$

Again, the corresponding $G^2 = 39.262$ on 6 d.f. This model also does not fit the data.

From the above analyses, the best models for the data are the parallel regression models fitted in (c) above. The sketch of the parallel models are displayed in Figure 8.9.

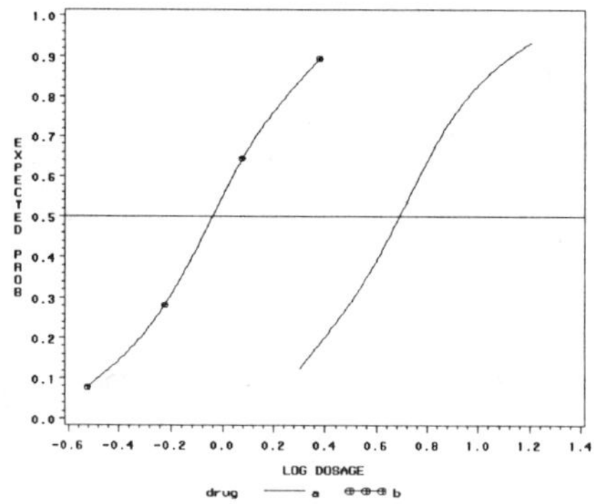

Figure 8.9: Predicted probabilities plot

Model 4 clearly indicates that there are significant differences between the two drugs as measured by the relative potency of the drugs.
The above analysis for the data in Table 8.8 can therefore be summarized as:

(i) A linear logistic model seems appropriate for the data.

(ii) The dosage effect is the same for both drugs.

(iii) Drug B is 5.313 times more potent than drug A.

8.5 Analyzing Data Arising from Cohort Studies

A cohort study involves selecting a sample of individuals free of the diseases under investigation. These individuals in the sample are then stratified according to exposure factors of interest, and followedup for a given period of time. Each individual is then classified according to whether or not he or she has developed the disease that is being studied. The relationship between the probability of disease occurrence and the exposure factors is then modeled. Such study has also been characterized as the *prospective* study.

8.5.1 Example 8.5: The Framingham Study Data

The data in Table 8.9 come from a cohort study where the relationship between the probability of disease occurrence and the exposure factor is to be modeled. Framingham is an industrial town.located some 20 miles west of Boston. In 1948, a cohort study was begun with the broad aim of determining which of a number of potential risk factors are related to the occurrence of coronary heart disease (CHD). At the start of the study, a large proportion of the town's inhabitants were examined for the presence of CHD. Measurements were also made on a number of other variables, including age, serum cholesterol level, systolic blood pressure, smoking history, and the result of an electrocardiogram. Those individuals found to be free of CHD at that time were followedup for 12 years and those who developed CHD during that period were identified. The resulting data set consisted of this binary response variable and information on the risk factors of 2187 men and 2669 women aged between 30 and 62. The summary data below are adapted from Truett, Cornfield, and Kannel (1967) and relate to the initial serum cholesterol level (in units of mg/100 ml) of these individuals, cross-classified according to their age and sex.

Sex	Age Group	Serum Cholesterol Level			
		< 190	$190 - 219$	$220 - 249$	≥ 250
Male	30-49	13/340	18/408	40/421	57/362
	50-62	13/123	33/176	35/174	49/183
Female	30-49	6/542	5/552	10/412	18/357
	50-62	9/58	12/135	21/218	48/395

Table 8.9: Proportions of cases of CHD, cross-classified by age, sex and initial cholesterol level

Source: *Journal of Chronic Diseases*, **20**, 511-524.

In the data in Table 8.9, the ratio 13/340 for males aged 30 to 49 for instance, denotes that of the 340 individuals in this category, 13 had CHD by the end of the study. Our goal here is to model the extent to which CHD is associated with initial serum cholesterol level, after adjusting for the effects of age group and sex (confounders) and whether the degree of association is similar for each sex and age group. The data have also been analyzed by Collet (1991). The exposure or risk factor here is the serum cholesterol level and the disease is coronary heart disease.

The model of interest here is the model given by:

$$\begin{aligned}\text{logit}(p_{ijk}) = \beta_0 + \beta_{1i}\,\text{sex}_i + \beta_{2j}\,\text{age}_j + \beta_{3k}\,\text{chol}_k + \beta_{4ij}\,(\text{sex} * \text{age})_{ij} \\ + \beta_{5jk}\,(\text{age} * \text{chol})_{jk} + \beta_{6ik}\,(\text{sex} * \text{chol})_{ik} + \text{higher terms}\end{aligned} \tag{8.17}$$

where

- sex_i: Is the effect of the i-th sex
- age_j: Is the effect of the j-th age group
- chol_k: Is the effect due to group k-th level of the cholesterol

The last three terms in (8.17) are the interaction terms. Note that because sex and age are confounders, we must include them and possibly their interaction (if significant) in the model. We give below the SAS software statements and the corresponding partial output for the analysis of the above data.

```
data cohort;
do sex=1 to 2;  do age=1 to 2;  do chol=1 to 4;
input r n @@; output; end; end; end;
datalines;
13 340 18 408 40 421 57 362 13 123 33 176 35 174 49 183
6 542 5 552 10 412 18 357 9 58 12 135 21 218 48 395
;
proc logistic; class sex age chol;
model r/n=sex|age|chol/selection=forward details;
run;
```

Type III Analysis of Effects

Effect	DF	Wald Chi-Square	Pr > ChiSq
sex	1	78.2833	<.0001
age	1	109.9376	<.0001
sex*age	1	5.5202	0.0188
chol	3	55.8450	<.0001
age*chol	3	14.5409	0.0023

Residual Chi-Square Test

Chi-Square	DF	Pr > ChiSq
8.8651	6	0.1813

Analysis of Effects Not in the Model

Effect	DF	Score Chi-Square	Pr > ChiSq
sex*chol	3	4.4495	0.2168

NOTE: No (additional) effects met the 0.05 significance level for entry into the model.

Summary of Forward Selection

Step	Effect Entered	DF	Number In	Score Chi-Square	Pr > ChiSq
1	age	1	1	142.8944	<.0001
2	sex	1	2	82.3854	<.0001
3	chol	3	3	62.2124	<.0001
4	age*chol	3	4	13.3407	0.0040
5	sex*age	1	5	5.5649	0.0183

We employ PROC LOGISTIC to conduct a forward selection procedure for the model above. The significant effects and interactions are those displayed in the summary of the forward selection procedure. Thus, other higher terms are not needed in the model above. A similar result from GENMOD gives the type3 analysis displayed below:

```
set cohort; proc genmod; class sex age chol;
model r/n=sex|age|chol/dist=b type3; run;

                    LR Statistics For Type 3 Analysis

                                        Chi-
            Source              DF     Square    Pr > ChiSq
            -----------------------------------------------
            sex                  1      55.93        <.0001
            age                  1     114.31        <.0001
            sex*age              1       9.50        0.0021
            chol                 3      47.44        <.0001
            sex*chol             3       3.94        0.2681 ns
            age*chol             3      15.92        0.0012
            sex*age*chol         3       3.47        0.3248 ns
```

Based on the above initial analysis, therefore, our reduced model is now of the form:

$$\begin{aligned}\text{logit}(p_{ijk}) = \beta_0 + \beta_{1i}\,\text{sex}_i + \beta_{2j}\,\text{age}_j + \beta_{3k}\,\text{chol}_k + \beta_{4ij}\,(\text{sex} * \text{age})_{ij} \\ + \beta_{5jk}\,(\text{age} * \text{chol})_{jk}\end{aligned} \tag{8.18}$$

We present the SAS software program and output from PROC LOGISTIC for implementing the reduced model in (8.18).

```
set cohort;
proc logistic;class sex (ref=last) age (ref=last) chol (ref=last)/param=ref;
model r/n=sex|age age|chol/scale=none aggregate covb; run;

              Deviance and Pearson Goodness-of-Fit Statistics

        Criterion         DF          Value     Value/DF     Pr > ChiSq
        ---------------------------------------------------------------
        Deviance           6         7.5847       1.2641         0.2701
        Pearson            6         8.8651       1.4775         0.1813

                         Number of unique profiles: 16

                         Type III Analysis of Effects

                                           Wald
               Effect          DF     Chi-Square     Pr > ChiSq
               ------------------------------------------------
               sex              1        26.1780         <.0001
               age              1        23.6540         <.0001
               sex*age          1         5.5202         0.0188
               chol             3         8.3190         0.0399
               age*chol         3        14.5409         0.0023

                   Analysis of Maximum Likelihood Estimates
                                          Standard          Wald
      Parameter          DF     Estimate       Error    Chi-Square     Pr > ChiSq
      ---------------------------------------------------------------------------
      Intercept           1      -1.8972      0.1309      210.0556         <.0001
      sex        1        1       0.7907      0.1545       26.1780         <.0001
      age        1        1      -1.1161      0.2295       23.6540         <.0001
      sex*age    1 1      1       0.5718      0.2434        5.5202         0.0188
      chol       1        1      -0.6667      0.2611        6.5198         0.0107
      chol       2        1      -0.3806      0.2019        3.5526         0.0595
      chol       3        1      -0.3009      0.1856        2.6291         0.1049
      age*chol   1 1      1      -0.8776      0.3715        5.5812         0.0182
      age*chol   1 2      1      -1.1075      0.3181       12.1190         0.0005
      age*chol   1 3      1      -0.3191      0.2676        1.4223         0.2330

                           Estimated Covariance Matrix

Variable       Intercept          sex1          age1     sex1age1          chol1          chol2
-----------------------------------------------------------------------------------------------
Intercept       0.017135      -0.01034      -0.01714     0.010337       -0.00888       -0.00991
sex1            -0.01034      0.023886      0.010337     -0.02389       -0.00874       -0.00636
age1            -0.01714      0.010337      0.052665     -0.03699       0.008881       0.009908
```

```
sexlage1     0.010337   -0.02389   -0.03699   0.059229   0.008737   0.006364
chol1        -0.00888   -0.00874   0.008881   0.008737    0.06818    0.01499
chol2        -0.00991   -0.00636   0.009908   0.006364    0.01499   0.040777
chol3         -0.0111    -0.0036   0.011102   0.003605   0.013981   0.013622
age1chol1    0.008881   0.008737   -0.02581   -0.00674   -0.06818   -0.01499
age1chol2    0.009908   0.006364   -0.02596   -0.00553   -0.01499   -0.04078
age1chol3    0.011102   0.003605   -0.02596   -0.00436   -0.01398   -0.01362

                  Estimated Covariance Matrix (continued)

Variable          chol3      age1chol1      age1chol2      age1chol3
--------------------------------------------------------------------
Intercept       -0.0111       0.008881       0.009908       0.011102
sex1            -0.0036       0.008737       0.006364       0.003605
age1           0.011102       -0.02581       -0.02596       -0.02596
sex1age1       0.003605       -0.00674       -0.00553       -0.00436
chol1          0.013981       -0.06818       -0.01499       -0.01398
chol2          0.013622       -0.01499       -0.04078       -0.01362
chol3           0.03444       -0.01398       -0.01362       -0.03444
age1chol1      -0.01398        0.13801       0.030461       0.029362
age1chol2      -0.01362       0.030461       0.101201       0.029029
age1chol3      -0.03444       0.029362       0.029029       0.071599
```

This model fits the data with a deviance of 7.5847 on 6 degrees of freedom.

8.5.2 Model Parameter Interpretations

Since the interaction term (age $*$ chol) is significant, we only need to concentrate on this rather than on the main effects of age and cholesterol level. It is important to note here that the main effect of a variable that is also in a two-way interaction can be interpreted as the effect of that variable when the other variable is 0. We note that the adjusted log-parameter estimates for both the chol and (age $*$ chol) are given below, respectively, under PROC LOGISTIC.

Parameters	Cholesterol levels (k)			
	1	2	3	4
$\hat{\beta}_{3k}$	-0.6667	-0.3806	-0.3009	0.0000
Odds	0.5134	0.6835	0.7402	1.0000
Odds2	1.000	1.3313	1.4418	1.9478

Log-parameters $\hat{\beta}_{5jk}$ for the interaction term

Age Group (j)	Cholesterol levels (k)			
	1	2	3	4
30-49 (1)	-0.8776	-1.1075	-0.3191	0.0000
50-62 (2)	0.0000	0.0000	0.0000	0.0000

Here, odds and odds2 refer respectively to the odds based on the last category reference (GENMOD or LOGISTIC with REF command) and the odds from the perspective of referencing the first lowest category. We have asked PROC LOGISTIC to code all variables with the cell referencing approach (this is accomplished above with the REF=LAST in the class statement). PROC GENMOD in any case uses this coding scheme. Thus here, variable age is coded as:

$$\text{age} = \begin{cases} 1 & \text{if 30-49} \\ 0 & \text{if 50-62} \end{cases}$$

Thus, the effects of cholesterol (chol) coefficients represent the effect of cholesterol when age = 0, that is, when the individual is aged between 50 and 62. For those individuals aged 30-49, we add the effect of the interaction coefficients to those of chol. These are similarly displayed below.

Age	Cholesterol levels			
	1	2	3	4
30-49	-1.5443	-1.4881	-0.6200	0.0000
50-62	-0.6667	-0.3806	-0.3009	0.0000

Here for instance, $-1.5443 = -0.6667 + (-0.8776)$ and $-0.6200 = -0.3009 + (-0.3191)$. That is, for individuals in the age group 50-62, the log-odds (Ω_{jk}) would be computed for $j = 1, 2$ and $k = 1, 2, 3, 4$ as:

$$\hat{\Omega}_{2k} = \widehat{\text{chol}}_k = \hat{\beta}_{3k}$$

That is, under model (8.18), for individuals in the j-th group, the odds ratio for an individual exposed to level k of cholesterol, relative to someone exposed who has been exposed to level k', is estimated by:

$$\begin{aligned}\log(\hat{\Omega}_{jkk'}) &= \text{logit}(p_{ijk}) - \text{logit}(p_{ijk'}) \\ &= \hat{\beta}_{3k}\,\text{chol}_k + \hat{\beta}_{5jk}\,(\text{age} * \text{chol})_{jk} \\ &- \hat{\beta}_{3k'}\,\text{chol}'_k + \hat{\beta}_{5jk'}\,(\text{age} * \text{chol})_{jk'}\end{aligned} \tag{8.19}$$

The expression in (8.19) can be written more succinctly as:

$$\log(\hat{\Omega}_{jkk'}) = (\hat{\beta}_{3k} - \hat{\beta}_{3k'}) + (\hat{\beta}_{5jk} - \hat{\beta}_{5jk'}) \tag{8.20}$$

Thus for comparison among the 50-62 age group, $\hat{\beta}_{5jk} = \hat{\beta}_{5jk'} = 0$. In this case the relative odds reduces to

$$\log(\hat{\Omega}_{jkk'}) = (\hat{\beta}_{3k} - \hat{\beta}_{3k'}) \tag{8.21}$$

Similarly, for comparisons among persons aged between 30 and 49, the relative odds is as given in (8.20).

Exponentiating these log-odds, we have the following estimated odds ratio relative to both levels 4 (GENMOD) and 1, respectively.

Age	Odds	Cholesterol levels			
		1	2	3	4
30-49	Odds_1	0.2135	0.2258	0.5379	1.0000
	Odds_2	1.000	1.0576	2.5194	4.6838
50-62	Odds_1	0.5134	0.6835	0.7402	1.0000
	Odds_2	1.000	1.3313	1.4418	1.9478

The above table shows that the relative risk of CHD increases more rapidly with increasing initial serum cholesterol level for persons in the 30-49 age group, compared to that for persons in the 50-62 age group. Further, the relative risk of CHD for persons with serum cholesterol at level 2 is greater for those in the older age group, whereas if the initial cholesterol level exceeds level 3, then the relative risk of CHD is greater for individuals in the younger age group.

Suppose we also wish to obtain confidence intervals for the true odds ratios then we note, for example, for the odds of CHD occurring in persons aged 50-62 with an initial cholesterol level 2 relative to those with initial cholesterol level 1 is estimated using (8.21) by:

$$\hat{\beta}_{32} - \hat{\beta}_{31} = -0.3806 - (-0.6667) = 0.2861$$

The estimated variance from the variance-covariance matrix above is given by

$$\begin{aligned}\text{Var}\{\hat{\beta}_{32} - \hat{\beta}_{31}\} &= \text{Var}(\hat{\beta}_{31}) + \text{Var}(\hat{\beta}_{32}) - 2\text{COV}(\hat{\beta}_{31}, \hat{\beta}_{32}) \\ &= 0.06818 + 0.04078 - 2(0.01499) \\ &= 0.07898\end{aligned}$$

These are given respectively by parameter variances and covariance of Chol1 and Chol2 in the SAS software output above. Consequently, a 95% confidence interval for the difference in the log odds ratios is therefore computed as:

$$0.2861 \pm 1.96\sqrt{0.07898} = 0.2861 \pm 0.5508 = [-0.2647, 0.8369]$$

the corresponding interval for the true odds ratio is (0.77,2.31). This interval includes 1, which suggest that the risk of CHD occurring in persons aged 50-62 is not significantly different for those with initial serum cholesterol levels less than 190 mg/100 ml compared to those with cholesterol levels between 190 and 210 mg/100 ml. We can implement the above in SAS software using PROC LOGISTIC with the following statements, since this contrast from (8.21) is $\hat{\beta}_{32} - \hat{\beta}_{31}$, with $k = 1$ and $k' = -1$, respectively. We present only the result pertaining to the contrast statements only below.

```
set cohort;  proc logistic;
class sex (ref=last) age (ref=last) chol (ref=last)/param=ref;
model r/n=sex|age age|chol/scale=none aggregate;
contrast '2 vs 1 (50-62)' chol -1 1 0 0/estimate=both;
run;
```

Contrast Test Results

Contrast	DF	Wald Chi-Square	Pr > ChiSq
2 vs 1 (50-62)	1	1.0365	0.3086

Contrast Rows Estimation and Testing Results

Contrast	Type	Row	Estimate	Standard Error	Alpha	Lower Limit	Upper Limit
2 vs 1 (50-62)	PARM	1	0.2861	0.2810	0.05	-0.2647	0.8369
2 vs 1 (50-62)	EXP	1	1.3312	0.3741	0.05	0.7674	2.3093

Contrast Rows Estimation and Testing Results

Contrast	Type	Row	Wald Chi-Square	Pr > ChiSq
2 vs 1 (50-62)	PARM	1	1.0365	0.3086
2 vs 1 (50-62)	EXP	1	1.0365	0.3086

The SAS software program requests SAS software to use the reference cell coding scheme. The contrast statement allows us to formally estimate the relative odds of cholesterol level 2 to cholesterol level 1 for individuals aged 50-62. The log contrast estimate of 0.2861 agrees with our earlier result and the actual odds ratio of 1.3312 with a 95% C.I. of (0.7674,2.3093). Similarly, the C.I. for persons aged 30-49 with initial cholesterol level of say 3 and 1 can also be obtained using (8.20). This is given as: $(\hat{\beta}_{33} - \hat{\beta}_{31}) + (\hat{\beta}_{513} - \hat{\beta}_{511})$, since $j = 1$. Again this is implemented in SAS software with the following program and a partial output.

```
set cohort;  proc logistic;
class sex (ref=last) age (ref=last) chol (ref=last)/param=ref;
model r/n=sex|age age|chol/scale=none aggregate;
contrast '3 vs 1 (30-49)' chol -1 0 1 0
                          age*chol -1 0 1 0 /estimate=both; run;
```

Contrast Test Results

Contrast	DF	Wald Chi-Square	Pr > ChiSq
3 vs 1 (30-49)	1	11.2089	0.0008

Contrast Rows Estimation and Testing Results

Contrast	Type	Row	Estimate	Standard Error	Alpha	Lower Limit	Upper Limit
3 vs 1 (30-49)	PARM	1	0.9243	0.2761	0.05	0.3832	1.4655
3 vs 1 (30-49)	EXP	1	2.5202	0.6958	0.05	1.4670	4.3296

Contrast Rows Estimation and Testing Results

Contrast	Type	Row	Wald Chi-Square	Pr > ChiSq
3 vs 1 (30-49)	PARM	1	11.2089	0.0008
3 vs 1 (30-49)	EXP	1	11.2089	0.0008

The estimate of 0.9243 agree with that obtained from the table above, namely, $-0.6200 + 1.5443 = 0.9243$. The standard error can be appropriately obtained, and confidence intervals obtained.

Similarly, for a given level of cholesterol, the true odds and corresponding confidence intervals for comparisons of the age*chol interaction terms at fixed cholesterol levels 1 and 4 ($k = 1$ and 4, respectively) are implemented again with the following contrast statements in PROC LOGISTIC.

```
set cohort; proc logistic;
class sex (ref=last) age (ref=last) chol (ref=last)/param=ref;
model r/n=sex|age age|chol/scale=none aggregate;
contrast 'age1 vs age2 at k=1'  age 1 -1
                          age*chol 1 0 0 0 -1 /estimate=both;
contrast 'age1 vs age2 at k=4'  age 1 -1
                          age*chol  0 0 0 1 0 0 0 -1 /estimate=both;  run;
```

Contrast Rows Estimation and Testing Results

Contrast	Type	Row	Estimate	Standard Error	Alpha	Lower Limit	Upper Limit
age1 vs age2 at k=1	PARM	1	-1.9938	0.3729	0.05	-2.7246	-1.2629
age1 vs age2 at k=1	EXP	1	0.1362	0.0508	0.05	0.0656	0.2828
age1 vs age2 at k=4	PARM	1	-1.1161	0.2295	0.05	-1.5659	-0.6663
age1 vs age2 at k=4	EXP	1	0.3275	0.0752	0.05	0.2089	0.5136

Contrast Rows Estimation and Testing Results

Contrast	Type	Row	Wald Chi-Square	Pr > ChiSq
age1 vs age2 at k=1	PARM	1	28.5863	<.0001
age1 vs age2 at k=1	EXP	1	28.5863	<.0001
age1 vs age2 at k=4	PARM	1	23.6540	<.0001
age1 vs age2 at k=4	EXP	1	23.6540	<.0001

We can obtain true confidence intervals for all the other odds ratios in a similar manner.

We can also employ PROC GENMOD to estimate some of the true comparison odds ratios produced above. For instance, for the contrast of "3 vs 1 (20-39)" can be implemented as:

```
set cohort;
proc genmod;
class sex age chol;
model r/n=sex|age age|chol/dist=b type3;
contrast '3 vs 1 (30-49)' chol -1 0 1 0
                          age*chol -1 0 1 0 /wald;
estimate '3 vs 1 (30-49)' chol -1 0 1 0
                          age*chol -1 0 1 0 /exp;  run;
```

Contrast Estimate Results

Label	Estimate	Standard Error	Alpha	Confidence Limits	
3 vs 1 (30-49)	0.9245	0.2761	0.05	0.3833	1.4656
Exp(3 vs 1 (30-49))	2.5206	0.6960	0.05	1.4671	4.3303

Contrast Results

Contrast	DF	Chi-Square	Pr > ChiSq	Type
3 vs 1 (30-49)	1	11.21	0.0008	Wald

8.5.3 Further Analysis

The pattern of the estimated odds ratios and hence the parameter values suggest that there seems to be a linear trend in the effect of the cholesterol level. We explore this further by considering the chol variable as a continuous variable having integer scores 1, 2, 3, 4, respectively. This therefore allows us to fit a linear effect in cholesterol level model to our data. Implementing this in SAS software with PROC GENMOD or LOGISTIC, we have the following output;

```
set cohort; proc genmod; class sex age;
model r/n= sex|age chol|age/dist=b type3; run;
proc logistic;
class sex (ref=last) age (ref=last)/param=ref;
model r/n=sex|age age|chol/scale=none aggregate;
        units chol=1 2 3 4 age=0;
run;
```

Criteria For Assessing Goodness Of Fit

Criterion	DF	Value	Value/DF
Deviance	10	11.0318	1.1032
Pearson Chi-Square	10	12.1454	1.2145

Analysis Of Parameter Estimates

Parameter			DF	Estimate	Standard Error	Wald 95% Confidence Limits		Chi-Square	Pr > ChiSq
Intercept			1	-2.7563	0.2725	-3.2903	-2.2222	102.32	<.0001
sex	1		1	0.7898	0.1544	0.4872	1.0924	26.16	<.0001
age	1		1	-2.6652	0.4006	-3.4503	-1.8801	44.27	<.0001
sex*age	1	1	1	0.5707	0.2431	0.0943	1.0471	5.51	0.0189
chol			1	0.2099	0.0748	0.0633	0.3566	7.87	0.0050
chol*age	1		1	0.3845	0.1111	0.1668	0.6022	11.98	0.0005

```
          LR Statistics For Type 3 Analysis

                                Chi-
Source              DF       Square     Pr > ChiSq
---------------------------------------------------
sex                  1        86.34         <.0001
age                  1        44.89         <.0001
sex*age              1         5.65         0.0175
chol                 1        56.20         <.0001
chol*age             1        12.07         0.0005
```

A test of whether the linear model is worthwhile is provided by G^2=11.0318 - 7.5847 = 3.4471 on 10 - 6 = 4 degrees of freedom (pvalue = 0.4860). This model not only fits the data but also indicate that a linear effect of chol will be very appropriate. Thus when age = 0, the estimated relative risk of CHD for the 50-62 age group is exp(0.2099) = 1.2336. That is, each 1 level increase in chol increases the relative risk of a CHD by 1.2336 among persons aged 50-62. A 2 level increase will make this relative risk be $1.2336^2 = 1.5218$. Similarly, for persons aged 30-49, the estimated relative risk is exp(0.2099 + 0.3845) = exp(0.5944) = 1.8119. Once again, each 1 level increase in chol increases the relative risk of a CHD by 1.8119 among persons aged 30-49. A 2 level increase will make this relative risk be $1.8119^2 = 3.2831$. The results are similar when PROC LOGISTIC is similarly employed. We note, however, that there are very slight differences in some of the parameter estimates. This does not in any way affect our overall conclusions above.

In the above analysis of a cohort study data, we have assumed that the follow-up time is the same for each person. However, if the follow-up time differed for each person, because they have been at risk for different periods before the study began, or at risk intermittently through the duration of the study, it would be sensible to take account of this variation in the analysis. Appropriate methods include Collet (1991) using the Poisson regression method with the offset being the person-years of exposure. One common approach is to compute the *person-years* of exposure for each individuals and the number of individuals who develop the disease in a particular group is then expressed as a proportion of the person-years of exposure. This rate of occurrence is then modeled using the *Poisson regression method.*

On the other hand, if the time from entry into the study until the occurrence of a particular disease is of interest, then models developed for survival analysis such as *Cox's proportional hazards model* can be used.

8.6 Example 8.6: Analysis of Data in Table 8.1

For the data in Table 8.1, our analysis started by fitting a saturated model involving the explanatory variables site, race, and their interaction using PROC GENMOD. The type3 partial analysis below indicate that the interaction between site and race is not significant, and neither is the site effect.

```
data tab81;
input site $ race $ r n @@;
datalines;
chton black 56 319 chton white 139 635
evans black 69 407 evans white 184 711
;
proc genmod order=data; class race site;
model r/n=site|race/dist=bin type3; run;
```

LR Statistics For Type 3 Analysis

Source	DF	Chi-Square	Pr > ChiSq
site	1	0.57	0.4508
race	1	12.24	0.0005
race*site	1	1.23	0.2675

Based on this information, we fitted a model involving site and race. This model although it fits the data with a $G^2 = 1.1178$ on 1 d.f., the type3 partial analysis again indicate that the effect of site with a partial pvalue of (0.1722) is again not significant. We therefore next fit a model involving only race.

```
set tab81;
proc genmod order=data;
class race site;
model r/n=race/dist=bin type3; run;
```

Criteria For Assessing Goodness Of Fit

Criterion	DF	Value	Value/DF
Deviance	2	2.9812	1.4906
Pearson Chi-Square	2	2.9722	1.4861

Analysis Of Parameter Estimates

Parameter		DF	Estimate	Standard Error	Wald 95% Confidence Limits		Chi-Square	Pr > ChiSq
Intercept		1	-1.1528	0.0638	-1.2779	-1.0277	326.27	<.0001
race	black	1	-0.4174	0.1172	-0.6472	-0.1877	12.68	0.0004

As seen from the SAS software output above, this model fits the data with $G^2 = 2.9812$ on 2 d.f. We notice that the odds that a Black person will die due to CHD are given by $e^{-0.3320} = 0.66$ lower than those for White men over the period of the study. This conclusion is consistent with those obtained by the authors in the original article. The estimated odds ratio has a 95% C.I. given by (0.529, 0.829).

We notice from the above output the ratio of "value/DF" is 1.4906. Ideally, we would want this value to be very close to 1.00. This value does affect our parameter estimates as well as their estimated standard errors, and we would reconsider this analysis later under discussion of over dispersion in binomial or Poisson based models.

8.7 Diagnostics

Influential observations and goodness-of-fit in logistic regression can be assessed by some of the residual statistics discussed in chapter 5 and some other measures that will be discussed in this section. These are effected by using the option *iplots* in PROC LOGISTIC. Let us give an example of this by revisiting the data in Table 8.6. In that section, we found that the complimentary log-log fits our data well. We reproduce part of the results below with the PROC LOGISTIC options **iplots** and **influence** invoked. The results below are from the PROC LOGISTIC output. We now discuss some of the terms in this output.

```
set tab86;
proc logistic;
model r/n=dose/link=cloglog scale=none aggregate influence iplots; run;
```

Deviance and Pearson Goodness-of-Fit Statistics

Criterion	DF	Value	Value/DF	Pr > ChiSq
Deviance	6	3.4464	0.5744	0.7511
Pearson	6	3.2947	0.5491	0.7711

Having fitted the model, the first diagnostic step is usually to examine both the chi-squared (χ_i) and deviance (d_i) residuals for identifying those observations that do not fit the model well. The two residuals are defined as:

$$\chi_i = \frac{y_i - n_i\hat{p}_i}{\sqrt{\{n_i\hat{p}_i(1-\hat{p}_i)\}}} \tag{8.22}$$

and

$$d_i = \begin{cases} -\left[2y_i \log\left(\frac{y_i}{\hat{y}_i}\right) + 2(n_i - y_i)\log\left(\frac{n_i - y_i}{n_i - \hat{y}_i}\right)\right]^{1/2} & \text{if } y_i < \hat{y}_i \\ \left[2y_i \log\left(\frac{y_i}{\hat{y}_i}\right) + 2(n_i - y_i)\log\left(\frac{n_i - y_i}{n_i - \hat{y}_i}\right)\right]^{1/2} & \text{if } y_i \geq \hat{y}_i \end{cases} \tag{8.23}$$

where $\hat{y}_i = n\hat{p}_i$; χ_i and d_i are appropriately referred to as the Pearson's and deviance residuals respectively. These two residuals are designated as **reschi** and **resdev** in SAS software and are produced below for our data. Standardized versions of these residuals are obtained by dividing the above raw residuals by the appropriate factors based on the leverages. The corresponding standardized residuals are designated as **streschi** and **stresdev** respectively. We may also note here that $G^2 = \sum d_i^2$. These are produced by PROC GENMOD in SAS®.

The chi-squared and deviance residuals from the complimentary log-log model applied to data in Table \ref{Ta:86}.

Obs	DOSE	Reschi	Streschi	Resdev	Stresdev	Reslik
1	1.6907	0.1825	0.2111	0.1806	0.2088	0.2094
2	1.7242	0.5681	0.6701	0.5577	0.6579	0.6614
3	1.7552	-0.7932	-0.9258	-0.8033	-0.9376	-0.9345
4	1.7842	-0.6355	-0.7132	-0.6344	-0.7120	-0.7122
5	1.8113	1.2430	1.4560	1.2888	1.5097	1.4953
6	1.8369	-0.5413	-0.6721	-0.5237	-0.6503	-0.6580
7	1.8610	-0.1212	-0.1445	-0.1188	-0.1416	-0.1425
8	1.8839	0.2298	0.2391	0.3250	0.3380	0.3315

While the raw residuals do not have the unit variances, the standardized residuals have unit variances, and although the standardized Pearson residuals are not closely approximated by a normal distribution, Collett (1991) has advocated the use of the standardized deviance residuals for routine model checking. Any standardized residual outside [-2,2] will be considered unsatisfactory. We see from the above results that none of the observations is unsatisfactory.

A check of influential observations is obtained by conducting influential diagnostics either with PROC LOGISTIC OR PROC GENMOD. The **Dfbeta** for both the intercept and dose parameters (usually for intercept and covariates parameters) generated indicates that none of the observations when removed (or deleted) from the model have any significant effects on the original parameter estimates of -39.5725 and 22.0412, respectively. The diagnostics measures **Delta deviance** and **Delta chi-square** indicate by what the goodness-of-fit test statistics G^2 and X^2 values would change if the designated observation were deleted from the model.

Plots of these against the case number is provided by **iplots** and displayed by the **Influence** options in PROC LOGISTIC. The results show that were we to remove observation 5 from the model, the G^2 and X^2 will change by 2.2360 and 2.1200, respectively: that is, by almost 65% and 64%, respectively.

```
                           Regression Diagnostics
                         Pearson Residual                  Deviance Residual
          Covariates
 Case                               (1 unit = 0.16)              (1 unit = 0.16)
Number        DOSE      Value    -8  -4  0 2 4 6 8      Value -8  -4  0 2 4 6 8

   1        1.6907     0.1825    |         |*      |     0.1806  |        |*   |
   2        1.7242     0.5681    |         |   *   |     0.5578  |        |*   |
   3        1.7552    -0.7932    |   *     |       |    -0.8033  |   *     |   |
   4        1.7842    -0.6355    |    *    |       |    -0.6344  |    *    |   |
   5        1.8113     1.2430    |         |      *|     1.2888  |         | *|
   6        1.8369    -0.5413    |     *   |       |    -0.5237  |     *   |   |
   7        1.8610    -0.1212    |        *|       |    -0.1188  |        *|   |
   8        1.8839     0.2298    |         |*      |     0.3249  |         | *|

                  Hat Matrix Diagonal
                                              Intercept
 Case                  (1 unit = 0.02)          DfBeta        (1 unit = 0.04)
Number      Value     0 2 4 6 8  12  16          Value      -8  -4  0 2 4 6 8

   1       0.2522     |          *      |        0.1093     |        |  *   |
   2       0.2813     |            *    |        0.3320     |        |     *|
   3       0.2660     |           *     |       -0.3340     |*       |      |
   4       0.2060     |        *        |       -0.0920     |      * |      |
   5       0.2712     |           *     |       -0.1039     |      * |      |
   6       0.3515     |               * |        0.1127     |        |  *   |
   7       0.2961     |            *    |        0.0173     |        *      |
   8       0.0757     |  *              |       -0.00848    |        *      |

                                           Confidence Interval Displacement C
            DOSE
 Case      DfBeta      (1 unit = 0.04)                       (1 unit = 0.05)
Number      Value     -8  -4  0 2 4 6 8          Value      0 2 4 6 8  12  16

   1      -0.1082     |    *   |       |         0.0150     |*              |
   2      -0.3274     |*       |       |         0.1758     |    *          |
   3       0.3263     |        |      *|         0.3106     |      *        |
   4       0.0862     |        | *     |         0.1320     |   *           |
   5       0.1152     |        |  *    |         0.7888     |              *|
   6      -0.1162     |    *   |       |         0.2448     |     *         |
   7      -0.0176     |        *       |         0.00878    |*              |
   8       0.00859    |        *       |         0.00468    |*              |

      Confidence Interval Displacement CBar             Delta Deviance
 Case                   (1 unit = 0.04)                         (1 unit = 0.14)
Number      Value     0 2 4 6 8  12  16          Value      0 2 4 6 8  12  16

   1       0.0112     |*               |         0.0439     |*              |
   2       0.1263     |   *            |         0.4374     |   *           |
   3       0.2280     |     *          |         0.8732     |      *        |
   4       0.1048     |  *             |         0.5072     |    *          |
   5       0.5749     |               *|         2.2360     |              *|
   6       0.1588     |   *            |         0.4330     |   *           |
   7       0.00618    |*               |         0.0203     |*              |
   8       0.00433    |*               |         0.1099     | *             |
```

```
                    Delta Chi-Square
   Case                   (1 unit = 0.13)
  Number        Value      0 2 4 6 8  12  16

     1         0.0446     |*                 |
     2         0.4491     |   *              |
     3         0.8570     |      *           |
     4         0.5087     |    *             |
     5         2.1200     |                 *|
     6         0.4517     |   *              |
     7         0.0209     |*                 |
     8         0.0571     |*                 |
```

These values are quite substantial and influential. We therefore find it attractive to remove this observation (5-th observation) from the data and refit the new revised model based on the remaining observations. Results from this implementation are displayed below. It gives a better fit and the diagnostics procedures reveal no further lack of fit or influential observations.

```
set tab86;
proc genmod; where dose ne 1.8113;
model r/n=dose/link=cloglog scale=none aggregate influence iplots; run;
```

Criteria For Assessing Goodness Of Fit

Criterion	DF	Value	Value/DF
Deviance	5	1.2364	0.2473
Pearson Chi-Square	5	1.1525	0.2305

Analysis Of Parameter Estimates

Parameter	DF	Estimate	Standard Error	Chi-Square	Pr > ChiSq
Intercept	1	-39.0409	3.2115	-45.3353	<.0001
DOSE	1	21.7096	1.7838	18.2134	<.0001
Scale	0	1.0000	0.0000		

Obs	DOSE	Reschi	Streschi	Resdev	Stresdev	Reslik
1	1.6907	0.2537	0.2924	0.2499	0.2881	0.2892
2	1.7242	0.7108	0.8404	0.6943	0.8209	0.8265
3	1.7552	-0.5728	-0.6793	-0.5784	-0.6859	-0.6840
4	1.7842	-0.3137	-0.3626	-0.3136	-0.3624	-0.3624
5	1.8369	-0.0640	-0.0844	-0.0637	-0.0841	-0.0842
6	1.8610	0.2311	0.2936	0.2395	0.3044	0.3003
7	1.8839	0.3144	0.3354	0.4445	0.4741	0.4595

In conclusion, care must be taken in accepting a logistic regression model without conducting diagnostic tests for lack of fit and influential observations. Collett (1991) has given a comprehensive diagnostics tests for linear logistic regression models. Interested readers are encouraged to consult this reference.

8.8 Overdispersion

A problem that often arises with modeling binomial and Poisson data is *overdispersion*, where data involving proportions or counts tend to be more variable than the underlying binomial or Poisson distributions can accommodate. This phenomenon in the binomial case is also known as the *extra binomial* or simply *overdispersion*. This problem usually arises when the y_i observations are correlated. For, instance, for the logistic model fitted to binomial data, we assume that the y_i are independent and follow a binomial distribution. However, if the y_i are positively correlated, then the variance of y_i will be greater than $n_i p_i(1-p_i)$, and in this case we would have overdispersion. On the other hand, when the y_i are negatively correlated, the variance of the binomial response variable will be less than the above expected variance, and in this case we would have *underdispersion.* This latter case is less frequently encountered than the former, and hence the former, that is, *overdispersion* is considered in this section.

Overdispersion also occurs in incorrectly specified models where some interactions or transformed variables have been omitted from the model. However, the former is more common. For positively correlated binomial observations, we show below that the ratio of the residual deviance (or G^2) to the corresponding degree of freedom, that is, the *residual mean deviance* **RMD** would take one of the three possible values below, which correspond respectively to overdispersion, no dispersion, and underdispersion. That is,

$$\text{RMD} = \begin{cases} >1 & \text{overdispersion} \\ 1 & \text{no dispersion} \\ <1 & \text{underdispersion} \end{cases}$$

If we consider the binomial case where, a random variable Y has probability π of success. Then, if Y given π has a binomial distribution $b(r/n, \pi)$ and π has a distribution of its own with mean μ and variance σ^2, we can write

$$E(Y) = E[E(Y \mid \pi)] = n\mu$$

and the conditional variance of Y can be shown to be:

$$\text{Var}\{Y\} = n\mu(1-\mu) + n(n-1)\sigma^2,$$

that is,

$$\text{Var}\{Y\} > n\mu(1-\mu)$$

If we let $\sigma^2 = \phi\mu(1-\mu)$, then the above becomes

$$\text{Var}\{Y\} = n\mu(1-\mu)[1+(n-1)\phi]$$

where $\phi \geq 0$ is an unknown scale parameter.

From the above, it follows that if the Y_i are uncorrelated, then, $\phi = 0$ and the above leads to Var{Y}=$n\mu(1-\mu)$. However, if $\phi > 0$, then Var(Y) will exceed the binomial variance $n_i p_i(1-p_i)$, and in this case we have overdispersion. Similarly, if $\phi < 0$, then Var(Y) $< n_i p_i(1-p_i)$, and in this case we have underdispersion.

The conditional variance of Y above is obtained from conditional probability property, where the unconditional expected value of a random variable Y can be obtained from the conditional expected value of Y given X. That is,

$$E(Y) = E\{E(Y \mid X)\} \text{ and}$$
$$V(Y) = E\{V(Y \mid X)\} + V\{E(Y \mid X)\}$$

8.8.1 Modeling Overdispersed Data

In this section, we shall consider methods that have been used to model overdispersion in binomial data. We shall in all cases reanalyze the data in Table 8.1 where the logistic model gives a $G^2 = 2.9812$ on 2 d.f. and a corresponding $X^2 = 2.9722$ on 2 d.f. The ratio, value/DF = Deviance/(n - p), an estimate of ϕ, for either case is greater than 1, indicating overdispersion for this data set. The first method, which is applicable for the case of equal n_i, estimates ϕ by the ratio of Pearson's X^2 to its corresponding degrees of freedom. First let us reproduce our earlier analysis but requesting a printout of the variance-covariance matrix of the parameters. This is displayed below.

```
set tab81;
proc logistic order=data;
class race(ref=last) site(ref=last)/param=ref;
model r/n=race/scale=none  covb;
run;
       Deviance and Pearson Goodness-of-Fit Statistics

Criterion          DF          Value     Value/DF      Pr > ChiSq
------------------------------------------------------------------
Deviance            2         2.9812       1.4906          0.2252
Pearson             2         2.9722       1.4861          0.2262

                      Analysis of Maximum Likelihood Estimates

                                            Standard           Wald
      Parameter            DF     Estimate     Error     Chi-Square     Pr > ChiSq
      ----------------------------------------------------------------------------
      Intercept             1      -1.1528    0.0638       326.2642         <.0001
      race        black     1      -0.4173    0.1172        12.6761         0.0004


                             Estimated Covariance Matrix

                      Variable         Intercept         raceblack
                      --------------------------------------------
                      Intercept         0.004073          -0.00407
                      raceblack         -0.00407          0.013736
```

An estimate of ϕ is provided by the deviance or Pearson's mean square (1.4906, 1.4861, respectively). Further, the estimated variance-covariance matrix under this model is:

$$\hat{C}_{ij} = \begin{bmatrix} 0.0041 & -0.0041 \\ -0.0041 & 0.0137 \end{bmatrix}$$

With the estimate of ϕ obtained above, a new model is then fitted to the data, and this is accomplished in SAS software in either PROC LOGISTIC or GENMOD by specifying the options SCALE=(Deviance or Pearson). Suppose we invoke SCALE=Deviance in PROC LOGISTIC to the data in Table 8.1 then a new estimate of ϕ is computed and the model refitted with this new value. We give below the SAS software output when this is invoked.

```
set tab81;
proc logistic order=data;
class race(ref=last)/param=ref;
model r/n=race/scale=deviance covb;
run;

       Deviance and Pearson Goodness-of-Fit Statistics

Criterion          DF          Value     Value/DF      Pr > ChiSq
------------------------------------------------------------------
Deviance            2         2.9812       1.4906          0.2252
Pearson             2         2.9722       1.4861          0.2262

Number of events/trials observations: 4


NOTE: The covariance matrix has been multiplied by the heterogeneity factor
      (Deviance / DF) 1.49058.
```

Analysis of Maximum Likelihood Estimates

Parameter		DF	Estimate	Standard Error	Wald Chi-Square	Pr > ChiSq
Intercept		1	-1.1528	0.0779	218.8835	<.0001
race	black	1	-0.4173	0.1431	8.5041	0.0035

Estimated Covariance Matrix

Variable	Intercept	raceblack
Intercept	0.006072	-0.00607
raceblack	-0.00607	0.020475

For these data, an estimated of ϕ is given by the column Value/DF= 1.49058. We note here that the estimates of the parameters β_0 and β_1 are unaffected through this scaling, but the variance covariance matrix of the parameters are now duly affected. The new variance-covariance matrix is now given from the output by:

$$\hat{C}'_{ij} = \begin{bmatrix} 0.0061 & -0.0061 \\ -0.0061 & 0.0205 \end{bmatrix}$$

This variance-covariance matrix has been adjusted to take into account the overdispersion present in the data. We note that

$$\hat{C}'_{ij} = \hat{\phi}\hat{C}_{ij} = 1.49058\,\hat{C}_{ij}$$

That is,

$$\hat{C}'_{ij} = \begin{bmatrix} 0.0061 & -0.0061 \\ -0.00611 & 0.0205 \end{bmatrix} = \hat{\phi}\hat{C}_{ij} = 1.49058 \begin{bmatrix} 0.0041 & -0.0041 \\ -0.0041 & 0.0137 \end{bmatrix}$$

As expected, this model has a deviance of 2.000 on 2 d.f., that is, a $\hat{\phi}$ value of 1.000, as would be expected, which now suggests no overdispersion. The standard errors of the parameter estimates are computed by $\sqrt{\hat{C}'_{ij}}$. We can obtain a similar result for Pearson's X^2 by also specifying SCALE=Pearson in the options line of the model statement in PROC LOGISTIC. In PROC GENMOD, you can either use the Pearson's X^2 for instance with a PSCALE option, which gives a $\phi = \sqrt{\text{Value/DF}}$ or with a SCALE=Pearson option, which gives the square root of ϕ. Corresponding options for the deviance (G^2) are DSCALE and SCALE=Deviance in the option statement. We give below a similar partial output when the option is DSCALE in PROC GENMOD.

```
set tab81;
proc genmod order=data; class race;
model r/n=race/dist=b dscale covb; run;
```

Criteria For Assessing Goodness Of Fit

Criterion	DF	Value	Value/DF
Deviance	2	2.9812	1.4906
Scaled Deviance	2	2.0000	1.0000
Pearson Chi-Square	2	2.9722	1.4861
Scaled Pearson X2	2	1.9940	0.9970

Estimated Covariance Matrix

	Prm1	Prm2
Prm1	0.006072	-0.006072
Prm2	-0.006072	0.02048

Analysis Of Parameter Estimates

Parameter		DF	Estimate	Standard Error	Wald 95% Confidence Limits		Chi-Square	Pr > ChiSq
Intercept		1	-1.1528	0.0779	-1.3056	-1.0001	218.88	<.0001
race	black	1	-0.4174	0.1431	-0.6979	-0.1370	8.51	0.0035
Scale		0	1.2209	0.0000	1.2209	1.2209		

NOTE: The scale parameter was estimated by the square root of DEVIANCE/DOF.

We observe for GENMOD that the scale parameter is 1.2209 instead of the customary or familiar 1.0. This value is simply the square root of $\hat{\phi}$, that is, $\sqrt{1.4906}$.

8.8.2 Williams' Procedure

While the above corrections for overdispersion give exact parameter estimates for both uncorrected and corrected overdispersion models, the difference only being the adjustment of the variance-covariance matrix of the parameter estimates, Williams (1975, 1982) proposed a procedure for fitting overdispersed data with several explanatory variables that not only modify the variance-covariance matrix (and hence standard errors) but also the parameter estimates of the model. This procedure also requires an estimate of ϕ, the *heterogeneity factor*, from the data. Then, if we consider each binomial observations y_i to have weights w_i, the Pearson's test statistic X^2 becomes

$$X^2 = \sum \frac{w_i(y_i - n_i\hat{p}_i)^2}{n_i\hat{p}_i(1-\hat{p}_i)}$$

which has approximately the expected value

$$\sum w_i(1 - w_i d_i v_i)[1 + \phi(n_i - 1)]$$

where $v_i = n_i p_i(1 - p_i)$ and d_i is the diagonal element of the variance-covariance matrix of the linear predictor, $\hat{g}_i = \sum \beta_j x_{ji}$. Williams's procedure can only be implemented in PROC LOGISTIC. GENMOD does not yet have options for Williams's procedure. Once ϕ is estimated from the data, the weights $[1 + (n_i - 1)\hat{\phi}]^{-1}$ are then used in fitting models that have fewer terms than our original full model. We present again SAS software implementation of Williams's procedure on the data in Table 8.1.

```
proc logistic order=data;
class race(ref=last) site(ref=last)/param=ref;
model r/n=race site/scale=williams  covb; run;
```

Model Information

Data Set	WORK.TAB81
Response Variable (Events)	r
Response Variable (Trials)	n
Number of Observations	4
Weight Variable	1 / (1 + 0.000522 * (n - 1))
Sum of Weights	1605.3396016
Model	binary logit
Optimization Technique	Fisher's scoring

Deviance and Pearson Goodness-of-Fit Statistics

Criterion	DF	Value	Value/DF	Pr > ChiSq
Deviance	1	0.9968	0.9968	0.3181
Pearson	1	0.9999	0.9999	0.3173

```
                 Number of events/trials observations: 4

NOTE: Since the Williams method was used to accommodate overdispersion, the Pearson
      chi-squared statistic and the deviance can no longer be used to assess the
      goodness of fit of the model.

                 Analysis of Maximum Likelihood Estimates

                                         Standard          Wald
Parameter            DF     Estimate        Error    Chi-Square    Pr > ChiSq
-----------------------------------------------------------------------------
Intercept             1      -1.0908       0.0934      136.3507        <.0001
race        black     1      -0.4197       0.1306       10.3307        0.0013
site        chton     1      -0.1353       0.1228        1.2139        0.2706
```

We observe here that we have fitted the full model involving site and race in order to implement Williams's overdispersed model. It has been suggested that it is better to use a full model when employing Williams's procedure, as this would reduce the risk of corrupting ϕ with a misspecified or incorrect model. In the analysis above, the estimate of ϕ is 0.000522 and is given at the beginning of the SAS software output under the formula for the WEIGHT variable. Because again the site effect is not significant, we can remove this variable from the model, giving a reduced model involving only race. In general, for multivariable explanatory models, the maximum likelihood will be examined to eliminate certain variables that are not significant at this stage, and a revised model will be fitted by specifying in the options statement, the value of ϕ estimated above, namely; Scale = Williams($\hat{\phi}$) in the model statement. That is, scale=williams(0.000522).

```
set tab81;
proc logistic order=data;
class race(ref=last)/param=ref;
model r/n=race/scale=williams(0.000522) covb;
run;
              Deviance and Pearson Goodness-of-Fit Statistics

        Criterion         DF           Value     Value/DF     Pr > ChiSq
        ----------------------------------------------------------------
        Deviance           2          2.2141       1.1070         0.3305
        Pearson            2          2.2086       1.1043         0.3314

                 Number of events/trials observations: 4

                 Analysis of Maximum Likelihood Estimates

                                         Standard          Wald
Parameter            DF     Estimate        Error    Chi-Square    Pr > ChiSq
-----------------------------------------------------------------------------
Intercept             1      -1.1544       0.0742      241.8578        <.0001
race        black     1      -0.4153       0.1305       10.1345        0.0015

                           Odds Ratio Estimates

                                  Point          95% Wald
            Effect             Estimate      Confidence Limits
            ----------------------------------------------------
            race black vs white   0.660      0.511       0.852

                        Estimated Covariance Matrix

                 Variable        Intercept       raceblack
                 ---------------------------------------
                 Intercept         0.00551        -0.00551
                 raceblack        -0.00551        0.017019
```

When this was applied to our data, parameter values not only changed, but their estimated standard errors are also lower than the uncorrected for dispersion model. This of course in turn affect the estimated odds ratio, as well as its corresponding 95% confidence intervals.

8.9 Other Topics: Case-Control Data Analysis

8.9.1 Modeling Case-Control Study Data

In what follows, we give examples of analysis of data arising from case-control studies having binary outcome. The first example relates to analyzing the general case-control study data while the second example analyzes data arising from a matched case-control study. The data for both examples are taken from Collett (1991), who offers similar analyses for both data sets.

8.9.2 Example 8.7: Diverticular Disease Data

Diverticular disease and carcinoma of the colon are among the most common diseases of the large intestine in the population of the UK and North America. Diverticular disease is estimated to be present in one-third of the population over 40 years of age, while carcinoma of the colon accounts for about 10,000 deaths in the UK alone per annum. It has been found that there is a geographical association between the prevalence of the two diseases in that their worldwide incidence tends to go hand in hand. In other to investigate whether the two conditions tend to be associated in individual patients, a case-control study was carried out and reported by Berstock, Villers, and Latto (1978).

A total of 80 patients, admitted to the Reading group of hospitals between 1974 and 1977 with carcinoma of the colon, were studied and the incidence of diverticular disease in each patient was recorded. A control group of 131 individuals were randomly selected from a population known not to have colonic cancer, and were similarly tested for the presence of diverticular disease. Table 8.10 gives the proportions of individuals with diverticular diseases, classified by age, sex, and according to whether or not they had colonic cancer.

To facilitate analysis, the data is first transformed to reflect the proportion of cases in the case-control study. In Table 8.11, for instance, are the case-control format necessary for the 40-49 M, 40-49 F, 80-89 F, and 80-89 M, respectively.

The original data with the other variables are now rearranged and the results presented in Table 8.12 together with the proportion of cases with or without the presence of diverticular disease (DD).

In this setup, if an individual has the diverticular disease, then DD takes the value 1 and 0 for those without the disease. That is,

$$\text{DD} = \begin{cases} 1 & \text{If diseased} \\ 0 & \text{Otherwise} \end{cases}$$

The following SAS software program and the accompanying partial outputs accomplish the above, together with the fitting of one of the many models applied to the data. The full output is available in appendix F.5.

```
data case;
input age $ sex $ r1 n1 r2 n2 @@;
```

Age interval	Midpoint of age age range	Sex	Proportion with DD Cancer patients	Controls
40-49	44.5	M	0/3	0/7
		F	0/6	1/15
50-54	52.0	M	1/2	1/7
		F	0/0	0/0
55-59	57.0	M	2/5	3/15
		F	1/7	4/18
60-64	62.0	M	1/5	5/18
		F	0/2	2/8
65-69	67.0	M	1/4	6/11
		F	0/5	7/17
70-74	72.0	M	0/5	1/4
		F	3/13	2/6
75-79	77.0	M	1/3	0/0
		F	5/9	0/0
80-89	84.5	M	1/2	4/5
		F	4/9	0/0

Table 8.10: Proportion of individuals with diverticular disease (DD) classified by age, sex and the presence of colonic cancer

```
rr1=n1-r1; rr2=n2-r2;
ndd=r1+r2; nndd=rr1+rr2;
rdd=r1; rndd=rr1;
drop rr1 rr2 r1 n1 r2 n2;
datalines;
40-49  m  0 3 0 7 40-49 f 0 6 1 15
50-54 m 1 2  1 7 50-54 f 0 0 0 0
55-59 m 2 5 3 15 55-59 f 1 7 4 18
60-64 m 1 5 5 18 60-64 f 0 2 2 8
65-69 m 1 4 6 11 65-69 f 0 5 7 17
70-74 m 0 5 1 4 70-74 f 3 13 2 6
75-79 m 1 3 0 0 75-79 f 5 9 0 0
80-89 m 1 2 4 5 80-89 f 4 9 0 0
;
proc print;
var rndd nndd rdd ndd;
run;
data new1 (rename=(rndd=r nndd=n));
set case; dd=0; output;
drop  rdd ndd; run;
data new2  (rename=(rdd=r ndd=n));;
set case; d=1; output;
drop rndd nndd; run;
data comb; set new1 new2;
proc print data=comb; run;
data comb2; set comb;
if r=0 and n =0 then delete;
proc print data=comb2; run;
proc genmod order=data data=comb2;
class age sex dd;
model r/n=age|sex dd/dist=binomial link=logit type3;
run;
proc logistic order=data data=comb2;
class age sex dd;
model r/n=age|sex dd/scale=none aggregate;
run;
```

DD	Cases	Controls	Total
1	0	0	0
0	3	7	10
Total	3	7	10

DD	Cases	Controls	Total
1	0	1	1
0	6	14	20
Total	6	15	21

DD	Cases	Controls	Total
1	1	4	5
0	1	1	2
Total	2	5	7

DD	Cases	Controls	Total
1	4	0	4
0	5	0	5
Total	9	0	9

Table 8.11: Transformations before analysis for 40-49 M, 40-49 F, 80-89 M, and 80-89 F respectively.

		Proportion of Cases	
Age interval	Sex	Without DD	With DD
40-49	M	3/10	0/0
	F	6/20	0/1
⋮	⋮	⋮	⋮
75-79	M	2/2	1/1
	F	4/4	5/5
80-89	M	1/2	1/5
	F	5/5	4/4

Table 8.12: Rearranged result for data in Table 8.10

Obs	age	sex	n	r	dd
1	40-49	m	10	3	0
2	40-49	f	20	6	0
3	50-54	m	7	1	0
4	50-54	f	0	0	0
5	55-59	m	15	3	0
6	55-59	f	20	6	0
7	60-64	m	17	4	0
8	60-64	f	8	2	0
9	65-69	m	8	3	0
10	65-69	f	15	5	0
11	70-74	m	8	5	0
12	70-74	f	14	10	0
13	75-79	m	2	2	0
14	75-79	f	4	4	0
15	80-89	m	2	1	0
16	80-89	f	5	5	0
17	40-49	m	0	0	1
18	40-49	f	1	0	1
19	50-54	m	2	1	1
20	50-54	f	0	0	1
21	55-59	m	5	2	1
22	55-59	f	5	1	1
23	60-64	m	6	1	1
24	60-64	f	2	0	1
25	65-69	m	7	1	1
26	65-69	f	7	0	1
27	70-74	m	1	0	1
28	70-74	f	5	3	1
29	75-79	m	1	1	1
30	75-79	f	5	5	1
31	80-89	m	5	1	1
32	80-89	f	4	4	1

Criteria For Assessing Goodness Of Fit

Criterion	DF	Value	Value/DF
Deviance	13	9.1927	0.7071
Pearson Chi-Square	13	8.5699	0.6592

The LOGISTIC Procedure

Model Information

Data Set	WORK.COMB2
Response Variable (Events)	r
Response Variable (Trials)	n
Number of Observations	29
Link Function	Logit
Optimization Technique	Fisher's scoring

Response Profile

Ordered Value	Binary Outcome	Total Frequency
1	Event	80
2	Nonevent	131

Class Level Information

		Design Variables						
Class	Value	1	2	3	4	5	6	7
age	40-49	1	0	0	0	0	0	0
	50-54	0	1	0	0	0	0	0
	55-59	0	0	1	0	0	0	0
	60-64	0	0	0	1	0	0	0
	65-69	0	0	0	0	1	0	0
	70-74	0	0	0	0	0	1	0
	75-79	0	0	0	0	0	0	1
	80-89	0	0	0	0	0	0	0
sex	f	1						
	m	0						
dd	0	1						
	1	0						

Deviance and Pearson Goodness-of-Fit Statistics

Criterion	DF	Value	Value/DF	Pr > ChiSq
Deviance	13	9.1929	0.7071	0.7582
Pearson	13	8.5700	0.6592	0.8046

Number of unique profiles: 29

Odds Ratio Estimates

Effect	Point Estimate	95% Wald Confidence Limits	
dd 0 vs 1	2.058	0.857	4.945

The analysis is based on 29 out of the 32 observations in the data set. This is because observations 4, 17, and 20 have both r and $n = 0$ and therefore not used in the analysis. We present in Table 8.13 other possible models applied to the data together with their deviances and corresponding degrees of freedom.

Model	d.f.	G^2
μ	28	73.49
{Age}	21	24.05
{Sex}	27	69.01
{Age, Sex}	20	22.73
{Age | Sex}	14	11.97
{Age | Sex, DD}	13	9.19
{Age | Sex, DD | Age}	6	3.71
{Age | Sex, DD | Sex}	12	8.13

Table 8.13: Results of models applied to the data

There are possible confounding effects of age and sex on the relationship between the occurrence of colon cancer and diverticular disease. Hence, the effect of the exposure factor DD needs to be assessed after due allowance for the effects of age and sex. Only age seems to have a potential confounding effect, while the sex effect is very light. There is interaction present between age and sex.

Once DD is added to the model, there does not seem to be significant interaction effects of DD with either age or sex, based on the differences in the G^2 and corresponding degrees of freedom. The DD parameter for DD in GENMOD is 0.7217 for DD=0 and 0 for DD=1. Hence the ratio of the odds of colonic cancer relative to one without is $\exp(-0.7217) = 0.49$. This implies that an estimate of the relative risk of colonic cancer occurring in patients with diverticular diseases, relative to those without is 0.49. That is, the occurrence of diverticular disease decreases the risk of colonic cancer, in other words, a patient with diverticular disease is less likely to be affected by colonic cancer than one without. Put another way, those without diverticular disease are 2.058=1/0.49 times more likely to have colonic cancer than those without.

If the cell referencing option is not employed in the class statement in PROC LOGISTIC, the estimate of the DD_0 parameter is given by 0.3609 and, the parameter $DD_1 = -0.3609$. Hence DD(1) versus DD(0) equals -0.3609 - $0.3609 = -0.7218$. The estimated odds ratio is that of DD(0) versus DD(1) = 2.058. Hence estimated relative risk of DD(1) versus DD(0) is $1/2.058 = 0.49$.

8.9.3 Matched Case-Control Study Data

An alternative to adjusting for confounding effects is to take account of their presence in the study design. This is achieved by what is known as *matched case-control study*, in which each diseased person included in the study as a case is matched to one or more disease-free persons who will be included as controls. The matching is usually based on potential confounding variables, such as gender, ethnicity, marital status, parity, etc., or by residential area, or place of work for those variables than cannot be easily quantified.

1. A design with M controls per case is known as a **1:M** matched study, and the individuals that constitute the one case and the M controls to which the case

has been matched are referred to as a **matched set**. Usually, in this case, $1 \leq M \leq 5$

2. A design in which $M = 1$ is called a **1:1** matching design. Here, the matched set consists of one case and one control from each stratum.

3. In a **m:n** matched study, the matched set consists of n cases with m controls, where usually $1 \leq (m, n) \leq 5$.

The appropriate form of logistic regression for these types of data is called *conditional logistic regression*, which is based on conditional likelihood estimation (Chamberlain, 1980). In constructing the likelihood function, we condition on the number of 1's and 0's that are observed for each individual. Stokes et al. (1995) has presented an elegant treatment of the derivation of this procedure. Suffice it to say here that for a randomized clinical trial with $i = 1, 2, \cdots, q$ centers that are randomly selected, then for two individuals from the same matched set, the i-th, say, the ratio of the odds of diseases (event) occurring in a person with covariate $\mathbf{x}_{i1}$ relative to one with covariate $\mathbf{x}_{i2}$ is given by

$$\log\left\{\frac{p(\mathbf{x}_{i1})/\{1-p(\mathbf{x}_{i1})\}}{p(\mathbf{x}_{i2})/\{1-p(\mathbf{x}_{i2})\}}\right\} = \beta_1(x_{i11} - x_{i21}) + \beta_2(x_{i12} - x_{i22}) + \cdots + \beta_k(x_{i1k} - x_{i2k}) \tag{8.24}$$

where $i = 1, 2, \cdots, q$. The process of estimating the β-parameters in the above model is often referred to as the *conditional logistic modeling.* A model that has no explanatory variable is referred to as the *null model* and in this case, $\beta_h = 0$ for $h = 1, 2, \cdots, k$, and the deviance reduces to $2q\log(1 + M)$. If in $1:1$ matching, then, the deviance reduces to $2q\log(2)$.

8.9.4 Example 8.8: An Etiological Study Data

Kelsy and Hardy (1975) described an etiological study to investigate whether driving of motor vehicles is a risk factor for low back pain caused by acute herniated lumbar invertebral discs. The design used was a matched case-control study in which one control was matched to each case. The cases were selected from persons between the ages of 20 and 64 living in the area of New Haven, Connecticut, who had X-rays taken of the low back between June 1971 and May 1973. Those diagnosed as having acute herniated lumbar invertebral discs, and who had only recently acquired symptoms of the disease, were used as the cases in the study. The controls were drawn from patients who were admitted to the same hospital as a case, or who presented at the same radiologist's clinic as a case, with a condition unrelated to the spine. Individual controls were further matched to a case on the basis of sex and age. The age matching was such that the ages of the case and control in each matched pair were within 10 years of one another. In total, 217 matched pairs were recruited, consisting of 89 female pairs and 128 male pairs.

After an individual had been entered into the study as either a case or a control, information on a number of potential risk factors, including place of residence and driving habits, was obtained. Those individuals who lived in New Haven itself were considered to be city residents, while others were classified as suburban residents. Data on whether or not each individual was a driver and their residential status are presented in the table below.

The number of matched sets of cases and controls according to driving and residence.

STATUS	Driver?	Suburban resident?	# of matched sets
case	no	no	9
control	no	no	
case	no	yes	2
control	no	no	
case	yes	no	14
control	no	no	
case	yes	yes	22
control	no	no	
case	no	no	0
control	no	yes	
case	no	yes	2
control	no	yes	
case	yes	no	1
control	no	yes	
case	yes	yes	4
control	no	yes	
case	no	no	10
control	yes	no	
case	no	yes	1
control	yes	no	
case	yes	no	20
control	yes	no	
case	yes	yes	32
control	yes	no	
case	no	no	7
control	yes	yes	
case	no	yes	1
control	yes	yes	
case	yes	no	29
control	yes	yes	
case	yes	yes	63
control	yes	yes	

Since this is a **1:1** matching design, i.e., $M = 1$ and there is no treatment except the explanatory variables, the conditional likelihood function can be written (Collett, 1991) as

$$\prod_{i=1}^{q}\left[1+\exp\left\{\sum_{h=1}^{k}\beta_h(x_{i1h}-x_{i2h})\right\}\right]^{-1}$$
$$\prod_{i=1}^{q}\left[1+\exp\left\{\sum_{h=1}^{k}\beta_h z_{ih}\right\}\right]^{-1} \tag{8.25}$$

where $z_{ih} = x_{i1h} - x_{i2h}$, so that z_{ih} is the value of the h-th explanatory variable for the cases minus that for the control, in the i-th matched set. Equation (8.25) is the likelihood for a linear logistic model with q binary observations that are all equal to unity, where the linear part of the model contains k explanatory variables $z_{i1}, z_{i2}, \cdots, z_{ik}$, with $i = 1, 2, \cdots, q$ and so no constant term. To implement this model in SAS software for the data in example 8.8, we first create the following indicator variables:

$$x_1 = \begin{cases} 0 & \text{if the individual is not a driver-No} \\ 1 & \text{if the individual is a driver-Yes} \end{cases}$$

and

$$x_2 = \begin{cases} 0 & \text{if the individual is not a suburban resident-No} \\ 1 & \text{if the individual is a suburban resident-Yes} \end{cases}$$

If we assume that interaction may be present between residence and driving, we can create variable $x_3 = x_1 \times x_2$ at this point. Note that both the cases and the controls are characterized by these three variables. From this, we now create the z's, which are the differences between corresponding x's for the cases and the controls. The following SAS software program fits the conditional logistic regression model to the data in example 8.8.

```
data cond;
input idd status $ driver $ res $ count @@;
if driver='yes' then x1=1;
else x1=0;
if res ='yes' then x2=1;
else x2=0;
x3=x1*x2;
datalines;
1 case no no 9 1 control no no 9 2 case no yes 2 2 control no no 2
3 case yes no 14 3 control no no 14 4 case yes yes 22 4 control no no 22
5 case no no 0 5 control no yes 0 6 case no yes 2 6 control no yes 2
7 case yes no 1 7 control no yes 1 8 case yes yes 4 8 control no yes 4
9 case no no 10 9 control yes no 10 10 case no yes 1 10 control yes no 1
11 case yes no 20 11 control yes no 20 12 case yes yes 32 12 control yes no 32
13 case no no 7 13 control yes yes 7 14 case no yes 1 14 control yes yes 1
15 case yes no 29 15 control yes yes 29 16 case yes yes 63 16 control yes yes 63
;
proc print; run;
data new;
set cond;
if status = 'control' then delete;
r=count;
drop  driver res count;
output; run;
data new1;
set cond;
if status = 'case' then delete;
y1=x1;y2=x2;y3=x3;n=count;
drop x1-x3 driver res count;
output; run;
data comb;
merge new new1;
by idd;
z1=x1-y1; z2=x2-y2; z3=x3-y3;
case=0; run;
proc print data=comb; run;
proc logistic data=comb;
weight n;
model case=z1-z2/noint details; run;
```

Obs	dd	status	x1	x2	x3	r	y1	y2	y3	n	z1	z2	z3	case
1	1	case	0	0	0	9	0	0	0	9	0	0	0	0
2	2	case	0	1	0	2	0	0	0	2	0	1	0	0
3	3	case	1	0	0	14	0	0	0	14	1	0	0	0
4	4	case	1	1	1	22	0	0	0	22	1	1	1	0
5	5	case	0	0	0	0	0	1	0	0	0	-1	0	0
6	6	case	0	1	0	2	0	1	0	2	0	0	0	0
7	7	case	1	0	0	1	0	1	0	1	1	-1	0	0
8	8	case	1	1	1	4	0	1	0	4	1	0	1	0
9	9	case	0	0	0	10	1	0	0	10	-1	0	0	0
10	10	case	0	1	0	1	1	0	0	1	-1	1	0	0
11	11	case	1	0	0	20	1	0	0	20	0	0	0	0
12	12	case	1	1	1	32	1	0	0	32	0	1	1	0
13	13	case	0	0	0	7	1	1	1	7	-1	-1	-1	0
14	14	case	0	1	0	1	1	1	1	1	-1	0	-1	0
15	15	case	1	0	0	29	1	1	1	29	0	-1	-1	0
16	16	case	1	1	1	63	1	1	1	63	0	0	0	0

```
                    Response Profile

       Ordered                     Total             Total
```

```
              Value          case      Frequency              Weight
            ----------------------------------------------------
                  1             0             15           217.00000

NOTE: 1 observation having zero frequency or weight was excluded since it does
      not contribute to the analysis.

                        Testing Global Null Hypothesis: BETA=0
                Test                   Chi-Square       DF      Pr > ChiSq
                --------------------------------------------------------
                Likelihood Ratio           9.5456        2          0.0085
                Score                      9.3130        2          0.0095
                Wald                       8.8484        2          0.0120

                       Analysis of Maximum Likelihood Estimates
                                          Standard
          Parameter    DF     Estimate       Error     Chi-Square     Pr > ChiSq
          -----------------------------------------------------------------
          z1            1       0.6576      0.2940         5.0043         0.0253
          z2            1       0.2554      0.2258         1.2792         0.2581

                                  Odds Ratio Estimates
                                    Point          95% Wald
                       Effect     Estimate       Confidence Limits
                       ---------------------------------------------
                       z1            1.930        1.085        3.434
                       z2            1.291        0.829        2.010

NOTE: Since there is only one response level, measures of association between
      the observed and predicted values were not calculated.
```

The results of some of the models applied to the data in this example are presented in Table 8.14, while Table 8.15 gives the conditional or partial tests based on the results displayed in Table 8.14.

Model	d.f.	G^2
z_1	14	292.57
z_2	14	296.54
$z_1 + z_2$	13	291.28
$z_1 + z_2 + z_3$	12	291.20

Table 8.14: Results of models applied to the data

Source of variation	d.f.	G^2	Comments
$z_3 \mid (z_1, z_2)$	1	0.08	Interaction adjusted for driving and residence
$z_2 \mid z_1$	1	1.29	Residence, adjusted for driving
$z_1 \mid z_2$	1	5.26	Driving, adjusted for residence

Table 8.15: Analysis of G^2 based on the models above

From the model fitted involving z_1 and z_2, the approximate relative risk of herniated disc in a driver ($x_1 = 1$) relative to a nondriver ($x_1 = 0$), after adjusting for the place of residence is $\exp(0.6576) = 1.93$. Similarly, the approximate relative risk of herniated disc occurring in a suburban resident ($x_2 = 1$) relative to a nonresident ($x_2 = 0$), adjusted for driving is $\exp(0.2554) = 1.291$.

Based on the above results therefore, we may conclude that the risk of herniated disc occurring in a driver is about twice that of a nondriver, but the risk is not affected by whether or not they are suburban or city resident.

8.9.5 Alternative Analysis of the Data in Example 8.8

Conditional logistic modeling can also be implemented by using the **Cox's proportional hazard** model, which utilizes PROC PHREG in SAS software. However, in order to do this, we must create subject specific covariates and dependent variable. Consequently, we read the data in for each individual giving a total of 434 obervations (217 pairs). The first and last 20 observations are reproduced below from implementation of the SAS software data step in the program below.

```
data cond;
infile 'c:\classdata\cl8\condiii.txt';
input idd status $ driver $ res $ count @@;
if driver='yes' then x1=1;
else x1=0;
if res ='yes' then x2=1;
else x2=0;
x3=x1*x2;
if status='case' then event=1;
else event=0;
drop count;
proc sort;
by idd;
run;
data new;
set cond;
event=2-event;
run;
proc print;
run;
proc phreg nosummary;
model event=x1 x2/ties=discrete;
strata idd;
run;
```

```
Obs    idd    status     driver    res    x1    x2    x3    event
-----------------------------------------------------------------
  1      1    case         no       no     0     0     0      1
  2      1    control      no       no     0     0     0      2
  3      2    case         no       no     0     0     0      1
  4      2    control      no       no     0     0     0      2
  5      3    case         no       no     0     0     0      1
  6      3    control      no       no     0     0     0      2
  7      4    case         no       no     0     0     0      1
  8      4    control      no       no     0     0     0      2
  9      5    case         no       no     0     0     0      1
 10      5    control      no       no     0     0     0      2
 11      6    case         no       no     0     0     0      1
 12      6    control      no       no     0     0     0      2
 13      7    case         no       no     0     0     0      1
 14      7    control      no       no     0     0     0      2
 15      8    case         no       no     0     0     0      1
 16      8    control      no       no     0     0     0      2
 17      9    case         no       no     0     0     0      1
 18      9    control      no       no     0     0     0      2
 19     10    case         no      yes     0     1     0      1
 20     10    control      no       no     0     0     0      2
..............................................................
..............................................................
..............................................................
413    207    case        yes      yes     1     1     1      1
414    207    control     yes      yes     1     1     1      2
415    208    case        yes      yes     1     1     1      1
```

```
416   208   control   yes   yes   1   1   1   2
417   209   case      yes   yes   1   1   1   1
418   209   control   yes   yes   1   1   1   2
419   210   case      yes   yes   1   1   1   1
420   210   control   yes   yes   1   1   1   2
421   211   case      yes   yes   1   1   1   1
422   211   control   yes   yes   1   1   1   2
423   212   case      yes   yes   1   1   1   1
424   212   control   yes   yes   1   1   1   2
425   213   case      yes   yes   1   1   1   1
426   213   control   yes   yes   1   1   1   2
427   214   case      yes   yes   1   1   1   1
428   214   control   yes   yes   1   1   1   2
429   215   case      yes   yes   1   1   1   1
430   215   control   yes   yes   1   1   1   2
431   216   case      yes   yes   1   1   1   1
432   216   control   yes   yes   1   1   1   2
433   217   case      yes   yes   1   1   1   1
434   217   control   yes   yes   1   1   1   2
```

The **DISCRETE** option is necessary because the data consist of matched pairs with each pair containing a 1 or a 0 on the dependent variable. For either $1 : m$ or $n : m$ matching designs, the DISCRETE option is essential. In the current setting, every matched pair has both of the two values of the dependent variable. In treatment-control matching, there are several ways of carrying out the analysis, but in case-control matching designs, the options are limited (e.g., conditional logit and Cox's approaches). The generalized estimating equations (GEE) is usually suitable for the treatment-control matching.

The event is recoded so that the 0's (the controls) become 2's. This ensures that the model variable is the probability of being a case. The result of implementing the above program is displayed in the partial output below.

```
The PHREG Procedure

        Model Information

Data Set                WORK.NEW
Dependent Variable      event
Ties Handling           DISCRETE

         Model Fit Statistics

              Without          With
Criterion    Covariates     Covariates
--------------------------------------
-2 LOG L        300.826        291.280
AIC             300.826        295.280
SBC             300.826        303.426

        Testing Global Null Hypothesis: BETA=0

Test                  Chi-Square      DF     Pr > ChiSq
-------------------------------------------------------
Likelihood Ratio          9.5456       2         0.0085
Score                     9.3130       2         0.0095
Wald                      8.8484       2         0.0120

                  Analysis of Maximum Likelihood Estimates

                 Parameter     Standard                                Hazard
Variable   DF     Estimate        Error   Chi-Square   Pr > ChiSq       Ratio
-----------------------------------------------------------------------------
x1          1      0.65761      0.29396       5.0043       0.0253       1.930
x2          1      0.25542      0.22583       1.2792       0.2581       1.291
```

We observe that the results presented from this alternate approach are identical with those obtained from using the differences of the variables for conditional logistic approach. Both procedures have their pros and cons. The PHREG approach requires data to be subject specific. This does not pose any problem if our data come in this form (as is usual), and it further has the advantage that no transformations to the z's are necessary.

The conditional logistic approach is also simple to use if the data are subject specific, but one would need to output the cases as well as the controls to take differences. It however has the advantage that it could easily be adapted to random or mixed effect models in which the centers constitute a random sample from a larger number of centers, i.e., the case in which the nuisance parameters α_i are random. In such cases, the SAS® macro **GLIMMIX**, which is available on the SAS® site will be appropriate.

8.10 A Five-Factor Response Example

The data for this example came from example 7.2, the Danish Welfare Study data previously analyzed in chapter 7. (see page 270. The data relate to the response to the question of whether there was a freezer in the household or not among subjects in the 1976 Danish Welfare Study (from Andersen, 1997).
The data are a five-way $2 \times 2 \times 3 \times 2 \times 2$ contingency table, with variables A sex, B age, C family taxable income, D employment sector, and E whether there is a freezer in the household. The age and income variables are defined as:

$$\text{Age} = \begin{cases} \text{Old} & \text{if} > 40 \\ \text{Young} & \text{if} \leq 40 \end{cases} \qquad \text{Income} = \begin{cases} \text{Low} & \text{if} < 60,000 \ \text{ D.kr} \\ \text{Medium} & \text{if } 60,000 - 100,000 \ \text{ D.kr} \\ \text{High} & \text{if} > 100,000 \ \text{ D.kr} \end{cases}$$

We wish to reanalyze this data by employing the subset selection strategies (the forward, backward, and stepwise) capability in PROC LOGISTIC. These strategies are discussed in turn below.

8.10.1 Model Based on the Forward Selection Procedure

This model is implemented in SAS software with the following program together with a partial output.

```
set tab77;
proc logistic;
class a (ref=last) b (ref=last)
      c (ref=last) d (ref=last)/param=ref;
weight count;
model e=a|b|c|d/scale=none aggregate=(a b c d)
        selection=forward details;
run;
```

```
Step  0. Intercept entered:
                                Analysis of Effects Not in the Model

                                                Score
                             Effect       DF    Chi-Square    Pr > ChiSq
                             -----------------------------------------
                             a             1       0.0341        0.8536
                             b             1       0.4058        0.5241
```

```
                     c              2        105.1919          <.0001
                     d              1          0.0655          0.7980
Step  1. Effect c entered:
                              Residual Chi-Square Test

                         Chi-Square        DF     Pr > ChiSq
                         ---------------------------------
                            23.4781        21         0.3190

                        Analysis of Effects Not in the Model

                                                Score
                     Effect        DF     Chi-Square    Pr > ChiSq
                     ------------------------------------------
                     a              1         0.0232        0.8791
                     b              1         0.0369        0.8476
                     d              1         0.3582        0.5495
NOTE: No (additional) effects met the 0.05 significance level for entry into the
      model.
                 Deviance and Pearson Goodness-of-Fit Statistics

          Criterion           DF           Value       Value/DF      Pr > ChiSq
          ---------------------------------------------------------------
          Deviance            21         23.4249         1.1155          0.3218
          Pearson             21         23.4781         1.1180          0.3190

                      Analysis of Maximum Likelihood Estimates

                                                 Standard          Wald
     Parameter              DF      Estimate        Error     Chi-Square      Pr > ChiSq
     -------------------------------------------------------------------------------
     Intercept               1        0.5209       0.0820        40.3436          <.0001
     c             1         1        1.0960       0.1111        97.2486          <.0001
     c             2         1        0.7865       0.1177        44.6221          <.0001

                               Odds Ratio Estimates

                                 Point            95% Wald
                    Effect     Estimate      Confidence Limits
                    ------------------------------------------
                    c 1 vs 3      2.992        2.407      3.720
                    c 2 vs 3      2.196        1.743      2.766
```

For the forward selection model, step 0 enters the intercept into the model. The analysis of the effect in the model indicates that variable C is the only one with a pvalue less than 0.05 (the cutoff point for inclusion of a variable from the model). Hence, at step 1, variable C was admitted into the model. The resulting analysis of effects not in the model indicates that no other variable meets the criterion for inclusion in the model. The model involving only variable C is therefore selected. The model has a $G^2 = 23.4781$ on 21 d.f. The model fits the data. We next consider the model selection strategy based on the backward procedure.

8.10.2 Model Based on the Backward Selection Procedure

This model is implemented in SAS software with the following program together with a partial output.

```
data new;
set tab77; proc logistic;
class a (ref=last) b (ref=last)
      c (ref=last) d (ref=last)/param=ref;
weight count;
model e=a|b|c|d/scale=none aggregate=(a b c d)
        selection=backward details; run;
```

```
Step  0. The following effects were entered:
Intercept  a  b  a*b  c  a*c  b*c  a*b*c  d  a*d  b*d  a*b*d  c*d  a*c*d  b*c*d a*b*c*d

Step  1. Effect a*b*c*d is removed:
Step  2. Effect a*b*c is removed:
Step  3. Effect b*c*d is removed:
Step  4. Effect a*c*d is removed:
Step  5. Effect c*d is removed:
Step  6. Effect b*c is removed:
Step  7. Effect a*b*d is removed:
Step  8. Effect a*b is removed:
Step  9. Effect a*d is removed:
```

Residual Chi-Square Test

Chi-Square	DF	Pr > ChiSq
11.7796	15	0.6956

NOTE: No (additional) effects met the 0.05 significance level for removal from the model.

Summary of Backward Elimination

Step	Effect Removed	DF	Number In	Wald Chi-Square	Pr > ChiSq
1	a*b*c*d	2	14	3.9139	0.1413
2	a*b*c	2	13	0.0619	0.9695
3	b*c*d	2	12	0.3052	0.8585
4	a*c*d	2	11	0.4001	0.8187
5	c*d	2	10	1.0766	0.5837
6	b*c	2	9	2.3672	0.3062
7	a*b*d	1	8	2.9163	0.0877
8	a*b	1	7	0.3077	0.5791
9	a*d	1	6	0.4656	0.4950

Deviance and Pearson Goodness-of-Fit Statistics

Criterion	DF	Value	Value/DF	Pr > ChiSq
Deviance	15	11.8149	0.7877	0.6930
Pearson	15	11.7796	0.7853	0.6956

Number of unique profiles: 24

Type III Analysis of Effects

Effect	DF	Wald Chi-Square	Pr > ChiSq
a	1	4.6557	0.0310
b	1	2.5730	0.1087
c	2	74.3851	<.0001
a*c	2	7.1950	0.0274
d	1	2.9882	0.0839
b*d	1	4.4481	0.0349

Analysis of Maximum Likelihood Estimates

Parameter		DF	Estimate	Standard Error	Wald Chi-Square	Pr > ChiSq
Intercept		1	0.1749	0.1521	1.3228	0.2501
a	1	1	0.3619	0.1677	4.6557	0.0310
b	1	1	0.2607	0.1625	2.5730	0.1087
c	1	1	1.4396	0.1707	71.1437	<.0001
c	2	1	1.0335	0.1853	31.1068	<.0001
a*c	1 1	1	-0.5933	0.2259	6.8963	0.0086
a*c	1 2	1	-0.4325	0.2411	3.2180	0.0728
d	1	1	0.2237	0.1294	2.9882	0.0839
b*d	1 1	1	-0.4213	0.1997	4.4481	0.0349

The analysis begins at step 0 with the inclusion of all effects in the model. The criterion for the removal of a variable in the model is that its type3 pvalue must be greater than 0.05 and for two effects with competing pvalues, the strategy would be to remove the higher order effect first. The backward selection procedure requires 10 steps (0-9) to obtain the best subset variables for this data. The effects removed at each stage or step are summarized in the Summary of Backward Elimination in the output above. For instance, at step 1, the $a*b*c*d$ effect was removed because its pvalue 0.1413 is greater than 0.05. At each stage, type3 analysis was carried out (not included in the output above) and effects were removed according to their type3 pvalues being greater than the cutoff point of 0.05. The final model is the logit model {AC,BD} which is equivalent to the log-linear model {ABCD,ACE,BDE}. The model is based on 15 d.f. with a G^2 value of 11.8149.

8.10.3 Model Based on the Stepwise Selection Procedure

This model is implemented in SAS software with the following program together with a partial output.

```
data new;
set tab77;
proc logistic;
class a (ref=last) b (ref=last)
      c (ref=last) d (ref=last)/param=ref;
weight count;
model e=c|b|c|d/scale=none aggregate=(a b c d)
        selection=stepwise; run;
```

```
Step  0. Intercept entered:
                              Analysis of Effects Not in the Model

                                                   Score
                          Effect        DF    Chi-Square     Pr > ChiSq
                          ---------------------------------------------
                          a              1        0.0341         0.8536
                          b              1        0.4058         0.5241
                          c              2      105.1919         <.0001
                          d              1        0.0655         0.7980

Step  1. Effect c entered:
                              Analysis of Effects Not in the Model

                                                   Score
                          Effect        DF    Chi-Square     Pr > ChiSq
                          ---------------------------------------------
                          a              1        0.0232         0.8791
                          b              1        0.0369         0.8476
                          d              1        0.3582         0.5495

NOTE: No (additional) effects met the 0.05 significance level for entry into the
      model.
                                   Summary of Stepwise Selection

                  Effect                          Number        Score          Wald
    Step    Entered    Removed       DF              In   Chi-Square    Chi-Square    Pr > ChiSq
    ---------------------------------------------------------------------------------------------
       1    c                         2               1     105.1919             .        <.0001

                       Deviance and Pearson Goodness-of-Fit Statistics

                Criterion           DF           Value      Value/DF      Pr > ChiSq
                -------------------------------------------------------------------
                Deviance            21         23.4249        1.1155          0.3218
                Pearson             21         23.4781        1.1180          0.3190
```

```
                 Number of unique profiles: 24

            Analysis of Maximum Likelihood Estimates

                                          Standard          Wald
Parameter              DF     Estimate       Error    Chi-Square    Pr > ChiSq
-----------------------------------------------------------------------------
Intercept               1       0.5209      0.0820       40.3436        <.0001
c          1            1       1.0960      0.1111       97.2486        <.0001
c          2            1       0.7865      0.1177       44.6221        <.0001

                       Odds Ratio Estimates

                             Point          95% Wald
              Effect      Estimate      Confidence Limits
              -------------------------------------------
              c 1 vs 3       2.992       2.407       3.720
              c 2 vs 3       2.196       1.743       2.766
```

Again for the stepwise approach, at step 0, the intercept is introduced into the model. At step 1, inclusion in the model is based on pvalue less than 0.05. The type3 analysis shows that effect C is the prime candidate at this stage. Effect C is therefore introduced at this stage. The type 3 analysis of the effects not in the model at this stage indicates that no other effect meets the entry criterion of 0.05. The entry criterion can be changed in SAS software as **SLE=value**. Value in this case, is the default value, 0.05. Similarly, to stay or remain in the model, the stay value can similarly be changed with the option **SLSTAY=value**. Again, here the default is 0.05. The model selected is the logit model {C}, which is equivalent to the log-linear model {ABCD,CE}. This is the same model selected by the forward selection procedure.

One obvious disadvantage of the forward and stepwise selection procedures is the lack of consideration for entry of two-factor, three-factor, or four-factor effects. We can change this by forcing the procedures to include, say, the first n of the $s = 2^r - 1$ effect terms, where r is the number of factor variables. In our case, $s = 2^4 - 1 = 15$. We present below a stepwise procedure with a starting value of 10. The model selected with this option corresponds to that selected by the backward selection procedure, namely, the logit model {AC,BD}. The partial output for the implementation is presented below.

```
proc logistic;
class a (ref=last) b (ref=last)
      c (ref=last) d (ref=last)/param=ref;
weight count;
model e=a|b|c|d/scale=none aggregate=(a b c d)
       selection=stepwise details start=10;
run;

Step  0. The following effects were entered:

Intercept  a  b  a*b  c  a*c  b*c  a*b*c  d  a*d  b*d

Step  1. Effect a*b*c is removed:
Step  2. Effect a*b is removed:
Step  3. Effect a*d is removed:
Step  4. Effect b*c is removed:

NOTE: No (additional) effects met the 0.05 significance level for entry into the
      model.

                     Summary of Stepwise Selection

            Effect                      Number        Score         Wald
```

Step	Entered	Removed	DF	In	Chi-Square	Chi-Square	Pr > ChiSq
1		a*b*c	2	9	.	0.1773	0.9151
2		a*b	1	8	.	0.3720	0.5419
3		a*d	1	7	.	0.3882	0.5332
4		b*c	2	6	.	2.7224	0.2563

Deviance and Pearson Goodness-of-Fit Statistics

Criterion	DF	Value	Value/DF	Pr > ChiSq
Deviance	15	11.8149	0.7877	0.6930
Pearson	15	11.7796	0.7853	0.6956

Type III Analysis of Effects

Effect	DF	Wald Chi-Square	Pr > ChiSq
a	1	4.6557	0.0310
b	1	2.5730	0.1087
c	2	74.3851	<.0001
a*c	2	7.1950	0.0274
d	1	2.9882	0.0839
b*d	1	4.4481	0.0349

8.10.4 Interpretations of Results

Based on the above analyses, the most parsimonious model selected is the logit model {AC,BD}, which is equivalent to the log-linear model {ABCD,ACE,BDE}. The model has a $G^2 = 11.8149$ on 15 d.f., and the ACE parameters under the logit model are presented below.

Parameter		DF	Estimate	Standard Error	Wald Chi-Square	Pr > ChiSq
a*c	1 1	1	-0.5933	0.2259	6.8963	0.0086
a*c	1 2	1	-0.4325	0.2411	3.2180	0.0728
b*d	1 1	1	-0.4213	0.1997	4.4481	0.0349

The ACE interaction terms indicates that given the response E_p, sex A_i and family income C_k of respondents are conditionally independent of age and employment sector. Thus if we keep sex constant we can compute and interpret relevant log odds ratios pertaining to family income and freezer status. But first, we form below the three-way ACE table as in Table 8.16 and we use this to obtain the estimated log of the odds ratios. These estimates are displayed in Table 8.17.

We observe that all the six estimates in Table 8.17 are positive. The overall pattern is that for the same values of k and k', females are more likely to have a freezer in the household. For females, the odds of having a freezer relative to not having one is $e^{1.419} = 4.1$ times higher for respondents with lower family income than those with high family income. For men, this value is 2.33 times higher. Similarly for females, the odds is $e^{1.029} = 2.80$ times higher for those on medium income than those with high family income. Again the corresponding odds for men is 1.82.

For the BDE interaction effect, we again present in Table 8.18 the BDE observed. The estimates of log odds ratios $\hat{\tau}^{DE.B}_{(ll')(pp').j}$ are -0.241 and 0.126 for $j = 1$ and $j = 2$, respectively. Thus for older respondents age > 40, the odds are 11% lower of owning a freezer for those in the private sector, relative to those in the public

Sex	Income	Freezer status	
A	C	Yes	No
i	k	$p = 1$	$p = 2$
1	1	594	129
	2	407	113
	3	239	121
2	1	479	84
	2	251	65
	3	160	116

Table 8.16: Observed ACE marginal table with A fixed at 1 and 2

		Family income (C)				Freezer status				
Sex	i	C	k	C	k'	Yes	p	No	p'	$\hat{\tau}^{CE.A}_{(kk')(pp').i}$
Male	1	Low	1	Medium	2	Yes	1	No	2	0.246
		Low	1	High	3	Yes	1	No	2	0.846
		Medium	2	High	3	Yes	1	No	2	0.601
Female	2	Low	1	Medium	2	Yes	1	No	2	0.390
		Low	1	High	3	Yes	1	No	2	1.419
		Medium	2	High	3	Yes	1	No	2	1.029

Table 8.17: Estimated log odds ratios for the interaction of family income and freezer status, given sex

sector. However, for the younger respondents under 40 years old, the corresponding odds are 13% higher for those in the private sector relative to those in the public sector of having a freezer.

8.11 Exercises

1. The data in the table below are reported in Woodward et al., (1941) and are reproduced from Christensen (1990). The data examine the relationship between exposure to chloroacetic acid and the death of mice. Ten mice were exposed at each dose level and the doses are measured in grams per kilogram of body weight.

Age B	Employment D	Freezer status Yes	No
j	l	$p = 1$	$p = 2$
1	1	533	177
	2	322	84
2	1	856	236
	2	419	131

Table 8.18: Observed BDE marginal table with B fixed at 1 and 2

Dose	Dead	Exposed
0.0794	1	10
0.1000	2	10
0.1259	1	10
0.1413	0	10
0.1500	1	10
0.1588	2	10
0.1778	4	10
0.1995	6	10
0.2239	4	10
0.2512	5	10
0.2818	5	10
0.3162	8	10

Fit the logistic regression model to the data and estimate the LD_{50}, LD_{90}, and LD_{999}. Discuss the possible danger of extrapolation to LD_{999}. Determine how well the model fits the data.

2. For the data in problem 1 above, fit the probit, complimentary log-log, and the extreme value and probability models. What can you say about your fits? Calculate the residuals and test whether there is lack of fit. (Note: A probability model has $\pi(x) = \beta_0 + \beta x$, where $\pi(x)$ is the proportion and is fitted by invoking the identity link.) Also obtain estimated proportions $\hat{\pi}(x)$ for all the models.

3. Show that the logistic distribution function:

$$\pi(x) = \frac{\exp(\beta_0 + \beta_1 x)}{1 + \exp(\beta_0 + \beta_1 x)}$$

has the steepest slope when $\pi(x) = \frac{1}{2}$. By rewriting $\pi(x)$ as a linear model, show that the LD_{50} is obtained as $-\frac{\beta_0}{\beta_1}$.

4. Find the LD_{50} for the complimentary log-log function.

5. The data in Table 8.19 compare male lung cancer patients with control patients having other diseases, according to the average number of cigarettes smoked daily over a period of 10-year period preceding the onset of lung cancer. The data are from a retrospective study of the incidence of lung cancer

and tobacco smoking among patients in hospitals in English cities (Agresti, 1990).

Daily average no. of cigarettes	Disease group Lung cancer patients	Control patients
0	7	61
3	55	129
9.5	489	570
19.5	475	431
37	293	154
55	38	12

Table 8.19: Incidence of cancer and smoking on patients

Fit a logit model to the data.

6. An antihistaminic drug was used at various doses to protect test animals against a certain lethal dose of histamine, with the results given below.

Dose μg/kg	Alive/ total
1000	8/8
500	7/8
250	4/8
125	4/8
62.5	1/8

Fit the logistic and probit models to the data above and compute the LD_{50} in each case. Comment about your models.

7. The proportion of individuals in a certain city in Zaire, with malaria antibodies present was recorded (Morgan, 1992) as in the following table:

Mean age	Sero-positive	Total	Mean age	Sero-positive	Total
1.0	2	60	22.0	20	84
2.0	3	63	27.5	19	77
3.0	3	53	32.0	19	58
4.0	3	48	36.8	24	75
5.0	1	31	41.6	7	30
7.3	18	182	49.7	25	62
11.9	14	140	60.8	44	74
17.1	20	138			

(a) Fit a logistic model to the above data.

(b) Repeat the analysis with a probit model.

(c) Which of the models fit the data better?

(d) With your chosen model, predict the expected probability for an individual with a mean age of 25 years.

(e) Find an ED_{90} from your chosen model and interpret it.

8. Two anticonvulsant drugs were compared by administering them to mice, which were then given electric shock under conditions that caused all control mice to convulse. The results of the experiment are displayed in the table below (Goldstein, 1965).

Drug A	
Dose mg/kg	Convulsed/ total
10	13/15
30	9/15
90	4/15

Drug B	
Dose mg/kg	Convulsed/ total
200	12/15
600	6/15
1800	2/15

(a) Fit separate regression lines for both drugs and hence obtain an estimate of the relative potency from estimates of their LD_{50}'s.

(b) Fit a combined regression line and test for equality slopes. Test whether there is dosage and/or drug effects. Summarize your conclusions.

9. The data below are reproduced from Collett (1991) and were originally reported by Strand (1930). The experiment was to assess the response of the confused flour beetle, Tribolium confusum, to gaseous carbon disulphide (CS_2). Prescribed volumes of liquid carbon disulphide were added to flasks in which a tubular cloth cage containing a batch of about thirty beetles was suspended. Duplicate batches of beetles were used for each concentration of CS_2. At the end of 5-hour period, the proportion killed was recorded and the actual concentration of gaseous CS_2 in the flask, measured in mg/liter, was determined by a volumetric analysis. The mortality data are presented below.

Number of beetles killed, y, out of n exposed to concentrations of CS_2

Concentration	Replicate I		Replicate II	
of CS_2	y	n	y	n
49.06	2	29	4	30
52.99	7	30	6	30
56.91	9	28	9	34
60.84	14	27	14	29
64.76	23	30	29	33
68.69	29	31	24	28
72.61	29	30	32	32
76.54	29	29	31	31

(a) Fit separate logistic models to the data in each replicate. How well do these models fit the data?

(b) Now combine the data and fit a third-degree polynomial logistic regression to the model. Discuss your findings and hence give the most parsimonious model for the combined data. Estimate both the LD_{50} and the LD_{90} under your assumed model and interpret.

10. Repeat the above analysis using the probit models.

11. The data in Table 8.20 from Breslow and Day (1980,) relate to the occurrence of esophageal cancer in Frenchmen. Potential risk factors related to the occurrence are age and alcohol consumption where any comsumption of wine more than one liter a day is considered high.

Age group	Alcohol consumption	Cancer Yes	Cancer No
25-34	High	1	9
	Low	0	106
35-44	High	4	26
	Low	5	164
45-54	High	25	29
	Low	21	138
55-64	High	42	27
	Low	34	139
65-74	High	19	18
	Low	36	88
75+	High	5	0
	Low	8	31

Table 8.20: Occurrence of esophageal cancer

(a) Fit a logistic model with the explanatory variables age, and alcohol consumption by first considering age as categorical variable and as a continuous variable.

(b) Consider fitting the interaction term in both situations above. Use the stepwise regression procedure to fins the most parsimonious model. Interpret your results.

12. For the data in the table below (reproduced from Collett, 1991):)

Mortality of the tobacco bud worm 72 hours after exposure to cypermethrin

(Holloway, 1989)

Gender of moth	Dosage of cypermethrin	Number affected out of 20
Male	1.0	1
	2.0	4
	4.0	9
	8.0	13
	16.0	18
	32.0	20
Female	1.0	0
	2.0	2
	4.0	6
	8.0	10
	16.0	12
	32.0	16

(a) Fit separate lines to male and female data.

(b) Fit parallel lines to both gender data.

(c) Fit a common line to both data and discuss your results in each case.

(d) Also fit a logistic model with gender and dose as the explanatory variable (you may also consider including their interaction term in the model). Again, discuss your results.

(e) Consider the fact that the dosage level are continuous and fit a logistic model with gender as an explanatory variable and including terms with powers of dosage level included (e.g., D^5)

13. Refer to problem 6.10:

(a) Fit a log-linear model that explains following politics regularly in terms of nationality and education level. Hint: We assume here that FP is a response variable.

(b) Repeat (a) but this time ignore the fact that FP is a response variable. That is, treat all variables as factor variables.

(c) Compare your results.

14. The following data from Finney (1941) and Pregibon (1981) relate to the occurrence of vasoconstriction in the skin of the fingers as a function of the rate and volume of air breathed. In times of stress, vasoconstriction restricts blood flow to the extremities (such as fingers and toes), forcing blood to the central vital organs. The data are reproduced below. A constriction value of 1 indicates that constriction occurred.

Constriction	Volume	Rate	Constriction	Volume	Rate
1	0.825	3.7	0	2.0	0.4
1	1.09	3.5	0	1.36	0.95
1	2.5	1.25	0	1.35	1.35
1	1.5	0.75	0	1.36	1.5
1	3.2	0.8	1	1.78	1.6
1	3.5	0.7	0	1.5	0.6
0	0.75	0.6	1	1.5	1.8
0	1.7	1.1	0	1.9	0.95
0	0.75	0.9	1	0.95	1.9
0	0.45	0.9	0	0.4	1.6
0	0.57	0.8	1	0.75	2.7
0	2.75	0.55	0	0.03	2.35
0	3.0	0.6	0	1.83	1.1
1	2.33	1.4	1	2.2	1.1
1	3.75	0.75	1	2.0	1.2
1	1.64	2.3	1	3.33	0.8
1	1.6	3.2	0	1.9	0.95
1	1.415	0.85	0	1.9	0.75
0	1.06	1.7	1	1.625	1.3
1	1.8	1.8			

(a) Use SAS software to plot the graph (a half page) of rate against volume for both constriction values on the same page. Are there any ouliers or influential observations?

(b) Fit a logistic regression to the data with rate and volume as explanatory variables. Discuss your results.

(c) Pregibon (1981) fits a logit response model having logs of volume and rate as explanatory variables. How does this model compares with your model in (b)? Pregibon suggests that the rate for observation 32 should be 0.3 rather than 0.03 as it appears in the table above. Is there any evidence for this suggestion? Identify any lack of fit by carrying out the necessary diagnostics procedures.

15. The Data below relate to a sample of patients with coronary heart disease (CHD) and a "normal" sample free of CHD (Lunneborg, 1994). A 1 indicates the patient has no CHD, while a 2 indicates that patient has CHD. Three risk factors are being evaluated. The risk factors are systolic blood pressure (SBP), blood-cholesterol level (Chol), and age of the patients.

group	sbp	chol	age
1	135	227	45
1	122	228	41
1	130	219	49
1	148	245	52
1	146	223	54
1	129	215	47
1	162	245	60
1	160	262	48
1	144	230	44
1	166	255	64
1	138	222	59
1	152	250	51
1	138	264	54

```
1      140      271      56
1      134      220      50
2      145      238      60
2      142      232      64
2      135      225      54
2      149      230      48
2      180      255      43
2      150      240      43
2      161      253      63
2      170      280      63
2      152      271      62
2      164      260      65
-----------------------------
```

If we define the variable Y to be

$$Y = \begin{cases} 1 & \text{if CHD} \\ 0 & \text{if no CHD} \end{cases}$$

fit a parsimonious linear logistics regression model to the data and interpret your results.

16. The data in Table 8.21 are reproduced from Slaton et al., (2000). They came from an experiment that examined the *in utero* damage in laboratory rodents after exposure to boric acid. The experiment uses four levels of boric acid, and records the number of rodents in the litter and the number of dead embryos.

Dose=0.0		Dose=0.1		Dose=0.2		Dose=0.3	
Dead	Litter size	Dead	Litter size	Dead	Litter size	Dead	Litter size
0	15	0	6	1	12	12	12
0	3	1	14	0	12	1	12
1	9	1	12	0	11	0	13
1	12	0	10	0	13	2	8
1	13	2	14	0	12	2	12
2	13	0	12	0	14	4	13
0	16	0	14	4	15	0	13
0	11	3	14	0	14	1	13
1	11	0	10	0	12	0	12
2	8	2	12	1	6	1	9
0	14	3	13	2	13	3	9
0	13	1	11	0	10	0	11
3	14	1	11	1	14	1	14
1	13	0	11	1	12	0	10
0	8	0	13	0	10	3	12
0	13	0	10	0	9	2	21
2	14	1	12	1	12	3	10
3	14	0	11	0	13	3	11
0	11	2	10	1	14	1	11
2	12	2	12	0	13	1	11
0	15	2	15	0	14	8	14
0	15	3	12	1	13	0	15
2	14	1	12	2	12	2	13
1	11	0	12	1	14	8	11
1	16	1	12	0	13	4	12
0	12	1	13	0	12	2	12
0	14	1	15	1	7		

Table 8.21: Damage in laboratory rodents after exposure to boric acid

Analyze the above data and discuss your results.

Chapter 9

Logit and Multinomial Response Models

9.1 Introduction

In logistic regression discussed in chapter 8, the emphasis has been on the conditional probabilities of a single response variable given one or several factors. However, in this chapter, we shall be concerned with data with a dependent variable and several other independent or factor variables. Specifically, we shall be interested in categorical type dependent and/or explanatory variables, that is, **ANOVA** type models. This class of log-linear models that utilize the binary nature of the dependent variables is called *logit models.* These logit models are examined in the first few sections of this chapter. We shall examine this methodology to situations where the the response variable has multiple outcomes. Specificically, we shall examine in turn cases when the multiple outcomes are either nominal or ordinal in nature.

However, before we go any further, let us consider the general $2 \times J$ table below.

	B					
A	1	2	$\cdots$	j	$\cdots$	J
1	$\hat{m}_{11}$	$\hat{m}_{12}$	$\cdots$	$\hat{m}_{1j}$	$\cdots$	$\hat{m}_{1J}$
2	$\hat{m}_{21}$	$\hat{m}_{22}$	$\cdots$	$\hat{m}_{2j}$	$\cdots$	$\hat{m}_{2J}$

For the data in the table above, the saturated log-linear model is given by:

$$\ln(\hat{m}_{ij}) = \ell_{ij} = \mu + \lambda_i^A + \lambda_j^B + \lambda_{ij}^{AB}$$

with $i = 1, 2, j = 1, 2, 3, \cdots, J$, and with the usual constraints.

However, since variable A is the dependent variable and is dichotomous (or binary), we can use logits that are given by:

$$\ln\left(\frac{\hat{m}_{1j}}{\hat{m}_{2j}}\right) = \ell_{1j} - \ell_{2j} = 2(\lambda_1^A + \lambda_{1j}^{AB})$$

In particular,

$$\ln\left(\frac{\hat{m}_{11}}{\hat{m}_{21}}\right) = 2(\lambda_1^A + \lambda_{11}^{AB})$$

and similarly,

$$\ln\left(\frac{\hat{m}_{12}}{\hat{m}_{22}}\right) = 2(\lambda_1^A + \lambda_{12}^{AB})$$

Thus, the logits are the functions of the same λ parameters that appear in the general log-linear model.

Let us now extend this further to a three-way $2 \times 2 \times 2$ contingency table of expected values (under some model) below:

		C	
A	**B**	$k = 1$	$k = 2$
$i = 1$	$j = 1$	$\hat{m}_{111}$	$\hat{m}_{112}$
	$j = 2$	$\hat{m}_{121}$	$\hat{m}_{122}$
$i = 2$	$j = 1$	$\hat{m}_{211}$	$\hat{m}_{212}$
	$j = 2$	$\hat{m}_{221}$	$\hat{m}_{222}$

In the above 2^3 table of expected values, let us suppose that A and B are factor variables and C is a binary response or dependent variable.

Let $\pi_{k.ij} = \pi_{k.ij}^{C.AB}$ denote the conditional probability that $C = k$ given that $A = i$ and $B = j$, and let

$$\pi_{k.ij}^{C.AB} = \left(\frac{\hat{m}_{ijk}}{\hat{m}_{ij}^{AB}}\right)$$

Further let us define

$$\omega_{ij} = \ln\left(\frac{\hat{m}_{ij1}}{\hat{m}_{ij2}}\right) \tag{9.1}$$

to be the log-odds (i.e., logit) that C is 1 rather than 2 given that $A = i$ and $B = j$, and since $\pi_{1.ij} + \pi_{2.ij} = 1$, we have

$$\omega_{ij} = \ln\left(\frac{p_{1.ij}}{1 - p_{1.ij}}\right)$$

where the $p's$ are the corresponding observed probabilities.

The three variable saturated log-linear model for the table is given by:

$$\ln(\hat{m}_{ijk}) = \mu + \lambda_i^A + \lambda_j^B + \lambda_k^C + \lambda_{ij}^{AB} + \lambda_{ik}^{AC} + \lambda_{jk}^{BC} + \lambda_{ijk}^{ABC} \tag{9.2}$$

But from (9.1)

$$\omega_{ij} = \ln(\hat{m}_{ij1}) - \ln(\hat{m}_{ij2}) \tag{9.3}$$

Hence, substituting (9.3) in (9.2), we have

$$\omega_{ij} = 2(\lambda_1^C + \lambda_{i1}^{AC} + \lambda_{j1}^{BC} + \lambda_{ij1}^{ABC})$$

which can succinctly be written in the *logit* model form:

$$\omega_{ij} = \eta + \eta_i^A + \eta_j^B + \eta_{ij}^{AB} \tag{9.4}$$

with $\eta_1^A + \eta_2^A = 0$ etc. and where

$$\eta = 2\lambda_1^C \qquad \eta_i^A = 2\lambda_{i1}^{AC}$$
$$\eta_j^B = 2\lambda_{j1}^{BC} \quad \text{and} \quad \eta_{ij}^{AB} = 2\lambda_{ij1}^{ABC}$$

We can now give in Table 9.1 equivalent logit models for the corresponding log-linear model for a three-way contingency table.

For any given logit model, the corresponding log-linear equivalent model can be obtained by noting the followings:

Model	Logit model	Log-linear
1	$\eta + \eta_i^A + \eta_j^B + \eta_{ij}^{AB}$	ABC
2	$\eta + \eta_i^A + \eta_j^B$	AB/AC/BC
3	$\eta + \eta_i^A$	AB/AC
4	$\eta + \eta_j^B$	AB/BC
5	η	AB/C

Table 9.1: Equivalence of logit and log-linear models

(a) The log-linear equivalent always includes the two-way interaction between the explanatory or factor variables, that is, AB in this case.

(b) The log-linear equivalent contains the interaction of the response variable C with the effects that are specified in the logit model.

For simplicity, we would like to write logit models succinctly. For example, the saturated model:

$$\eta + \eta_i^A + \eta_j^B + \eta_{ij}^{AB}$$

would be written succinctly as {AB}. The logit model {B} is equivalent to the log-linear model given by {AB,BC} using (a) and (b) above.

If we extend this to a four-way 2^4 contingency table having A, B, and C as factor variables and D as the response variable, we similarly for example have the following logit and log-linear model equivalents.

Models	Logit	Log-linear
1	ABC	ABCD
2	B/C	ABC/BD/CD
3	A/C	ABC/AD/CD
4	A/B	ABC/AD/BD
5	A/B/C	ABC/AD/BD/CD
6	BC/A	ABC/BCD/AD
7	AC/B	ABC/ACD/BD
8	AB/C	ABC/ABD/CD
9	AC/BC	ABC/ACD/BCD
10	AB/BC	ABC/ABD/BCD
11	AB/AC	ABC/ABD/ACD
12	AB/AC/BC	ABC/ABD/ACD/BCD
13	η	ABCD

Table 9.2: Equivalent models for four-way tables

We note here again that by (a) above, each log-linear equivalent models contains the three-way factor interaction ABC. The remaining terms are D multiplied by the other terms in the logit model formulation.

9.1.1 Gun Registration Data Revisited: Example 9.1

In chapter 6, we analyzed the gun registration data presented in that chapter in Table 6.10. There, we have a 2^3 contingency table with a response variable R response to gun registration(favors, opposes), Y year of survey (1975, 1976) and Q the form of questionnaire administered (Q1, Q2). Since R is a response variable, we start by fitting the saturated logit model (YQ), which is equivalent to the log-linear model (RYQ). This can be accomplished by either using SAS® PROC CATMOD or PROC LOGISTIC or PROC GENMOD. Let us use PROC LOGISTIC in this example. Preliminary analysis gives the following SAS software output.

```
data tab521;
INPUT R $ Y $ Q $ COUNT @@;
DATALINES;
OPP 1975 Q1 126 OPP 1975 Q2 141 OPP 1976 Q1 152 OPP 1976 Q2 182
FAV 1975 Q1 319 FAV 1975 Q2 290 FAV 1976 Q1 463 FAV 1976 Q2 403
;
PROC LOGISTIC;
CLASS Y Q; WEIGHT COUNT; MODEL R=Y|Q/SCALE=NONE AGGREGATE; RUN;

         Type III Analysis of Effects
                          Wald
Effect       DF    Chi-Square     Pr > ChiSq
-------------------------------------------
Y             1        1.7466         0.1863
Q             1        7.2378         0.0071
Y*Q           1        0.3220         0.5704
```

We see that only the effect of Q on R is important. We also note here that we are modeling the response category "favors gun registration." Had we wanted to model "oppose gun registration," we would have included the statement **descending** after the PROC LOGISTIC statement. We next therefore fit the logit model (Q), which is equivalent to the log-linear model (YQ,RQ). We use PROCs CATMOD, LOGISTIC and GENMOD to implement this model. The following SAS software program and partial outputs illustrate these fits.

```
proc catmod; population y q;
weight count; model r=q;  run;
                                The CATMOD Procedure

                       Analysis of Maximum Likelihood Estimates
                                              Standard         Chi-
          Effect        Parameter    Estimate      Error       Square    Pr > ChiSq
          ------------------------------------------------------------------------
          Intercept          1         0.8988     0.0485       343.22        <.0001
          Q                  2         0.1354     0.0485         7.79        0.0052

proc logistic; class y q; weight count;
model r=q/scale=none aggregate=(y q); output out=aa p=pred; run;

                 Deviance and Pearson Goodness-of-Fit Statistics

         Criterion           DF           Value      Value/DF      Pr > ChiSq
         ---------------------------------------------------------------------
         Deviance             2          2.0154        1.0077          0.3651
         Pearson              2          2.0228        1.0114          0.3637

                    Analysis of Maximum Likelihood Estimates

                                          Standard
      Parameter          DF    Estimate      Error     Chi-Square    Pr > ChiSq
      -------------------------------------------------------------------------
      Intercept           1      0.8988     0.0485       343.2217        <.0001
      Q            Q1     1      0.1354     0.0485         7.7922        0.0052

                             Odds Ratio Estimates

                                  Point          95% Wald
                Effect         Estimate      Confidence Limits

                Q Q1 vs Q2        1.311        1.084        1.586
```

```
proc genmod; class r y q; model count=y|q r|q/dist=poi link=log type3;
run;
                              Analysis Of Parameter Estimates

                                          Standard  Wald 95% Confidence   Chi-
Parameter              DF  Estimate  Error     Limits                Square Pr > ChiSq
------------------------------------------------------------------------------------
R*Q          FAV    Q1    1   0.2709    0.0970    0.0807     0.4610     7.79   0.0052
R*Q          FAV    Q2    0   0.0000    0.0000    0.0000     0.0000      .
```

The statement population Y Q in CATMOD induces a sampling scheme with the YQ fixed. The same is accomplished in LOGISTIC with the aggregate statement. Because the coding schemes in PROC CATMOD and LOGISTIC are effect coding scheme, the parameter estimate for Q and its corresponding standard error is half of that given by PROC GENMOD and the odds of Q1 against Q2 equal $e^{0.2709} = e^{2(0.1354)} = 1.311$.The logit model (Q) gives a $G^2 = 2.0154$ on 2 d.f. ($p = 0.3651$), a good fit. The parameter estimates under the GENMOD model equal $\hat{\lambda}^{RQ} = 0.2709$, the equivalent parameter estimated in CATMOD is 0.1354. Thus the odds ratio for Q1 versus Q2 (for those who oppose gun registration) is $e^{0.2709} = 1.31$ under the GENMOD model and is equal to $e^{2(.1354)} = 1.31$ under the CATMOD model. Thus those that are administered the form of questionnaire 1 are 1.31 times more likely to respond favorably to gun registration than those that were administered the form of questionnaire 2. This conclusion is consistent with that obtained earlier in chapter 6. The following prediction probabilities (pred) for favoring gun registration also confirms our result regardless of year of survey.

```
  Y        Q          pred
  ------------------------
  1975     Q1        0.7377
  1975     Q2        0.6821
  1976     Q1        0.7377
  1976     Q2        0.6821
```

9.1.2 Example 9.2

The data in Table 9.3 are for a 2^4 contingency table from Demo and Parker (1987). They relate to the effect of academic achievement on self-esteem among black and white college students.

SEX(i)	GPA(j)	RACE(k)	ESTEEM(l) High	ESTEEM(l) Low
Males	High	Black	15	9
		White	17	10
	Low	Black	26	17
		White	22	26
Females	High	Black	13	22
		White	22	32
	Low	Black	24	23
		White	3	17

Table 9.3: A 2^4 table from Demo and Parker (1987)

If we regard the variable "self-esteem" as a response variable, we can employ the logit model formulation to find a model that adequately describes the data. We give below the relevant SAS software program for fitting the saturated logit model to this data, using PROCs CATMOD, LOGISTIC, and GENMOD.

```
data tab82;
do s=1 to 2; do g=1 to 2; do r=1 to 2; do e=1 to 2;
input count @@; output; end; end; end; end;
datalines;
15 9 17 10 26 17 22 26 13 22 22 32 24 23 3 17
;
*** Use CATMOD***;
proc catmod; weight count; population s g r; model e=s|g|r; run;
*** Use PROC LOGISTIC ***;
proc logistic; class s g r; weight count; model e=s|g|r\scale=none aggregate=(s g r); run;
***Use PROC GENMOD ***;
proc genmod; class s g r e; model count= s|g|r|e/dist=poi link=log type3; run;
```

We start modeling the above data by first fitting the saturated logit model to the data using PROC CATMOD and LOGISTIC (version 8.0) and the equivalent log-linear model using PROC GENMOD.

In the use of PROC CATMOD above, we fit the logit saturated model {SGR} to the data, which produces the analysis of variance table below that enables us to determine which effect or effects need to be included in a future reduced model. We present a modified output from PROC CATMOD below.

```
The CATMOD Procedure

     Maximum Likelihood Analysis of Variance
Source                 DF   Chi-Square    Pr > ChiSq
-------------------------------------------------
s                       1        12.69        0.0004 *
r                       1         4.05        0.0443 *
g*r                     1         5.40        0.0201 *
```

Of the effects and interactions {S,G,SG,R,SR,GR,SGR} in the model, only the interaction term GR is significant, with a G^2 value of 5.40 on 1 d.f. (pvalue $= 0.020$). None of the other two-factor interactions SG and SR is significant. The main effect terms S and R are also significant. Thus our reduced model would be the logit model {S,GR}. This model corresponds to the log-linear model {SGR,ES,EGR}. Below are modified SAS software outputs from the saturated models based on ML analysis (or Type 3) of effects and interactions from PROC LOGISTIC and GENMOD. In the GENMOD output, we are only looking for significant interactions involving E (the response variable). Note that we have only reported those that are significant.

```
The LOGISTIC Procedure

        Type III Analysis of Effects
                            Wald
Effect      DF    Chi-Square    Pr > ChiSq
-----------------------------------------
s            1       12.6839       0.0004 *
r            1        4.0442       0.0443 *
g*r          1        5.4027       0.0201 *

The GENMOD Procedure

     LR Statistics For Type 3 Analysis
                                 Chi-
Source              DF         Square     Pr > ChiSq
-------------------------------------------------
s*e                  1          13.82        0.0002 *
r*e                  1           4.22        0.0398 *
g*r*e                1           5.68        0.0171 *
```

Having identified our reduced model, that is the logit model {S,GR} with corresponds to the log-linear model {SGR, ES, EGR}, we present below the SAS software program to implement this model for the three procedures together with the modified outputs again from the three procedures. The reduced model fits the data with $G^2 = 2.5165$ on 3 d.f. (pvalue = 0.4723).

```
set tab82; proc catmod; weight count; population s g r; model e=g|r s; run;
proc logistic; class s g r; weight count; model e=g|r s/scale=none aggregate=(s g r);
run;
proc genmod; class s g r e; model count=s|g|r g|r|e s|e/dist=poi link=log type3; run;
```

The CATMOD Procedure

Maximum Likelihood Analysis of Variance

Source	DF	Chi-Square	Pr > ChiSq
Intercept	1	0.84	0.3600
g	1	1.14	0.2847
r	1	3.41	0.0647
g*r	1	4.84	0.0278
s	1	11.52	0.0007
Likelihood Ratio	3	2.52	0.4723

Analysis of Maximum Likelihood Estimates

Effect	Parameter	Estimate	Standard Error	Chi-Square	Pr > ChiSq
Intercept	1	-0.1109	0.1212	0.84	0.3600
g	2	0.1336	0.1249	1.14	0.2847
r	3	0.2254	0.1220	3.41	0.0647
g*r	4	-0.2722	0.1237	4.84	0.0278
s	5	0.4282	0.1262	11.52	0.0007

The LOGISTIC Procedure

Deviance and Pearson Goodness-of-Fit Statistics

Criterion	DF	Value	Value/DF	Pr > ChiSq
Deviance	3	2.5165	0.8388	0.4723
Pearson	3	2.4482	0.8161	0.4847

Type III Analysis of Effects

Effect	DF	Wald Chi-Square	Pr > ChiSq
g	1	1.1446	0.2847
r	1	3.4120	0.0647
g*r	1	4.8429	0.0278
s	1	11.5162	0.0007

Analysis of Maximum Likelihood Estimates

Parameter		DF	Estimate	Standard Error	Chi-Square	Pr > ChiSq
Intercept		1	-0.1109	0.1212	0.8379	0.3600
g	1	1	0.1336	0.1249	1.1446	0.2847
r	1	1	0.2254	0.1220	3.4120	0.0647
g*r	1 1	1	-0.2722	0.1237	4.8429	0.0278
s	1	1	0.4282	0.1262	11.5162	0.0007

Odds Ratio Estimates

Effect	Point Estimate	95% Wald Confidence Limits	
s 1 vs 2	2.355	1.436	3.861

```
The GENMOD Procedure

          Criteria For Assessing Goodness Of Fit

Criterion                   DF            Value        Value/DF
----------------------------------------------------------------
Deviance                     3           2.5165          0.8388
Pearson Chi-Square           3           2.4482          0.8161

                      Analysis Of Parameter Estimates

                                       Standard       Chi-
Parameter             DF   Estimate       Error     Square  Pr > ChiSq
-----------------------------------------------------------------------
Intercept              1     2.7255      0.2361     133.26      <.0001
g*e          1  1      1     0.8116      0.3591       5.11      0.0238 *
r*e          1  1      1     0.9951      0.3443       8.35      0.0038 *
g*r*e        1  1  1   1    -1.0887      0.4947       4.84      0.0278 **
s*e          1  1      1     0.8564      0.2524      11.52      0.0007 **
Scale                  0     1.0000      0.0000

     LR Statistics For Type 3 Analysis

                            Chi-
Source            DF      Square      Pr > ChiSq
-------------------------------------------------
e                  1        0.84          0.3595
g*e                1        1.15          0.2831
r*e                1        3.44          0.0636
g*r*e              1        4.92          0.0265
s*e                1       11.91          0.0006
```

The three procedures each gives a model fit of $G^2 = 2.5165$ on 3 d.f. (pvalue = 0.4723), a very good fit. Again the parameter estimates from both PROC CATMOD and LOGISTIC are the same because of the same coding scheme employed by these two procedures. The parameter estimates from GENMOD, being from the cell reference coding scheme; are much easier to interpret. Let us therefore concentrate on the parameter estimates from PROC GENMOD for a moment. Since we are only concerned with those interactions involving the response variable E, we notice that the following effects appear to be significant: GE, RE, GRE, SE. Of these four terms, only SE and GRE belong to the logit model formulation. However, when all the above terms are examined in the light of their type 3 contributions to the model, only the GRE and SE terms are important. These results are consistent with those from PROC CATMOD and LOGISTIC being the only ones significant at $\alpha = 0.05$. These are the two terms that will be employed in interpreting the data in Table 9.3. In Table 9.4 are the expected values obtained under this model.

			ESTEEM(l)	
SEX(i)	GPA(j)	RACE(k)	High	Low
Males	High	Black	14.392	9.608
		White	16.792	10.208
	Low	Black	28.553	14.447
		White	20.264	27.736
Females	High	Black	13.608	21.392
		White	22.208	31.792
	Low	Black	21.447	25.553
		White	4.736	15.264

Table 9.4: Estimated expected values under the logit model S/GR

The estimated odds ratio from the logit model of a high self-esteem to a low self-esteem for a male with high GPA and who is Black = 14.392/9.608 = 1.498. The above is obtained from the table of estimated expected cell values. The estimated odds ratios for all cells are given in Table 9.5.

		Race	
Sex	GPA	Black	White
Males	High	1.498	1.645
	Low	1.9764	0.7306
Females	High	0.6361	0.6985
	Low	0.8393	0.3103

Table 9.5: Estimated odds ratios of high to low self-esteem perception change under logit model S/GR

From Table 9.5, we can show that the estimated odds of having a high self-esteem to low-esteem are 2.355 times greater for males than for females. This is computed as:

$$2.355 = \frac{1.498}{0.6361} = \frac{1.9764}{0.8393} = \frac{1.645}{0.6985} = \frac{0.7306}{0.3103}$$

In other words,

$$\ln\left(\frac{\hat{m}_{1jk1}\hat{m}_{2jk2}}{\hat{m}_{1jk2}\hat{m}_{2jk1}}\right) = 2.355$$

The parameter estimate for the SE interaction from the above logit model (CATMOD and LOGISTIC) are:

Sex	Male	Female
$\hat{\lambda}^{SE}$:	0.4282	−0.4282

Since high self-esteem was modeled, the parameter estimates indicate that high self-esteem is higher among males than females, with the odds of a male having a higher self-esteem to a female being $2^{0.4282-(-0.4282)} = e^{2(0.4282)} = 2.355$, as obtained from above. Similarly, from PROC GENMOD output, we can obtain this from the parameter estimate for the SE interaction effect, which is $e^{(0.8569)} = 2.356$. The 2.355 measures the main effect for sex in the logit model, and it involves the GPA-race interaction. PROC LOGISTIC gives us this estimate in its output displayed above.

Because the logit term GR is significant in the logit model, we present below the parameter estimates from this term (which corresponds to a three-factor interaction λ^{GRE}), we present these estimates below, again from either PROC CATMOD or LOGISTIC. The results indicate that high self-esteem is lower among the combination of high GPA-Blacks and low GPA-Whites than among high GPA-Whites and low GPA-Blacks.

		Race	
	GPA	Black	White
$\hat{\lambda}_{GRE}$:	High	−0.2722	0.2722
	Low	0.2722	−0.2722

We also present the prediction probabilities $\hat{\pi}_{ijk}$ for the high self-esteem category. This probability is designated as PHAT in the following SAS software output.

```
s   g   r        PHAT
-----------------------
1   1   1      0.5997
1   1   2      0.6219
1   2   1      0.6640
1   2   2      0.4222
2   1   1      0.3888
2   1   2      0.4113
2   2   1      0.4563
2   2   2      0.2368
```

It is immediately obvious from the output presented above that the males have higher prediction probabilities for high self-esteem and that the highest prediction probability for high self-esteem is obtained among Black males with low GPA, while the lowest prediction probability is obtained among White females with low GPA.

It is therefore obvious from the above analyses that the chance of having a high self-esteem change are much higher among males than females. Further, White females tend to have the lowest chance of changing their high self-esteem, even when they have low GPA. Within each gender, however, Whites with low GPA tend to have less propensity for changing their self-esteem from high to low.

9.1.3 Another Example: Example 9.3

We re-analyze here the 2^4 table example in chapter 7. The table presented in Table 7.1 relates to the study of marijuana use among adults in a large city and a suburb. The variables are G (geographical location, 1 for San Francisco and 2 for Contra Costa), family status F (married with children & un married or without children), religion R (Protestant, Catholic, or others). The response variable is M (used marijuana, did not use marijuana). The data were analyzed in chapter 7 as a log-linear model. We present here the reanalysis of these data using logit models.

Our analysis starts by employing SAS® PROC LOGISTIC to select the most parsimonious logit model using the forward, backward and stepwise selection strategies. We present below the SAS software program and a partial output for the implementation of the forward selection procedure.

```
data chap93;
do M=1 to 2;
 do R=1 to 2;
   do G=1 to 2;
     do F=1 to 2;
input count @@;
  output;
end; end; end; end;
datalines;
52 3 23 12 37 9 69 17 35 15 23 35 130 67 109 136
;
proc logistic order=data; class R (ref=last) G (ref=last) F (ref=last)/param=ref;
weight count; model M=R|G|F/scale=none aggregate=(R G F)
         selection=forward details;  run;
```

```
                 Analysis of Effects Not in the Model

                                 Score
              Effect      DF    Chi-Square    Pr > ChiSq
              -------------------------------------------
              R*F          1        0.0112        0.9155
              G*F          1        0.4942        0.4821

NOTE: No (additional) effects met the 0.05 significance level for entry into the
      model.
```

Summary of Forward Selection

Step	Effect Entered	DF	Number In	Score Chi-Square	Pr > ChiSq
1	F	1	1	50.8415	<.0001
2	R	1	2	32.6132	<.0001
3	G	1	3	4.2012	0.0404
4	R*G	1	4	4.7890	0.0286

Deviance and Pearson Goodness-of-Fit Statistics

Criterion	DF	Value	Value/DF	Pr > ChiSq
Deviance	3	4.3403	1.4468	0.2270
Pearson	3	4.5169	1.5056	0.2108

Number of unique profiles: 8

Type III Analysis of Effects

Effect	DF	Wald Chi-Square	Pr > ChiSq
R	1	6.2577	0.0124
G	1	8.3019	0.0040
R*G	1	4.7637	0.0291
F	1	45.5430	<.0001

Analysis of Maximum Likelihood Estimates

Parameter		DF	Estimate	Standard Error	Wald Chi-Square	Pr > ChiSq
Intercept		1	-1.8923	0.1929	96.2575	<.0001
R	1	1	0.6546	0.2617	6.2577	0.0124
G	1	1	-0.6163	0.2139	8.3019	0.0040
R*G	1 1	1	0.8077	0.3701	4.7637	0.0291
F	1	1	1.3626	0.2019	45.5430	<.0001

Odds Ratio Estimates

Effect	Point Estimate	95% Wald Confidence Limits	
F 1 vs 2	3.906	2.630	5.802

The forward selection procedure selects the logit model {RG,F} for the above data. The model fits with $G^2 = 4.3403$, (pvalue = 0.22). This model is equivalent to a log-linear model {RGF,RGM,FM}, which is the model arrived at in chapter 7. The backward and stepwise procedures came up with the same logit model {RG,F} in this example.

9.1.4 Interpretations of Selected Model

First we interpret the significance of F. The odds of 3.906 indicates that for a given religion and geographical location, the odds are 3.91 times higher of having smoked marijuana as against not having smoked marijuana among those individuals having children versus those not having children.

Since the R$*$G interaction is significant, we can interpret the interaction effects by constructing relevant contrasts in SAS software. These contrasts are implemented in the SAS software program below together with an output from their implementations.

```
set tab93;
proc logistic order=data;
class R (ref=last)  G (ref=last) F (ref=last)/param=ref;
weight count; model M=R|G F/scale=none aggregate=(R G F);
CONTRAST 'R1 vers 0,G=1' R 1 -1   R*g 1 0 -1 0/ESTIMATE=EXP;
CONTRAST 'R1 vers 0, G=0' R 1 -1  R*g 0 1 0 -1/ESTIMATE=EXP;
CONTRAST 'G1 vers 0, R=1' G 1 -1  R*g 1 -1/ESTIMATE=EXP;
CONTRAST 'G1 vers 0, R=0' G 1 -1  R*g 0 0 1 -1/ESTIMATE=EXP; run;
```

Odds Ratio Estimates

Effect	Point Estimate	95% Wald Confidence Limits	
F 1 vs 2	3.906	2.630	5.802

Contrast Test Results

Contrast	DF	Wald Chi-Square	Pr > ChiSq
R1 vers 0,G=1	1	31.2547	<.0001
R1 vers 0, G=0	1	6.2577	0.0124
G1 vers 0, R=1	1	0.3876	0.5336
G1 vers 0, R=0	1	8.3019	0.0040

Contrast Rows Estimation and Testing Results

Contrast	Type	Row	Estimate	Standard Error	Alpha	Lower Limit	Upper Limit
R1 vers 0,G=1	EXP	1	4.3158	1.1289	0.05	2.5848	7.2062
R1 vers 0, G=0	EXP	1	1.9243	0.5035	0.05	1.1522	3.2137
G1 vers 0, R=1	EXP	1	1.2110	0.3723	0.05	0.6628	2.2124
G1 vers 0, R=0	EXP	1	0.5399	0.1155	0.05	0.3550	0.8211

For a given family status, the odds are 4.32 times higher that an individual will respond to having smoked to not having smoked for a San Francisco respondent among Protestants and Catholics than among those with other religion. This odds ratio is significant. Corresponding odds ratio for those respondents living in Contra Costa is 1.92 times higher among Protestants and Catholics than among other religious groups. This odds value is also significant.

For a given family status, the odds are 47% lower for having smoked to not having smoked among San Francisco residents than among Contra residents, given that the respondent has other religious belief. This odds value is also very significant. Its corresponding odds value for the Protestant or Catholic respondents is not significant.

9.2 Poisson Regression, and Models for Rates

9.2.1 The Poisson Regression

Suppose the observed counts n_i ($n_i \geq 0$) follow a Poisson distribution with parameter λ. Then from Poisson distribution properties, we have

$$E(n_i) = \lambda \quad \text{and}$$
$$\text{Var}(n_i) = \lambda$$

We assume here that observed counts occur over a fixed interval, and because these counts are nonnegative, a Poisson regression model is defined in terms of log of expected counts ($\hat{m}_i$) as:

$$\ell_i = \mathbf{x}^{'}\boldsymbol{\beta} \tag{9.5}$$

where the **x** represents the explanatory variables. The above is equivalent to modeling the intensity ($\boldsymbol{\lambda}$) as:

$$\boldsymbol{\lambda} = \exp(\mathbf{x}^{'}\boldsymbol{\beta})$$

The latter is the multiplicative version of (9.5). We are often concerned with the rate or intensity with which our events occur and whether this intensity is constant or changing over time. Events with constant intensity are described as *homogeneous Poisson processes*, while those with varying intensities are appropriately described as *nonhomogeneous processes*. We now consider an approach for modeling these kinds of data.

As an example, consider the data below from Zeger (1988), reproduced from Lindsey (1995), relating the monthly numbers of cases of poliomyelitis over 14 years in the United States.

	Months											
Year	J	F	M	A	M	J	J	A	S	O	N	D
1970	0	1	0	0	1	3	0	2	3	5	3	5
1971	2	2	0	1	0	1	3	3	2	1	1	5
1972	0	3	1	0	1	4	0	0	1	6	14	1
1973	1	0	0	1	1	1	1	0	1	0	1	0
1974	1	0	1	0	1	0	1	0	1	0	0	2
1975	0	1	0	1	0	0	1	2	0	0	1	2
1976	0	3	1	1	0	2	0	4	0	2	1	1
1977	1	1	0	1	1	0	2	1	3	1	2	4
1978	0	0	0	1	0	1	0	2	2	4	2	3
1979	3	0	0	2	7	8	2	4	1	1	2	4
1980	0	1	1	1	3	0	0	0	0	1	0	1
1981	1	0	0	0	0	0	1	2	0	2	0	0
1982	0	1	0	1	0	1	0	2	0	0	1	2
1983	0	1	0	0	0	1	2	1	0	1	3	6

A homogeneous Poisson process model contains a common mean for all the cases. When this model is implemented, $G^2 = 326.2621$ on 167 d.f. with $\hat{\lambda} = e^{0.2467} = 1.28$. The SAS software output for implementing this and other models in GENMOD is displayed below:

```
data poiss;
do year=1970 to 1983; do month=1 to 12;
input count @@; output; end; end;
datalines;
0 1 0 0 1 3 0 2 3 5 3 5 2 2 0 1 0 1 3 3 2 1 1 5
0 3 1 0 1 4 0 0 1 6 14 1 1 0 0 1 1 1 1 0 1 0 1 0
1 0 1 0 1 0 1 0 1 0 0 2 0 1 0 1 0 0 1 2 0 0 1 2
0 3 1 1 0 2 0 4 0 2 1 1 1 1 0 1 1 0 2 1 3 1 2 4
0 0 0 1 0 1 0 2 2 4 2 3 3 0 0 2 7 8 2 4 1 1 2 4
0 1 1 1 3 0 0 0 0 1 0 1 1 0 0 0 0 0 1 2 0 2 0 0
0 1 0 1 0 1 0 2 0 0 1 2 0 1 0 0 0 1 2 1 0 1 3 6
;
***Fit homogeneous model***;
proc genmod order=data; model count=/dist=poi; run;
***Fit non-homogeneous year model***;
proc genmod order=data; class year month; model count=year/dist=poi type3;
```

```
contrast 'yy' year 0 0 0 1 -1 0 0 0 0 0 0 0 0 0,
              year 0 0 0 0 1 -1 0 0 0 0 0 0 0 0; run;

                          Analysis Of Parameter Estimates

                                     Standard       Chi-
Parameter           DF  Estimate        Error     Square  Pr > ChiSq
---------------------------------------------------------------------
Intercept            1    0.2231       0.2582       0.75      0.3875
year          1970   1    0.4274       0.3319       1.66      0.1978
year          1971   1    0.3365       0.3381       0.99      0.3196
year          1972   1    0.7259       0.3145       5.33      0.0210
year          1973   1   -0.7621       0.4577       2.77      0.0959
year          1974   1   -0.7621       0.4577       2.77      0.0959
year          1975   1   -0.6286       0.4378       2.06      0.1510
year          1976   1   -0.0000       0.3651       0.00      1.0000
year          1977   1    0.1252       0.3542       0.12      0.7238
year          1978   1   -0.0000       0.3651       0.00      1.0000
year          1979   1    0.8183       0.3100       6.97      0.0083
year          1980   1   -0.6286       0.4378       2.06      0.1510
year          1981   1   -0.9163       0.4830       3.60      0.0578
year          1982   1   -0.6286       0.4378       2.06      0.1510
year          1983   0    0.0000       0.0000        .           .

                    Contrast Results

                               Chi-
Contrast            DF       Square     Pr > ChiSq     Type
-------------------------------------------------------------
yy                   2         0.09         0.9562      LR
```

If we allow the process to vary with the years (that is, different intensities for each year), then a nonhomogeneous model gives $G^2 = 260.1858$ on 154 d.f. The difference in $G^2 = 66.08$ on 13 d.f. has (pvalue = <0.0001), which is highly significant.

From examination of the parameter estimates for this model, we observe that the first three years (1970-1972) have positive contrasts of log intensities, while the next three years (1973-1975) all have negative contrasts of log intensities. Further, the magnitudes of these contrasts of log intensities are very similar for the years 1973-1975. Could it be then that these three years have the same intensity?. This is equivalent to equating the three parameters for these years.

$$H_0 : \lambda_{1973} = \lambda_{1974} = \lambda_{1975}$$

We tested the above hypotheses with the contrast statement in GENMOD above, and the results gives a $G^2 = 0.09$ on 2 d.f. Clearly this hypothesis is tenable. That is, the intensities of the cases of poliomyelitis are not significantly different for years 1973-1975. The next four years (1976-1979) all have positive intensities and the last three years (1980-1982) all have negative intensities.

On the other hand, we may also allow the process to vary with the months (that is, different intensities for each month): then a nonhomogeneous model gives $G^2 = 271.0113$ on 156 d.f. The difference in $G^2 = 55.2508$ on 11 d.f. has (pvalue = <.0001), which is highly significant. Below is the GENMOD statements and a modified output.

```
proc genmod order=data; class month; model count=month/dist=poi;
run;
                       Analysis Of Parameter Estimates

                               Standard        Chi-
Parameter      DF   Estimate      Error      Square   Pr > ChiSq
-----------------------------------------------------------------
Intercept       1     0.9445     0.1667       32.11       <.0001
```

```
month        1   1   -1.3863    0.3727       13.84        0.0002
month        2   1   -0.9445    0.3150        8.99        0.0027
month        3   1   -2.1972    0.5270       17.38        <.0001
month        4   1   -1.2809    0.3575       12.84        0.0003
month        5   1   -0.8755    0.3073        8.12        0.0044
month        6   1   -0.4925    0.2706        3.31        0.0688
month        7   1   -1.0186    0.3236        9.91        0.0016
month        8   1   -0.4480    0.2669        2.82        0.0933
month        9   1   -0.9445    0.3150        8.99        0.0027
month       10   1   -0.4055    0.2635        2.37        0.1239
month       11   1   -0.1495    0.2450        0.37        0.5417
month       12   0    0.0000    0.0000         .             .
```

One of the main difference between log-linear modeling and Poisson regression is that unlike log-linear modeling techniques in which margins, rows, or columns are fitted, in Poisson regression, rows, columns, or other margins of the data do not come into play. These margins take the role of "explanatory variables" in Poisson regression. In other words, we do not fit the marginal totals as in log-linear modeling. However, we have observed for these data that both the month and years are important in explaining the number of cases of poliomyelitis during this period. Consequently, we incorporate these two factors into our model in the GENMOD statement above, but asking for types 1 and 3 tests. The model when implemented has $G^2 = 204.9351$ on 143 d.f. Results from both the type 1 and type 3 tests indicate that both factor variables are important in our model.

```
proc genmod order=data; class year month;
model count=year month/dist=poi type1 type3; run;

               LR Statistics For Type 1 Analysis
                                                  Chi-
Source          Deviance        DF       Square     Pr > ChiSq
--------------------------------------------------------------
Intercept       326.2621
year            260.1858        13        66.08         <.0001
month           204.9351        11        55.25         <.0001

        LR Statistics For Type 3 Analysis

                                Chi-
Source              DF        Square     Pr > ChiSq
-----------------------------------------------------
year                13        66.08         <.0001
month               11        55.25         <.0001
```

From the model, the intensities can be obtained from the **xbeta** output from PROC GENMOD.

9.2.2 Another Example

The data below from Upton and Fingleton (1985) relate to hypothetical quadrant data where there is a possibility of north-south or east-west trends in the data.

Quadrant counts displaying possible north-south trend

0	0	0	1	0
3	3	5	2	0
2	3	6	2	5
1	5	4	6	7
6	2	4	3	4

Following Upton and Fingleton, to model for the north-south and east-west possible trend in the data, suppose we let X_1 and X_2 be quadrant coordinates ranging from (1,1) in the north-west corner to (5,5) in the south-east corner. A constant intensity model gives $G^2 = 50.1493$ on 24 d.f. A poor fit. However, when the continuous explanatory variables X_1 and X_2 are introduced into the model, PROC GENMOD produces the following summary results for the parameters of the model:

$$E(n_{ij}) = \exp(\beta_0 + \beta_1 X_{1i} + \beta_2 X_{2j})$$

```
data upton;
do x1=1 to 5;do x2=1 to 5;
input count @@; output;end; end;
datalines;
0 0 0 1 0 3 3 5 2 0 2 3 6 2 5 1 5 4 6 7
6 2 4 3 4
;
proc genmod order=data; model count=x1 /dist=poi type1 type3;
run;
```

Criteria For Assessing Goodness Of Fit

Criterion	DF	Value	Value/DF
Deviance	22	34.9906	1.5905
Pearson Chi-Square	22	29.7345	1.3516

Analysis Of Parameter Estimates

Parameter	DF	Estimate	Standard Error	Chi-Square	Pr > ChiSq
Intercept	1	-0.1788	0.4252	0.18	0.6742
x1	1	0.3248	0.0878	13.68	0.0002
x2	1	0.0609	0.0824	0.55	0.4598
Scale	0	1.0000	0.0000		

The model has a G^2 of 34.991 on 22 d.f., which is a good fit. The estimated cell counts for row i and column j are given by:

$$\exp(\hat{\beta}_0)(\exp(\hat{\beta}_1))^{X_{1i}}(\exp(\hat{\beta}_2))^{X_{2j}} = (0.8363)(1.3838)^{X_{1i}}(1.0628)^{X_{2i}}$$

Thus the estimated Poisson parameter λ for, say, the quadrant in the third row and fourth column is $0.8363 \times 1.3838^3 \times 1.0628^4 = 2.8273$. Thus a movement down r rows of the table implies an increase in λ by the multiplicative factor 1.3838^r, and similarly for movement across the columns, we have a multiplicative factor of 1.0628^c. The type3 analysis suggest that β_2 is not significant ($p = 0.4592$) and can well be removed from the model. Removing this parameter from the model now gives a model with parameter estimates $\hat{\beta}_0 = 0.0077$ (a.s.e. $= 0.3386$) and $\hat{\beta}_1 = 0.3248$, (a.s.e. $= 0.0878$). The model has G^2 of 35.538 on 23 d.f., which is again a good fit. The above results indicate that the data exhibit only the north-south trend. This is not surprising as the original artificial data were randomly generated from a Poisson distribution having $\lambda = i$, where i is the row of the table.

9.2.3 Weighted Log-Linear Models

The data in Table 9.6 are from a stratified sampling that resulted in a disproprpor-tionate number of population size across age groups for the two cities. The data relate to incidence of nonmelanoma skin cancer among women in Minneapolis-St. Paul and Dallas-Ft. Worth (Le, 1992). We shall therefore analyze the data using as weights the population size in each category. The advantage of the weighted analysis is that it removes the bias due to the unequal population sizes. Such an analysis gives rise to what has been dubbed *Poisson regression models.* The justification for a Poisson regression in this case is that few responses are observed out of a very large number of possible responses or large number of population size.

	Minn-St. Paul		Dallas-Ft. Worth	
Age Group (yr)	No of cases	Population Size	No of Cases	Population Size
15-24	1	172,675	4	181,343
25-34	16	123,065	38	146,207
35-44	30	96,216	119	121,374
45-54	71	92,051	221	111,353
55-64	102	72,159	259	83,004
65-74	130	54,722	310	55,932
75-84	133	32,185	226	29,007
85+	40	8,328	65	7,538

Table 9.6: Incidence of nonmelanoma skin cancer among women

We can model the incidence of nonmelanoma skin cancer on age (A) and city (C), if we let n_{ij} be the observed number of cases reported for age group i and city j in a total population size of W_{ij}. Then, if we also let the corresponding expected cell value be $\hat{m}_{ij}$, our model becomes (Agresti, 1990):

$$\ln\left(\hat{m}_{ij}/W_{ij}\right) = \mu + \lambda_i^{\text{city}} + \lambda_j^{\text{age}} \tag{9.6}$$

where the term on the left-hand side is the log-weighted frequencies. Estimates of the λ parameters are obtained by conditioning on the cell weights. The model assumes that there is no interaction between age and cities. The above model seems like the familiar log-linear model formulation except for the fact that the left-hand side equals $\ln(\hat{m}_{ij}) - \ln(W_{ij})$ instead of the familiar $\ln(\hat{m}_{ij})$. The term $\ln(W_{ij})$, which is referred to as the *adjustment* term, is called an *offset.* The above model can be implemented in SAS® GENMOD by specifying in the option statement that log(popl) is an offset, that is, off $= \ln(\text{popl})$.

```
proc genmod order=data;
class city age;
model cases=city age/dist=poi link=log offset=off type3 obstats; run;
```

```
                                Standard        Chi-
Parameter         DF  Estimate     Error      Square  Pr > ChiSq
-----------------------------------------------------------------
Intercept          1   -4.6754    0.0991     2225.55      <.0001
city       msp     1   -0.8043    0.0522      237.34      <.0001
age        15-24   1   -6.1782    0.4577      182.17      <.0001
age        25-34   1   -3.5480    0.1675      448.76      <.0001
age        35-44   1   -2.3308    0.1275      334.36      <.0001
age        45-54   1   -1.5830    0.1138      193.38      <.0001
age        55-64   1   -1.0909    0.1109       96.75      <.0001
age        65-74   1   -0.5328    0.1086       24.06      <.0001
age        75-84   1   -0.1196    0.1109        1.16      0.2809
age        85+     0    0.0000    0.0000         .           .
Scale              0    1.0000    0.0000
```

The model above considers age and city as factor variables. When this model is fitted to the data in Table 9.6, it gives G^2 value of 8.1950 on 7 d.f. with a city estimate (Minneapolis/St. Paul) of −0.8043. Consequently, the relative risk for women from Minneapolis/St. Paul of contracting the skin cancer given the age group of an individual (that is, after adjusting for age group) is $e^{(-0.8043)} = 0.4429$ times that of Fort Worth Dallas. Put in another way, the relative risk of contracting non-melanoma skin cancer among women is $\frac{1}{.4429} = 2.235$ times higher in Fort Worth Dallas (after adjusting for age group) than for women in the Twin Cities of Minneapolis-St. Paul. Figure 9.1 gives the cancer rates for these two cities for varying age groups. The figure confirms the higher rate for Dallas and also indicate the gradual rise in the rate as individuals get older. The rates seem to decline between the ages of 65 to 70.

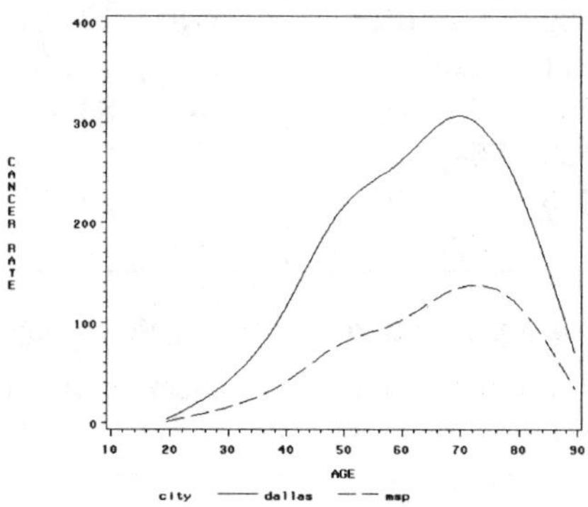

Figure 9.1: Plot of Rates against age

An alternative analysis for the data in Table 9.6 is to assume that age is not a categorical variable. Kleinbaum et al. (1998) suggested a transformation of the form:

$$U_i = \frac{(\text{Midpoint of } i\text{-th age group}) - 15}{35}, \quad i = 1, 2, \cdots, 8$$

and then fitted a Poisson model that has $\ln(U)$ as a covariate: that is, the model with

$$\ln(\hat{m}_{ij}/W_{ij}) = \beta_0 + \beta_1 \ln(U_i) + \beta_2 \text{city}_j \tag{9.7}$$

When this model is implemented in GENMOD, we have the following summary results:

```
U=(AGE-15)/35; T=LOG (U); off=log(popl);
proc genmod order=data; class city;
model cases=city TT/dist=poi offset=off type3 obstats; run;
```

Criteria For Assessing Goodness Of Fit

Criterion	DF	Value	Value/DF
Deviance	13	14.2877	1.0991
Pearson Chi-Square	13	14.1568	1.0890

Analysis Of Parameter Estimates

Parameter		DF	Estimate	Standard Error	Wald 95% Confidence Limits		Chi-Square	Pr > ChiSq
Intercept		1	-6.2354	0.0324	-6.2989	-6.1718	36988.0	<.0001
city	msp	1	-0.8027	0.0522	-0.9050	-0.7004	236.67	<.0001
city	dallas	0	0.0000	0.0000	0.0000	0.0000	.	.
T		1	2.2493	0.0621	2.1276	2.3710	1312.44	<.0001
Scale		0	1.0000	0.0000	1.0000	1.0000		

This model fits the data with a deviance of 14.287 on 13 d.f, with an adjusted odds ratio of $1/(\exp{-0.8027}) = 2.2316$, indicating again that the risk is about 2.2316 higher for women in Fort. Worth than in Minneapolis/St. Paul after adjusting for the effect of age. Figure 9.2 gives the sketch of the rates under this model. We observe that the graph is smoother this time around.

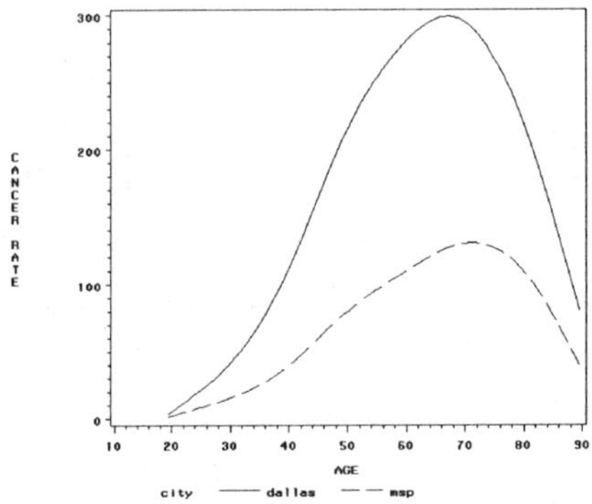

Figure 9.2: Plot of rates against age

9.2.4 Log-Linear Models for Rates

There is a clear relationship between *odds* and *rates* and hence between models for rates and logit models. Odds and rates are approximately equal (Clogg & Eliason, 1988) when the Poisson approximation to the binomial can be used, that is, whenever rare events are considered. However, when analyzing rare events (that is, when p is very small), logit models for a binary response variable will be virtually identical to log-linear models for rates.

Consider an $I \times J \times 2$ contingency table with observed frequency n_{ijk}, where the third variable is the response binary variable (sucess or failure). The logit model has:

$$\ln\left(\hat{m}_{ij1}/\hat{m}_{ij2}\right) = 2\lambda_1^C + 2\lambda_{i1}^{AC} + 2\lambda_{j1}^{BC} \tag{9.8}$$

which represents a model with additive effects of the factors on the log-odds of success. Since logit models always fit the "group totals," in this case the AB interaction or n_{ij+}, the marginal distribution of the joint variable composed of all independent variables. Suppose we define the expected rate for sucess as:

$$R_{ij}^{AB} = \hat{m}_{ij1}/\hat{m}_{ij+}$$

Then if we condition on the observed group totals, n_{ij+}, the following *log-rate model* can be used:

$$\begin{aligned} \ln\left(R_{ij}^{AB}\right) &= \ln\left(\hat{m}_{ij1}/\hat{m}_{ij+}\right) \\ &= \eta + \eta_i^A + \eta_j^B \end{aligned} \tag{9.9}$$

The η parameters relate to the parameters of the model. The model above is equivalent to the log-linear model {AB,AC,BC}.

9.2.5 Example 9.3

The data in Table 9.7 relate to a $2 \times 5 \times 2$ table (Clogg & Eliason, 1988) involving the variables industry, age, and lung functioning (abnormal versus normal), since

Independent variables		Response		
Industry	Age group	Abnormal	Normal	Total
Manufact.	20-29	9	394	403
	30-39	22	666	688
	40-49	15	668	683
	50-59	37	502	539
	60+	17	116	133
Service	20-29	12	244	256
	30-39	17	508	525
	40-49	17	582	599
	50-59	30	423	453
	60+	14	141	155

Table 9.7: Lung functioning by age and industry

the rate of abnormal lung functioning is a rare event; indeed, observed rates are below 0.10 for all age-industry combinations.

The data have been displayed to reflect the marginal total n_{ij+}, the number of observations in category i of variable A and category j of variable B averaged over the response variable C. Following Clogg and Eliason (1988), let $N_{ij} = n_{ij+} = n_{ij1} + n_{ij2}$. For example, for the data in the Table 9.7, $N_{11} = 403$ and the observed proportion of abnormal lung functioning in the manufacturing industry among the age-group 1 (20-29) is thus $9/403 = 0.0223$, giving an observed odds of $0.0223/(1.0-0.0223) = 0.0228$ and a resulting observed logit of -3.7808. The three quantities *proportions, odds*, and *logit* will be used to model contingency tables in this section. To fit the logit model, we would use the observed 20 cell counts, whereas to fit the log-rates model, we would use only the 10 success counts (abnormal) and the marginal totals n_{ij+} as the weights.

Three models are considered for modeling the data in Table 9.7. These are:

(a) The linear-probability model (Grizzle et al., 1969), where if we let $p_{ij} = \hat{m}_{ij1}/N_{ij}$ denote the expected probability that level of response = 1, when industry and age are at levels i and j respectively, then the saturated model is

$$p_{ij} = \alpha + \beta_i + \gamma_j + \delta_{ij} \tag{9.10}$$

where α is a constant, β_i and γ_j pertain to main effects of industry and age, and δ_{ij} denotes the interaction term if significant or the departure from additivity (see Clogg & Eliason, 1988).

(b) The logit model (Goodman, 1978) has the formulation:

$$\ln\left[p_{ij}/(1-p_{ij})\right] = \alpha + \beta_i + \gamma_j + \delta_{ij} \tag{9.11}$$

where $\ln\left[p_{ij}/(1-p_{ij})\right] = \ln\left(n_{ij1}/n_{ij2}\right)$ is the observed logit when variable C takes the value 1 (abnormal) and industry and age are at levels i and j respectively. The parameters are explained in the previous case, although estimates from both models are not necessarily the same.

(c) When the proportion (p) is referred to as the *rate*, then the model in **(a)** above can be regarded as a linear model for rates. The relevant model here is the *log-rate model* or *Poisson regression*, which is defined as:

$$\ln(\hat{m}_{ij1}/N_{ij}) = \alpha + \beta_i + \gamma_j + \delta_{ij} \tag{9.12}$$

The linear-probabilty model can be implemented using weighted least squares (as indicated in chapter 8), since the variances of the p_{ij} are not constant, being functions of the N_{ij}, that is, the marginal totals. We will concentrate our attention on the other two models in **(b)** and **(c)** above. The saturated logit and log-rate models are first fitted to the data, and type 3 partial tests are examined to find important explanatory terms. That is, we wish to fit a log-linear model of the form

$$\ln(\hat{m}_{ij}/w_{ij}) = \alpha_j + \sum_{j=1}^{K} \beta_k x_{ijk}$$

where $\hat{m}_{ij}$ are the expected cell counts under some model and w_{ij} are the weights(or offsets). Thus to fit the log-rate model, we specify the weights to be the N_{ij}. The logit model is also fitted by specifying the fixed marginal totals with the *fixed* command. That is, the logit model is fitted by conditioning on the industry-age marginal.

```
data tab96;
input Ind $ age1 $ abn total@@; off-log(total);
datalines;
mf 20-29 9 403 mf 30-39 22 688 mf 40-49 15 683 mf 50-59 37 539
mf 60+ 17 133 sv 20-29 12 256 sv 30-39 17 525 sv 40-49 17 599
sv 50-59 30 453 sv 60+ 14 155
;
***Fit saturated logit Model***;
proc genmod order=data; class ind age;
model abn/total=ind|age/dist=bin link=logit type3; run;

     LR Statistics For Type 3 Analysis

                            Chi-
Source           DF       Square     Pr > ChiSq
-----------------------------------------------
Ind               1         0.58         0.4448
age               4        50.28         <.0001
Ind*age           4         4.40         0.3546
```

The result of the saturated logit model indicate that only the effect of age is important in explaining abnormal lung functioning. We next fit the saturated log-rate model and we further display partial results below:

```
set tab96;
proc genmod order=data;
class ind age;
model abn=ind|age/dist=poi offset=off type3;
run;
     LR Statistics For Type 3 Analysis

                            Chi-
Source           DF       Square     Pr > ChiSq
-----------------------------------------------
Ind               1         0.64         0.4236
age               4        47.71         <.0001
Ind*age           4         4.18         0.3817
```

The results from the saturated log-rate model also indicate that only age is important for a proper explanation of the data in Table 9.7. The χ^2 values are very close in both cases. We note here that the second model (the log-rate model) employs the weights $(\ln(N_{ij}))$ as offsets in this model.

The above results indicate that the type of industry does not seem significant, as for the interaction effects between age and industry. Thus a reduced model without the interaction terms is next fitted to the data. Both logit and log-rate models now fit the data. They have G^2 values of 4.3999 and 4.1837 on 4 degrees of freedom, respectively. We also notice that the estimates of the parameters are very similar for both models, indicating that for rare events, both models can be considered equivalent. However, examination of the parameter estimates in both cases again confirms that the type of industry is again not significant in both models. The industry parameter has G^2 values of 0.17 and 0.16, respectively on 1 d.f. for both the logit and log-rate models (pvalue >0.8816).

Since age has been found to be signifant, we next consider models involving only age as the explanatory variable. In this case, the logit and log-rate models give G^2 values of 4.5683 and 4.3427 on 5 degreees of freedom, respectively. The parameter estimates under both models are displayed in Table 9.8.

	Logit		Log-rate	
Parameters	Estimates	ASE	Estimates	ASE
$\hat{\alpha}$	-2.1151	0.1901	-2.2290	0.1796
Age:				
$\hat{\gamma}_1$	-1.2987	0.2921	-1.2172	0.2826
$\hat{\gamma}_2$	-1.2895	0.2503	-1.2083	0.2406
$\hat{\gamma}_3$	-1.5501	0.2611	-1.4615	0.2520
$\hat{\gamma}_4$	-0.5100	0.2284	-0.4661	0.2172
$\hat{\gamma}_5$	0	0	0	0

Table 9.8: Parameter estimates, log-rate models, with only age as the explanatory variable

The contrast between age 1 and age 5 (i.e., between the 20-29 and 60+) under the log-rate model is -1.2172, hence, the estimated relative risk ratio is given by $e^{-1.217} = 0.296$. That is, the risk of abnormal lung functioning for the 20-29 years old is about 29.6 % times those for the 60+ years old. In other words, those 60+ have relative risk that is $\frac{1}{.296} = 3.38$ times higher than those 20-29 years old. Based on the above analyses, the response is very well explained by only the explanatory variable, age. The equivalent log-linear model {IA,AR} has $G^2 = 5.3001$ on 5 degrees of freedom. The corresponding interaction AR parameter estimates are $\{-1.2457, -1.2365, -1.4970, -0.4889, 0\}$ which are again close to those presented above.

Because of the ordinal nature of the age variable, we consider next fitting a model with a linear trend and a quadratic response in age, using the centers of the age classes (ignoring industry), to the data. The results of the quadratic model are displayed in the partial SAS software output below.

```
set tab96;
if age1='20-29' then age=.5*(20+29);
else if age1='30-39' then age=0.5*(30+39);
else if age1='40-49' then age=.5*(40+49);
```

```
else if age1='50-59' then age=.5*(50+59);
else age=.5*(60+69);
age2=age*age;
proc genmod; model abn=age age2/dist=poi offset=off type3; run;

                    Analysis Of Parameter Estimates

                               Standard      Chi-
Parameter    DF    Estimate       Error    Square    Pr > ChiSq
----------------------------------------------------------------
Intercept     1     -1.9041      0.9281      4.21        0.0402
age           1     -0.0991      0.042       5.38        0.0204
age2          1      0.0015      0.0005     10.14        0.0014
Scale         0      1.0000      0.0000
```

Under the log-rate model, the model gives a G^2 value of 11.3355 on 7 degrees of freedom. When we next consider a cubic term in the model, although the model fits the data, the extra cubic parameter is not significant, and we therefore conclude that an appropriate model for the data is the log-rate or logit model with a quadratic age effect. The estimated model is given by the equation below and the plot of this response is given in Figure 9.3.

$$\ln(\hat{m}_{ij}/\omega_{ij}) = -1.9041 - 0.0991(age) + 0.0015(age)^2 \tag{9.13}$$

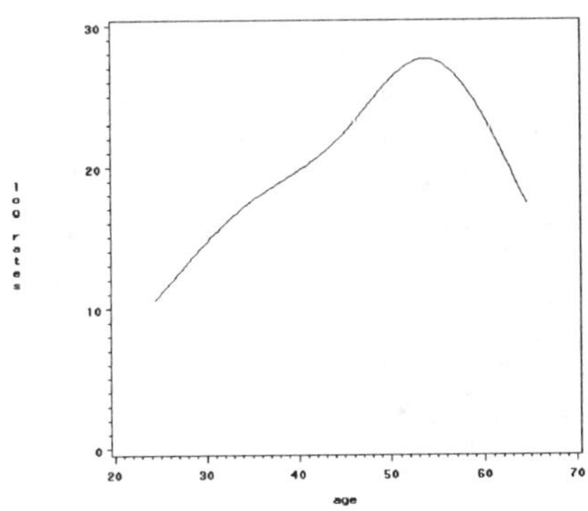

Figure 9.3: Plot of rates against age

It should however be noted that when we have nonrare events, the two models (logit and log-rate) can give quite different paramcter estimates. In this situation, we would then need "to use both statistical and substantive criteria to choose between the two" (Clogg & Eliason, 1988).

9.2.6 Example 9.4

As our last example in this section, the data in Table 9.9 are reproduced from Clogg and Eliason (1988). The data arise from a longitudinal study where exposures are obtained by following individuals through time and the frequencies of occurrence of events obtained over time. The data in the table relate to death counts among subjects who had received kidney transplants from donors (cadavar or living), time, and match grade (number of matched antigens out of a maximum of four). This resulting table is a $2 \times 5 \times 3$ contingency table. If we designate the exposure in cell (i, j, k) by E_{ijk}, then a saturated Poisson regression model for the data is given by:

$$\ln\left(\hat{m}_{ijk}/E_{ijk}\right) = \alpha + \lambda_i^D + \lambda_j^T + \lambda_k^G + \lambda_{ij}^{DT} + \lambda_{ik}^{DG} + \lambda_{jk}^{TG} + \lambda_{ijk}^{DTG}$$

where D refers to donors, T to time, and G to Match grade.

		Deaths: Match grade			**Exposures:** Match grade		
Donor	Time	0-2	3	4	0-2	3	4
Cadaver	0-1 mo.	204	30	1	27390.5	3471	521
	1-3 mo.	170	16	4	43470	5340	930
	3-6 mo.	94	10	2	54000	7110	1080
	6-12 mo.	54	6	0	94860	12600	1980
	> 1 yr.	91	10	3	351090	48600	6570
Living relative	0-1 mo.	35	19	5	13677	4665.5	4839
	1-3 mo.	64	9	6	24510	8325	9300
	3-6 mo.	31	8	4	32715	11880	13590
	6-12 mo.	26	4	3	60210	22590	26595
	> 1 yr.	34	5	6	242100	104940	115380

Table 9.9: Graft failures following kidney transplants

The fit of the saturated weighted log-linear model {DTG} with the weights being the log of exposures indicates that only the interaction term TG is significant, while a a partial test of the two-factor interaction terms (DT, DG, TG) indicates that only the interaction term DG is important. This model is implemented for instance with PROC GENMOD in SAS software with the following:

```
data tab98;
do donor=1 to 2; do time = 1 to 5; do grade=1 to 3;
input death exps @@; off=log(exps); output; end; end; end;
datalines;
204 27390.5 30 3471 1 521 170 43470 16 5340 4 930
94 54000 10 7110 2 1080 54 94860 6 12600 0 1980
91 351090 10 48600 3 6570 35 13677 19 4665.5 5 4839
64 24510 9 8325 6 9300 31 32715 8 11880 4 13590
26 60210 4 22590 3 26595 34 242100 5 104940 6 115380
;
run;
proc genmod order=data; class donor time grade;
model death=donor|time donor|grade time|grade/dist=poi link=log offset=off type3;
run;
```

Results with various models applied to the data in Table 9.9 give contradictory conclusions, and we therefore decided to fit various combinations of these two-factor interactions to the data. These results are summarized in Table 9.10.

Model (i) fits the data well, and we therefore seek a more parsimonious model with fewer number of parameters than model (i). Models (ii), (iii), and (iv) enable us to test the hypotheses whether each of the interaction terms DT, DG, and TG is respectively zero. The results of these tests indicate that only the DG interaction term is worthy of inclusion in our model. We therefore fit model {T,DG} in (v), and the pvalue for this model of 0.1104 indicates that the model fits the data. As expected, models (vi) and (vii) do not fit the data in view of our earlier conclusion. We again sought to fit a reduced model to (v), by fitting the linear components of variables T and G, which are ordinal in nature. To implement this, we used

Number	Model	G^2	d.f.	pvalue
(i)	DT,DG,TG	9.1353	8	0.3307
(ii)	DG,TG	16.0698	12	0.1882
(iii)	DT,TG	15.7920	10	0.1059
(iv)	DG,DT	22.3274	16	0.1334
(v)	T,DG	27.9643	20	0.1104
(vi)	D,TG	23.4127	14	0.0537
(vii)	G,DT	29.8700	18	0.0385
(viii)	G,T,D,D*G(1)	28.2226	21	0.1342
(ix)	G(1),T,D,D*G(1)	28.6098	22	0.1566
(x)	G,T(1),D,D*G(1)	42.5739	24	0.0111
(xi)	G(1),T(1),D,D*G(1)	42.9260	25	0.0142

Table 9.10: Results of fitting various models to the data in Table 9.9

the midpoints of variable T as scores and integer scores {1,2,3} for variable G. Our attempts at fitting these reduced models give models (viii) to (xi), from which model (ix) (T, DG(1)) emerged as the most parsimonious model. This model has a G^2 value of 28.6098 and is based on 22 degrees of freedom. We present in the next SAS software output, the parameter estimates from this model.

Clogg and Shockey (1988) also give an equivalent log-rate model utilizing cell covariates. Their model has the saturated log-linear model formulation:

$$\begin{aligned}\ln(\hat{m}_{ijk}) &= \alpha + \lambda_i^D + \lambda_j^T + \lambda_k^G + \lambda_{ij}^{DT} \\ &\quad + \lambda_{ik}^{DG} + \lambda_{jk}^{TG} + \lambda_{ijk}^{DTG} + \gamma[\ln(E_{ijk})].\end{aligned}$$

A fit of model (ix) with the covariate gives G^2=28.4356 on 21 degrees of freedom. This is the model {G(1),T,D,D*G(1),X}, where X in this case is the covariate, that is, the log of the exposures E_{ijk}. For this data, there is only one covariate. However, it is common to have situations involving several covariates. We give the SAS software statements and partial output from the Poisson regression model applied to the above data.

```
set tab98;
proc genmod order=data; class donor time;
model death=donor|grade time/dist=poi offset=off type3;
contrast 'donor1 vs 2' donor 1 -1   donor*grade 1 -1;
estimate 'donor1 vs 2' donor 1 -1   donor*grade 1 -1/exp;  run;
```

```
                         Analysis Of Parameter Estimates

                                 Standard        Wald 95%          Chi-
Parameter          DF  Estimate     Error   Confidence Limits    Square  Pr > ChiSq
-----------------------------------------------------------------------------------
Intercept           1   -8.3446    0.1614   -8.6610   -8.0282   2672.33      <.0001
donor          1    1    0.1599    0.1861   -0.2048    0.5246      0.74      0.3902
donor          2    0    0.0000    0.0000    0.0000    0.0000       .          .
grade               1   -0.5410    0.0946   -0.7263   -0.3557     32.73      <.0001
grade*donor    1    1    0.4461    0.1376    0.1764    0.7158     10.51      0.0012
grade*donor    2    0    0.0000    0.0000    0.0000    0.0000       .          .
time           1    1    3.3464    0.1007    3.1489    3.5438   1103.41      <.0001
time           2    1    2.7644    0.1022    2.5641    2.9647    731.49      <.0001
time           3    1    1.9274    0.1159    1.7002    2.1545    276.58      <.0001
time           4    1    0.8747    0.1322    0.6157    1.1337     43.80      <.0001
time           5    0    0.0000    0.0000    0.0000    0.0000       .          .
Scale               0    1.0000    0.0000    1.0000    1.0000
```

Based on these estimates, the estimated log-rate model becomes:

$$\begin{aligned}\ln(\hat{m}_{ijk}/E_{ijk}) &= 0.1599\,\text{donor}_i - 0.5410\,\text{grade}_j \\ &+ 0.4461\,(\text{donor} * \text{grade})_{ij} + 3.3464\,\text{time}_1 \\ &+ 2.7644\,\text{time}_2 + 1.9274\,\text{time}_3 + 0.8747\,\text{time}_4\end{aligned}$$

For the second model employing the cell covariates, we would normally expect the value of γ to be one (Clogg & Shockey, 1988). Our results in the above table indicate that $\hat{\gamma} = 0.884$ in our model. Of course if $\hat{\gamma} = 1$, the parameter estimates from both models would be expected to be identical. Our results further indicate that the log-rate model utilizing the cell weights is more parsimonious for our data than the second model, because it has a smaller AIC (Akaike information criterion; see chapter 7). Further, parameter estimates from this model all have smaller asymptotic standard errors (a.s.e.) Hence, we could conclude that the explanatory variable **Time** is important for a proper explanation of the data, as well as the type of donor. However, only the linear component of match grade helps to explain the variation in the data.

The linear match grade and donor interaction is also very significant in the model. Controlling for the time effects, an estimate of the overall parameter for interpreting this interaction can be formulated as follows:

$$\hat{\beta} = \begin{cases} -0.5410 * \text{grade}, & \text{if donor=0} \\ 0.1599 - 0.0949 * \text{grade}, & \text{if donor=1} \end{cases}$$

Thus when donor is zero (that is, living relative donors), the odds are $\exp(-0.5410)$, $\exp(-0.5410*2), \exp(-0.5410*3)$ for match grades 0-2, 3, and 4, respectively. That is, the odds of rejection are $\{0.582, 0.339, 0.197\}$ for the match grades, respectively. Thus for living relative donors, the least odds are those with match grade 4.

Similarly, when donor $= 1$, that is, those with cadaver donors, the odds are $\exp(0.065), \exp(-0.0299)$, and $\exp(-0.1249)$ for match grades 0-2, 3, and 4, respectively. That is, the odds of rejection are $\{1.067, 0.971, 0.883\}$ for the match grades, respectively. Thus for those who received kidney transplants from cadaver donors, the least odds of rejection are again those with match grade 4. In comparing the two results, its quite clear here that those that received living relative organs are relatively less likely to reject the organ than those that receive cadaver organs for the same match grade. It should be noted that the main effect of donor in itself is not significant. It only becomes important when match grade is incorporated into it. The results of the "contrast" and "estimate" statements in the SAS software program are displayed below.

Contrast Estimate Results

Label	Estimate	Standard Error	Alpha	Confidence Limits		Chi-Square	Pr > ChiSq
donor1 vs 2	0.6060	0.0815	0.05	0.4463	0.7657	55.30	<.0001
Exp(donor1 vs 2)	1.8331	0.1494	0.05	1.5625	2.1505		

Contrast Results

Contrast	DF	Chi-Square	Pr > ChiSq	Type
donor1 vs 2	1	60.54	<.0001	LR

The results indicate that the odds of rejection for match grade (0-2) are 1.83 times higher among those who received cadaver organs than among those who received living relative organs. This odds increase to $1.833^2 = 3.36$ for those with match grade 3 and to $1.833^3 = 6.15$ for those receiveing match grade 4 organs. In all the comparisons, theses differences are statistically significant.

The effect of time is very significant. Thus the odds of rejection are exp(3.3464) = 28.40 times higher for 0-1 months survived patients than those that received the transplant and lived for a year or more. Similar odds relative to those with one year or more after transplantation are 15.87, 6.87, 2.40 for those who had received transplants for between 1-3, 3-6, and 6-12 months, respectively, while the odds for 1-3 months patients are $\exp(2.7644 - 1.9274) = 2.31$ times higher than those of 3-6 months patients. These results indicate that the odds of rejection decreases as the elapsed time between transplantion increases after adjusting for donor and match grade.

9.2.7 Survival-Time Models

The log-rate or Poisson regression model discussed above can easily be adapted for fitting survival-time models. Here, if we let $\hat{m}_i$ denote the expected number of deaths (or events) for subject i, then

$$\ln(\hat{m}_i/t_i) = \ln(\lambda) + \boldsymbol{\beta}' x_i$$

where λ is the hazard function, which for the negative exponential equals a constant; T is the time to death, x_i is the explanatory variable i, and $\boldsymbol{\beta}$ are the parameters of the model. Further, $(\sum t_i)$ is the total exposure time at each setting. A log-linear rate model is therefore usually of the form:

$$\ln(\hat{m}_{ijk}/E_{ijk}) = \mu + \lambda_i^A + \lambda_j^B + \cdots$$

where A, B, $\cdots$, are the explanatory variables. Agresti (1990, p. 195) analyzed an example of a lung cancer survival data for 539 males diagonised as having lung cancer in 1973. The data are reproduced in Table 9.11.
The prognostic factors are histology (3 levels), stage of disease (3 levels), and follow-up period which was divided into seven 2-month intervals. Models of the form:

$$\ln(\hat{m}_{ijk}/E_{ijk}) = \mu + \lambda_i^H + \lambda_j^S + \lambda_k^T + \cdots$$

where E_{ijk}, the offset, is the log of exposure time, are considered. Results from partial tests analysis indicate that only the effect of stage (S) is important. All other terms turn out not to be significant.

```
         LR Statistics For Type 3 Analysis

                             Chi-
Source              DF     Square    Pr > ChiSq
-----------------------------------------------
t                    6      10.97        0.0892
s                    2      47.92        <.0001
t*s                 12      15.22        0.2295
h                    2       2.78        0.2486
t*h                 12      10.12        0.6051
h*s                  4       2.73        0.6042
```

	Histology								
	I Stage			II Stage			III Stage		
Time Interval Stage:	1	2	3	1	2	3	1	2	3
0-	9	12	42	5	4	28	1	1	19
	(157	134	212	77	71	130	21	22	101)
2-	2	7	26	2	3	19	1	1	11
	(139	110	136	68	63	72	17	18	63)
4-	9	5	12	3	5	10	1	3	7
	(126	96	90	63	58	42	14	14	43)
6-	10	10	10	2	4	5	1	1	6
	(102	86	64	55	42	21	12	10	32)
8-	1	4	5	2	2	0	0	0	3
	(88	66	47	50	35	14	10	8	21)
10-	3	3	4	2	1	3	1	0	3
	(82	59	39	45	32	13	8	8	14)
12-	1	4	1	2	4	2	0	2	3
	(76	51	29	42	28	7	6	6	10)

Table 9.11: Number of deaths and total follow-up for the time periods, histology, and stage of disease

```
              Criteria For Assessing Goodness Of Fit

Criterion                    DF           Value        Value/DF
-----------------------------------------------------------------
Deviance                     60         57.4021          0.9567
Pearson Chi-Square           60         54.3456          0.9058

                   Analysis Of Parameter Estimates

                                    Standard          Chi-
Parameter        DF    Estimate       Error         Square    Pr > ChiSq
--------------------------------------------------------------------------
Intercept         1     -1.7010      0.0676         633.66        <.0001
s            1    1     -1.3758      0.1477          86.80        <.0001
s            2    1     -0.8929      0.1331          44.98        <.0001
```

The model {S}, therefore, when applied to the data gives a parsimonious fit with $G^2 = 57.4021$ on 60 d.f. and the following estimates of the parameters of S based on PROC GENMOD.

$$\hat{\lambda}_1^S - \hat{\lambda}_3^S = -1.3758$$

$$\hat{\lambda}_2^S - \hat{\lambda}_3^S = -0.8929$$

From the above, we have for instance $\hat{\lambda}_2^S - \hat{\lambda}_1^S = 0.4829$ and we can therefore estimate the corresponding sum to zero parameters by first adding the two expressions and imposing the sum to zero constraints to give $\hat{\lambda}_1^S = -0.6196, \hat{\lambda}_2^S = -0.1367$, hence, $\hat{\lambda}_3^S = 0.7562$ and $\hat{\lambda}_2^S - \hat{\lambda}_1^S = -0.1367 + 0.6196 = 0.4829$. For instance, regardless of the follow-up time and histology levels, we estimate the risk to be $e^{0.4829}$=1.62 times higher at the second stage of disease than at the first stage. Similarly, $\hat{\lambda}_3 - \hat{\lambda}_2 =$ 0.8929 with a risk of 2.44 times higher at the third stage of the disease than at the second stage. This result is consistent with fitting a more complicated model of the form (H,T,S) to the model. The latter model would give a relative risk ratio of about 2.349.

Below are the relevant SAS software statements used in implementing the model discussed above for the data in Table 9.11.

```
data surv;
do t=0 to 12 by 2; do h=1 to 3;  do s=1 to 3;
input count tot @@; off=log(tot);
output; endl; end; end;
datalines;
9 157 12 134 42 212 5 77 4 71 28 130
....................................
4 51 1 29 2 42 4 28 2 7 0 6 2 6 3 10
;
proc genmod order=data; class t h s;
model count=s/dist=poi offset=off type3; run;
```

9.3 Multicategory Response Models

So far, we have discussed cases in which the response or outcome variables have two categories. But what happens if the response category has more than two categories? Do we collapse the many categories to two, so that we might be able to use the theory and methods developed for binary response situations? As pointed out by Lawal (1980), collapsing categories together may sometimes detract from the main import of the investigation. What happens for those situations where collpasing categories does not really make sense? A typical example of this is the response "party affiliation" with categories (*Republican, Democrat, Independent*): how do we collapse these categories to only two categories without detracting from the differences (even if philosophically) among the three parties? We shall develop in this section the necessary methodology required to analyzing data in which the outcome variable has many categories.

When a response variable has many categories, we often refer to that variable as multicategory or we may simply describe it as a *polychotomous* response variable. In this section we shall examine the various multinomial logit models that have been proposed for analyzing data in which a polychotomous outcome variable is either nominal or ordinal. Let us first consider the case when the outcome variable is multicategory but nominal.

9.3.1 Baseline Category Model

Consider a two-way $I \times J$ contingency table indexed by variables A and R, respectively. Suppose variable R is a response variable. We have shown earlier that for the case when $J = 2$, the logit model can be written as

$$\ln\left(\frac{\pi_1|i}{\pi_2|i}\right) = \lambda + \lambda_i^A \tag{9.14}$$

with the usual restrictions on λ_i^A. We see that the model of independence implies that λ_i^A be zero. Now suppose $J > 2$, and let j and j' be any two columns of the response variable R. Then,

$$\ln\left(\frac{\pi_j \mid i}{\pi_{j'} \mid i}\right) = \lambda_j^R + \lambda_{ij}^{AR} \tag{9.15}$$

where $j^{'}$ is a fixed reference point and j takes all values except $j^{'}$ and with $\sum_{i=1}^{I} \lambda_{ij}^{AR} = 0$ being the $(J-1)$ linear constraints imposed on the parameters. Then, we have

$$\lambda_j^R = \alpha_j^R - \alpha_{j'}^R$$
$$\lambda_{ij}^{AR} = \alpha_{ij}^{AR} - \alpha_{ij'}^{AR}$$

The model described by (9.15) has been referred to as the *baseline category* model. The model treats R as a nominal variable and category $j^{'}$ as the baseline category. It does not matter which category is chosen as the baseline. The parameter estimates under any other choice of j', say $j*$, are linear transformations of the baseline model with $j = j'$, and the goodness-of-fit statistics will be identical in both situations in which $j = j'$ or $j = j*$. Generally, the last category of variable R is often used as the baseline, and the baseline category logit model with categorical explanatory variable A becomes

$$\ln\left(\frac{\pi_j}{\pi_J}\right) = \lambda_j^R + \lambda_{ij}^{AR} \tag{9.16}$$

which corresponds to the log-linear model

$$\ell_{ij} = \mu + \lambda_i^A + \lambda_j^R + \beta\,\nu_j\,\beta_i^A \tag{9.17}$$

where the ν_j are scores such that $\nu_1 \leq \nu_2 \leq \cdots, \leq \nu_J$ and $\beta = (\beta_i^A - \beta_{i'}^A)$. Similarly, the baseline category logit model with continuous explanatory variable(s) can be written in the form:

$$\ln\left(\frac{\pi_j}{\pi_J}\right) = \alpha_j + \beta_j\mathbf{X}, \quad j = 1, 2, \cdots, (J-1) \tag{9.18}$$

9.3.2 Example 9.6: Severity of Pneumoconiosis

The data in Table 9.12 are reproduced from Lindsey (1995) and show the severity of pneumoconiosis as related to the number of years working at a coal face.

	Pneumoconiosis		
Years	Normal	Mild	Severe
0.5-11	98	0	0
12-18	51	2	1
19-24	34	6	3
25-30	35	5	8
31-36	32	10	9
37-42	23	7	8
43-49	12	6	10
50-59	4	2	5

Table 9.12: Years of employment and frequency of pneumoconiosis examination outcome among coal miners

In this example, $J = 3$ and for a specific year i, let

π_{1i} = the probability that pneumoconiasis is severe for year i

π_{2i} = the probability that pneumoconiasis is mild for year i

π_{3i} = the probability that pneumoconiasis is normal for year i

The baseline category logit model or multinomial logit model, with the last category being the baseline, sought to model $\ln(\pi_{1i}/\pi_{3i})$ and $\ln(\pi_{2i}/\pi_{3i})$, where $\pi_{ji}, j =$

1, 2, 3 refer respectively to severe, mild, and normal cases of pneumoconiosis for a given number of years (i) on the coal surface. Here, we have used the normal category as the reference, because we are assuming that workers are more likely to fall ill with increasing number of years of exposure to the coal face than otherwise. That is, a worker's health would rather deteriorate from normal to severe than the reverse with increasing years of exposure. To model these data, we could consider years as either a factor or categorical variable with 8 categories and try to fit a saturated baseline category model to the data. To implement this, however, we notice that for years group (0.5-11), two of the observations are zero. This would result in an infinite number of estimates for years (Agresti, 1996, p. 222). We can overcome this by simply adding some constant (0.1 or 0.5) to the two observations. The comparisons of the first year category will definitely depend on the choice of constants so employed.

The second approach that we would explore here is to consider a linear trend in age by utilizing the midpoints of the classes for years. In this case, the response R is the severity of pneumoconiosis and X is the number of years at the coal face. That is, we wish to fit the model,

$$\ln\left(\frac{\pi_j}{\pi_J}\right) = \alpha_j + \beta_j x_i, \quad j = 1, 2; i = 1, 2, \cdots, 8 \tag{9.19}$$

It has been suggested (McCullagh & Nelder, 1989) that a logarithmic transformation of years will explain the data better. We explore both options (years and log-years) in what follows. With explanatory variable years, the baseline-category logit model gives $G^2 = 13.90$ on 12 d.f. (pvalue = 0.3073), while with the explanatory variable being the log of years (designated as lyear here), we have $G^2 = 5.32$ on 12 d.f. (pvalue = 0.9461). Clearly, the log-year model fits better and we will therefore adopt this model. Basically, the baseline category model reduces for our example to the following:

$$\ln\left[\frac{\text{Severe} \mid \text{year}}{\text{Normal} \mid \text{year}}\right] = \alpha_1 + \beta_1 \text{lyear} \quad \text{and} \tag{9.20a}$$

$$\ln\left[\frac{\text{Mild} \mid \text{year}}{\text{Normal} \mid \text{year}}\right] = \alpha_2 + \beta_2 \text{lyear} \tag{9.20b}$$

The model in (9.20) has four parameters to be estimated. Below we give the parameter estimates under this model as well as the ML analysis of variance table. This model fits the data with a $G^2 = 5.33$ on 12 d.f. The effects of lyear are also significant (pvalue < 0.0001).

```
data base;
input years $ rep $ year count @@;
if rep eq 'sev' then resp='asever';
else if rep eq 'mild' then resp='bmild';
else resp='normal';
lyear=log(year);
datalines;
1 norm 5.75 98 1 mild 5.75 0 1 sev 5.75 0
.........................................
8 norm 54.5 4 8 mild 54.5 2 8 sev 54.5 5
;
run;
proc catmod;
weight count;
direct lyear; model resp=lyear;
run;
```

```
Maximum Likelihood Analysis of Variance

Source                DF   Chi-Square    Pr > ChiSq
--------------------------------------------------
Intercept              2        60.45        <.0001
lyear                  2        45.26        <.0001

Likelihood Ratio      12         5.33        0.9461


                 Analysis of Maximum Likelihood Estimates

           Function                  Standard       Chi-
Parameter  Number      Estimate        Error      Square    Pr > ChiSq
-----------------------------------------------------------------------
Intercept     1        -11.8366       1.9533       36.72        <.0001
              2         -8.8549       1.5558       32.39        <.0001
lyear         1          3.0249       0.5514       30.09        <.0001
              2          2.1402       0.4501       22.61        <.0001
```

Notice that to implement this model, we do not need to add constants to the two obervations whose average age on the coal face is 5.75 years because we are now not fitting a saturated model. The data=order statement asks that the severity, which is now recoded as asever, be used as the first category. Generally, CATMOD always makes the category with the highest value of the response variable the reference category or the last in alphabetical order if the categories of the response variable are read in as alphanumeric.

From the above result, we have, where $x = \ln(years)$,

$$\ln(\hat{\pi}_1/\hat{\pi}_3) = -11.8366 + 3.0249\, x_i \quad \text{and}$$
$$\ln(\hat{\pi}_2/\hat{\pi}_3) = -8.8549 + 2.1402\, x_i$$

Consequently, the estimated log odds that the response is "severe" rather than "mild" equals for a given x_i,

$$\begin{aligned}\ln(\hat{\pi}_1/\hat{\pi}_2) &= (-11.8366 + 8.8549) + (3.0249 - 2.1402)\, x_i \\ &= -2.9817 + 0.8847\, x_i\end{aligned} \tag{9.21}$$

The SAS software output shows that the Wald statistics for the four parameters are all significant ($p < .0001$).

The odds of severe versus normal pneumoconiosis are estimated to be

$$\left[\frac{\text{Severe}}{\text{Normal}}\right] = \exp[-11.8366 + 3.0249 * lyears]$$

For coal miners employed for two years, the model estimates the odds of pneumoconiosis to be $\frac{1}{17{,}000}$ and is at $\frac{1}{131}, \frac{1}{38}, \frac{1}{16}$ at 10, 15, and 20 years, respectively. That is, one of every 17 miners working on the coal face for 20 years will be at severe risk of the illness.

Similarly, the odds of a mild pneumoconiosis is estimated to be

$$\left[\frac{\text{Mild}}{\text{Normal}}\right] = \exp[-8.8549 + 2.1402(\text{lyears})]$$

The odds of a mild form of pneumoconiosis for miners who have been employed for 2, 10, 15, and 20 years on the coal face are estimated to be $\frac{1}{1590}, \frac{1}{51}, \frac{1}{21}, \frac{1}{12}$ respectively. That is, 1 in 13 coal face miners who have been employed for 20 years will develop a mild illness of pneumoconiosis.

9.3.3 Obtaining Response Probabilities

Estimated response probabilities (Agresti, 1990) can be obtained from the following expression

$$\hat{\pi}_j = \frac{\exp(\hat{\alpha}_j + \hat{\beta}_j x)}{\sum_k \exp(\hat{\alpha}_k + \hat{\beta}_k x)}, \quad j = 1, 2, \cdots, (J-1) \tag{9.22}$$

Using (9.22), we can obtain the estimated response probabilities of the outcomes (severe, mild and normal) as:

$$\hat{\pi}_1 = \frac{\exp(-11.8366 + 3.0249x)}{1 + \exp(-11.8366 + 3.0249x) + \exp(-8.8549 + 2.1402x)}$$

$$\hat{\pi}_2 = \frac{\exp(-8.8549 + 2.1402x)}{1 + \exp(-11.8366 + 3.0249x) + \exp(-8.8549 + 2.1402x)}$$

$$\hat{\pi}_3 = \frac{1}{1 + \exp(-11.8366 + 3.0249x) + \exp(-8.8549 + 2.1402x)}$$

where again $x =$ lyears is the log of years. When $j = J$ under the baseline category model, then, $\hat{\alpha}_J = \hat{\beta}_J = 0$. Consequently, $\exp(\hat{\alpha}_3 + \hat{\beta}_3) = 1$. This accounts for the 1 in the expression for the numerator in the expression for $\hat{\pi}_3$ above. For year $= 46$, then, $x = \ln(46) = 3.8286$ and therefore,

$$\hat{\pi}_2 = \frac{\exp[-8.8549 + 2.1402(3.8286)]}{1 + \exp[-11.8366 + 3.0249(3.8286)] + \exp[-8.8549 + 2.1402(3.8286)]}$$
$$= 0.2254$$

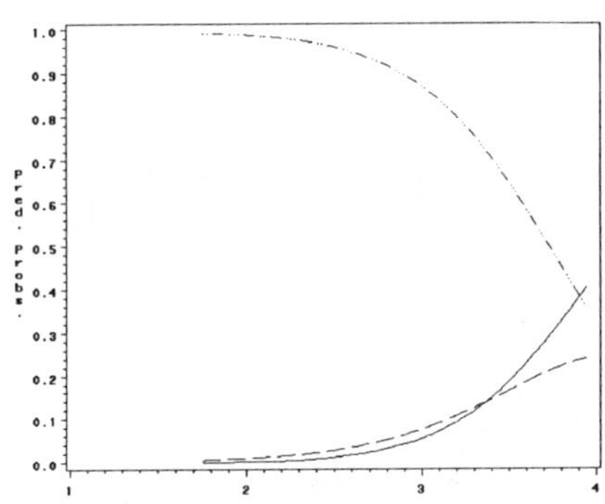

Figure 9.4: Predicted probabilities for severity of sickness

In Figure 9.4 is the plot of these predicted probabilities against log of years on the coal face. Here, the solid line represents those for mild, the normal has the top dotted line while the severe category has the elongated dotted line.

If we wish to test the hypothesis that all the parameter estimates for the $\ln(\pi_{1i}/\pi_{3i})$ model are equal to those of the $\ln(\pi_{2i}/\pi_{3i})$ model in equations (9.20), that is,

$$H_0 : \alpha_1 = \alpha_2 \quad \text{and} \quad \beta_1 = \beta_2$$

we can accomplish this with the following statements in SAS software by including the keyword **_RESPONSE_** as a variable in the model statement. In this case, a single set of parameters rather than separate sets of parameters at each cutpoint will be produced.

```
set base;
proc catmod; weight count; direct lyear;
model resp=_response_ lyear/noiter; run;

                    Maximum Likelihood Analysis

           Maximum likelihood computations converged.

     Maximum Likelihood Analysis of Variance

Source               DF   Chi-Square    Pr > ChiSq
--------------------------------------------------
Intercept             1        59.87        <.0001
_RESPONSE_            1         0.44        0.5080
lyear                 1        44.77        <.0001

Likelihood Ratio     13         7.21        0.8908


          Analysis of Maximum Likelihood Estimates

                               Standard        Chi-
Parameter          Estimate       Error      Square    Pr > ChiSq
-----------------------------------------------------------------
Intercept          -10.2032      1.3187       59.87        <.0001
_RESPONSE_  1        0.0733      0.1107        0.44        0.5080
lyear                2.5445      0.3803       44.77        <.0001
```

The corresponding partial output is also displayed and if we compare the G^2 for this model with the earlier model with separate sets of parameters for each cutpoint, we have $G^2 = 7.21 - 5.33 = 1.88$ on 1 d.f., which indicates that this hypothesis is tenable. Thus the pair of coefficients do not differ across the two models. We can also accomplish the above test by including a contrast statement in the original statements:

```
model resp=lyear; contrast 'test' @1 lyear 1 @2 lyear -1;
```

The contrast statement on implementation gives a Wald's Q = 1.88 on 1 degree of freedom. Again, there is no reason to doubt the hypothesis of equal set of parameter values for the two cutpoints. In other words, we can fit a single model of the form:

$$\ln(\hat{\pi}_1/\hat{\pi}_3) = -10.2032 + 2.5445x_i \quad \text{and}$$
$$\ln(\hat{\pi}_2/\hat{\pi}_3) = -10.2032 + 2.5445x_i$$

from which the relevant estimated response probabilities can be computed.

9.3.4 Example 9.7: Breast Examination Data

The data below are the results of a survey of women, relating frequency of breast self-examination and age (Senie et al., 1981).

	Freq. of breast self-examination		
Age	Monthly	Occasionally	Never
< 45	91	90	51
45-59	150	200	155
60+	109	198	172

Table 9.13: Frequency of self-examination of breast by age

For the data in Table 9.13, the explanatory variable is categorical and we would want to fit a baseline category logit model of the form:

$$\ln\left(\frac{\pi_j}{\pi_3}\right) = \alpha_j + \beta_j \text{ age} \quad j = 1, 2 \tag{9.23}$$

Because variable age is categorical with three levels, two dummy variables Z_1 and Z_2 would need to be created. PROC CATMOD creates its dummy variable by utilizing the *effect coding* scheme, where

$$Z_1 = \begin{cases} 1 & \text{if age } <45 \\ -1 & \text{if age } 60+ \\ 0 & \text{Otherwise} \end{cases} \qquad Z_2 = \begin{cases} 1 & \text{if age } 45\text{-}59 \\ -1 & \text{if age } 60+ \\ 0 & \text{Otherwise} \end{cases}$$

In this setup the baseline category logit model in (9.23) takes the form:

$$\ln\left(\frac{\pi_j}{\pi_3}\right) = \alpha_j + \beta_{1j} Z_1 + \beta_{2j} Z_2, \quad j = 1, 2 \tag{9.24}$$

Of course, it would be much simpler, especially if further analysis would be necessary to use the *reference coding* scheme where

$$Z_1 = \begin{cases} 1 & \text{for age } <45 \\ 0 & \text{elsewhere} \end{cases} \qquad Z_2 = \begin{cases} 1 & \text{for age } 45\text{-}59 \\ 0 & \text{elsewhere} \end{cases}$$

If we adopt this last coding scheme, then a baseline category logit model where the order of the response variable is {monthly, occasionally, never} has Never as the baseline category. We display below the SAS software output from the fit of this model to the data in Table 9.13. The ages are coded 1 to 3 for age groups <45, 45-59, and 60+, respectively.

```
DATA breast;
INPUT AGE EXAM $ COUNT @@;
Z1=(AGE= 1);
Z2=(AGE= 2);
X=AGE;
DATALINES;
1 MONTH 91 1 OCCASS 90 1 NEVER 51
2 MONTH 150 2 OCCASS 200 2 NEVER 155
3 MONTH  109 3 OCCASS 198 3 NEVER 172
;
PROC PRINT;
RUN;
PROC CATMOD ORDER=DATA; WEIGHT COUNT; DIRECT Z1 Z2;
MODEL EXAM=Z1 Z2/WLS NOITER; RUN;
```

Maximum Likelihood Analysis of Variance

Source	DF	Chi-Square	Pr > ChiSq
Intercept	2	25.61	<.0001
Z1	2	24.33	<.0001
Z2	2	6.67	0.0356
Likelihood Ratio	0	.	.

Analysis of Maximum Likelihood Estimates

Effect	Parameter	Estimate	Standard Error	Chi-Square	Pr > ChiSq
Intercept	1	-0.4561	0.1224	13.88	0.0002
	2	0.1408	0.1042	1.82	0.1768
Z1	3	1.0352	0.2135	23.51	<.0001
	4	0.4272	0.2039	4.39	0.0362
Z2	5	0.4234	0.1677	6.38	0.0116
	6	0.1141	0.1494	0.58	0.4449

The model is saturated and indicates that the effect of age group is very evident in the frequency of self-examination of breasts by women in the survey. The above estimates of the parameters lead to the following estimated logit equations:

$$\ln(\hat{\pi}_1/\hat{\pi}_3) = -0.4561 + 1.0352 z_1 + 0.4234 z_2$$
$$\ln(\hat{\pi}_2/\hat{\pi}_3) = 0.1408 + 0.4272 z_1 + 0.1141 z_2$$

The age effect seems to be concentrated on the <45 group relative to the over 60+ group for the "monthly" versus "never" equation and similarly on the 45-59 group relative to the 60+ group for the "monthly" versus "never" equation with pvalues of <0.0001 and 0.0116, respectively. We can therefore say that those who are under 45 years of age are exp(1.0352) = 2.82 times more likely to examine their breasts monthly rather than never than in the 60+ years old group. Similarly, the 45-59 age group has odds that are exp(0.4234) = 1.53 times higher than the over 60+ years group of monthly breast examination rather than never breast examination. The analysis also show that the group under 45 years are exp(0.4272) = 1.53 times more likely to examine their breasts occassionally than the 60+ age group.

The predicted probabilities of each of our responses are designated here as pi1 to pi3. For instance, the estimated probability for π_2="occassionally" when age group 45-59 is 0.3960, which is computed as:

$$\frac{\omega_2}{1 + \omega_1 + \omega_2}$$

where

$$\omega_1 = \exp[-0.4561 + (1.0352 * z_1) + (0.4234 * z_2)]$$
$$\omega_2 = \exp[0.1408 + (0.4272 * z_1) + (0.1141 * z_2)]$$

and substituting $z_1 = 0$ and $z_2 = 1$ in the above expression.

In the table below are these estimated probabilities for values of the explanatory variables.

AGE	Z1	Z2	pi1	pi2	pi3
1	1	0	0.3923	0.3879	0.2198
1	1	0	0.3923	0.3879	0.2198
1	1	0	0.3923	0.3879	0.2198
2	0	1	0.2970	0.3960	0.3069
2	0	1	0.2970	0.3960	0.3069
2	0	1	0.2970	0.3960	0.3069
3	0	0	0.2276	0.4134	0.3591
3	0	0	0.2276	0.4134	0.3591
3	0	0	0.2276	0.4134	0.3591

The model that assumes equal slope for $Z1$ and $Z2$, that is, $H_0 : \beta_{1j} = \beta_{2j}, j = 1, 2$ in equation (9.24) gives $G^2 = 10.52$ on 2 d.f. (pvalue = 0.0052), indicating that the model is not tenable.

Suppose, instead of the above analysis, we have assumed that the variable age group has a linear effect. That is, the effect from age group "<45" to "45-59" is the same as the effect from "45-59" to "60+." If we designate these levels by integer scores {X=1, 2, 3}, respectively, and we apply this in our model, we have a model with $G^2 = 0.53$ on 2 d.f. ($p = 0.7674$). The model fits well the data. The parameter estimates under this model are given below.

```
set breast;
PROC CATMOD ORDER=DATA;
WEIGHT COUNT; DIRECT X; MODEL EXAM=X/ML NOITER; RUN;

                Analysis of Maximum Likelihood Estimates

                                              Standard        Chi-
Effect         Parameter     Estimate           Error       Square    Pr > ChiSq
-------------------------------------------------------------------------------
Intercept              1       1.0139          0.2367        18.35        <.0001
                       2       0.6878          0.2283         9.08        0.0026
X                      3      -0.4985          0.1025        23.64        <.0001
                       4      -0.1904          0.0955         3.97        0.0463
```

Again, a model that assumes equal slopes for the X variable gives a $G^2 = 11.18 - 0.53 = 10.65$ on 1 d.f. This again indicates that this assumption is not tenable. Hence, the estimated baseline category models are:

$$\ln(\hat{\pi}_1/\hat{\pi}_3) = 1.0139 - 0.4985\,x$$
$$\ln(\hat{\pi}_2/\hat{\pi}_3) = 0.6878 - 0.1904\,x$$

The baseline category logit model can also be fitted to situations involving either several explanatory categorical variables or a mixture of explanatory categorical variables and continuous type variables. Some of the exercises at the end of this chapter are examples of these cases.

9.4 Models for Ordinal Response Variables

We shall consider here the situation in which the multinomial response variable has J ordered categories. In this case, it is always possible to reduce these J categories to a two-category variable by collapsing over some categories: for example, category 1 versus others or category i versus category j such that $i \neq j$ and where category j implies all other categories (except i) combined. There are $\binom{J}{2}$ such possible pairs of categories, and it is quite possible to model each pair by using either the complimentary log-log model or the logit model, where analysis is carried out in turn on each $\binom{J}{2}$ pairs of categories.

The above approach, however, is not without its drawbacks, namely, that the resulting $\binom{J}{2}$ models for the proportions in the J categories may not in general give fitted probabilities that sum to 1, and that the models themselves might involve different functions of the explanatory variable or different link functions (see Aitken et al., 1989). The above drawbacks can be overcome by the use of the multinomial logit model. We shall next discuss some of the specialized models that are available for the ordinal response variable.

9.4.1 Cumulative Logit Model

The cumulative logits are defined for observed counts f_{ij} as:

$$L_j = \ln\left(\sum_{j=i}^{i} f_{ij} / \sum_{j=i+1}^{J} f_{ij}\right) = \alpha_j + \beta X_i$$

for $j = 1, 2, \cdots, (J-1)$. We illustrate this concept with an example below where for a given i, the ordinal response variable has $J = 4$ categories.

i	j			
	1	2	3	4
	Low	Medium	High	Very high
	f_{i1}	f_{i2}	f_{i3}	f_{i4}

Then, for a particular level i of the explanatory variable X, we have the following decomposition of the cumulative model:

$$\ln\left(\frac{f_{i1}}{f_{i2}+f_{i3}+f_{i4}}\right) = \alpha_1 + \beta X_i$$
$$= \text{logit}\left(\frac{\text{Low}}{\geq \text{Medium}}\right)$$
$$\ln\left(\frac{f_{i1}+f_{i2}}{f_{i3}+f_{i4}}\right) = \alpha_2 + \beta X_i$$
$$= \text{logit}\left(\frac{\leq \text{Low}}{\geq \text{High}}\right)$$
$$\ln\left(\frac{f_{i1}+f_{i2}+f_{i3}}{f_{i4}}\right) = \alpha_3 + \beta X_i$$
$$= \text{logit}\left(\frac{\leq \text{High}}{\text{V. High}}\right)$$

The above model thus has one slope (β) but three intercepts (cut points): α_1, α_2, and α_3 respectively.

The simplest cumulative logit model has

$$L_j = \alpha_j, \quad j = 1, 2, \cdots, (J-1) \tag{9.25}$$

The model implies that the response variable is independent (simultaneously) of the explanatory variable X. In this case, the $\{\alpha_j\}$, the cut-point parameters, are nondecreasing in j. An application of the cumulative logit model to the data in Table 9.12 is implemented in PROC GENMOD as follows:

```
set base;
proc genmod; freq count;
model resp=lyear/dist=multinomial
                link=cumlogit
                aggregate=lyear type1;   run;

              Model Information
Data Set                              WORK.BASE
Distribution                        Multinomial
Link Function                  Cumulative Logit
Dependent Variable                         resp
Frequency Weight Variable                 count
Observations Used                            22
Sum Of Frequency Weights                    371
Missing Values                                2

        Response Profile
 Ordered                   Total
   Value    resp       Frequency
       1    asever            44
       2    bmild             38
       3    normal           289

PROC GENMOD is modeling the probabilities of levels of resp having LOWER Ordered
Values in the response profile table. One way to change this to model the
probabilities of HIGHER Ordered Values is to specify the DESCENDING option in the
PROC statement.
```

Criteria For Assessing Goodness Of Fit

Criterion	DF	Value	Value/DF
Deviance	13	5.0007	0.3847
Scaled Deviance	13	5.0007	0.3847
Pearson Chi-Square	13	4.6806	0.3600
Scaled Pearson X2	13	4.6806	0.3600
Log Likelihood		-204.2611	

Analysis Of Parameter Estimates

Parameter	DF	Estimate	Standard Error	Wald 95% Confidence Limits		Chi-Square	Pr > ChiSq
Intercept1	1	-10.5729	1.3446	-13.2081	-7.9376	61.83	<.0001
Intercept2	1	-9.6673	1.3241	-12.2625	-7.0721	53.30	<.0001
lyear	1	2.5943	0.3812	1.8472	3.3414	46.32	<.0001
Scale	0	1.0000	0.0000	1.0000	1.0000		

LR Statistics For Type 1 Analysis

Source	Deviance	DF	Chi-Square	Pr > ChiSq
Intercepts	101.6406			
lyear	5.0007	1	96.64	<.0001

The **aggregate=lyear** in the model statement asked that lyear be used to define the multinomial populations for computing the goodness-of-fit test statistics. A test of the significance of the covariate **lyear** is provided by the type1 options. This test is highly significant. It is important to make sure that the response profile is such that the level we wish to model is indeed the one being modeled. PROC GENMOD has several options to implement this reordering should this be the case. From the above results we have the following estimated cumulative logit equations

$$\ln\left(\frac{\hat{\pi}_1}{1-\hat{\pi}_1}\right) = -10.5729 + 2.5943\,\text{lyear} \tag{9.26a}$$

$$\ln\left(\frac{\hat{\pi}_1+\hat{\pi}_2}{1-(\hat{\pi}_1+\hat{\pi}_2)}\right) = -9.6673 + 2.5943\,\text{lyear} \tag{9.26b}$$

We consider in the next sections some of the specialized cumulative models that have been applied to categorical data having ordinal response variable(s).

9.4.2 The Proportional Odds Model

For this model, each category of the J ordinal category response variable is considered in turn and the frequency of response at least up to that point on the ordinal scale is compared to the frequency for all points higher on the scale. That is, the first category is compared to all the rest combined, then the first and the second combined are compared to all the rest combined, and so on, which results in the original J-response table being converted into a series of J − 1 subtables, each with a binary categorization, lower or higher than the point on the scale. The model can be formally written for a single explanatory variable X as:

$$\ln\left[\frac{P(Y \le j \mid x_i)}{P(Y > j \mid x_i)}\right] = \alpha_j + \beta x_i \quad j = 1, 2, \cdots (J-1);\; i = 1, 2, \cdots, I \tag{9.27}$$

We see that there are $(J-1)$ separate proportional odds equations for each possible cut point j, with each having a common slope β and different intercepts α_j. In this formulation, the α_j's are not themselves important, but the slope parameter β is.

As an example of the application of this model, let us consider again the data in Table 9.12 relating years of employment on the coal face to the development of pneumoconiosis.

If we let $\pi_{j|i}$ be the probability that the response is in category j given that the explanatory variable is x_i, then

$$\sum_{j=1}^{J} \pi_{j|i} = 1, \quad i = 1, 2, \cdots, I$$

Thus there is a linear restriction on the $\pi_{j|i}$, and there would therefore be $I(J-1)$ parameters of the conditional multinomial distribution. For the data in Table 9.12, $J = 3$, and we would therefore have $(J - 1) = 2$ cut points in our model. The proportional odds model for this data can thus be formulated as follows:

$$\ln\left[\frac{P(\text{Severe} \mid x_i)}{P(\text{Mild or Normal} \mid x_i)}\right] = \alpha_1 + \beta_1 \text{ lyear}$$

$$\ln\left[\frac{P(\text{Severe or Mild} \mid x_i)}{P(\text{Normal} \mid x_i)}\right] = \alpha_2 + \beta_2 \text{ lyear}$$

Although there are four parameters to be estimated in the above model, the proportional odds model assumes that $\beta_1 = \beta_2 = \beta$. This assumption will be tested with what is called the *score test for the proportional odds assumption.* If this assumption holds, then the model has three parameters to be estimated from the data. We give below the SAS software statements to implement this model together with a modified SAS software output from the implementation.

```
set base; proc logistic; weight count;
model resp=lyear/scale=none aggregate; run;

Score Test for the Proportional Odds Assumption

Chi-Square        DF      Pr > ChiSq
-----------------------------------
   0.1387          1          0.7096

        Deviance and Pearson Goodness-of-Fit Statistics
Criterion          DF          Value      Value/DF      Pr > ChiSq
------------------------------------------------------------------
Deviance           13         5.0007        0.3847          0.9752
Pearson            13         4.6806        0.3600          0.9816
Number of unique profiles: 8

                Analysis of Maximum Likelihood Estimates
                                    Standard
Parameter      DF     Estimate        Error     Chi-Square      Pr > ChiSq
---------------------------------------------------------------------------
Intercept       1     -10.5728       1.3463        61.6776          <.0001
Intercept2      1      -9.6672       1.3249        53.2392          <.0001
lyear           1       2.5943       0.3813        46.2850          <.0001

               Odds Ratio Estimates
               Point             95% Wald
Effect      Estimate          Confidence Limits
-----------------------------------------------
lyear         13.387         6.340          28.268
```

The score test for the proportional odds assumption gives a $G^2 = 0.1387$ on 1 d.f. (p $= 0.7096$). Thus the assumption that $\beta_1 = \beta_2$ is well justified. Its degree of freedom is obtained as $P(J-2)$, where P is the number of explanatory variables in the model.

In our case, P = 1 and hence the d.f. $= 1(3-2) = 1$. The goodness-of-fit test statistic for this model is the deviance or $G^2 = 5.0007$ on $I(J-1) - r = 8(3-1) - 3 = 13$ degrees of freedom. Here r is the number of parameters being estimated and I is the number of cell combinations generated by the product of the levels of the explanatory variables. In this example, $I = 8$ and $r = 3$.

The odds of severe pneumoconiosis is estimated to be

$$\left[\frac{\text{Severe}}{\text{Mild or Normal}}\right] = \exp(-10.5728 + 2.5943\,\text{lyear})$$

For coal miners employed for 2 years, the model estimates the odds of pneumoconiosis to be $\frac{1}{6468}$, and $\frac{1}{100}$, $\frac{1}{35}$, and $\frac{1}{17}$ at 10, 15, and 20 years, respectively. That is, one of every 18 miners working on the coal face for 20 years will be at severe risk of the illness.

Similarly, the estimates of the odds of severe or mild pneumoconiosis are estimated to be

$$\left[\frac{\text{Sever or Mild}}{\text{Normal}}\right] = \exp(-9.6672 + 2.5943\,\text{lyear})$$

The odds of pneumoconiosis in this case, for 2, 10, 15, and 20 years on the coal face, are again estimated to be $\frac{1}{2615}, \frac{1}{40}, \frac{1}{14}$, and $\frac{1}{7}$, respectively. That is, 1 in 8 coal face miners who have been employed for twenty years will develop a severe or mild illness of pneumoconiosis. The results here are very consistent with our earlier results, which utilized the baseline category logit model. However, the proportional odds model is more parsimonious than the baseline category model. The above analysis is contingent upon the proportional odds assumption being true. The assumption assumes the same slope for each odds.

9.4.3 Breast Examination Example Revisited

We shall next consider, in the following sections, the fitting of the proportional odds model when the explanatory variable has more than two categories or when there are several explanatory variables, each having two or more categories. We shall illustrate the first with the data in Table 9.13. The proportional odds model can be formulated again as:

$$\ln\left[\frac{P(\text{monthly} \mid \text{age})}{P(\text{occasionally or never} \mid \text{age})}\right] = \alpha_1 + \beta\,\text{age}$$

$$\ln\left[\frac{P(\text{monthly or occasionally} \mid \text{age})}{P(\text{never} \mid \text{age})}\right] = \alpha_2 + \beta\,\text{age}$$

Since the explanatory variable is categorical, we again create two dummy variables with the last category as the reference category. The two indicator variables are

$$Z_1 = \begin{cases} 1 & \text{for age} < 45 \\ 0 & \text{elsewhere} \end{cases} \qquad Z_2 = \begin{cases} 1 & \text{for age 45-59} \\ 0 & \text{elsewhere} \end{cases}$$

Specifically in this problem, we have the following 6 equations:

$$\ln\left[\frac{P(\text{monthly} \mid <45)}{P(\text{occasionally or never} \mid <45)}\right] = \alpha_1 + \beta_1$$

$$\ln\left[\frac{P(\text{monthly} \mid 45\text{-}59)}{P(\text{occasionally or never} \mid 45\text{-}59)}\right] = \alpha_1 + \beta_2$$

$$\ln\left[\frac{P(\text{monthly} \mid 60+)}{P(\text{occasionally or never} \mid 60+)}\right] = \alpha_1$$

$$\ln\left[\frac{P(\text{monthly or occasionally} \mid <45)}{P(\text{never} \mid <45)}\right] = \alpha_2 + \beta_1$$

$$\ln\left[\frac{P(\text{monthly or occasionally} \mid 45\text{-}59)}{P(\text{never} \mid 45\text{-}59)}\right] = \alpha_2 + \beta_2$$

$$\ln\left[\frac{P(\text{monthly or occasionally} \mid 60+)}{P(\text{never} \mid 60+)}\right] = \alpha_2$$

Here $J = 3, I = 3$. The model has a total of 4 parameters to be estimated from the data. We give below some SAS software output from this analysis.

```
set breast; proc logistic order=data; freq count;
MODEL EXAM=z1 z2/scale=none aggregate; test: test z1=z2; run;

Score Test for the Proportional Odds Assumption

Chi-Square        DF      Pr > ChiSq
------------------------------------
    0.7077         2          0.7020

       Deviance and Pearson Goodness-of-Fit Statistics

Criterion         DF           Value      Value/DF      Pr > ChiSq
-------------------------------------------------------------------
Deviance           2          0.7124        0.3562          0.7003
Pearson            2          0.7127        0.3563          0.7002

Number of unique profiles: 3
              Analysis of Maximum Likelihood Estimates

                                       Standard
Parameter      DF     Estimate            Error     Chi-Square     Pr > ChiSq
-----------------------------------------------------------------------------
Intercept       1      -1.1783           0.0943       156.1188         <.0001
Intercept2      1       0.5514           0.0888        38.5888         <.0001
Z1              1       0.7314           0.1494        23.9616         <.0001
Z2              1       0.2895           0.1181         6.0063         0.0143

           Odds Ratio Estimates
             Point            95% Wald
Effect     Estimate       Confidence Limits

Z1            2.078        1.550       2.785
Z2            1.336        1.060       1.684
```

The proportional odds model assumption that the slope is the same across the response profiles is tested with a score test of 0.7077 on 2 d.f. ($p = 0.7020$). Here $P = 2$, hence d.f. $= 2(3 - 2) = 2$. This assumption is well satisfied. The model gives $G^2 = 0.7124$ on $3(3 - 1) - 4 = 2$ d.f. ($p = 0.7003$), which indicates that the model again fits well the data. The proportional odds or cumulative odds model can also be implemented for the Breast examination data in Table 9.13 via PROC CATMOD and GENMOD with the following statements.

```
set breast; PROC CATMOD order=data; WEIGHT COUNT;
DIRECT Z1 Z2; response clogits;
MODEL EXAM=_response_ z1 z2/WLS NOITER;  RUN;
proc genmod rorder=data; class age; freq count;
model exam=age/dist=multinomial
                link=cumlogit
                aggregate=age
                type1;
estimate 'odds1' age 1 -1/exp; estimate 'odds2' age 1 0 -1/exp;
estimate 'odds3' age 0 1 -1/exp;  run;
                              Contrast Estimate Results

                          Standard                                      Chi-
Label          Estimate      Error    Alpha    Confidence Limits    Square    Pr > ChiSq
-----------------------------------------------------------------------------------------
odds1            0.4419     0.1477     0.05     0.1523      0.7315      8.95        0.0028
Exp(odds1)       1.5557     0.2298     0.05     1.1645      2.0782
odds2            0.7315     0.1493     0.05     0.4388      1.0242     23.99        <.0001
Exp(odds2)       2.0781     0.3104     0.05     1.5508      2.7848
odds3            0.2896     0.1181     0.05     0.0582      0.5210      6.02        0.0142
Exp(odds3)       1.3359     0.1577     0.05     1.0599      1.6837
```

Notice that the implementation of this model in PROC GENMOD instructs genmod to model the response profiles as MONTHLY, OCCASIONALLY, and NEVER in that order. This is accomplished with the **rorder** option in the PROC GENMOD statement line. The odds computed agree with those obtained under the logistic model approach. We also observe here that we have have used the **CLASS** statement in GENMOD, because the coding here is reference coding. We accomplish same in PROC LOGISTIC with the **REF=last** statement.

A further test of the equality of the β parameters, that is, $\beta_1 = \beta_2$, gives a Wald test statistic of 8.9872 on 1 d.f. ($p = 0.0027$) in PROC LOGISTIC, indicating the rejection of this hypothesis. This hypothesis is tested in PROC GENMOD with the **odds1** estimate with a pvalue of 0.0028. The same conclusion is obtained.

Similarly, a test of the hypothesis $\beta_1 = \beta_2 = 0$, which is equivalent to the model of independence, also gives a Wald test statistic of 24.1866 on 2 d.f. ($p < .0001$), again indicating that the model of independence is not tenable and that the model permitting an effect fits better than the model of independence. These tests are implemented in PROC LOGISTIC with the following for this model.

```
proc logistic order=data; freq count;
model exam=z1 z2/aggregate scale=none; test: test z1=z2;
test2: test z1, z2; output out=aa p=phat; run;

    Linear Hypotheses Testing Results
                 Wald
 Label     Chi-Square      DF     Pr > ChiSq
 -------------------------------------------
 test          8.9872       1         0.0027
 test2        24.1866       2         <.0001
```

Estimated Response Probabilities

Estimated response probabilities can easily be calculated. For instance, the cumulative probabilities are computed from

$$P(R \leq j) = \frac{\exp(\hat{\alpha}_j + \hat{\beta}_1 z_1 + \hat{\beta}_2 z_2)}{1 + \exp(\hat{\alpha}_j + \hat{\beta}_1 z_1 + \hat{\beta}_2 z_2)}$$

The second estimated cumulative probability for respondents who are in the age group 45-59 and are monthly or occasionally examining their breasts ($z_1 = 0, z_2 = 1, j = 2$) equals

$$\frac{\omega_2}{1+\omega_2} = 0.2914 \qquad \text{where}$$
$$\omega_2 = \exp[-1.1783 + 0.2895\,(1)]$$

These probabilities are obtained in PROC LOGISTIC with the option statement **output out=aa p=phat** in the program above.

Sometimes if the proportional odds model does not fit well the data of interest, a partial proportional odds model (Koch et al., 1985) is suggested as an alternative. We can of course seek other multinomial logit models that might fit the data better.

In what follows, we again consider age as a continuous variable having a linear effect with categories again assigned integer scores {1, 2, 3}. In this case, the cumulative logit model again fits the data with $G^2 = 1.1839$ on 3 d.f. ($p = 0.7569$).

The hypothesis of common slope, which is tested by the score test, has $p = 0.4736$, indicating that the proportional odds model assumption of equal slopes is satisfied. We present below the result of this analysis.

```
set breast; proc logistic order=data; weight count;
model exam=x/scale=none aggregate; run;
```

Score Test for the Proportional Odds Assumption

Chi-Square	DF	Pr > ChiSq
0.5135	1	0.4736

Deviance and Pearson Goodness-of-Fit Statistics

Criterion	DF	Value	Value/DF	Pr > ChiSq
Deviance	3	1.1839	0.3946	0.7569
Pearson	3	1.1904	0.3968	0.7553

Analysis of Maximum Likelihood Estimates

Parameter	DF	Estimate	Standard Error	Chi-Square	Pr > ChiSq
Intercept	1	-0.1394	0.1684	0.6853	0.4078
Intercept2	1	1.5899	0.1751	82.4020	<.0001
X	1	-0.3535	0.0726	23.7230	<.0001

We have for a given age $= x$, $\exp(-0.3535) = 0.7022$, implying that each level unit increase in the level of that variable multiplies the odds of monthly versus (occasionally or never), (monthly or occasionally) versus (never) by 0.7022. Consequently, the odds of examination monthly are 0.7022 times greater when the respondent is in the age group <45 than when the respondent is in age group 45-59. The odds are $0.7022^2 = 0.4931$ times greater when the respondent is in age group <45 or 45-59 than when the respondent is in age group 60+. In other words, those in the 60+ group are 2.03 times more likely to monthly examine their breasts than those in the $\leq$59 year age group.

It is obvious from the two examples examined in this section that the proportional odds model is equivalent to the cumulative logit model. While PROC GENMOD and CATMOD can be employed to implement the cumulative logit model, PROC LOGISTIC can similarly be employed to implement the equivalent proportional odds model.

9.4.4 Adjacent-Category Logits

For an ordinal response variable with J categories, comparing adjacent categories (that is, each category to the next category) leads to what has been described (defined here in terms of expected frequencies) as the *adjacent category logits* (Agresti, 1990). It is defined for an individual or subject i to be:

$$\ln\left(\frac{\hat{m}_{ij}}{\hat{m}_{i,j+1}}\right) = \alpha_j + \boldsymbol{\beta}_j^{'}\mathbf{X}, \quad j = 1, 2, \cdots, (J-1) \tag{9.28}$$

If we let the expected frequencies be modeled as in (9.28), then the above model is also equivalent to:

$$\ln\left(\frac{\hat{m}_{i,j+1}}{\hat{m}_{i,j}}\right) = \alpha_j + \boldsymbol{\beta}_j^{'}\mathbf{X}_i \quad j = 1, 2, \cdots, (J-1) \tag{9.29}$$

In the above two formulations, the adjacent-category model is obtained by the imposition of a constraint on the set of $(J-1)$ equations in either (9.28) or (9.29). Specifically, we impose the constraint $\beta_j = \beta$ for all j. This is accomplished in CATMOD with the **_response_** in the model statement.

The above does define logits for all $\binom{J}{2}$ pairs of response categories, and can be interpreted as the odds of getting category j relative to category $j+1$. They can also be viewed as the conditional odds given that either category j or $j+1$ occurs.

Let us reanalyze the breast examination survey data data in Table 9.13 using the adjacent category model approach. This model has indicator variables Z_1 and Z_2 as defined previously. To fit adjacent-category models in PROC CATMOD in SAS software, we shall define the response as **ALOGITS** or **ALOGIT**. The adjacent-category model when fitted to the data has the partial SAS software output displayed below. The goodness-of-fit test statistic for this model has $G^2 = 0.63$ on 2 d.f. ($p = 0.7308$). The model fits the data very well.

```
set breast;
proc catmod order=data; weight count; direct z1 z2 x;
response alogits; model exam=_response_ z1 z2/ml noiter;
contrast  'three' z1 1 z2 -1; run;
```

Analysis of Weighted Least Squares Estimates

Effect	Parameter	Estimate	Standard Error	Chi-Square	Pr > ChiSq
Intercept	1	0.2226	0.0608	13.39	0.0003
RESPONSE	2	0.3112	0.0590	27.86	<.0001
Z1	3	-0.5158	0.1059	23.71	<.0001
Z2	4	-0.2059	0.0835	6.08	0.0137

The ML estimates of the age group effects are $\hat{\beta}_1 = -0.5158$ and $\hat{\beta}_2 = -0.2059$. The effects of βs are significant. Based on the above parameter estimates, therefore, the estimated odds that an individual in age group <45 does do self-examination of breast as being in category $j+1$ rather than in j are $\exp(-0.5158) = 0.597$ times the estimated odds for those in age group 60+. Further, the estimated odds of "monthly" instead of "occasionally" self-examination of breast of the <45 age group is about 60% of the odds of the 60+ group. In general, the estimated odds for any pair of response categories $(c1, c2), c1 > c2$ is $\exp[\hat{\beta}_1(c1 - c2)]$ for the <45 age group and $\exp[\hat{\beta}_2(c1 - c2)]$ for the 45-59 age group (all relative to the last category). For instance, the estimated odds of "Monthly" (category 1) instead of

"Never" (category 3) is $e^{[-0.5158(3-1)]} = 0.356$. That is, those in the age group <45 have odds that are about 36% of the odds of those in the 60+ group in the response to self-examination of breast. In the above model, we see that the effect of Z_2 is also significant and similar interpretations can be made for the 45-59 group relative to the 60+ group. A test of whether the effect of Z_1 is significantly different from that of Z_2 is provided by the contrast statement in the program above. This test yields a Wald $W_Q = 8.95$ on 1 d.f., with a pvalue of 0.0028.

An equivalent log-linear model formulation of the adjacent category model can be implemented with the following SAS software statements:

```
data breast2;
input age exam  count @@;
examl=exam;
datalines;
1 1 91 1 2 90 1 3 51 2 1 150 2 2 200 2 3 155
3 1  109 3 2 198 3 3 172
;
proc genmod; class exam age;
model count=age exam age*examl/dist=poi type3; run;
```

Criteria For Assessing Goodness Of Fit

Criterion	DF	Value	Value/DF
Deviance	2	0.6268	0.3134
Scaled Deviance	2	0.6268	0.3134
Pearson Chi-Square	2	0.6274	0.3137
Scaled Pearson X2	2	0.6274	0.3137

Analysis Of Parameter Estimates

Parameter		DF	Estimate	Standard Error	Wald 95% Confidence Limits		Chi-Square	Pr > ChiSq
Intercept		1	5.1664	0.0717	5.0257	5.3070	5185.60	<.0001
.....								
examl*age	1	1	-0.5160	0.1060	-0.7237	-0.3083	23.71	<.0001
examl*age	2	1	-0.2060	0.0833	-0.3693	-0.0428	6.12	0.0134

Note that, "exam" and "age" group are declared as categorical variables, while a derived variable **examl** from a data step is declared as a quantitative (integer scored) variable. The log parameter estimates for examl*age agree with those obtained from PROC CATMOD.

Again if we consider age as a continuous variable, then a fit of this model has the following results:

```
set breast;
proc catmod order=data; weight count; direct x;
response alogits; model exam=_response_ x/wls noiter; run;
```

Analysis of Weighted Least Squares Estimates

Effect	Parameter	Estimate	Standard Error	Chi-Square	Pr > ChiSq
Intercept	1	-0.5112	0.1190	18.45	<.0001
RESPONSE	2	0.3107	0.0589	27.78	<.0001
X	3	0.2498	0.0516	23.47	<.0001

The WLS estimate for this model is $\hat{\beta} = 0.2498$. The model fits the data with Wald statistic $Q = 1.07$ on 3 d.f. ($p = 0.7832$). Based on this model, the estimated odds that an individual in age group <45 does do self-examination of breast as being in category $j+1$ rather than j are exp(0.2498) = 1.284 times the estimated odds for those in age group 45-59. This implies that a percentage change in the odds for each 1-unit increase (group change e.g. <45 to 45-59) in age group is associated with a 28% increase in the odds of being in $j+1$-th rather than in the j-th response category.

The estimated odds of "Monthly" (category 1) instead of "Never" (category 3) is $e^{[0.2498(3-1)]} = 1.65$. That is, those in the age group <45 have odds that are about 65% higher than those in the 45-59 age group. Further, those in the age group <45 have odds that are about $(1 - 1.65^2) \times 100 = 172\%$ higher than those in the 60+ group in the response to self-examination of breast. The latter result is as a result of the linear assumption for the effect of age.

When the adjacent-odds ratio model is applied to the coal miners data (with 0.1 added to the cells with zero counts), the model gives a Wald statistic of $Q = 4.70$ on 13 d.f. ($p = 0.9812$). The model clearly fits the data with an estimated equation:

$$\ln\left(\frac{\hat{m}_{ij}}{\hat{m}_{i,j+1}}\right) = 6.0428 - 1.5490 \text{ lyear} \quad \text{for } j = 1, 2 \tag{9.30}$$

For a miner who has spent 10 years on the coal face, the odds that the individual would have a severe pnemoconiosis rather than a mild pnumoconiosis are exp(2.476)=11.9 times higher. Similarly, the predicted odds that the individual would have mild pnemoconiosis rather than normal are also 11.9 times higher. In other words, whenever we compare adjacent pneumoconiosis categories, miners have about 11.9 times more chance of being in the worse category for those that have been employed for 10 years. Similar odds can be obtained for those with 15 or 20 years of employment on the coal face.

9.4.5 Continuation Ratio Model

For an ordered response variable, the continuation ratio logits are defined as

$$\begin{aligned} L_j &= \ln\left(\frac{\hat{m}_j}{\hat{m}_{j+1} + \cdots + \hat{m}_J}\right), \quad j = 1, 2, \cdots (J-1) \\ &= \ln\left(\frac{\hat{m}_i}{\sum_{j=i+1}^{J} \hat{m}_j}\right), \quad j = 1, 2, \cdots (J-1) \end{aligned} \tag{9.31}$$

which in terms of probabilities, becomes

$$\ln\left(\frac{\pi_1}{\pi_2 + \cdots + \pi_J}\right) + \ln\left(\frac{\pi_2}{\pi_3 + \cdots + \pi_J}\right) + \cdots + \ln\left(\frac{\pi_{(J-1)}}{\pi_J}\right) \tag{9.32}$$

The model can alternatively be defined as:

$$\ln\left(\frac{\pi_1}{\pi_2}\right) + \ln\left(\frac{\pi_1 + \pi_2}{\pi_3}\right) + \cdots + \ln\left(\frac{\pi_1 + \cdots + \pi_{(J-1)}}{\pi_J}\right) \tag{9.33}$$

where the first category is compared to the second, the first and second combined to the third, and so on. The two model formulations lead to different parameter estimates as well as different goodness-of-fit test statistics. We shall adopt the form in (9.33) in this text.

The continuation ratios model resembles the proportional odds model, except that for each category of the ordinal table considered in turn, the frequency of the response variable at least up to that point on the ordinal scale is compared only to the frequency for the immediately following category. The implementation of this model involves creating a series of subtables and fitting individual logistic regression models to these subtables. We illustrate the fitting of the continuation ratio model by applying this model to the data in Tables 9.12 and 9.13 again. First for the data in Table 9.12, we form two separate subtables of successes/failures (a) and (b) in the first subtable and (a)+(b) and (c) in the second subtable. The first subtable (ratio 1, Table 1) corresponding to the ratio $\left(\frac{\text{severe}}{\text{mild}}\right)$, while the second ratio (Table 2) corresponds to $\left(\frac{\text{severe+mild}}{\text{normal}}\right)$. The results from forming these subtables are displayed in the Table 9.14.

Table 1			Table 2			
severe	Mild		Severe+mild	Normal		
(a)	(b)	n	(a)+(b)	(c)	n	years
0	0	0	0	98	98	5.75
1	2	3	3	51	54	15.00
3	6	9	9	34	43	21.50
8	5	13	13	35	48	27.50
9	10	19	19	32	51	33.50
8	7	15	15	23	38	39.50
10	6	16	16	12	28	46.00
5	2	7	7	4	11	51.50

Table 9.14: Subtables formed from the original data

To fit the continuation ratio model to these data, we now fit separate logistic regression models to each subtable using the log of years as the explanatory variables. Goodness-of-fit statistics for these models as well as the parameter estimates are displayed below.

```
data crm;
input normal mild severe year @@;
lyear=log(year);
datalines;
98 0 0 5.75 51 2 1 15.0
.......................
12 6 10 46.0 4 2 5 51.50
;
run;
data new1;
set crm;
table=1; r=severe; n=severe+mild; output;
table=2; r=severe+mild; n=normal+mild+severe; output; run;
proc print; run;

SUB-TABLE ONE
-------------
proc logistic; where table=1;
model r/n=lyear/scale=none aggregate;
run;
       Deviance and Pearson Goodness-of-Fit Statistics

Criterion         DF          Value      Value/DF      Pr > ChiSq
-----------------------------------------------------------------
Deviance           5         1.7471        0.3494          0.8829
Pearson            5         1.7356        0.3471          0.8844
```

```
              Analysis of Maximum Likelihood Estimates

                                 Standard
Parameter    DF     Estimate        Error     Chi-Square     Pr > ChiSq
-----------------------------------------------------------------------
Intercept     1      -3.8639       2.6880         2.0663         0.1506
lyear         1       1.1363       0.7588         2.2424         0.1343

SUB-TABLE TWO
-------------
proc logistic;
where table=2;
model r/n=lyear/scale=none aggregate;
run;
       Deviance and Pearson Goodness-of-Fit Statistics

Criterion          DF          Value      Value/DF      Pr > ChiSq
--------------------------------------------------------------
Deviance            6         3.1045        0.5174          0.7956
Pearson             6         2.5801        0.4300          0.8594

Number of unique profiles: 8

              Analysis of Maximum Likelihood Estimates

                                 Standard
Parameter    DF     Estimate        Error     Chi-Square     Pr > ChiSq
-----------------------------------------------------------------------
Intercept     1      -9.5997       1.3399        51.3293         <.0001
lyear         1       2.5734       0.3866        44.3199         <.0001
```

The G^2 for the full continuation model is the sum of the separate G^2 for each subtable and corresponds to the G^2 for simultaneous fitting of the two models. In this case our overall $G^2 = 1.7471 + 3.1045 = 4.8516$ on $5 + 6 = 11$ degrees of freedom. The model certainly fits the data. The odds can be computed for each ratios as before, and we see here that the continuation ratio has two intercepts and two slopes, in contrast to the proportional odds model with a common slope.

Lindsey (1995) has suggested that the full continuation ratio should include the subtables as a factor variable in the model. When we implement this, we have the following results from SAS software. Note the parameterization of tables in this case. The table variable is coded 1 and 0 for Tables 1 and 2 respectively. This is accomplished with the **ref=** in the class statement.

```
proc logistic;
class table (ref=last)/param=ref;
model r/n=table|lyear/scale=none aggregate; run;

      Deviance and Pearson Goodness-of-Fit Statistics

Criterion          DF          Value      Value/DF      Pr > ChiSq
--------------------------------------------------------------
Deviance           11         4.8516        0.4411          0.9381
Pearson            11         4.3156        0.3923          0.9598

                Analysis of Maximum Likelihood Estimates

                                     Standard
Parameter          DF     Estimate       Error     Chi-Square     Pr > ChiSq
--------------------------------------------------------------------------
Intercept           1      -9.5995      1.3399        51.3293         <.0001
table         1     1       5.7355      3.0034         3.6467         0.0562
lyear               1       2.5734      0.3865        44.3198         <.0001
lyear*table 1       1      -1.4370      0.8516         2.8473         0.0915
```

The result of fitting the model with table, lyear, and lyear and table interaction terms gives a deviance value that agrees with the sum of deviances obtained from fitting the individual models as outlined above. However, we observe that the interaction term is not significant in this model. Removing this term and refitting the continuation ratio model gives the result displayed below.

```
         Deviance and Pearson Goodness-of-Fit Statistics

Criterion          DF          Value      Value/DF     Pr > ChiSq
------------------------------------------------------------------
Deviance           12         7.5899        0.6325         0.8163
Pearson            12         6.9192        0.5766         0.8629

                 Analysis of Maximum Likelihood Estimates

                                      Standard
Parameter        DF     Estimate         Error     Chi-Square     Pr > ChiSq
-----------------------------------------------------------------------------
Intercept         1      -8.7251        1.1288        59.7468         <.0001
table       1     1       0.6824        0.2750         6.1585         0.0131
lyear             1       2.3189        0.3270        50.3053         <.0001

                 Odds Ratio Estimates

                    Point          95% Wald
Effect           Estimate      Confidence Limits
--------------------------------------------------
table 1 vs 2        1.979        1.154       3.392
lyear              10.165        5.356      19.293
```

The model still fits the data with $G^2 = 7.5899$ on 12 d.f. The continuation ratio therefore provides an alternative to both the baseline category logit and proportional odds (cumulative logit) models.

To fit the continuation ratio model to the breast data, we need to first generate the following two subtables of successes/failures.

	Table1		Table2	
Age	(a)	(b)	(a+b)	(c)
1	91	181	181	232
2	150	350	350	505
3	109	307	307	479

The above is accomplished by the following SAS software staments and partial output:

```
options nodate nonumber ls=85 ps=66;
data breast;
do age= 1 to 3;
input month occass never @@;
output; end;
datalines;
91 90 51 150 200 155 109 198 172
;
run;
data new1;
set breast;
table=1; r=month; n=month+occass; output;
table=2; r=month+occass; n=month+occass+never; output;
run; proc print; run;

Obs    age    month    occass    never    table      r      n
---------------------------------------------------------------
 1      1       91        90       51       1       91    181
 2      1       91        90       51       2      181    232
 3      2      150       200      155       1      150    350
 4      2      150       200      155       2      350    505
 5      3      109       198      172       1      109    307
 6      3      109       198      172       2      307    479
```

A logistic regression fitted to subtable 1 gives a$G^2 = 0.0013$ on 1 d.f. with $\hat{\beta} = -0.3045$, a.s.e. (0.0943). Similarly, the logistic model applied to subtable 2 gives a $G^2 = 0.6601$ on 1 d.f. with $\hat{\beta} = -0.3199$, a.s.e. (0.0865). Consequently, the combined continuation ratio is $G^2 = 0.0013 + 0.6601 = 0.6614$ on 2 d.f. Each of our models assumes a linear trend in age group where integer scores are assigned. We are exploiting the intrinsic ordering of this variable. The continuation ratio model therefore fits the breast data with a pvalue of 0.7184.

For each of the above subtables with three observations per subtable, a logistic model that assumes that age group is categorical with three categories, as an explanatory variable would have been a saturated model. Consequently, we assume in the previous analysis that the variable "age" has a linear effect by assigning integer scores (1,2,3) to the three categories. An alternative method of fitting the continuation ratio (Lindsey, 1995) employs fitting a single model simultaneously to the subtables so created by assuming the subtables form a factor variable and does not in any way assume the linear trend effect of the original explanatory variable.

A logistic regression fitted to the above data with age and subtable as factor variables give $G^2 = 0.3649$ on 2 d.f. ($p = 0.8332$).

```
proc logistic order=data;
class table (ref=last)  age (ref=last)/param=ref;
model r/n=table age/scale=none aggregate;
run;
```

Deviance and Pearson Goodness-of-Fit Statistics

Criterion	DF	Value	Value/DF	Pr > ChiSq
Deviance	2	0.3649	0.1824	0.8332
Pearson	2	0.3643	0.1822	0.8335

Analysis of Maximum Likelihood Estimates

Parameter		DF	Estimate	Standard Error	Chi-Square	Pr > ChiSq
Intercept		1	0.5730	0.0830	47.6906	<.0001
table	1	1	-1.1599	0.0946	150.4801	<.0001
age	1	1	0.6431	0.1315	23.9054	<.0001
age	2	1	0.2672	0.1038	6.6270	0.0100

Odds Ratio Estimates

Effect	Point Estimate	95% Wald Confidence Limits	
table 1 vs 2	0.314	0.260	0.377
age 1 vs 3	1.902	1.470	2.462
age 2 vs 3	1.306	1.066	1.601

If we assume that the variable "age" has an equal interval scale, then the continuation ratio model is implemented as:

```
proc logistic order=data;
class table (ref=last)/param=ref;
model r/n=table|age|/sacle=none aggregate;
run;
```

Deviance and Pearson Goodness-of-Fit Statistics

Criterion	DF	Value	Value/DF	Pr > ChiSq
Deviance	2	0.6614	0.3307	0.7184
Pearson	2	0.6609	0.3304	0.7186

Analysis of Maximum Likelihood Estimates

Parameter		DF	Estimate	Standard Error	Chi-Square	Pr > ChiSq
Intercept		1	1.5117	0.2060	53.8560	<.0001
table	1	1	-1.1934	0.2959	16.2681	<.0001
agel		1	-0.3200	0.0865	13.6877	0.0002
agel*table	1	1	0.0155	0.1280	0.0146	0.9038

Since the interaction term is not significant, a model that fits the linear effect of age and the effects of the subtables has:

Deviance and Pearson Goodness-of-Fit Statistics

Criterion	DF	Value	Value/DF	Pr > ChiSq
Deviance	3	0.6760	0.2253	0.8788
Pearson	3	0.6741	0.2247	0.8793

Analysis of Maximum Likelihood Estimates

Parameter		DF	Estimate	Standard Error	Chi-Square	Pr > ChiSq
Intercept		1	1.4957	0.1574	90.2701	<.0001
table	1	1	-1.1595	0.0945	150.4096	<.0001
agel		1	-0.3129	0.0637	24.1118	<.0001

Odds Ratio Estimates

Effect	Point Estimate	95% Wald Confidence Limits	
table 1 vs 2	0.314	0.261	0.377
agel	0.731	0.645	0.829

The advantage of the continuation ratio model is that it is readily more interpretable than baseline category model with the same number of parameters. Further, the link functions can readily be changed to the complimentary log-log or probit.

9.4.6 Mean Response Models

Sometimes, we wish to be able to obtain regression type models because of the ease of interpretations of its parameters. In contingency table analysis employing an ordinal response variable, this can be accomplished by the use of the *mean response model* where the response variable is assigned scores (usually integer scores) and the mean of these scores is then used as a response function.

For example, consider again the Breast Self-Examination Survey data in Table 9.13, which relate to the frequency of breast self-examination in individual age groups. In these data, both the explanatory and the response variables are ordinal. Let us assign scores $\{u_i\}, i = 1, 2, 3$ and $\{v_j\}, j = 1, 2, 3$ to the two variables (age and examination), respectively. We discuss the possibility of other forms of scores in the next chapter. Thus within each level of the factor variables, the conditional mean of the response variable is

$$\mathrm{M}_i = \sum_j v_j \hat{m}_{ij}/n_{i+}, \quad i = 1, 2, 3 \tag{9.34}$$

where n_{i+} is the marginal observed total and $\hat{m}_{ij}$ is the expected count for cell (i, j).

The regression model therefore becomes (Agresti, 1984)

$$\mathrm{M}_i = \mu + \beta(u_i - \bar{u}) \tag{9.35}$$

where μ is the average of the conditional means, and β is the change in the conditional mean for a unit change in X (the explanatory variable). Since only two parameters (μ, β) are being estimated in this case, the model would therefore be based on $(I-2)$ degrees of freedom. Hence, $I \geq 3$ is required to obtain an unsaturated model. The parameter estimates can only be obtained by using a weighted least squares (WLS) approach (Bhapkar, 1966; Grizzle et al., 1969; Williams & Grizzle, 1972). The WLS solution is applicable only when the explanatory variable(s) are categorical.

We fit the model described in (9.35) to the Breast Self-Examination Survey data with frequency of self-examination as the response variable and age as a continuous explanatory variable. We obtain, using PROC CATMOD in SAS software the output below. $\hat{\mu} = 1.6983$, (a.s.e. $= 0.0690$,) and $\hat{\beta} = 0.1472$, (a.s.e. $= 0.0295$), for a WLS solution based on integer scores (1,2,3). The predicted increase in the mean opinion response is 0.1472 categories for every age group <45 to 45-59 and 45-59 to 60+ changes. The hypothesis that $H_0 : \beta = 0$ is tested by Wald's value of 24.83 on 1 d.f., which is not tenable. The residual G^2 of 0.43 on 1 d.f indicates that the model adequately fits the data and that there is a very strong association between the two classificatory variables. The predicted mean responses are {1.8276, 2.0099, 2.1315}. It does appear that the frequencies of breast examination are different in the three age group categories, and that individuals who are 60+ exhibit a more positive self-examination of breast.

```
data breast2;
INPUT AGE EXAM $ COUNT @@;
AGEL=AGE;
DATALINES;
1 MONTH 91 1 OCCASS 90 1 NEVER 51 2 MONTH 150 2 OCCASS 200 2 NEVER 155
3 MONTH  109 3 OCCASS 198 3 NEVER 172
;
PROC PRINT; RUN;
PROC CATMOD ORDER=DATA;WEIGHT COUNT; DIRECT X;
RESPONSE 1 2 3; MODEL EXAM=AGEL/FREQ PROB; RUN;
```

Analysis of Variance

Source	DF	Chi-Square	Pr > ChiSq
Intercept	1	606.00	<.0001
AGEL	1	24.83	<.0001
Residual	1	0.43	0.5099

Analysis of Weighted Least Squares Estimates

Effect	Parameter	Estimate	Standard Error	Chi-Square	Pr > ChiSq
Intercept	1	1.6983	0.0690	606.00	<.0001
AGEL	2	0.1472	0.0295	24.83	<.0001

9.4.7 Multidimensional Tables

For multidimensional tables, we give in Table 9.15, the results of a study to examine the relationship between drinking habits of subjects living in group quarters to the location in New York, and the length of time they have been there (Upton, 1978).

Number of years in quarters	Location	Drinking Habits Light	Moderate	Heavy
0		25	21	26
1-4	Bowery	21	18	23
5+		20	19	21
0		29	27	38
1-4	Camp	16	13	24
5+		8	11	30
0		44	19	9
1-4	Park Slope	18	9	4
5+		6	8	3

Table 9.15: Drinking habits of subjects living in New York neighborhood

Below is the SAS software output for our preliminary analysis. Notice that because the response variable is read in as a character variable, we specify the scores assigned to each category in the response statement line. The saturated model fitted indicate that the interaction term between length of time spent and location is not significant. ($p = 0.4327$). Hence, a reduced model can now be fitted to the data.

```
data mean2; input years $ loc $ habit $ count @@;
datalines;
0 bowery light 25 0 bowery mod 21 0 bowery heavy 26
1-4 bowery light 21 1-4 bowery mod 18 1 4 bowery heavy 23
5+ bowery light 20 5+ bowery mod 19 5+ bowery heavy 21
0 camp light 29 0 camp mod 27 0 camp heavy 38
1-4 camp light 16 1-4 camp mod 13 1-4 camp heavy 24
5+ camp light 8 5+ camp mod 11 5+ camp heavy 30
0 park light 44 0 park mod 19 0 park heavy 9
1-4 park light 18 1-4 park mod 9 1-4 park heavy 4
5+ park light 6 5+ park mod 8 5+ park heavy 3
;
proc catmod order=data; weight count;
response 1 2 3; model habit=years|loc/wls noiter; run;

           Analysis of Variance

Source         DF   Chi-Square    Pr > ChiSq
--------------------------------------------
Intercept       1      2633.35        <.0001
years           2         5.96        0.0508
loc             2        38.36        <.0001
years*loc       4         3.81        0.4327
Residual        0            .             .
```

The reduced model fits the data with a $Q_W = 3.81$ on 4 d.f. ($p = 0.4327$). The main effect of location is highly significant, while the effects of length of time as measured in years is on the border at a specified 0.05 α level. We nonetheless decide to keep it in the model. In order to make comparisons among the locations and length of time, we constructed the following contrasts, which are self-explanatory. From the analysis below, it is indicative that those with 0 years (new comers) are significantly different from the over 5 years residents, but the newcomers are not in any way different from those with (1-4) years length of stay. However, for the location, all the three pairs are significantly different with the major difference between those residing in the Camp neighborhood and those at Park Slope.

```
set mean2;
proc catmod order=data;
weight count; response 1 2 3; model habit=years loc;
contrast '0 versus (1-4)' years 1 -1;    contrast '0 versus 5+' years 2 1;
contrast '(1-4) versus 5+' years 1 2;    contrast 'Bowery versus camp' loc 1 -1;
contrast 'Bowery versus Park' loc 2 1;   contrast 'Camp versus Park' loc 1 2; run;
```

Analysis of Contrasts

Contrast	DF	Chi-Square	Pr > ChiSq
0 versus (1-4)	1	0.30	0.5844
0 versus 5+	1	5.62	0.0177
(1-4) versus 5+	1	2.79	0.0946
Bowery versus camp	1	5.67	0.0172
Bowery versus Park	1	21.19	<.0001
Camp versus Park	1	48.63	<.0001

For instance, the contrast "0 versus 5+" is constructed in CATMOD as "years 2 1." This is so because the contrast is $\beta_1 - \beta_3$. But $\beta_3 = -(\beta_1 + \beta_2)$ because of the usual sum-to-zero constraints employed in CATMOD. Substituting this in the contrast gives the given contrast equation. Here, we have assumed that $\beta_1, \beta_2, \beta_3$ refer respectively to the parameters for year level 0, 1-4, and 5+, respectively.

It should be 'noted here that the mean response model is most appropriate for making inference about an underlying continuous set of variables, where the categories of the variables reflect some underlying continuous characteristics.

9.4.8 Further Analysis of the Data in Table 9.15

So far, we applied the multicategory logit models to the pnemoconiasis data as well as a two-way contingency table with an ordered response variable (the breast self-examination frequency data). In order to further explain how to implement all the above multicategory logit models to other multidimensional contingency tables, we now show the application of these models to the data in Table 9.15, which has three response categories and two factor or explanatory variables each at three levels.

We now fit the proportional odds, the adjacent category, and continuation ratio odds models to this data set. First, we create dummy variables LOC1, LOC2 (Location 3 is the reference), and YY1 and YY2 to represent dummy variables for the year. Note that the response category ''heavy" is coded to have the highest value on the drinking scale. The following SAS software statements and selected output for the proportional odds model indicate that the proportional odds assumption is satisfied based on the score test ($p = 0.5089$). The model also fits the data with a deviance value of 8.6706 on 12 d.f. ($p = 0.7308$). Examination of the parameter estimates and the corresponding Wald tests indicates that location 1 and location 2 are highly significant, as well as the year 1 of residence relative to year level 3. Based on the estimated odds-ratios below, we can conclude that a resident living in Bowery is .391 less likely than those living in Park Slope to have light drinking habits. Put another way, those leaving in Bowery are $\frac{1}{.391} = 2.554$ or 125% more likely to have heavy drinking habits than those living in Park Slope. Similarly, those who have recently moved into this neighborhood (less than 1 year) are 59% more likely than those that have lived in the neighborhood for at least 5 years to have light rather than heavy drinking habits.

On the other hand, those residents who are living in Camp are $\dfrac{1}{0.250} = 4$ times more likely to have heavy drinking habits rather than moderate drinking habits than those living in Park Slope. Since the effect of year of living is not significant for this case, we decide not to give any interpretation to the corresponding odds ratio. We can just take differences if interest centers on comparing those with moderate and light habits, as explained earlier.

```
data tab917;
if lo eq 'bowery' then loc=1;
else if lo eq 'camp' then loc=2;
else loc=3;
loc1=loc eq 1;      loc2=loc eq 2;
yy1=year eq 1;      yy2=year eq 2;
if habit eq 'light' then habit=1;
else if habit eq 'mod' then habit=2;
else habit=3;
proc logistic descending;
freq count; model habit=loc1 loc2 yy1 yy2/scale=n aggregate; run;
```

Score Test for the Proportional Odds Assumption

Chi-Square	DF	Pr > ChiSq
3.3005	4	0.5089

Deviance and Pearson Goodness-of-Fit Statistics

Criterion	DF	Value	Value/DF	Pr > ChiSq
Deviance	12	8.6706	0.7226	0.7308
Pearson	12	8.9117	0.7426	0.7105

Number of unique profiles: 9

Analysis of Maximum Likelihood Estimates

Parameter	DF	Estimate	Standard Error	Chi-Square	Pr > ChiSq
Intercept	1	-0.0228	0.2460	0.0086	0.9260
Intercept2	1	1.2402	0.2520	24.2126	<.0001
loc1	1	-0.9378	0.2276	16.9797	<.0001
loc2	1	-1.3849	0.2276	37.0231	<.0001
yy1	1	0.4662	0.2105	4.9036	0.0268
yy2	1	0.3491	0.2285	2.3333	0.1266

Odds Ratio Estimates

Input	Odds Ratio
loc1	0.391
loc2	0.250
yy1	1.594
yy2	1.418

We next implement the adjacent category model on this set of data to again produce the following select SAS software output. Here again, this model fits with $G^2 = 8.06$ on 12 d.f. ($p = 0.7805$). The Wald tests also indicate that only LOC1, LOC2, and YY1 are significant ($p < 0.05$). Again for those living in Bowery versus those in Park Slope, the odds of having a heavy drinking habit rather than moderate drinking habit are exp(0.6472)=1.910 times higher. For those with less than 1 year versus those with more than 5 years, the odds of heavy drinking habit to moderate drinking habit are exp(-0.2921)=0.747. Again, for those living in the camp versus those living at Park Slope, the odds of having a moderate drinking habit rather than a light drinking habit is exp(0.9237)= 2.519 higher. These conclusions are very similar to the earlier conclusions under the proportional odds model.

```
proc catmod;
weight count;
direct loc1 loc2 yy1 yy2;
response alogit;
model habit= _response_ loc1 loc2 yy1 yy2/wls; run;

The CATMOD Procedure

              Analysis of Variance

Source          DF    Chi-Square     Pr > ChiSq
-----------------------------------------------
Intercept        1          6.38         0.0115
_RESPONSE_       1          2.49         0.1146
loc1             1         16.55         <.0001
loc2             1         33.76         <.0001
yy1              1          4.25         0.0392
yy2              1          2.13         0.1448

Residual        12          8.06         0.7805

                 Analysis of Weighted Least Squares Estimates

                                         Standard          Chi-
Effect           Parameter    Estimate      Error        Square     Pr > ChiSq
------------------------------------------------------------------------------
Intercept                1     -0.4364     0.1728          6.38         0.0115
_RESPONSE_               2     -0.1591     0.1009          2.49         0.1146
loc1                     3      0.6472     0.1591         16.55         <.0001
loc2                     4      0.9237     0.1590         33.76         <.0001
yy1                      5     -0.2921     0.1417          4.25         0.0392
yy2                      6     -0.2220     0.1523          2.13         0.1448
```

To implement the continuation ratio model, we first construct the tables Light versus Moderate and (Light and Moderate) versus Heavy. The SAS software output below gives the fit of the continuation ratio model to the data in Table 9.15. The LOGISTIC model for each of the table gives deviances of 2.0894 and 3.2870, respectively, on 4 d.f. The continuation ratio model therefore gives a deviance value of 5.3764 (2.0894+3.2870) on 8 d.f. The model fits the data. Instead of fitting individual logistic models for each table, we can fit a single model to the data that incorporates the interaction terms of tables with LOC1, LOC2, YY1, and YY2. This model produces the 8 d.f deviance in the result below. However, this model indicates that the interaction terms involving LOC1 and LOC2 with tables are not significant (Wald tests not shown). Deleting these terms from the model leads to a more parsimonious continuation ratio model with a deviance of 5.5163 on 10 d.f.

```
data cont;
do year=1 to 3;
do loc=1 to 3;
input r s table @@;
yy1=year eq 1;          yy2=year eq 2;
loc1=loc eq 1;          loc2=loc eq 2;
tot=r+s; output; end; end;
datalines;
25 21 1 21 18 1 20 19 1 29 27 1 16 13 1 8 11 1
44 19 1 18 9 1 6 8 1 46 26 2 39 23 2 39 21 2
56 38 2 29 24 2 19 30 2 63 9 2 27 4 2 14 3 2
;
proc genmod;
where table=1;
model r/tot=loc1 loc2 yy1 yy2/dist=b type3;
run;
proc genmod;
class table;
model r/tot=loc1 loc2 table|yy1 table|yy2/dist=b type3;
run;
```

year	loc	r	s	table	yy1	yy2	loc1	loc2	tot
1	1	25	21	1	1	0	1	0	46
1	2	21	18	1	1	0	0	1	39
1	3	20	19	1	1	0	0	0	39
2	1	29	27	1	0	1	1	0	56
2	2	16	13	1	0	1	0	1	29
2	3	8	11	1	0	1	0	0	19
3	1	44	19	1	0	0	1	0	63
3	2	18	9	1	0	0	0	1	27
3	3	6	8	1	0	0	0	0	14
1	1	46	26	2	1	0	1	0	72
1	2	39	23	2	1	0	0	1	62
1	3	39	21	2	1	0	0	0	60
2	1	56	38	2	0	1	1	0	94
2	2	29	24	2	0	1	0	1	53
2	3	19	30	2	0	1	0	0	49
3	1	63	9	2	0	0	1	0	72
3	2	27	4	2	0	0	0	1	31
3	3	14	3	2	0	0	0	0	17

Criteria For Assessing Goodness Of Fit

Criterion	DF	Value	Value/DF
Deviance	4	2.0894	0.5223-Table 1
Deviance	4	3.2870	0.8217-Table 2

Criteria For Assessing Goodness Of Fit

Criterion	DF	Value	Value/DF
Deviance	8	5.3764	0.6720
Pearson Chi-Square	8	5.3626	0.6703

Criteria For Assessing Goodness Of Fit

Criterion	DF	Value	Value/DF
Deviance	10	5.5163	0.5516
Scaled Deviance	10	5.5163	0.5516
Pearson Chi-Square	10	5.5053	0.5505

Analysis Of Parameter Estimates

Parameter		DF	Estimate	Standard Error	Chi-Square	Pr > ChiSq
Intercept		1	1.5522	0.3036	26.13	<.0001
loc1		1	0.4072	0.1847	4.86	0.0275
loc2		1	0.3194	0.2002	2.54	0.1106
table	1	1	-1.2432	0.3391	13.44	0.0002
yy1		1	-1.2296	0.3097	15.77	<.0001
yy1*table	1	1	0.7991	0.4125	3.75	0.0527
yy2		1	-1.7107	0.3052	31.42	<.0001
yy2*table	1	1	1.1318	0.4175	7.35	0.0067
Scale		0	1.0000	0.0000		

LR Statistics For Type 3 Analysis

Source	DF	Chi-Square	Pr > ChiSq
loc1	1	4.86	0.0275
loc2	1	2.55	0.1103
table	1	14.46	0.0001
yy1	1	16.71	<.0001
yy1*table	1	3.85	0.0498
yy2	1	32.52	<.0001
yy2*table	1	7.59	0.0059

The most obvious significant parameter here is that involving the interaction between YY2 and the table. For the continuation ratio model to hold, the effects of YY1 and YY2 are very important. Based on the significant YY2 and table interaction, the odds that an individual will have a light or moderate drinking to a heavy drinking habit is $\exp(-1.7107) = 0.181$ times for those who have lived 1-4 years than for those who have lived at least 5 or more years after adjusting for neighborhood residence. In other words, the 5+ years residents have odds that are 5.5 times higher than those who have lived 1-4 years of going from moderate or light drinking to heavy drinking habit. Similarly, the odds are $\exp(-1.7107+1.1318) = 0.56$ times greater for those who have lived 1-4 years than those who have lived 5+ years of going from light to moderate drinking habit. That is, the 5+ years residents have odds that are 1.78 times higher than those who have lived 1-4 years of going from light to moderate drinking habits, again having controlled for the neighborhood residence.

9.5 Exercises

1. For the data in exercise 5 chapter 6, we plan to construct a log-linear model that corresponds to a logit model in which intercourse is the response. Based on the odds ratios obtained in that exercise, which log-linear model seems appropriate?

2. For a three-way table with binary response C, give the equivalent log-linear and logit models for which:
(a) C is jointly independent of A and B.
(b) C is conditionally independent of B.
(c) There is no interaction between A and B in their effects on C.

3. For a four-way table with binary response D, give the equivalent log-linear and logit models for which:
(a) D is jointly independent of A, B, and C.
(b) D is jointly independent of B and C, given A.
(c) There are main effects of A and B on D, but D is conditionally independent of C, given A and B.
(d) There is interaction between A and B in their effects on D, and C has main effects.

4. Refer to exercise 7 in chapter 8. The limiting distribution for the binomial is the Poisson. Reanalyze these data by using a Poisson regression model. Discuss the differences from your chosen model in that exercise.

5. Radelet (1981) gives data on the relationship between race and the imposition of the death penalty. The data are given in the table below. Analyze the data using logit models by considering death penalty as a response variable.

6. For the data in Table 4.16, fit a suitable mean response model. Interpret your results.

Defendant's race	Victim's race	Death penalty Yes	No
Black	Black	6	97
	White	11	52
White	Black	0	9
	White	19	132

7. The following data is taken from the 1984 General Social Survey of the National Data program in the United States as reproduced by Agresti (1990).

	Job satisfaction			
Income (US$)	Very dissatisfied	Little dissatisfied	Moderately satisfied	Very satisfied
< 6000	20	24	80	82
6000-15,000	22	38	104	125
15,000-25,000	13	28	81	113
$> 25,000$	7	18	54	92

Table 9.16: Cross-classification of job satisfaction by income

Treating job satisfaction as the response variable, fit a cumulative logit model to the data that gives a good fit and interpret the estimated effect.

8. The data in the table below present (Hedlund, 1978) the relationship between an ordinal variable, political ideology, and a nominal variable, party affiliation, for a sample of voters in the 1976 presidential primary election in Wisconsin.

	Political ideology		
Party affiliation	Liberal	Moderate	Conservative
Democrat	143	156	100
Independent	119	210	141
Republican	15	72	127

Analyze the above data using cumulative logits, treating political ideology as the response. Test the adequacy of the model fit and interpret parameter estimates. What can you deduce from your analysis as being the influence of party affiliation on ideology?

9. The data below relate to patients undergoing chemotherapy who were categorized by the severity of nausea and by whether or not they received cisplatinum (treatment) (Farewell, 1982).

	Response				
Treatment	None	Mild	Moderate	Severe	Total
No	43	39	50	29	161
Yes	7	7	30	14	58

Analyze the above data.

10. The data below are from Agresti (1984) and relate to attitude toward abortion and schooling of a sample of people.

Education	Attitude		
	Disapprove	Middle	Approve
$<$ High school	209	101	237
High school	151	126	426
$>$ High school	16	21	138

Fit a proportional odds model to the data and draw your conclusions.

11. The data below are from Christen (1990) and relate to 1237 men between the ages of 40 and 59 (who did not develop coronary heart attack) taken from a study conducted in Massachusetts. The men were cross-classified according to their serum cholesterol and systolic blood pressure.

Cholesterol (in mg/100 cc)	Blood Pressure (in mm Hg)			
	$<$ 127	127-148	147-166	167+
$<$ 200	117	121	47	22
200-219	85	98	43	20
220-259	119	209	68	43
$\geq$ 260	67	99	46	33

Fit the adjacent category, proportional odds, and continuation odds ratio models to the above data. Based on your analyses, which of these models would you recommend and why?

12. The data below relate to 3 year survival of breast cancer patients according to nuclear grade and diagnostic center (Whittaker, 1990, p. 220).

Center	Malignant		Benign	
	Died	Survived	Died	Survived
Boston	35	59	47	112
Glamorgan	42	77	26	76

Fit an appropriate logit model to this data. What would be the equivalent log-linear model?

13. The table below relate opinions on whether one agrees or not that grocery shopping is tiring to availability of a car, obtained in a survey in Oxford, England (Lindsey, 1995).

Opinions from the Oxford Shopping Survey

Car available	Grocery shopping is tiring				
	Disagree	Tend to disagree	In between	Tend to agree	Agree
No	55	11	16	17	100
Sometimes	101	7	18	23	103
Always	91	20	25	16	77

Fit the proportional odds, adjacent category, and continuation ratio models to the above data. Which model is the most parsimonious? Interpret your choice model.

Chapter 10

Models in Ordinal Contingency Tables

10.1 Introduction

We shall in this chapter explore the special analyses of the general $I \times J$ contingency table when one or both classificatory variables are ordered. For this class of tables, the usual independence or quasi-independence analyses may not be adequate enough for a proper explanation of the variation in the data. As an example, the data below in Table 10.1 (Christensen, 1990) relate to a sample of 1237 men between the ages of 40 and 59 (who did not develop coronary heart attack) taken from the Framingham longitudinal study. The men were cross-classified according to their serum cholesterol and systolic blood pressure. In this example, $I = J = 4$, though genuine cases when $I \neq J$ such as the very well-analyzed Manhattan 4×6 Manhattan mental health by parents' socioeconomic status data (Goodman, 1979a), is also presented in Table 10.2.

Cholesterol	Blood pressure (in mm Hg)			
(in mg/100 cc)	< 127	127-148	147-166	167+
< 200	117	121	47	22
200-219	85	98	43	20
220-259	119	209	68	43
≥ 260	67	99	46	33

Table 10.1: Classification by SCL and BP

For the data in Table 10.1 we see that both row and column variables can be assumed to be ordered, leading to a 4×4 ordered table. In Table 10.2, we can also assume that the response variable "Mental health status" has some intrinsic order in its categories. In Table 10.3, too, both rows and columns can also be assumed to have ordinal ordering about their categories.

Ee let f_{ij} be the observed frequency in the ij-th cell of a general $I \times J$ table with at least one variable ordered and also let $\hat{m}_{ij}$ be the corresponding expected frequency under some model. Then, the saturated log-linear model formulation for this table becomes:

Mental health status	Parents' socioeconomic status					
	A	B	C	D	E	F
Well	64	57	57	72	36	21
Mild symptom Formation	94	94	105	141	97	71
Moderate symptom Formation	58	54	65	77	54	54
Impaired	46	40	60	94	78	71

Table 10.2: Cross classification og mental health status by Socioeconomic status

$$\ln(\hat{m}_{ij}) = \mu + \lambda_i^X + \lambda_j^Y + \lambda_{ij}^{XY} \tag{10.1}$$

for $i = 1, 2, \cdots, I$ and $j = 1, 2, \cdots, J$ where X and Y relate to the row and column variables respectively. Further, we assume the usual identifiability constraints on the parameters as discussed in chapter 6. The corresponding independence model, which Goodman called the *null* or (O) model, has the formulation

$$\ln(\hat{m}_{ij}) = \mu + \lambda_i^X + \lambda_j^Y \tag{10.2}$$

which can be written in the multiplicative form as

$$\hat{m}_{ij} = \alpha_i \beta_j$$

Periodontal Condition	Level			
	1	2	3	4
A	5	3	10	11
B	4	5	8	6
C	26	11	3	6
D	23	11	1	2

Table 10.3: Cross-classification of 135 women according to their periodontal condition and calcium intake level (Goodman, 1979a)

The model of independence is based on $(I-1)(J-1)$ degrees of freedom. For the data in Tables 10.1 and 10.3, the model of independence yields $G^2 = 20.38$ and 46.817 on 9 degrees of freedom, respectively, which clearly indicates that the model of independence does not fit both data. We will now explore the possibility of exploiting the fact that one or both of the classificatory variables are ordered into modeling the tables by assigning known scores to the categories of the variables. We also note here that the model of independence being inadequate suggests that there is a significant interaction or association term between the row variable X and the column variable Y.

Suppose we assign known scores x_i and y_i pertaining to row and column categories respectively to the table, and let us now introduce these known scores to the model in (10.1) (that is, rewriting the interaction term) to give a revised model of the form:

$$\ln(\hat{m}_{ij}) = \mu + \lambda_i^X + \lambda_j^Y + \phi x_i y_j \tag{10.3}$$

where ϕ is a parameter describing the intrinsic association between variables X and Y. The model in (10.3) is known as the *linear-by-linear* association model. The scores x_i and y_i are either specified or known in advance. Possible values of the scores are:

(i) Integer scores: that is,

$$x_i = i \quad \text{and} \quad y_j = j$$

(ii) Centered Scores: where

$$x_i = i - \left(\frac{I+1}{2}\right) \quad \text{and} \quad y_j = j - \left(\frac{J+1}{2}\right)$$

which centers the scores.

For a 4 × 4 table, the integers scores assume the values $(i, j) = 1, 2, 3, 4$, while the second set of scores assumes the values $(x_i, y_j) = -1.5, -0.5, 0.5, 2.5$.

The model formulation in (10.3) assumes that the categories are equally spaced (that is, interval scaled) and that the λ_{ij}^{XY} interaction in (10.1) has been modeled or smoothed (or restricted) by the term $\phi x_i y_i$.

If we consider the 2 × 2 subtable formed from adjacent rows (i.e., rows i and $i+1$) and adjacent columns (i.e. columns j and $j+1$) for the general $I \times J$ table, then let θ_{ij} denote the corresponding odds-ratio for $i = 1, 2, \cdots, (I-1)$ and $j = 1, 2, \cdots, (J-1)$ based on the expected cell frequencies $\hat{m}_{ij}$. Then,

$$\theta_{ij} = \left(\frac{\hat{m}_{ij}, \hat{m}_{i+1,j+1}}{\hat{m}_{i,j+1}, \hat{m}_{i+1,j}}\right) \tag{10.4}$$

and there are $(I-1)(J-1)$ of such odds ratios in such a table.

If we take the natural logarithm of θ_{ij} above and we define the log odds as: $\Phi_{ij} = \ln \theta_{ij}$, then from (10.4), we have

$$\Phi_{ij} = \ln \hat{m}_{ij} + \ln \hat{m}_{i+1,j+1} - \ln \hat{m}_{i,j+1} - \ln \hat{m}_{i+1,j} \tag{10.5}$$

Re writing (10.5) by changing the subscripts appropriately using (10.3) we have:

$$\Phi_{ij} = \phi(x_{i+1} - x_i)(y_{j+1} - y_j); \quad = \phi\Delta_i^x\Delta_j^y \tag{10.6}$$

where $\Delta_i^x = (x_{i+1} - x_i)$ is the distance under the postulated set of scores between the i-th and the $(i+1)$-th categories of the row variable X, and where Δ_j^y is similarly defined.

We see that the log-odds ratio in the 2 × 2 subtable is a function of the intrinsic association ϕ, the distance between the row categories Δ_i^x, and the distance between the column categories Δ_j^y. Obviously, if $\Delta_i^x = \Delta_j^y = 1$, then $\Phi_{ij} = \phi$. That is, $\Phi_{ij} = \phi$ if and only if the adjacent row and column categories are one unit apart.

10.1.1 Properties of ϕ

1. ϕ is unaffected by a location shift.

   For instance, if we change x_i to $x_i^* = x_i + a$ and y_j to $y_j^* = y_i + b$ (where a and b are constants), while these location shifts will produce changes in the main effect parameters in (10.3), they would not change the value of ϕ since in this case

$$\begin{aligned}\Phi_{ij} &= \phi(x_{i+1}^* - x_i^*)(y_{j+1}^* - y_j^*) \\ &= \phi\{x_{i+1} + a - (x_i + a)\}\{y_{j+1} + b - (y_j + b)\} \\ &= \phi(x_{i+1} - x_i)(y_{j+1} - y_j) \\ &= \phi\Delta_i^x\Delta_j^y\end{aligned}$$

   The above is equivalent to (10.6), which indicates that ϕ is unchanged.

2. A scale change (or unit point restriction) of the known scores x_i and y_j, however, does have an effect on ϕ in addition to producing changes in the main effect parameters. To show this consider $x_i^* = ax_i$ and $y_j^* = by_j$. If we let ϕ^* be the new intrinsic association parameter, then we have

$$\begin{aligned}\Phi_{ij} &= \phi^*(ax_{i+1} - ax_i)(by_{j+1} - by_j) \\ &= \phi^* ab(x_{i+1} - x_i)(y_{j+1} - y_j) \\ &= \phi^* ab\Delta_i^x \Delta_j^y\end{aligned}$$

Comparing this with (10.6), we have $\phi^* ab = \phi$ and consequently, we see that in this case $\phi^* = \phi/ab$. The new intrinsic association parameter is thus a fraction of the original ϕ being ϕ/ab.

Under integer scoring, $\Delta_i^x = \Delta_j^y = 1$; hence, we have, using (10.6),

$$\Phi_{ij} = \phi, \quad \text{or,} \tag{10.7a}$$

$$\theta_{ij} = e^{\phi} \qquad = \theta \tag{10.7b}$$

The model described above has been called the *uniform association* (U) model by Goodman (1979) . Either ϕ or θ can be used as a measure of association between the X and Y variables if the model holds. Specifically, we can transform the measure to well known $[-1, +1]$ scale by taking $Q = \dfrac{\theta - 1}{\theta + 1}$ so that there is no difficulty at all in summarizing the association.

The multiplicative form of the model is given by

$$\hat{m}_{ij} = \alpha_i \beta_j \theta^{ij}$$

The (U) model has one parameter (namely, ϕ) more than the model of independence; hence, it will be based on $(I-1)(J-1) - 1 = (IJ - I - J)$ degrees of freedom. The model of independence, which Goodman described as the null or O model, of course is a special case of the linear-by-linear association model and is obtained when $\phi = 0$. For our data in Table 10.1 the uniform association (U) model gives $G^2 = 7.4291$ and Pearson's $X^2 = 7.4628$ on 8 d.f., which indicates that the model is very satisfactory and that the integer scoring system appears reasonable. Here, the MLE estimate of $\hat{\phi} = 0.1044$, which gives $\hat{\theta} = e^{\hat{\phi}} = 1.110$. Thus when the U model holds, then each 2×2 subtable formed from adjacent rows and adjacent columns has an odds ratio of this magnitude, which gives a Yule's Q of 0.0521, which represents a modest but significant positive association between the two variables.

Model (U) is implemented in SAS software for the data in Table 10.1 with the following program and partial output.

```
data tab101;
do chol=1 to 4;
  do bp=1 to 4;
input count @@;
u=chol*bp; /* creates product of integer scores *\;
output;
end; end;
datalines;
117 121 47 22 85 98 43 20 119 209 68 43 67 99 46 33
;
proc genmod; class chol bp;
model count=chol bp u/dist=poi link=log type3;
run;
```

```
Criterion                 DF          Value        Value/DF
-----------------------------------------------------------
Deviance                   8         7.4291          0.9286
Pearson Chi-Square         8         7.4628          0.9328

                      Analysis Of Parameter Estimates

                                 Standard   Wald 95% Confidence        Chi-
Parameter       DF   Estimate       Error          Limits            Square   Pr > ChiSq
-----------------------------------------------------------------------------------------
Intercept        1     1.7591      0.4120     0.9515     2.5667       18.23       <.0001
.........................................................................................
u                1     0.1044      0.0293     0.0470     0.1617       12.73       0.0004
```

In general, the odds ratio between arbitrary rows i and i' $(i' > i)$ and columns j' and j $(j' > j)$ is given by

$$\theta_{(ii')(jj')} = e^{rc\hat{\phi}} = \hat{\theta}^{rc} \tag{10.8}$$

where $r = i' - i$ and $c = j' - j$, respectively are the differences between the levels of the X and Y variables. That is, the expression in (10.8) can be written in terms of scores as

$$(x_{i'} - x_i)\,(y_{j'} - y_j)\,\hat{\theta}$$

If $r = c = 1$, we have the earlier result above. For instance, the odds ratio between the third and first categories of Y and the first and second categories of X is given by $\hat{\theta} = e^{2(0.1044)} = 1.232$, since $r = 1$ and $c = 2$.

	Y			
X	1	2	3	4
1	112.034	130.954	43.088	20.930
2	81.276	105.450	38.512	20.764
3	130.141	187.416	75.974	45.467
4	64.549	103.180	46.427	30.840

Table 10.4: Expected values under model U for the data in Table 10.1

Thus,

$$\hat{\theta}_{(1,1)(2,3)} = \left[\frac{(112.034)(38.512)}{(81.276)(43.088)}\right] = 1.2320$$

while for $r = 2$ and $c = 2$ we have $\hat{\theta} = e^{4(0.1044)} = 1.518$. We can demonstrate this again from the table of expected values as:

$$\hat{\theta}_{(1,1)(3,3)} = \left[\frac{(112.034)(75.974)}{(130.141)(43.088)}\right] = 1.518 \quad \text{and}$$

$$\hat{\theta}_{(2,1)(4,3)} = \left[\frac{(81.276)(46.427)}{(64.549)(38.512)}\right] = 1.518$$

If the model of independence holds, then $\theta = 1$, and this implies that $G^2(O) - G^2(U)$ has a χ^2 distribution with 1 d.f which tests independence conditional on the (U) model holding true.

Usually, the (U) association model always fits the data of interest, but in case this is not the case, we explore below some other association models that can be used to further examine the variability in the table and to test whether the (U) model is sufficient for a proper explanation of the row-column association in the data.

10.2 The Row Association (R) Model

If the column variables is ordinal, then suppose we assign scores $\{\nu_j\}$ to these column categories. With this set up, the row association model (R) can be defined in terms of the odds-ratios as

$$\theta_{ij} = \theta_{i+} \qquad \text{for } i = 1, 2, \cdots, I-1 \tag{10.9}$$

The model has $(I-1)$ more parameters, namely, θ_{i+}, than the (O) model, and is therefore based on $(I-1)(J-1)-(I-1) = (I-1)(J-2)$ degrees of freedom. The model assumes that the row categories are not necessarily ordinal. The model can be written in the log-linear form (Agresti, 1984) as:

$$\ln(\hat{m}_{ij}) = \mu + \lambda_i^X + \lambda_j^Y + \tau_i(\nu_j - \bar{\nu}) \tag{10.10}$$

where $\bar{\nu} = \frac{\sum_j \nu_j}{J}$ and $\sum \lambda_i^X = \sum \lambda_j^Y = \sum \tau_i = 0$.

The model is equivalent to the model of independence when $\tau_i = 0$ for all i. The row association parameter $\{\tau_i\}$ is interpreted as being the deviation within a particular row of $\ln(\hat{m}_{ij})$ from row independence of a known function of the ordinal variable with slope τ_i.

For arbitrary rows i and i' ($i' > i$) and columns j and j' ($j' > j$), we have the log odds ratios for the model above as

$$\begin{aligned}
\ln\left(\frac{\hat{m}_{ij}\hat{m}_{i'j'}}{\hat{m}_{ij'}\hat{m}_{i'j}}\right) &= \ln(\hat{m}_{ij}) + \ln(\hat{m}_{i'j'}) - [\ln(\hat{m}_{ij'}) + \ln(\hat{m}_{i'j})] \\
&= (\tau_{i'} - \tau_i)(\nu_{j'} - \bar{\nu}) - (\tau_{i'} - \tau_i)(\nu_j - \bar{\nu}) \\
&= (\tau_{i'} - \tau_i)(\nu_{j'} - \nu_j)
\end{aligned}$$

That is, the log odds ratio is proportional to the distance between the columns and is always positive whenever $(\tau_{i'} - \tau_i) > 0$. If $\{\nu_j = j\}$, that is, integer scores, then the log odds ratio is constant and equals $(\tau_{i'} - \tau_i)$ for all $(J-1)$ pairs of adjacent columns.

The row association model is naturally suited for the general $I \times J$ table having nominal row variable and ordinal column variable since the model has the same form (and produces the same G^2) if rows of the table are permuted. That is, permuting the row categories would not produce any changes in the value of G^2. In other words, the rows are permutation proof.

Goodman (1979a) has formulated an alternative form of the row-association model (R). His formulation is of the form:

$$\theta_{ij} = \theta\xi_{i+} \tag{10.11}$$

where θ_{i+} is equated to $\theta\xi_{i+}$ and where

$$\prod_{i=1}^{I-1} \xi_{i+} = 1 \tag{10.12}$$

Taking natural logarithms then, (10.11) and (10.12) are equivalent on the additive scale to

$$\ln\theta_{ij} = \psi + \eta_{i+} \quad \text{with} \tag{10.13}$$

$$\sum_{i=1}^{I-1} \eta_{i+} = 0 \tag{10.14}$$

where $\psi = \ln(\theta)$ and $\eta_{i+} = \ln(\xi_{i+})$. The above implicitly assumed that $\zeta_{I+} = 1$ and $\eta_{I+} = 0$.

10.2.1 Example 10.1: Analysis of Data in Table 10.3

We consider in the table below the 4×4 data relating to periodontal condition and calcium intake level of 135 women (Goodman, 1986a) described previously in Table 10.3. The SAS software program for implementing the row-association model is presented below with partial outputs from PROC GENMOD.

```
data tab103;
do cond=1 to 4; do level=1 to 4;
input count @@;
nu=level; /* create integer column scores*\;
mu=cond; /* creates integer row scores*\;
output; end; end;
datalines;
5 3 10 11 4 5 8 6 26 11 3 6 23 11 1 2
;
proc print;
proc genmod; class cond level;
model count= cond|nu level/dist=poi; run; /* fits (R) model*\;
```

```
                Criteria For Assessing Goodness Of Fit

Criterion                    DF             Value          Value/DF
-------------------------------------------------------------------
Deviance                      6            9.8761            1.6460
Pearson Chi-Square            6            9.2897            1.5483

              Analysis Of Parameter Estimates

                                       Standard      Chi-
Parameter         DF    Estimate          Error    Square   Pr>ChiSq
------------------------------------------------------------------
Intercept          1      3.4416         1.1800      8.51     0.0035
nu                 1     -0.7634         0.3896      3.84     0.0501
nu*cond       1    1      1.2701         0.2812     20.40     <.0001
nu*cond       2    1      1.0824         0.2850     14.42     0.0001
nu*cond       3    1      0.3101         0.2543      1.49     0.2227
```

In order to find the sum to zero parameter estimates in terms of the $\eta' s$, we note the following from the above SAS software output:

$$\hat{\eta}_1 - \hat{\eta}_4 = 1.2701 \quad \hat{\eta}_2 - \hat{\eta}_4 = 1.0824 \qquad \text{and}$$
$$\hat{\eta}_3 - \hat{\eta}_4 = 0.3101$$

Adding the above and remembering that $\sum_{i=1}^{4} \hat{\eta}_i = 0$ to get the equivalent sum to zero constraint parameters, we have,

$$\hat{\eta}_1 = 0.6045 \qquad \hat{\eta}_2 = 0.4167$$
$$\hat{\eta}_3 = -0.3556 \quad \text{and} \quad \hat{\eta}_4 = -0.6657$$

We can compute estimated odds ratios from values of either the sum to zero parameter estimates or the GENMOD estimates (last parameter set to zero constraint). In either case, for example, $\hat{\eta}_3 - \hat{\eta}_1 = -0.9600$ and the estimated odds ratio is given by exp {-0.9600}=0.383. This could have been obtained from the GENMOD parameter estimates as $\exp\{0.3101 - 1.2701\} = -0.9600$ as before.

10.2.2 Estimating the Log Odds Ratios

We can estimate the odds ratios (or log of them) based on the above parameter estimates. If we assume the row association model with integer scores for the column variable, then we should expect constant odds ratios for adjacent column categories in the multiplicative model, which is reduced to constant difference of log odds ratios for adjacent categories.

$$\hat{\Phi}_{(i,j)(i',j')} = \hat{\eta}_{i'} - \hat{\eta}_i \quad \text{hence} \quad \hat{\theta}_{(i,j)(i',j')} = \exp(\hat{\eta}_{i'} - \hat{\eta}_i) \tag{10.15}$$

To illustrate, $\hat{\eta}_3 - \hat{\eta}_1 = -0.960$, with corresponding odds-ratio $\hat{\theta} = 0.383$, which means that the odds of calcium intake being 4 rather than 3, or 3 rather than 2 or 2 rather than 1 are 0.383 lower for periodontal condition C than condition A. These odds ratios can be obtained from the table of expected cell frequencies presented below under model (R):

Periodontal condition	Calcium intake level 1	2	3	4
A	5	3	10	11
	4.365	5.321	7.261	12.052
B	4	5	8	6
	4.867	4.918	5.563	7.652
C	26	11	3	6
	24.657	11.508	6.013	3.822
D	23	11	1	2
	24.110	8.253	3.163	1.474

Table 10.5: Table of expected values under model R

* Second-row values denote expected values under the R association model.

For instance, $\hat{\eta}_3 - \hat{\eta}_1 = -0.960$ can be obtained from the table of expected values as:

$$\ln\left[\frac{4.365 \times 11.508}{5.321 \times 24.657}\right] = \ln\left[\frac{5.321 \times 6.013}{7.261 \times 11.508}\right] = \ln\left[\frac{7.261 \times 3.822}{12.052 \times 6.013}\right] = -0.960$$

Similarly,

$$\hat{\eta}_2 - \hat{\eta}_1 = -0.1877 \qquad \hat{\eta}_3 - \hat{\eta}_2 = -0.7723$$
$$\hat{\eta}_4 - \hat{\eta}_3 = -0.3092 \quad \text{while} \quad \hat{\eta}_4 - \hat{\eta}_1 = -1.2701$$

which are the log odds ratios for comparing A with B, B with C, C with D and A with D conditions respectively.

A conditional test of independence for the data in Table 10.3 given that the row association model holds true, is accomplished by testing the hypothesis under the multiplicative (τ)s or log (η)s-scales, respectively, as

$$\tau_1 = \tau_2 = \cdots = \tau_I = 1 \qquad \eta_1 = \eta_2 = \cdots = \eta_I = 0$$

respectively. If the model holds, then the homogeneous row association model corresponds to the model of independence as stated before. And the conditional test is based on

$$G^2(O \mid R) = G^2(O) - G^2(R)$$

and will be based on $(I-1)(J-1) - (I-1)(J-2) = (I-1)$ degrees of freedom. For the data in Table 10.3, we present the results of this analysis below:

Model	d.f.	G^2
O	9	46.887
R	6	9.876
O | R	3	37.011

The analysis above indicates that the conditional test gives a G^2 value of 37.011 on 3 d.f. which shows very strong evidence of a row association. It should be noted here that the row association model can also be employed not only for situations when the row category is nominal, but when such category is also ordinal.

10.2.3 Effect of Changing Scoring

In the above analysis, we have assumed integer scoring for the column categories. That is, we assume that $\nu_j = \{1, 2, 3, 4\}$. We present below, the analysis based on centered scoring scheme, where $\nu_j = j - \left(\frac{J+1}{2}\right) = \{-1.5, -0.5, 0.5, 1.5\}$ in this case. This is accomplished in SAS software with the following together with a partial output.

```
set tab103;
if level eq 1 then score=-1.5;   else if level eq 2 then score=-0.5;
else if level eq 3 then score=0.5;  else score=1.5;
proc genmod; class cond level;
model count= cond|score level/dist=poi link=log type3; run;
```

```
            Criteria For Assessing Goodness Of Fit
Criterion                    DF           Value           Value/DF
-----------------------------------------------------------------
Deviance                      6          9.8761             1.6460
Pearson Chi-Square            6          9.2897             1.5483

                          Analysis Of Parameter Estimates
                                Standard         Wald 95%              Chi-
Parameter         DF  Estimate     Error   Confidence Limits     Square  Pr > ChiSq
-----------------------------------------------------------------------------------
Intercept          1    1.5331    0.4049    0.7396    2.3266      14.34      0.0002
...................................................................................
score              1   -0.7634    0.3896   -1.5270    0.0002       3.84      0.0501
score*cond    1    1    1.2701    0.2812    0.7189    1.8212      20.40      <.0001
score*cond    2    1    1.0824    0.2850    0.5238    1.6410      14.42      0.0001
score*cond    3    1    0.3101    0.2543   -0.1884    0.8086       1.49      0.2227
score*cond    4    0    0.0000    0.0000    0.0000    0.0000        .           .
```

The parameter changes in this case only affect the row and column parameters estimates. The estimates of the ηs are not affected and neither are the goodness-of-fit test statistics affected.

10.3 The Column Association (C) Model

The column association model (C) is defined in terms of the odds ratios as

$$\theta_{ij} = \theta_{+j} \quad \text{for } j = 1, 2, \cdots, J-1 \tag{10.16}$$

The model also has $(J-1)$ more parameters, namely, θ_{+j}, than the (O) model, and is also based on $(I-1)(J-1) - (J-1) = (I-2)(J-1)$ degrees of freedom. The model can similarly be written in the log-linear model form as:

$$\ln(\hat{m}_{ij}) = \mu + \lambda_i^X + \lambda_j^Y + \rho_j(\mu_i - \bar{\mu}) \tag{10.17}$$

where $\sum \lambda_i^X = \sum \lambda_j^Y = \sum_j \rho_j = 0$.

The model is equivalent to the model of independence when $\rho_j = 0$. The column association parameters $\{\rho_j\}$ is interpreted as being the deviation within a particular column of $\ln(\hat{m}_{ij})$ from column independence of a known function of the ordinal variable with slope ρ_j.

Again, for arbitrary rows i and i' ($i' > i$) and columns j and j' ($j' > j$), we can show that the log odds are equivalent to:

$$\ln\left(\frac{\hat{m}_{ij}\hat{m}_{i'j'}}{\hat{m}_{ij'}\hat{m}_{i'j}}\right) = (\rho_{j'} - \rho_j)(\mu_{i'} - \mu_i)$$

That is, the log odds ratio is again proportional to the distance between the rows and is always positive whenever $(\rho_{j'} - \rho_j) > 0$. If $\{\mu_i = i\}$, that is, integer scores, then the log odds ratio is constant and equals $(\rho_{j'} = \rho_j)$ for all $(I - 1)$ pairs of adjacent rows.

Like the row association model, the column association model is naturally suited for the general $I \times J$ table having nominal column variable and ordinal row variable since the model has the same form (and produces the same G^2) if columns of the table are again permuted.

All the models considered so far, namely, the O, U, R, and C association models, are nested in the sense that O implies U and U implies both R and C. However, R and C are not themselves nested. We can thus carry out conditional tests of the form $G^2(O \mid U)$,$G^2(U \mid R)$ or $G^2(O \mid C)$. These can be used to conditionally test the significance of the row or column parameters or simply to decompose the baseline G^2 value for the model of independence into parts corresponding to contributions from the factors to the total.

10.4 The R+C Association Model

We next consider a generalization of the row and column association models that will include an overall effect plus the effects of the rows and columns association. This model can be written as:

$$\begin{aligned} \theta_{ij} &= \theta\, \theta_{i+}\theta_{+j} \quad \text{or} \\ \Phi_{ij} &= \phi + \phi_{i+} + \phi_{+j} \end{aligned} \tag{10.18}$$

The former is the multiplicative form of the model, while the latter is the additive form of the model. In the above model, θ_{i+} and θ_{+j} are unspecified and so are the $\phi's$.

Models (10.18) describes the $(I - 1)(J - 1)$ odds ratios θ_{ij} and log odds ratios $\ln(\theta_{ij})$ in terms of the row and column effects, and it is based on $(I - 2)(J - 2)$ degrees of freedom. The model is called the R+C model because both the row effect and column effect are added to the overall effect ϕ. The model is log-linear or additive in the log odds ratios and is sometimes referred to as *model I.*

The R+C model requires both ordering of the two classificatory variables. That is, it assumes that both the row and column variables are ordinal. Thus, changing the order of any two rows or columns categories changes the model structurally. We recollect that changing the order of rows in the R model or the order of columns in the C model does not in any way change the model structurally. The R+C model would therefore be naturally suited for contingency tables with doubly ordered categories and where the spacings of the categories are also assumed known. The model is based on $(I - 2)(J - 2)$ degrees of freedom.

10.5 The RC Association Model

Because of the restrictions imposed on the R+C model above, namely,

(i) known row and column integer scores and

(ii) both rows and columns being ordinal,

the model RC, which is the multiplicative form of the additive log odds ratio model for the R+C model, is therefore proposed by Goodman (1979a). The RC model has the log odds ratios Φ_{ij} modeled as:

$$\Phi_{ij} = \phi' \phi'_{i+} \phi'_{+j} \tag{10.19}$$

where the ϕ' values here should not be confused with the ϕ values in the R+C model. It is sometimes referred to as *model II* in comparison to the R+C model, which is model I.

This model when written in terms of the linear-by-linear model has row score parameters as μ_i and column score parameters as ν_j where the μ_i and the ν_j are unknown and need to be estimated from the data, and the corresponding log of the expected frequencies can be written in the form:

$$\ln(\hat{m}_{ij}) = \mu + \lambda_i^X + \lambda_j^Y + \phi \mu_i \nu_j \tag{10.20}$$

with the following constraints imposed:

$$\sum_i \mu_i = \sum_j \nu_j = 0 \tag{10.21a}$$

$$\sum_i \mu_i^2 = \sum_j \nu_j^2 = 1 \tag{10.21b}$$

$$\sum_i \lambda_i^X = \sum_j \lambda_j^Y = 0 \tag{10.21c}$$

Constraints in (10.21a) and (10.21b) are often referred to as properties that ensure that a scale has to have a zero and unit variance properties. It would therefore be desirable in order to make scale comparison meaningful to invoke the properties that a scale has to have a zero and unit properties.

The constraints in (10.21a) and (10.21b) are due to Goodman (1979b) where the first constraint ensures that the scores are centered at zero, while the second also ensures that the length of each vector is normed to 1. The above is analogous to the standardized normal Z variable with mean 0 and unit variance.

The above method has been described (Goodman, 1991) as the *unweighted solution* or marginal independent constraints. Goodman (1981, 1985) also proposed instead the *marginal-weighted scores* where (10.21a) and (10.21b) are now defined as:

$$\sum_i \mu_i P_{i+} = \sum_j \nu_j P_{+j} = 0 \tag{10.22a}$$

$$\sum_i \mu_i^2 P_{i+} = \sum_j \nu_j^2 P_{+j} = 1 \tag{10.22b}$$

Here the score parameters are normed with the marginal distribution where $P_{i+} = \sum_j p_{ij}$, $P_{+j} = \sum_i p_{ij}$ and the p_{ij} are the observed probabilities.

A more general rule is to define row weights g_i and column weights h_j (see Becker & Clogg, 1989), which may or may not sum to 1. If $g_i > 0$ for all i and $\sum_i g_i = 1$, then we may regard g_i as a probability distribution for the rows, and with similar arguments for the h_i) for the columns. With these weights, we have:

$$\sum_i g_i\mu_i = \sum_j h_j\nu_j = 0 \tag{10.23a}$$

$$\sum_i g_i\mu_i^2 = \sum_j h_j\nu_j^2 = 1 \tag{10.23b}$$

In all cases, the first constraint centers scores at zero or the mean score at zero or the weighted score at zero. The second constraints adjusts the length of score vectors in all cases.

The unweighted scores discussed above is a special case of the above when $g_i = h_j = 1$ for all i and j. Similarly, the marginal-weighted solution is obtained if we let $g_i = P_{i+}$ and $h_j = P_{+j}$. Other choices have $g_i = 1/I$ and $h_j = 1/J$, which has been referred to as the *uniform-weighted* solution. Other weights that have been employed include $g_i = i - (I+1)/2, \quad i = 1, 2, \cdots, I$ and $h_j = j - (J+1)/2, \quad j = 1, 2, \cdots, J$ which adjusts the sums of the weights to zero.

The corresponding log odds ratio from (10.21) thus becomes

$$\Phi_{ij} = \phi'(\mu_{i+1} - \mu_i)(\nu_{j+1} - \nu_j) = \phi\, \phi'_{i+}\phi'_{+j} \tag{10.24}$$

We notice immediately that this model is not log-linear but rather multiplicative in the log of odds ratios, and this generally complicates estimation procedure, but the ANOAS algorithm by Goodman or the ANOAS module in CDAS by Eliason (1990) makes the estimation and fitting of this model very simple.

With μ and ν defined as scores, it thus becomes obvious that ϕ'_{i+} denotes the distance between rows i and $i+1$ and that ϕ'_{+j} similarly denotes the distance between columns j and $j+1$. We also note that the ordering of the categories has been made redundant because of the integer nature of the score parameters, and it can be shown that this model is unchanged by any permutation of rows or columns. Because of this property, it does not matter whether we model the RC model in terms of the log odds ratios or in terms of log of expected odds ratios. The RC association model, like its additive R+C counterpart, is also based on $(I-2)(J-2)$ degrees of freedom.

Models O, U, R, and C are all special cases of the RC model. For instance, the (O) model is obtained when $\phi = 0$. For a more detailed discussion of this model, readers are advised to refer to the following references: Goodman (1979a), Haberman (1981), and Clogg (1982). Various scores have been advocated. These scores range from marginal weighted scores (Goodman, 1981, 1985) and row weights (Becker & Clogg, 1989).

10.6 Homogeneous R+C or RC Models

The homogeneous row-column effect models imply the following additional constraints:

$$(R+C)_H : \mu_1 = \nu_1; \quad \mu_2 = \nu_2; \quad \cdots, \quad \mu_I = \nu_I$$
$$(RC)_H : \mu_1 = \nu_1; \quad \mu_2 = \nu_2; \quad \cdots, \quad \mu_I = \nu_I$$

We note that both the R+C and the RC homogeneous models can only be employed for square contingency tables in which $I = J$ and that there are additional I constraints on the homogeneous model parameters. But because of the location and scale constraints, there are, however, $(I-2)$ non-redundant, identifiable or estimable parameters and the degrees of freedom is given by $(I-2)(I-2)+(I-2) = (I-2)(I-1)$. Table 10.6 gives the degrees of freedom for all the models discussed in the preceding sections.

Models	d.f.
O	$(I-1)(J-1)$
U	$(IJ-I-J)$
R	$(I-1)(J-2)$
C	$(I-2)(J-1)$
R+C	$(I-2)(J-2)$
$(\text{R+C})_H$	$(I-2)(I-1)$
RC	$(I-2)(J-2)$
$(\text{RC})_H$	$(I-2)(I-1)$

Table 10.6: Models considered in this chapter

while Table 10.7 gives the degrees of freedom for the conditional association tests together with their corresponding G^2.

Effects on association	Models used	d.f.	G^2
1. General Effect	O-U	1	$G^2(O \mid U) = G^2(O) - G^2(U)$
2. Row effects	U-R	$I-2$	$G^2(R \mid U)$
3. Column effects	U-C	$J-2$	$G^2(C \mid U)$
4. Column effects given Rows $C \mid R$	RC-R	$J-2$	$G^2(RC \mid R)$
5. Row effects given Columns $R \mid C$	RC-C	$I-2$	$G^2(RC \mid C)$
6. Residual	RC	$(I-2)(J-2)$	$G^2(RC)$
7. Total	O	$(I-1)(J-1)$	$G^2(O)$

Table 10.7: Degrees of freedom for conditional association tests

10.6.1 General Comments

- When the classificatory variables are both ordered, that is, when we have a *doubly ordered* contingency table, then the R+C model should be used instead of the RC model.

- If only the rows variables are ordered but not the column variable, then the U model would not be relevant nor would either of R (since it assumes ordering of columns) or the R+C model. In this case, only the O, C, and RC models would be appropriate. This would also be the case even if the column category were partially ordered.

- When neither row nor column are ordered as in case of a nominal-nominal table, then only the O and the RC models would be appropriate, since they are the only models un-affected by permutation of either row categories or column categories.

10.6.2 Example 10.2: Analysis of Data in Table 10.1

We give below the results of applying the models listed in Table 10.6 to the data in Table 10.1.

Model	d.f.	G^2	BIC	AIC
O	9	20.378	-43.71	-12.38
U	8	7.429	-49.53	-8.57
R	6	7.404	-35.32	-4.60
C	6	5.534	-37.19	-6.47
R+C	4	5.488	-22.99	-2.51
$(R+C)_H$	6	6.236	-36.49	-5.76
RC	4	4.914	-23.57	-3.09
$(RC)_H$	6	6.619	-36.10	-5.38

The above models are implemented in SAS software with the following SAS software program.

```
set tab101;
***create uniform, column and row integer scores ***;
u=chol*bp;
cl=bp;
rl=chol;
*** INDEPENDENCE MODEL ***;
proc genmod; class chol bp;
model count=chol bp/dist=poi type3; run;
*** (U)-ASSOC. MODEL ***;
proc genmod; class chol bp;
model count=chol bp u/dist=poi; run;
*** R-ASSOC. MODEL ***;
proc genmod; class chol bp;
model count=chol|cl bp/dist=poi type3; run;
*** (C)-ASSOC. MODEL ***;
proc genmod; class chol bp;
model count=chol bp|rl/dist=poi type3; run;
*** R+C-ASSOC. MODEL ***;
proc genmod; class chol bp;
model count=chol|cl bp|rl/dist=poi type3; run;
```

Apart from the null model (O), all the other models considered adequately fit the data. Again, using the parsimony consideration in this case, model U (the uniform association) seems to be the most parsimonious of all the models. This model gives a BIC (Bayesian information criterion) and an AIC (Akaike information criterion) of -49.534 and -8.571, respectively. Both criteria agree in this case on the choice of the most parsimonious model. For these data, the sample size is 1237. Model RC which is implemented in CDAS, for instance, has parameter estimates displayed in the following table.

$\hat{\phi}$	0.51962			
	$i=1$	2	3	4
$\hat{\mu}_i$:	-0.64330	-0.32248	0.39399	0.57179
	$j=1$	2	3	4
$\hat{\nu}_j$:	-0.72656	0.06975	-0.02623	0.68305

Table 10.8: MLE under model RC for the data in Table 10.1

Similarly, the ML estimates (MLE) under the homogeneous (RC) model are presented in Table 10.9.

$\hat{\phi}$	0.52582			
	$i = 1$	2	3	4
$\hat{\mu}_i$:	-0.71443	-0.12398	0.17145	0.66696
	$j = 1$	2	3	4
$\hat{\nu}_j$:	-0.71443	-0.12398	0.17145	0.66696

Table 10.9: MLE under model $(\text{RC})_H$ for the data in Table 10.1

Similarly for the data in Table 10.2, we give the fits of the various models described in the preceding sections to these data.

Model	d.f.	G^2
O	15	47.418
U	14	9.895
R	12	6.281
C	10	6.829
R+C	8	3.045
RC	8	3.571

Using the independence model as the baseline model, the U model accounts for about 80% of the G^2 under independence and fits the data quite well. That the U model fits the data well indicates that there is positive association between the two classificatory variables.

10.7 The General Association Model

The general RC(**M**) association model (Goodman, 1986a, 1996) is defined in terms of log expected frequencies as:

$$\ln(\hat{m}_{ij}) = \lambda + \lambda_i^X + \lambda_j^Y + \sum_{m=1}^{M} \phi_m \mu_{im} \nu_{jm} \tag{10.25}$$

where $M = \min(I, J) - 1$. Here again, the parameters μ_{im} and ν_{jm} are scores to be estimated from the data, while the parameter ϕ_m is again defined to be a measure of the intrinsic association (in the **m** dimensions) that will be estimated from the data. We may assume without loss of generality that the ϕ_m are ordered so that:

$$\phi_1 \geq \phi_2 \geq \cdots \geq \phi_m.$$

When $M = 0$, the model in (10.25) reduces to the model of independence, that is, the null model (O). Similarly, when $M = 1$, the model defined by RC(1) is also equivalent to the (RC) model discussed in the previous section. And cases where $M > 1$ are of interest in this section as they relate to the dimensionality of the association which can be shown to be analogous to the usual correspondence analysis. Further, the fit of models in which $M > 1$ would be necessary for those data in which the RC model proves inadequate. The RC(**M**) model is based on $(I - M - 1)(J - M - 1)$ degrees of freedom.

The log odds ratio formed from rows i and i' and columns j and j' can be decomposed as:

$$\Phi_{ij(i'j')} = \sum_{m=1}^{M} \phi_m (\mu_{im} - \mu_{i'm})(\nu_{jm} - \nu_{j'm}) \tag{10.26}$$

Thus, the log odds ratio for any 2×2 subtable formed from rows i and i' and columns j and j' is represented by a sum of M components, one component for each dimension allowed under the model. Further, each component reflects the intrinsic level of association in that dimension, the distance between rows i and i', and also the distance between columns j and j'.

For identifiability purposes, constraints would need to be imposed on the score parameters (the usual zero-point and unit-point restrictions) because the RC(**M**) model has M ϕ's, MI μ's and MJ ν's, for a total of M$(1 + I + J)$ parameter values. For example, for the 4×4 data in Table 10.1, $I = J = 4$ and model RC(**M**) has $M = 3$, which would give a total of 27 parameter values for this model, in contrast to only $(4 - 1)(4 - 1) = 9$ nonredundant possible interaction parameters.

Consequently, if we let g_i and h_j be sets of weights, then we have for the zero-point restriction

$$\sum_{i=1}^{I} g_i \mu_{im} = \sum_{j=1}^{J} h_j \nu_{jm} \quad = 0 \tag{10.27a}$$

Similarly, the unit-point restrictions are:

$$\sum_{i=1}^{I} g_i (\mu_{im})^2 = \sum_{j=1}^{J} h_j (\nu_{jm})^2 = 1 \tag{10.27b}$$

where in both cases $\mathbf{m} = 1, 2, \cdots, M$. The above restrictions impose respectively 2M constraints and thus a total of 4M restrictions. This will in effect reduce the total number of parameters to $M(1 + I + J) - 4M = M(I + J - 3)$. Thus for our 4×4 table, we would now have for the RC(3) model 15, which are still too many parameters. We next impose orthogonal constraints in addition to the constraints imposed above. That is:

$$\sum_{i=1}^{I} g_i \mu_{im} \mu_{im*} = \sum_{j=1}^{J} h_j \nu_{jm} \nu_{jm*} = 0 \quad \text{for all } m \neq m* \tag{10.28}$$

Goodman (1986b, 1991) has described the general association model RC(**m**) for various weighted situations. Goodman (1985, 1996) has also discussed extensively correlation models in the general contingency tables. These models will not be covered in this text, and interested readers are referred to the references cited above.

These impose $\binom{M}{2}$ restrictions on the μ and ν parameters, respectively. We thus now have a total of $M(I + J - 3) - \frac{M(M-1)}{2} - \frac{M(M-1)}{2} = M(I + J - M - 2)$ interaction parameters. The degrees of freedom for the RC(**M**) model is therefore given by:

$$(I - 1)(J - 1) - M(I + J - M - 2) = (I - M - 1)(J - M - 1)$$

that is,

$$\text{d.f.} = \begin{cases} (I - M - 1)(J - M - 1) & \text{if } M < \min(I, J) - 1 \\ 0 & \text{if } M = \text{in}(I, J) - 1 \end{cases}$$

The case when $M = \min(I, J) - 1$ refers to the saturated situation. All other cases in which $M < \min(I, J) - 1$ refer to unsaturated cases.

We may also note here that the weights g_i and h_j may take any of the three (*unweighted*, *marginally weighted*, and *uniformly weighted*) weighted scores discussed in the previous section.

The RC(**M**) model can be implemented by the use of program RCDIM in the CDAS group of programs by Scott Eliason (1990).

10.7.1 Example 10.3: Application to Table 10.1

Again, let us consider the set of data in Table 10.1. For this data we have the following results when we fit either the unsaturated RC model or the RC(M) to the data when $M^* = 1$. Both models are equivalent in terms of estimates of parameters etc only if marginal weighted scores are employed. That is, ANOAS uses the row marginal scores $P_{i+} = \{0.2482, 0.1987, 0.3549, 0.1981\}$ and column marginal scores $P_{+j} = \{0.3137, 0.4260, 0.1649, 0.1981\}$. And it is only in this case that the estimated correlation can be obtained.

If we were to fit the RC(1) association model to the data for the five weighted situations listed below, then:

1. Unweighted, that is, $g_i = 1$,$h_j = 1$ for all (i, j).

2. Uniformly weighted, that is, $g_i = \frac{1}{I}$ and $h_j = \frac{1}{J}$

3. Integer-weighted,that is, $g_i = i$ and $h_j = j$; $i = 1, 2, \cdots, I$ and $j = 1, 2, \cdots, J$.

4. Marginally weighted, that is, $g_i = P_{i+}$ and $h_j = P_{+j}$.

5. Centered scores, where $g_i = i - (I + 1)/2$ and $h_j = j - (J + 1)/2$.

We would find that in all the five situations above, the G^2 values would remain the same indicating that the choice of scores do not affect the magnitude of the relevant test statistic (Becker & Clogg, 1989). However, the μ and ν estimates would be different for all the five scores, as well as the estimates of the intrinsic ϕ association parameter. We give an example to demonstrate this general theory. Again, we consider the data in Table 10.1 for analyses using scores in 1, 2, 3, and 4 as defined above. We have in Table 10.10 the following results when model RC(1) is fitted. A more detailed treatment is provided in Becker and Clogg (1989).

	Unweighted SCORES(1)		Uniform SCORES(2)		Integer SCORES(3)		Marginal SCORES(4)	
ROWS	$\hat{\mu}_i$	$\hat{\mu}_i'$	$\hat{\mu}_i$	$\hat{\mu}_i'$	$\hat{\mu}_i$	$\hat{\mu}_i'$	$\hat{\mu}_i$	$\hat{\mu}_i'$
1	-0.643	-0.464	-1.287	-0.464	-0.622	-0.626	-1.366	-0.458
2	-0.323	-0.232	-0.645	-0.232	-0.390	-0.393	-0.715	-0.239
3	0.394	0.284	0.788	0.284	0.127	0.128	-0.741	0.249
4	0.572	0.412	1.143	0.412	0.255	0.257	1.101	0.369
COLUMNS	$\hat{\nu}_i$	$\hat{\nu}_i'$	$\hat{\nu}_i$	$\hat{\nu}_i'$	$\hat{\nu}_i$	$\hat{\nu}_i'$	$\hat{\nu}_i$	$\hat{\nu}_i'$
1	-0.727	-0.524	-1.453	-0.524	-0.663	-0.668	-1.340	-0.449
2	0.070	0.050	0.140	0.050	-0.097	-0.098	0.471	0.158
3	-0.026	-0.019	-0.053	-0.019	-0.166	-0.167	0.252	0.085
4	0.683	0.492	1.366	0.492	0.339	0.341	1.865	0.626
ϕ	0.5195		0.1299		1.0125		0.1125	
ρ	NA		NA		NA		0.1116	
G^2	4.9138		4.9138		4.9138		4.9138	

Table 10.10: Estimated scores from fitting the RC(1) model under four different scoring systems

In all the four cases above, we notice that the G^2 values are the same and the only changes are in the parameter estimates as well as the estimates for the intrinsic association parameter ϕ. The correlation between the scores μ and ν is only obtainable for the case when the scores are marginally weighted, that is, case 4 in which $g_i = P_{i+}$ and $h_j = P_{+j}$. The models above are each based on 4 d.f. The adjusted scores estimates $\mu_i^{'}$ and $\nu_j^{'}$ are obtained by multiplying by the corresponding $\sqrt{\hat{\phi}}$.

10.8 Correlation Models

For the $I \times J$ contingency table, let P_{ij} denote the probability that an observation falls in cell $(i, j)\{i = 1, 2, \cdots, I; j = 1, 2, \cdots, J\}$. Then the model of independence is given by:

$$P_{ij} = P_{i+}P_{+j}$$

where $P_{i+} = \sum_j P_{ij}$ and $P_{+j} = \sum_i P_{ij}$.

Goodman (1985) has introduced a generalization of the above model, which he described as the *correlation* or *canonical correlation* model. This generalization takes the form:

$$P_{ij} = P_{i+}P_{+j}(1 + \sum_{m=1}^{M} \rho_m x_{im} y_{jm}) \tag{10.29}$$

where $M = \min(I - 1, J - 1)$, and where x_{im} are the row scores. Similarly, the y_{jm} are the column scores and the parameter ρ_m is a measure of the correlation between the row score x_{im} and the column score y_{jm}. Further, the scores satisfy the following constraints for $m \neq m'$:

$$\sum_{i=1}^{I} x_{im} P_{i+} = 0 \qquad \sum_{j=1}^{J} y_{jm} P_{+j} = 0$$

$$\sum_{i=1}^{I} x_{im}^2 P_{i+} = 1 \qquad \sum_{j=1}^{J} y_{jm}^2 P_{+j} = 1$$

$$\sum_{i=1}^{I} x_{im} x_{im'} P_{i+} = 0 \quad \sum_{j=1}^{J} y_{jm} y_{jm'} P_{+j} = 0$$

From the above, we notice that the x_{im} and the y_{jm} scores pertaining to the m-th component have been standardized and that for $m \neq m'$, the corresponding row and column scores are uncorrelated with each other.

With the above constraints, it is not too difficult to show that

$$\sum_{i=1}^{I}\sum_{j=1}^{J} x_{im} y_{jm} P_{ij} = \rho_m \quad \text{for} \quad m = 1, 2, \cdots, M$$

If we consider the 2×2 subtable formed from rows i and i' and from columns j and j', then the correlation coefficient is given by:

$$\rho = (P_{ij}P_{i'j'} - P_{ij'}P_{i'j})/(P_{i+}P_{i'+}P_{+j}P_{+j'})^{\frac{1}{2}}$$

Specifically, for a 2×2 contingency table, the above reduces to

$$\rho = (P_{11}P_{22} - P_{12}P_{21})/(P_{1+}P_{2+}P_{+1}P_{+2})^{\frac{1}{2}}$$

and the usual odds ratio becomes

$$\theta = (P_{11}P_{22}/P_{12}P_{21})$$

We see that ρ is marginal dependent, while θ is marginal free. θ is therefore useful for modeling unweighted association models while ρ is most useful for modeling correlation models.

We see from (10.29) that

$$(P_{ij} - P_{i+}P_{+j})/P_{i+}P_{+j} = \sum_{m=1}^{M} \rho_m x_{im} y_{jm} \tag{10.30}$$

Squaring (10.30) and summing over i and j, we have

$$\begin{aligned}
\sum_{i=1}^{I}\sum_{j=1}^{J}(P_{ij} - P_{i+}P_{+j})^2/(P_{i+}P_{+j}) &= \sum_{i=1}^{I}\sum_{j=1}^{J}\left(\sum_{m=1}^{M} \rho_m x_{im} y_{jm}\right)^2 P_{i+}P_{+j} \\
&= \sum_{m=1}^{M} \rho_m^2 \left(\sum_{i=1}^{I} x_{im}^2 P_{i+}\right) \\
&= \sum_{m=1}^{M} \rho_m^2 \left(\sum_{j=1}^{J} y_{jm}^2 P_{+j}\right) \\
&= \sum_{m=1}^{M} \rho_m^2
\end{aligned}$$

and the Pearson's X^2 like quantity:

$$(P_{ij} - P_{i+}P_{+j})^2/P_{i+}P_{+j} = \sum_{m=1}^{M} \rho_m^2 = \frac{X^2}{N} \tag{10.31}$$

If we multiply both sides of the above by N, the sample size, we have

$$X^2 = N\sum_{m=1}^{M} \rho_m^2$$

For a 4×4 table, $M = 3$ and from the above, we see that

$$\frac{\rho_1^2}{X^2/N} \qquad \frac{\rho_2^2}{X^2/N} \qquad \frac{\rho_3^2}{X^2/N}$$

and these correspond to the first inertia, second inertia and third inertia quantities respectively in the usual correspondence analysis. We note that these inertia quantities add up to 1. That is,

$$\frac{\rho_1^2}{X^2/N} + \frac{\rho_2^2}{X^2/N} + \frac{\rho_3^2}{X^2/N} = 1$$

As noted by Clogg, the scores parameters derive from principal components (or singular value decomposition) so that the correlations ρ_m represent correlations between principal components. Thus a singular value decomposition algorithm can be used to calculate the eigenvalues ($\rho_m' s$), the left (row) eigenvectors (x_{im}) and the right (column) eigenvectors (y_{jm}) based on the estimate of Δ_{ij}, where

$$\Delta_{ij} = (P_{ij} - P_{i+}P_{+j})/(P_{i+}P_{+j})$$

10.8.1 Example 10.4

We now fit all the models that have been described in the preceding sections to the data in Table 10.2.

Model	d.f.	G^2	$\hat{\phi}$	$\hat{\rho}$
O	15	47.418	-	-
U	14	9.895	0.152	0.150
R	12	6.281	0.160	0.157
C	10	6.829	0.158	0.156
R+C	8	3.045	0.004	0.100
RC	8	3.571	0.166	0.161

The results above are obtained by using the program ANOAS by Leo Goodman. We note here that for all the models, the centered scores $g_i = i - (I+2)/2$ and $h_j = j - (J+2)/2$ are employed. Under model U, for instance, the estimated values of the row and column scores are respectively:

$$\hat{\mu}_i = \{-1.439, -0.481, 0.477, 1.436\} \quad \text{and}$$

$$\hat{\nu}_j = \{-1.539, -0.918, -0.298, 0.323, 0.9441.565\} \quad \text{with } \hat{\phi} = 0.152$$

To convert the estimate of the intrinsic association parameter to those based on equally spaced rows and column scores, we use the relationship

$$\hat{\phi}' = \hat{\phi}(\hat{\mu}_i - \hat{\mu}_{i+1})(\hat{\nu}_j - \hat{\nu}_{j+1})$$

In our example above, $\hat{\phi}' = 0.153(-1.439 + 0.481)(-1.539 + .918) = 0.091$. Consequently, $\hat{\theta} = \exp(\hat{\phi}') = 1.095$. Also, using the independence model as the baseline model, the U model accounts for about 80% of the G^2 under independence and fits the data quite well. That the U model fits the data well indicates that there is positive association between the two classificatory variables.

Similarly, the RC(1) models for the data under various weights have as expected a G^2 value of 3.571 on 8 d.f. However, the estimates of the intrinsic association ϕ and the row and column scores differ in each of the cases considered. For instance, we have $\hat{\phi} = \{0.9649, 0.1970, 2.4081, 0.1665\}$ respectively for the uniform weights $g_i = h_j = 1$, uniform marginal weights, $g_i = 1/4, h_j = 1/6$; integer scores $g_i = i, h_j = j$ and marginal weighted $g_i = P_{i+}, h_j = P_{+j}$. In the latter case, the estimated correlation $\hat{\rho} = 0.1611$.

We present in Table 10.11, estimates of the parameters when models RC, RC(1), and the correlation model are applied to the data in Table 10.2.

10.9 Grouping of Categories in Two-Way Tables

A reasonable grouping of the rows and column categories of a two-way contingency table can sometimes simplify the analysis of association between the two classificatory variables. Thus by grouping categories, we may "get a more parsimonious and compact summary of the data" (Fienberg, 1980, p. 154). These will lead to a considerable reduction in the number of parameters under the hypothesized model. Thus, we are interested in *collapsing* our original $I \times J$ table to an $I^* \times J^*$ table, where I^* and J^* represent lower dimensions, that is, $I^* \leq I$ and $J^* \leq J$.

Parameter	Estimates		
	RC	RC(1)	CORR.
$\hat{\mu}_1$	-1.678	-1.678	1.637
$\hat{\mu}_2$	-0.140	-0.140	0.102
$\hat{\mu}_3$	0.137	0.137	-0.105
$\hat{\mu}_4$	1.414	1.414	-1.428
$\hat{\nu}_1$	-1.112	-1.112	1.201
$\hat{\nu}_2$	-1.121	-1.121	1.289
$\hat{\nu}_3$	-0.371	-0.371	-0.067
$\hat{\nu}_4$	0.027	0.027	0.040
$\hat{\nu}_5$	1.010	1.010	-0.934
$\hat{\nu}_6$	1.818	1.818	-1.743

Table 10.11: Results from correlation analysis

From chapter 6, the independence model, when it holds, is always collapsible. For other models, we need to establish reasonable criteria for collapsibility. Goodman (1981) has introduced the *homogeneity* criterion (amongst others), which assumes that "particular rows or columns can be combined if these particular rows or columns are homogeneous" (Gilula, 1986). This translates to:

> Two distinct columns a and b are said to be homogeneous if $P_{ia}/P_{+a} = P_{ib}/P_{+b}$ for all $1 \leq i \leq I$ and similarly for rows. Using the results of Gilula (1986), the above implies that a necessary and sufficient condition for two distinct columns a and b to be homogeneous is that the scores $\nu_{a\alpha} = \nu_{b\alpha}$ where $1 \leq \alpha \leq \min(I-1, J-1)$. In our case $\alpha = 1$. The case for the rows is similarly defined.

If we let $G^2(O)$ be the likelihood ratio statistic under the model of independence, then using the Williams (1952) and Goodman (1985) procedure, one can use $G_d^2 = G^2(O) - G^2(S)$ as a test statistic to judge whether the grouping that leads to model S is justified. The statistic is distributed asymptotically as χ^2 with $(I-1)(J-1) - (I_1-1)(J_1-1)$ d.f., where I_1 and J_1 are the reduced dimensions of the table. We would use the estimated row and column scores obtained above to explore the homogeneity of the rows and columns for the data in Table 10.2.

Based on the correlation scores in the above Table 10.11, Gilula (1986) has suggested that rows 2 and 3 are homogeneous as are columns 1 and 2, and columns 3 and 4. Thus combining rows 1 and 2, columns 1 and 2, and columns 3 and 4, we have the following new 3×4 observed table, that is, Table 10.12.

Mental health status	Socioeconomic status			
	A+B	C+D	E	F
Well	121	129	36	21
Mild+ moderate	300	388	151	125
Impaired	86	154	78	71

Table 10.12: Table 10.2 with some rows and columns combined

Clogg and Shihadeh (1994) has examined the case where rows 2 and 3 are combined and columns 1 and 2, columns 3 and 4, and columns 5 and 6 are also combined, leading to the new 3×3 table given as Table 10.13.

Mental health	Socioeconomic status		
status	A+B	C+D	E+F
Well	121	129	57
Mild+moderate	300	388	276
Impaired	86	154	149

Table 10.13: Table 10.2 with Clogg's collapsibility rule

Based on the row and column scores obtained from fitting models U, R ,C, and RC to the data in Table 10.2, I consider that only rows 2 and 3 need to combined, as well as columns 1 and 2 only. These are the only ones that can be justified from the use of the above association models. The correlation model of course leads to the 3×4 table suggested by Gilula. Our approach would therefore lead to a 3×5 table displayed in Table 10.14.

Mental health	Socioeconomic status				
status	A+B	C	D	E	F
Well	121	57	72	36	21
Mild+moderate	300	170	210	151	125
Impaired	86	60	95	78	71

Table 10.14: Table 10.2 with Gilula's rule

We now give the results of fitting the independence and uniform association models to the data in Tables 10.12 to 10.14. These results are displayed in Table 10.15.

Tables	Models	d.f.	G^2	$\hat{\phi}$	$\hat{\theta}$
4×6	O	15	47.418	-	-
	U	14	9.895	0.091	1.095
3×5	O	8	45.100	-	-
	U	7	2.390	0.181	1.198
3×4	O	6	43.437	-	-
	U	5	1.270	0.256	1.290
3×3	O	4	41.449	-	-
	U	3	1.119	0.318	1.374

Table 10.15: Results of fitting both the independence and uniform association models to the collapsed tables

The independence model for all the three collapsed tables gives G^2 values of 45.100, 43.437, and 41.449 on 8, 6, and 4 degrees of freedom, respectively. These values compare very favorably to the previous value of 47.418 on 15 d.f. Each of these, therefore, gives values of $(47.418-45.100) = 2.318$ (7 d.f.), $(47.418-43.437) = 3.981$ (9 d.f.) and $(47.418-41.449) = 5.969$ (11 d.f.), respectively, for the *grouping error*. None of these values is significant.

The estimates of the θ parameter for the three cases under the (U) model are 1.198, 1.290, and 1.374 respectively. When these are compared with the original value of $\hat{\theta} = 1.095$ for the 4×6 table, all the three uniform association models indicate that the U model is adequate or satisfactory for the collapsed tables; the estimates of the associations, as measured by θ, did change appreciably for the three tables. This implies that inferences drawn from a full table and a collapsed table may not necessarily be the same since the parameter estimates are not the same.

In order to reconcile these differences, Clogg and Shihadeh (1994) has advocated that since for the full table the estimates of the parameters are based on integer scores $\mu_i = i$ and $\nu_j = j$, a near equivalent result for a collapsed table can be obtained by modifying the scoring system for comparability. For instance, for the 3×3 collapsed table, if we score rows and columns as $\mu_1 = 1$, $\mu_2 = 2.5$, and $\mu_3 = 4$ and $\nu_1 = 1.5$, $\nu_2 = 3.5$, and $\nu_3 = 5.5$, then, these would be more consistent with the scoring for the full table. In this case, $G^2 = 1.12$, the same as before but with an estimate of θ given by $\hat{\theta} = 1.1118$, which is very close to the original 1.095 for the full table.

10.9.1 Association Versus Correlation Models

The question that readily comes to mind is, "When is the correlation model equivalent to the association model?" The correlation association models dates back to the earlier works of Pearson, while association models can similarly be linked to the earlier works of Yule. To answer the above question, Goodman (1996) suggests that in order to compare the two schools of thought, first consider, $R(.)$ to be a monotonic increasing function of x. Then, for the correlation models, define $R(x)$ as:

$$R(x) = P_{i+}P_{+j}(1 + \sum_{m=1}^{M} \rho_m x_{im} y_{jm})$$

Similarly, we can define

$$\ln R(x) = P_{ij} = \alpha_i \beta_j \text{exp}\left(\sum_{i=1}^{M} \phi_m \mu_{im} \nu_{jm}\right)$$

where ϕ is unweighted. Of course, ϕ can be weighted.

If $R(x) = x$, then we would have the correspondence (correlation or canonical correlation) analysis, that is, Pearsonian approach. On the other hand, if $R(x) = \ln(x)$, then we would have the association model.

Consequently, if we define a family of monotonic increasing function as

$$R(x) = \frac{x^c}{c}, \quad 0 \leq c \leq 1$$

then

$$R(x) = \begin{cases} x & \text{if} \quad c = 1 \\ \ln x & \text{as} \quad c \longrightarrow 0 \end{cases}$$

where $R(x) = \ln x$ is obtained as a limiting form of $R(x)$ as x approaches zero. That is,

$$\lim_{c \to 0} \frac{x^c}{c} = \ln x$$

Goodman (1996) has suggested that while $c = 1$ gives the correlation model, which is equivalent to the correspondence analysis of Pearson (1947), $c = 0$, gives the usual association (Yule's) models. A middle of the road can be taken such that $c = 1/2$ (where the two approaches will be equivalent). He describes such a case as the *mid-way association models.*

In general, Goodman (1986b) has suggested that in most situations, the association models seem better than the correlation model. With association models, you can think about rows and columns of the table being symmetric.

As an example, for the data in Table 10.1, we have $\hat{\rho} = 0.112$, which is not too high; consequently, the equivalent RC model can be employed for this data. Goodman (1991) has advocated that the RC model be used in these cases.

The analysis of association models in multi-way contingency tables when one or more classificatory variable are ordered is fully discussed in the next section.

10.10 Higher Dimensional Tables

In this section, we extend the concept of association models to higher dimensional contingency tables. First, consider a three-way contingency table having variables A, B, and C with I, J, and K categories respectively. If one, two, or all of these variables have ordinal categories, then we can exploit the ordinal nature of these categories in our analysis. Our interest here is to model the association in such a three-way table. Three types of models that exploit these ordering of categories of the variables will be discussed here.

(i) The first of such models relates to the case in which two of the variables (say A and B) are nominal and the third variable C has ordinal categories. Such models are very common and we will illustrate such a case with the data in Table 10.16 in the next section.

(ii) The second of such models relate to the case in which two of the variables (say, A and B) are ordinal and the third variable C has nominal categories (or groups or layers). Such models are referred to by Clogg (1982) as *conditional association models.* An alternative approach to modeling this class of tables is to fit K row-column association models in **m** dimensions, that is, we fit the RC(**m**) to the two ordinal variables at various levels of the third (nominal) variable. This type of models are fully discussed in Becker and Clogg (1989).

(iii) The third type of models to be discussed, considers the three variables A, B, and C to have ordinal categories. Again, following, Clogg (1982), models associated with these are described as the *partial association* models.

In all our discussions in this section, we would suppose that we have a three-way $I \times J \times K$ table. In addition, if we let n_{ijk} be the observed count in the table, then $\hat{m}_{ijk}$ would be the corresponding expected cell count under some model, for $i = 1, 2, \cdots, I$, $j = 1, 2, \cdots, J$, and $k = 1, 2, \cdots, K$.

10.10.1 Type I Group of Models

The models in this class have two variables that are nominal, with the third variable (response variable) having ordinal categories. An example is given in Table 10.16, which is reproduced from Clogg and Shockey (1988) and relates to job satisfaction on a 4-point ordinal scale and explanatory variables race (White, other) and degree attained (high school or less, more than high school), resulting in a $2 \times 2 \times 4$ contingency table. Let us assume for the sake of our analysis here that the variable "degree attained" is nominal.

		Job satisfaction			
Race	Degree attained	Very satisfied	Moderately satisfied	A Little dissatisfied	Very dissatisfied
White	≤HS	404	319	81	52
	>HS	112	81	16	10
Other	≤HS	48	46	21	7
	>HS	8	14	1	0

Table 10.16: Job satisfaction data

Here, if we let the response variable (job satisfaction) be represented by S, degree attained by D, and race by R, then a series of log-linear models can be fitted to this data. The models fitted with their corresponding G^2 and degrees of freedom are presented in Table 10.17.

Model	d.f.	G^2	pvalue
{RD S}	9	17.4553	0.0412
{RD, RS}	6	12.9673	0.0435
{RD, DS}	6	9.8122	0.1326
{RD,RS,DS}	3	5.5650	0.1348

Table 10.17: Results of fitting various logit models

The above models are implemented in SAS software with the following statements:

```
options nodate nonumber nocenter ls=85 ps=66;
data tab1013;
do r=1 to 2; do d=1 to 2; do s=4 to 1 by -1;
input count @@; sl=s; output; end; end; end;
datalines;
404 319 81 52 112 81 16 10 48 46 21 7 8 14 1 0
;
proc genmod;
class r d s; model count=r|d s/dist=poi link=log; run;
proc genmod;
class r d s; model count=r|d s|d/dist=poi link=log; run;

proc genmod;
class r d s; model count=r|d s|r/dist=poi link=log; run;
proc genmod;
class r d s; model count=r|d s|d s|r/dist=poi link=log; run;
```

For the four models fitted above, their log-linear model formulations can be written. For instance, for the first model {RD,S}, we have:

$$\ln\left(\hat{m}_{ijk}\right) = \mu + \lambda_i^R + \lambda_j^D + \lambda_k^S + \lambda_{ij}^{RD}$$

We see that the first two models {RD,S} and {RD,R} do not fit the data. Model {RD,S} states that job satisfaction is jointly independent of both race and degree attained. The last two models however provided better fits. Since the response variable has four levels, and if we assume integer scores, we can examine the linear effect component of this variable on the other factor variables. We therefore incorporate next, the linear components of S into each of the four models resulting in the results below in Table 10.18.

Model	d.f.	G^2	pvalue
RD, S(1)	11	91.5181	0.0000
RD, S, R*S(1)	8	14.3779	0.0724
RD, S, D*S(1)	8	14.3037	0.0741
RD, S, R*S(1), D*S(1)	7	11.4406	0.1204

Table 10.18: Results of fitting equivalent log-linear models with only the linear components of S

Both model which suggest linear association between the response variable S and either race or degree attained have G^2 in the neighborhood of 14.3, and each in turn indicates that the interaction components are significantly different from zero. Since neither can be eliminated from the model, the last model {RD, S, R*S(1), D*S(1)} gives a G^2 value of 11.4406 on 7 degrees of freedom. This model fits the data very well and is given by the log-linear formulation:

$$\ln(\hat{m}_{ijk}) = \mu + \lambda_i^R + \lambda_j^D + \lambda_k^S + \lambda_{ij}^{RD} + \lambda_{ik}^{RS}(u_k - \bar{u}) + \lambda_{jk}^{DS}(u_k - \bar{u})$$

where $\{u_k\}$ are the scores (integer) relating to the response variable S. Note that we have scored the response variable from 1 to 4 with a score of 4 corresponding to "very satisfied" and a score of 1 corresponding to "very dissatisfied." The ML estimates of the RS and DS associations in this case are 0.1715 (a.s.e. = 0.0985) and −0.1473 (a.s.e. = 0.0885), respectively. Since this model fits the data, we can therefore conclude that being White is associated with more job satisfaction (less job dissatisfaction) than being nonwhite. Similarly, having at most an high school degree is associated with less job satisfaction (more job dissatisfaction) than having more than a high school degree. The model is implemented in SAS software with the following, where **sl** are the integer scores for the levels of S.

```
set tab1013;
sl=s;
proc genmod;
class r d s; model count=r|d s r|sl d|sl/dist=poi link=log; run;
```

10.11 Conditional Association Models

The models in this class relate to situations in which two ordinal variables (say A and B) are observed for each of K groups (or K levels of variable C). We are interested in possible sources of between-group heterogeneity in the association. The sampling scheme in this case is the product multinomial for the different IJ-dimensional multinomials.

We shall assume that variables A and B have scores $\{u_i\}$ and $\{v_j\}$, respectively. Then for the three-way contingency table, the conditional odds ratios are defined as:

$$\theta_{ij(k)} = (\hat{m}_{ijk}\hat{m}_{i+1,j+1,k})/(\hat{m}_{i,j+1,k}\hat{m}_{i+1,j,k}) \tag{10.32}$$

where, $(i, j) = 1, 2, \cdots, (I - 1)$, and $k = 1, 2, \cdots, K$.

The data in Table 10.19 is an example of data relating to this class of models.

10.11.1 Example 10.6

The data in Table 10.19 relate to the $2 \times 3 \times 3$ table of cross-classification of popularity ratings of two groups of children over two occasions (von Eye & Spiel, 1996).

Groups	Ratings at Time 1	Ratings at Time 2			Total
		1	2	3	
	1	1	3	0	4
1	2	4	43	18	65
	3	0	4	13	17
Total		5	50	31	86
	1	7	4	0	11
2	2	3	36	9	48
	3	1	7	25	33
Total		11	47	34	92

Table 10.19: Cross-tabulation of popularity ratings of two groups of children over two occasions

If we designate ratings at time 1 to be variable A, ratings at time 2 to be variable B, and the groups to be variable C, then variables A, B, and C are indexed by $i = 1, 2, 3$, $j = 1, 2, 3$, and $k = 1, 2$, respectively.

Under the product multinomial sampling scheme, the baseline model is the log-linear model {AC,BC}, which is usually described as the null conditional association model. The following models of interest will be considered for the data in Table 10.19. These models are defined in terms of the odds ratios in (10.32) and the corresponding equivalent log-linear model formulations in terms of $l_{ijk} = \ln \hat{m}_{ijk}$.

(i) The *null conditional association* model has

$$\theta_{ij(k)} = 1 \tag{10.33a}$$

$$l_{ijk} = \mu + \lambda_i^A + \lambda_j^B + \lambda_k^C + \lambda_{ik}^{AC} + \lambda_{jk}^{BC} \tag{10.33b}$$

The model is based on $K(I-1)(J-1)$ degrees of freedom and is equivalent to the log-linear model {AC,BC}. In other words, A and B are conditional independent given the levels of variable C.

(ii) The *homogeneous conditional uniform association* model has

$$\theta_{ij(k)} = \theta \tag{10.34a}$$

$$\begin{aligned} l_{ijk} = \mu &+ \lambda_i^A + \lambda_j^B + \lambda_k^C + \lambda_{ik}^{AC} \\ &+ \lambda_{jk}^{BC} + \lambda_{ij}^{AB}(u_i v_j) \end{aligned} \tag{10.34b}$$

where $u_i = \mu_i - \bar{\mu_i}$ and $v_i = \nu_j - \bar{\nu_j}$. The model states that variables A and B are uniformly associated for every level of variable C. The model has $K(I-1)(J-1)-1$ d.f. The model can be fitted in SAS software as the model {AC, BC, A(1)*B(1)}. That is, the model containing the linear-by-linear component of the AB interaction term.

(iii) The *heterogeneous uniform association* has:

$$\theta_{ij(k)} = \theta_{++(k)} \tag{10.35a}$$

$$\begin{aligned} l_{ijk} &= \mu + \lambda_i^A + \lambda_j^B + \lambda_k^C + \lambda_{ij}^{AB}(u_i)(v_j) + \lambda_{ik}^{AC} \\ &+ \lambda_{jk}^{BC} + \lambda_{ijk}^{ABC}(u_i)(v_j) \end{aligned} \tag{10.35b}$$

Here, the strength of the uniform association between variables A and B changes across the levels of variable C. The model has $K(IJ-I-J)$ degrees of freedom and can be fitted in SAS software as the model: {AC, BC, A(1)*B(1), A(1)*B(1)*C }.

(iv) A *homogeneous row effect* model under (PM) has:

$$\theta_{ij(k)} = \theta_{i++} \tag{10.36a}$$

$$l_{ijk} = \mu + \lambda_i^A + \lambda_j^B + \lambda_k^C + \lambda_{ik}^{AC} + \lambda_{ik}^{BC} + \lambda_{ij}^{AB}(v_j) \tag{10.36b}$$

The model implies that there are only row effects on the association and that they are homogeneous for each category of variable C. The model can be implemented as: {AB, AC, A*B(1)} and has $(I-1)(JK-K-1)$ degrees of freedom.

(v) The *simple heterogeneous row effect* model under the (PM) sampling has:

$$\theta_{ij(k)} = \theta_{i++}\theta_{++k} \tag{10.37a}$$

$$\begin{aligned} l_{ijk} &= \mu + \lambda_i^A + \lambda_j^B + \lambda_k^C + \lambda_{ik}^{AC} + \lambda_{ik}^{BC} \\ &+ \lambda_{ij}^{AB}(v_j) + \lambda_{ijk}^{ABC}(u_i)(v_j) \end{aligned} \tag{10.37b}$$

and is implemented as: {AC, BC, A*B(1), A(1)*B(1)*C}. The model implies that there are only row effects on the association but the overall effects on the association is different for each level of variable C. The model has $K(IJ-I-K)-(I-2)$ degrees of freedom.

(vi) A *heterogeneous row effect* model, which allows for heterogeneity both in the overall effects on the association and in the row effects on the association, similarly has:

$$\theta_{ij(k)} = \theta_{i.(k)} \tag{10.38a}$$

$$\begin{aligned} l_{ijk} &= \mu + \lambda_i^A + \lambda_j^B + \lambda_k^C + \lambda_{ik}^{AC} + \lambda_{ik}^{BC} \\ &+ \lambda_{ij}^{AB}(v_j) + \lambda_{ijk}^{ABC}(v_j) \end{aligned} \tag{10.38b}$$

The model which is based on $K(IJ-2I+J+2)$ degrees of freedom is implemented as model {AC, BC, A*B(1), A*B(1)*C }.

Corresponding column models of the above row models $(iv)-(vi)$ can similarly be defined and the various interpretations for the row effects models discussed therein can be extended to the column effects models.

10.11.2 The Conditional Association RC Models under Product Multinomial Sampling Scheme

The row-column conditional association models under the product multinomial sampling S(PM) scheme are defined as follows:

(vii) The *homogeneous RC effects* model has:

$$\theta_{ij(k)} = \theta_{i++}\theta_{+j+} \tag{10.39a}$$

$$\begin{aligned} l_{ijk} &= \mu + \lambda_i^A + \lambda_j^B + \lambda_k^C + \lambda_{ik}^{AC} + \lambda_{jk}^{BC} \\ &\quad + \lambda_{ij}^{AB}(u_i) + \lambda_{ij}^{AB}(v_j) \end{aligned} \tag{10.39b}$$

This model is based on $[K(IJ+1) - (I+J)(K+1) - 3]$ degrees of freedom and is implemented in SAS software as {AC, BC, A(1)*B, A*B(1)}.

(viii) The *heterogeneous row, RC effects* model has:

$$\theta_{ij(k)} = \theta_{i+(k)}\theta_{+j+} \tag{10.40a}$$

$$\begin{aligned} l_{ijk} &= \mu + \lambda_i^A + \lambda_j^B + \lambda_k^C + \lambda_{ik}^{AC} + \lambda_{jk}^{BC} \\ &\quad + \lambda_{ij}^{AB}(u_i) + \lambda_{ij}^{AB}(v_j) + \lambda_{ijk}^{ABC}(v_j) \end{aligned} \tag{10.40b}$$

The model is based on $(J-2)(IK-K-1)$ degrees of freedom is implemented as: {AC, BC, A(1)*B, A*B(1), AC*B(1)}.

(ix) The *heterogeneous column, RC effects* model has:

$$\theta_{ij(k)} = \theta_{i++}\theta_{+j(k)} \tag{10.41a}$$

$$\begin{aligned} l_{ijk} &= \mu + \lambda_i^A + \lambda_j^B + \lambda_k^C + \lambda_{ik}^{AC} + \lambda_{jk}^{BC} \\ &\quad + \lambda_{ij}^{AB}(u_i) + \lambda_{ij}^{AB}(v_j) + \lambda_{ijk}^{ABC}(u_i) \end{aligned} \tag{10.41b}$$

The model is based on $(I-2)(JK-K-1)$ degrees of freedom and is implemented as model {AC, BC, A(1)*B, A*B(1), A(1)*BC}.

(x) The *heterogeneous row-column, RC effects* model has:

$$\theta_{ij(k)} = \theta_{i+(k)}\theta_{+j(k)} \tag{10.42a}$$

$$\begin{aligned} l_{ijk} &= \mu + \lambda_i^A + \lambda_j^B + \lambda_k^C + \lambda_{ik}^{AC} + \lambda_{jk}^{BC} \\ &\quad + \lambda_{ij}^{AB}(u_i) + \lambda_{ij}^{AB}(v_j) + \lambda_{ijk}^{ABC}(u_i) + \lambda_{ijk}^{ABC}(v_j) \end{aligned} \tag{10.42b}$$

The model is based on $K(I-2)(J-2)$ degree of freedom and can be implemented in SAS software as: {AC, BC, A(1)*B, A*B(1), A(1)*BC, AC*B(1)}.

The above models are implemented in SAS® using PROC GENMOD with the following statements.

```
data cond;
do gp=1 to 2; do time1=1 to 3; do time2=1 to 3;
 input count @@;  rl=time1;  cl=time2;
output; end; end; end;
datalines;
1 3 0 4  43 18 0 4 13 7 4 0 3 36 9 1 7 25
;
proc print;
RUN;
proc genmod; class time1 time2 gp;
(1)  model count=time1|gp time2|gp/dist=poi;run;
(2a) model count=time1|gp time2|gp rl|cl/dist=poi;run;
(2b) model count=time1|gp time2|gp rl|cl|gp/dist=poi;run;
(3a) model count=time1|gp time2|gp time1|cl/dist=poi;run;
(3b) model count=time1|gp time2|gp time1|cl rl|cl|gp/dist=poi;run;
(3c) model count=time1|gp time2|gp time1|cl time1|cl|gp/dist=poi;run;
(4a) model count=time1|gp time2|gp time2|rl/dist=poi;run;
```

```
(4b) model count=time1|gp time2|gp time2|rl rl|cl|gp/dist=poi;run;
(4c) model count=time1|gp time2|gp time2|rl|gp/dist=poi;run;
(5a) model count=time1|gp time2|gp time1|cl time2|rl/dist=poi;run;
(5b) model count=time1|gp time2|gp time1|cl|gp time2|rl/dist=poi;run;
(5c) model count=time1|gp time2|gp time1|cl time2|rl|gp/dist=poi;run;
(5d) model count=time1|gp time2|gp time1|cl|gp time2|rl|gp /dist=poi;run;
```

The models have been renumbered from 1 to 5d, where **rl** and **cl** are integer scores for the linear effects at Time1 and Time2, respectively. We have displayed the model statements in SAS software together for brevity, actual implementation calls using PROC GENMOD.

10.11.3 Models Under the Multinomial Sampling Scheme

The preceding models were obtained under the product-multinomial sampling scheme. Agresti (1984) considers fitting similar models to the cholesterol data in his example. In the product multinomial (PM) scheme, the baseline model is the null conditional association model {AC,BC}, with $K(I-1)(J-1)$=8 d.f. for the data in Table 10.19. However, in the multinomial (M) scheme, the baseline model would be the model of mutual or pairwise independence {A,B,C}, with $(IJK-I-J-K+2)$ which would be 12 d.f. for the data in Table 10.19. Equivalent models discussed in the previous section can be implemented under the multinomial sampling scheme with the following models (without the actual model formulation) in SAS software.

Model #	Model Description	Model Implementation
(1)	*Null association*	{A, B, C}
(2a)	*Homogeneous uniform*	{A, B, C, A(1)*C, B(1)*C, A(1)*B(1)}
(2b)	*Heterogeneous uniform*	{A, B, C, A(1)*C, B(1)*C, A(1)*B(1),A(1)*B(1)*C}
(3a)	*Homogeneous row effect*	{A, B, C, A(1)*C, B(1)*C, A(1)*B}
(3b)	*Simple heterogeneous row effect*	{A, B, C, A(1)*C, B(1)*C, A(1)*B, A(1)*B(1)*C}
(3c)	*Heterogeneous row effect*	{A, B, C, A(1)*C, B(1)*C, A(1)*B, A(1)*BC}
(5a)	*Homogeneous RC effects*	{A, B, C, A(1)*C, B(1)*C, A(1)*B,A*B(1)}
(5b)	*Heterogeneous row RC effects*	{A, B, C, A(1)*C, B(1)*C, A(1)*B,A*B(1), A(1)*BC}
(5c)	*Heterogeneous column RC effects*	{A, B, C, A(1)*C, B(1)*C, A(1)*B,A*B(1), AC*B(1)}
(5d)	*Heterogeneous row-column RC effects*	{A, B, C, A(1)*C, B(1)*C, A(1)*B,A*B(1), A(1)*BC, AC*B(1)}

We present the results of employing these models under both the product multinomial and multinomial sampling schemes to the data in Table 10.19 in Table 10.20.

Based on these results, the most parsimonious model would be model (2a) with G^2 value of 5.155 on 7 d.f. This is the model of homogeneous conditional uniform association, which gives $\hat{\theta}_{ij(k)} = 1.100$ for $i = 1, 2$, $j = 1, 2$, and all $k = 1, 2$. The parameter θ is estimated as exp(2.2565) $=$ 9.5496 where 2.2565 is the log-parameter estimate of $\lambda_{ij}^{AB}(u_i - \bar{u})(v_i - \bar{v})$ with (a.s.e. $=$ 0.3352) obtained from PROC GENMOD.

Models for the case when all three variables are ordinal have been described by Clogg (1982) as the *partial association* models. These models are not considered in this text but such models can easily be implemented in SAS software.

Other models that have been suggested for handling the kind of data presented in the latter sections of this chapter include the RC(**m**)-G model discussed in Becker and Clogg (1989). A general class of symmetry models is fully discussed in Clogg (1982) and in Clogg et al. (1990).

In addition to the models considered above, some of the models discussed earlier, namely, the uniform association, row, and column association models, are sometimes

#	Models	Product Multinomial d.f.	G^2	X^2	Multinomial Scheme d.f.	G^2	X^2
1.	Null	8	71.897	79.854	12	85.205	118.319
2.	Uniform:						
(2a)	Homogeneous	7	5.155	8.377	9	15.071	19.550
(2b)	Heterogeneous	6	4.737	10.278	8	7.728	19.930
3.	Row Effects:						
(3a)	Homogeneous	6	5.070	8.138	8	15.033	18.859
(3b)	Simple Heterogeneous	5	4.677	9.940	7	7.504	16.170
(3c)	Heterogeneous	4	4.275	9.025	5	7.336	19.208
4.	Column Effects:						
(4a)	Homogeneous	6	4.909	6.704	8	14.710	18.273
(4b)	Simple Heterogeneous	5	4.413	7.653	7	7.551	18.334
(4c)	Heterogeneous	4	4.365	8.105	5	4.287	8.650
5.	RC Effects:						
(5a)	Homogeneous	5	2.361	2.132	7	11.916	11.091
(5b)	Heterogeneous Row	3	1.142	0.802	4	4.496	4.126
(5c)	Heterogeneous Col	3	1.714	1.329	4	1.263	0.963
(5d)	Heterogeneous Row-Col	2	0.457	0.255	2	0.457	0.255

Table 10.20: Conditional association models for the data in Table 10.19

very useful for modeling contingency tables having one, two, or three ordinal classificatory variables. However, the RC(**m**)-G model has an advantage over most other models because of its ability to fit models in more than one dimension and estimating the row and column scores in individual groups or for the combined group. For more detailed explanations and merits of the RC(**m**)-G models, the reader is referred to the following references: Clogg (1982), Clogg and Goodman (1984), Becker and Clogg (1989), Gilula and Haberman (1988), and Gilula, Krieger, and Ritov (1988).

10.12 Exercises

1. The data below measure the joint distribution of years of education of married couples in the United States as obtained from the 1972 General Social Survey (Haberman, 1978).

Distribution of years of education of married couples

Years of education of husband	Years of education of wife 0-11	12	13-15	16+
0-11	283	141	25	4
12	82	180	43	14
13-15	20	104	43	20
16+	4	52	41	69

Fit the model of independence and the uniform association model to this data. For the uniform association model, estimate $\theta_{(1,2),(3,4)}$. Comment.

2. Refer to the table above:

(a) Fit the row association model to the data and estimate the τ's. Summarize the difference in the years of education distribution of husbands for the different years of education of wives.

(b) Fit a column effects model to describe how the years of education distribution of wives differs for the years of education of husbands.

(c) Fit the uniform association and R+C association model. Compare these with those in (a) and (b).

3. Refer again to the data in exercise 5.4; fit the R, RC, and the R+C models to the data and comment on your results.

4. The data in the table below relate to the initial and follow-up blood pressure according to hypertension status in the WHO definition (BPD).

Blood Pressure	Normal	Borderline	Elevated
Normal	105	9	3
Borderline	10	12	1
Elevated	3	2	7

Fit both the row and column associations models to the data and comment.

5. The data in the table below are from Clogg and Shockey (1988) and relate to voting for President Reagan by color and political views.

		Political view						
Vote	Color	1	2	3	4	5	6	7
Reagan	White	1	13	44	155	92	100	18
	Nonwhite	0	0	2	1	0	2	0
Other	White	12	57	71	146	61	41	8
	Nonwhite	6	16	23	31	8	7	4

The political views, coded 1-7 range from extremely liberal to extremely conservative. Analyze these data by using scores for the variable "political view." If you consider vote as a response variable, fit a suitable logit model to these data based on the results of your earlier analysis.

6. Bachman, Johnson, and O'Malley (1980) present Table 10.21 for two ordinal variables. A sample of high school seniors in 1979 gave responses to the classificatory variables: A, how often he or she rides a car for fun (1 = almost every day, 2 = at least once a week, 3 = once or twice a month, 4 = a few times a year and 5 = never) and B; drug use (1 = none, 2 = marijuana only, 3 = a few pills, 4 = more pills, 5 = heroin)

   Fit the row, column, and RC models to these data. Interpret the estimated scores for each model and test whether each of them would give a better fit than the uniform association model.

7. Bachman, Johnson, and O'Malley (1980) also present Table 10.22 for two ordinal variables. Here again, a sample of high school seniors in 1980 gave responses to the classificatory variables: A, attitudes toward school (1 = like very much, 2 = like quite a lot, 3 = like some, 4 = don't like very much, 5 =

A	Drug Use-B 1	2	3	4	5
1	290	296	196	395	29
2	402	342	148	216	10
3	206	118	58	62	3
4	157	72	27	37	2
5	81	48	27	27	1

Table 10.21: Source: Clogg & Shockey (1988)

don't like at all) and B, drug use (1 = none, 2 = marijuana only, 3 = a few pills, 4 = more pills, 5 = heroin) as in above.

Again, fit the row, column, and RC models to these data. Interpret the estimated scores for each model and test whether each of them would give a better fit than the uniform association model.

A	Drug use-B 1	2	3	4	5
1	199	95	44	54	1
2	398	256	119	162	6
3	372	343	152	291	6
4	72	91	34	85	1
5	20	24	13	51	1

Table 10.22: Source: Clogg & Shockey (1988)

8. Fit relevant models discussed in this chapter to the rheumatoid arthritis data in exercise 7 in chapter 5.. Repeat for the data in exercise 11 in chapter 6.

9. Duncan and McRae (1979) present the following data, which relate to the evaluations of performances of radio and TV networks in 1959 and 1971. The data previously appeared in Duncan et al., (1973).

	Respondents's	Performance networks		
Year	race	Poor	Fair	Good
1959	White	54	253	325
	Black	4	23	81
1971	White	158	636	600
	Black	24	144	224

Table 10.23: Radio and TV network evaluation

Analyze this data and draw your conclusions.

10. Refer to the data in exercise 5, chapter 5, treating smoking level as an ordinal variable. Fit ordinal models to the data and draw conclusions on the nature of association present in the data.

11. Refer to the data in Table 6.31, treating strenous work as an ordinal variable. Reanalyze these data using models discussed in this chapter.

12. The data below from Ku and Kullback (1974), which have also been analyzed by Agresti (1984), relate to a sample of male residents of Framingham, MA,

aged 40 to 59, which was classified on blood pressure (BP) and serum cholesterol (CH) levels. During a six-year follow up period, they were observed and classified according to whether they developed coronary heart disease (D), which is a binary response variable. The variables BP and CH are ordinal factor variables.

Coronary heart disease	Serum cholesterol (mg/100cc)	Systolic Blood Pressure (mm Hg)			
		< 127	127-146	147-166	167+
Present	< 200	2	3	3	4
	200-219	3	2	0	3
	220-259	8	11	6	6
	≥ 260	7	12	11	11
Absent	< 200	117	121	47	22
	200-219	85	98	43	20
	220-259	119	209	68	43
	≥ 260	67	99	46	33

Table 10.24: Classification of men by blood pressure, serum cholesterol, and heart failure

(a) Fit a log-linear model that the log odds of heart disease depends on blood pressure and cholesterol, treated as categorical variables.

(b) Fit a log-linear model that says that the log odds of heart disease depends on blood pressure and cholesterol, treated as covariates.

(c) Obtain confidence intervals for the effects of unit increases in blood pressure and cholesterol levels on the odds of heart disease.

Chapter 11

Analysis of Doubly Classified Categorical Data

11.1 Introduction

In chapter 10, we discussed association models that provide insight into associations present in the general $I \times J$ contingency tables having ordered (one or both) categories. These have been described as either the uniform, row, column, and RC association models for the two-way tables.

To what extent can we say that a variable is ordinal? In other words, what is the degree of ordinality of a variable? The variable *age*, for instance, may have categories that represent a discretized version of an underlying continuous distribution, while the variable *family size* has categories that truly reflect an underlying discrete distribution. On the other hand, we may have nebulous ordinal variables like the variable having categories: {*strongly agree, disagree,...,strongly disagree*} or a variable with categories: {*social class 1, social class 2,...*}. In the former, a definite ordering does exist even if respondents all have different conception of the locations of the relevant cutoff points. In the latter however, the existence of an underlying continuum for the purportedly ordinal variable is itself open to question.

While the association models discussed in chapter 10 may be appropriate for the general $I \times J$ table, there remains, as Upton (1985) puts it, "one distinct class of data for which these models are inappropriate. These are the square $I \times I$ tables (ordered or unordered) in which the classificatory variables are intimately related." Such data usually arise from repeated measures or from longitudinal studies. Square tables may arise in several different ways. I give below a few cases that may give rise to such tables.

1. When a sample of individuals or subjects is cross-classified according to two essentially similar categorical variables (e.g., vision of right and left eyes; strength of right hand and strength of left hand). In Table 11.1 is an example which relates unaided distance vision of 7477 women aged 30-39 employed in the Royal Ordnance factories from 1943 to 1946 (Stuart, 1955).

2. When samples of pairs of matched individuals or subjects (arising from matched-pair design) such as husbands and wives, or fathers and sons are each classified

	Left Eye Grade				
Right Eye Grade	Best (1)	Second (2)	Third (3)	Worst (4)	Total
Best (1)	1520	266	124	66	1976
Second (2)	234	1512	432	78	2256
Third (3)	117	362	1772	205	2456
Highest (4)	36	82	179	492	789
Total	1907	2222	2507	841	7477

Table 11.1: Unaided distant vision as reported in Stuart (1955)

according to some categorical variable of interest, e.g., mobility, migration, religious affiliation, attained highest qualification, etc. Two data examples that fall into this category are displayed in Tables 11.2 and 11.3.

Father's Status	Son's Status					
	(1)	(2)	(3)	(4)	(5)	Total
(1)	50	45	8	18	8	129
(2)	28	174	84	154	55	495
(3)	11	78	110	223	96	518
(4)	14	150	185	714	447	1510
(5)	3	42	72	320	411	848
Total	106	489	459	1429	1017	3500

Table 11.2: British Occupational mobility data, (Glass, 1955)

where,

1. Professional and high administrative
2. Managerial, executive and high supervisory
3. Low inspectional and supervisory
4. Routine non-manual and skilled manual
5. Semi and unskilled manual).

	Current residence				
Residence at age 16	North East	South	North Central	West	Total
NE	263	22	14	13	312
South	26	399	36	30	491
NC	10	41	368	46	465
West	1	8	5	148	162
Total	300	470	423	237	1430

Table 11.3: Migration data (Haberman, 1978)

3. In panel studies where each individual or subject in a sample is classified according to the same criterion at two different points in time. (e.g., party affiliation, party loyalty, religious affiliation, etc.). Again two data examples are presented in Tables 11.4 and 11.5.

1966	**1970** C	LIB	L	IND	Total
C	68	1	1	7	77
LIB	12	60	5	10	87
L	12	3	13	2	30
IND	8	2	3	6	19
Total	100	66	22	25	213

Table 11.4: British Election Study, Upton (1978)

where, C = Conservative, LIB = Liberal, L = Labor, and IND = Independents.

Table 11.5 refers to the religious mobility data for Great Britain (Breen & Hayes, 1996).

Affiliation at age 16	**Religious Affiliation Now** 1	2	3	4	Total
1	863	30	1	52	946
2	50	320	0	33	403
3	1	1	28	1	31
4	27	8	0	33	68
Total	941	359	29	119	1448

Table 11.5: Subjects' religious affiliation (1991)

where: 1 = Protestant, 2 = Catholic, 3 = Jewish, 4 = None or others.

4. In rating experiments in which a sample of N individuals or subjects is rated independently by the same two raters into one of I nominal or ordinal categories. Thus, the entries in such a resulting $I \times I$ table relate to individuals that are jointly classified into category i by the first rater and category j by the second rater. The example in Table 11.6 arose from diagnosis of multiple sclerosis (MS) from Landis and Koch (1977a): 149 Winnipeg patients were examined by two neurologists, one from New Orleans, and the other from Winnipeg. The two neurologists classified each patient into one of the following classes: (1 = Certain MS, 2 = Probable MS, 3 = Possible MS, 4 = Doubtful, unlikely, or definitely not MS).

New Orleans neurologist	Winnipeg neurologist 1	2	3	4	Total
1	38	5	0	1	44
2	33	11	3	0	47
3	10	14	5	6	35
4	3	7	3	10	23
Total	84	37	11	17	149

Table 11.6: Diagnostic classification regarding multiple sclerosis for the Winnipeg patients

11.2 Symmetry Models

When the single classificatory variable is nominal rather than ordinal, but individuals are jointly classified with this variable over time, like the data in Table 11.5, where, strictly speaking, the categories of the variable cannot be truly considered ordinal, we are usually interested in fitting models of symmetry and all its associated decompositions or derivatives.

Specifically, we are concerned with associations that exhibit symmetric or pattern about the main diagonal of the table. I shall illustrate this group of models with the models of complete symmetry, marginal symmetry or homogeneity, quasi-symmetry, quasi-independence, and conditional symmetry.

11.2.1 The Complete Symmetry Model

If we let the joint distribution for the square table be given by $\{\pi_{ij}\}$ for $1 \leq (i,j) \leq I$, then the model of *complete symmetry* (S) has the hypothesized form:

$$H_S : \pi_{ij} = \pi_{ji} \quad \text{for } 1 \leq i \leq j \leq I \tag{11.1}$$

The log-linear model formulation for this model can be written in the form:

$$\ln(\hat{m}_{ij}) = \mu + \lambda_i + \lambda_j + \lambda_{ij} \tag{11.2}$$

with $\lambda_{ij} = \lambda_{ji}$ and $\sum_i \lambda_i = \sum_j \lambda_j = 0$ and $\sum_i \lambda_{ij} = 0 \quad j = 1, \cdots, I.$

The MLE under the model of symmetry (S) has

$$\hat{m}_{ij} = \begin{cases} (f_{ij} + f_{ji})/2 & \text{for all } i \neq j \\ f_{ii} & \text{for } i = j \end{cases}$$

where f_{ij} denote the corresponding joint observed frequency in the i-th row and j-th column. The model S is based on $I(I-1)/2$ degrees of freedom. Both the Pearson's and likelihood ratio statistics reduce respectively to the following under this model:

$$X^2 = \sum_{i=2}^{I} \sum_{j=1}^{i-1} \frac{(f_{ij} - f_{ji})^2}{(f_{ij} + f_{ji})} \quad \text{and} \quad G^2 = 2 \sum_{i=1}^{I} \sum_{j=1}^{I} f_{ij} \ln\left(\frac{2f_{ij}}{f_{ij} + f_{ji}}\right)$$

For the case when $I = 2$, Pearson's test reduces to McNemar's symmetry statistic,

$$X_M^2 = \frac{(f_{21} - f_{12})^2}{f_{12} + f_{21}}$$

which is used to test the symmetry hypothesis in a 2×2 table. The statistic is based on 1 d.f.

11.2.2 The Quasi-Symmetry Model

The symmetry model (S) rarely fits the data of interest because of its highly structured form. A less restrictive form of the hypothesis in (11.1) assumes that the symmetric model would have held if it were not for the distorting effect of the marginal totals. The model of *quasi-symmetry* (QS) has the log-linear formulation of the form:

$$\ln(\hat{m}_{ij}) = \mu + \lambda_i^R + \lambda_j^C + \lambda_{ij}^{RC}, \quad \text{for} \quad i \neq j \tag{11.3}$$

with $\lambda_{ij}^{RC} = \lambda_{ji}^{RC}$; $\sum_i \lambda_i^R = \sum_j \lambda_j^C = 0$ and $\sum_i \lambda_{ij}^{RC} = 0, \quad j = 1, \cdots, I.$ The model has the additional constraints $\lambda_i^R = \lambda_j^C$ for $i \neq j$.

Model QS was first introduced by Caussinus (1965). The symmetry model is the special case in which $\lambda_i^R = \lambda_j^C$ for $i = 1, 2, \cdots, I$ and where R and C relate to the row and column variables respectively. The model is based on $(I-1)(I-2)/2$ d.f and has the likelihood equations:

$$\begin{aligned} \hat{m}_{i+} &= f_{i+}, \quad i = 1, 2, \cdots, I \\ \hat{m}_{+j} &= f_{+j}, \quad j = 1, 2, \cdots, I \\ \hat{m}_{ij} + \hat{m}_{ji} &= f_{ij} + f_{ji}, \quad \text{for } i \neq j \end{aligned}$$

For both models S and QS, it should be obvious that $\hat{m}_{ii} = f_{ii}$. We now propose a procedure for implementing these and other symmetry models in SAS® using PROC GENMOD in the following sections. Our approach here is to fit these models using the generalized linear modeling capability provided in SAS® PROC GENMOD by employing factor and regression generated variables (Lawal, 2001). All models that will be considered in this chapter shall be applied to the data in Table 11.1 as an example.

11.2.3 A Nonstandard Log-Linear Model

A nonstandard log-linear model (von Eye & Spiel, 1996) can be written in the generalized linear model form (Clogg et al., 1990):

$$\ell = \mathbf{X}\boldsymbol{\lambda} \tag{11.4}$$

where $\mathbf{X}$ is a design matrix consisting of 0s and 1s that are derived from the factor or regression variable, $\boldsymbol{\lambda}$ is a vector of parameters, $\ell = \ln(\mathbf{m})$, and $\mathbf{m}$ is a I^2 vector of expected values under some model. The formulation above allows us to incorporate various contrasts of interest in the factor variable as well as several other possible models (von Eye & Niedermeier, 1999).

To implement the symmetry and other similar models, we need to be able to generate the appropriate factor or regression variables necessary for their implementation. Kateri (1993) and Kutylowski (1989) have discussed the generation of factor variables required for the implementation of some of the models being considered in this chapter. Our implementation of the symmetry model here for instance, is consistent with the procedure proposed in Friedl (1995) except that our factor variable for the symmetry model is defined differently. Both ours and Friedl have $\frac{I(I+1)}{2}$ levels. The implementation of the symmetry model therefore would involve only this single factor variable, whereas the approach by Kateri and Kutylowski involves two such factor variables designated as **sc_** and **ss_** in both their papers. Further, their programs are written for GLIM.

The factor variable for implementing the symmetry model in our case is generated for the general $I \times I$ table from the recurrence relation (Lawal & Sundheim, 2002) as:

$$S_h^k = S_{h-1}^k + (I + 2 - h), \quad \text{for } h = 2, \cdots, (I - k) \tag{11.5}$$

where $k = \mid i - j \mid = 0, 1, \cdots, (I-1)$, k is the k-th diagonal and $S_1^k = k + 1$. For a 4×4 table for instance, $k = \mid i - j \mid = 0, 1, 3$. The main diagonal elements have $k = 0$ and $h = 2, 3, 4$. Lawal and Sundheim (2002) have developed a SAS® macro for implementing all the models that are being considered in this chapter. In their programming in SAS software for instance, the above recurrence relation and hence the entries for the factor variable **S** are generated with the following expressions for all (i, j):

$$\mathbf{S}_{ij} = \begin{cases} (k+1) - (i+1)(\frac{1}{2}i+1) + (I+3)(i+1) - 3 - 2I & \text{if } i \leq j \\ (k+1) - (j+1)(\frac{1}{2}j+1) + (I+3)(j+1) - 3 - 2I & \text{if } i < j \end{cases} \tag{11.6}$$

where k and I are as defined above. We note here that when $i = j$, then $k = 0$ in (11.6). The S factor variable has levels that equal $I(I+1)/2 = 10$ for the case when $I = 4$. Hence, the resulting vector (this is indicated as a factor variable in SAS software) necessary for implementing the complete symmetry model which is generated from the above expression for the 4×4 table example is:

$$\mathbf{S} = \begin{bmatrix} 1 & 2 & 3 & 4 \\ 2 & 5 & 6 & 7 \\ 3 & 6 & 8 & 9 \\ 4 & 7 & 9 & 10 \end{bmatrix}$$

We may note here that the factor variable defined for S has entries that do not exactly match those generated in Friedl (1995), but the common feature of both vectors here is that both have 10 levels as expected. The equivalent S' in Friedl is generated from the expression:

$$S' = 2^{(i-1)} + 2^{(j-1)} \quad \text{for } (i, j) = 1, 2, \cdots, I$$

Both the complete symmetry (S) and the quasi-symmetry (QS) models can be implemented in SAS software to the data in Table 11.1 by employing the GENMOD procedure as follows:

```
data tab1;
do r=1 to 4;   do c=1 to 4;
input count @@; output; end; end; end;
datalines;
1520 266 124 66 234 1512 432 78 117 362 1772 205 36 82 179 492
;
proc genmod; class r c s; model count=s/dist=poi; /* fits the S model*/;
proc genmod; class r c s ; model count =r c s/dist=poi; /* fits the QS model */; run;
```

The above SAS software statements for the complete symmetry model for example translate into the log-linear model:

$$l_{ij} = \mu + \lambda_{ij}^S$$

where $l_{ij} = \ln(m_{ij})$ and we will impose the usual last parameter set to zero identifiability constraint on the λs. The above can also be rewritten as:

$$\begin{aligned} l_{ij} = \mu &+ \lambda_1^S Z_{1ij} + \lambda_2^S Z_{2ij} + \lambda_3^S Z_{3ij} + \lambda_4^S Z_{4ij} + \lambda_5^S Z_{5ij} \\ &+ \lambda_6^S Z_{6ij} + \lambda_7^S Z_{7ij} + \lambda_8^S Z_{8ij} + \lambda_9^S Z_{9ij} + \lambda_{10}^S Z_{10ij} \end{aligned}$$

where

$$Z_{1ij} = \begin{cases} 1 & \text{if } (i,j) = (1,1) \\ 0 & \text{elsewhere} \end{cases} \qquad Z_{2ij} = \begin{cases} 1 & \text{if } (i,j) = (1,2), (2,1) \\ 0 & \text{elsewhere} \end{cases}$$

The other indicator variables are similarly defined. However, because of the structure of S, there are only 10+1 = 11 such parameters including the common intercept. Thus the equivalent log-linear model in this case reduces to:

$$
\begin{bmatrix} l_{11} \\ l_{12} \\ l_{13} \\ l_{14} \\ l_{21} \\ l_{22} \\ l_{23} \\ l_{24} \\ l_{31} \\ l_{32} \\ l_{33} \\ l_{34} \\ l_{41} \\ l_{42} \\ l_{43} \\ l_{44} \end{bmatrix} = \begin{bmatrix} 1 & 1 & 0 & 0 & 0 & 0 & 0 & 0 & 0 & 0 & 0 \\ 1 & 0 & 1 & 0 & 0 & 0 & 0 & 0 & 0 & 0 & 0 \\ 1 & 0 & 0 & 1 & 0 & 0 & 0 & 0 & 0 & 0 & 0 \\ 1 & 0 & 0 & 0 & 1 & 0 & 0 & 0 & 0 & 0 & 0 \\ 1 & 0 & 1 & 0 & 0 & 0 & 0 & 0 & 0 & 0 & 0 \\ 1 & 0 & 0 & 0 & 0 & 1 & 0 & 0 & 0 & 0 & 0 \\ 1 & 0 & 0 & 0 & 0 & 0 & 1 & 0 & 0 & 0 & 0 \\ 1 & 0 & 0 & 0 & 0 & 0 & 0 & 1 & 0 & 0 & 0 \\ 1 & 0 & 0 & 1 & 0 & 0 & 0 & 0 & 0 & 0 & 0 \\ 1 & 0 & 0 & 0 & 0 & 0 & 1 & 0 & 0 & 0 & 0 \\ 1 & 0 & 0 & 0 & 0 & 0 & 0 & 0 & 1 & 0 & 0 \\ 1 & 0 & 0 & 0 & 0 & 0 & 0 & 0 & 0 & 1 & 0 \\ 1 & 0 & 0 & 0 & 1 & 0 & 0 & 0 & 0 & 0 & 0 \\ 1 & 0 & 0 & 0 & 0 & 0 & 0 & 1 & 0 & 0 & 0 \\ 1 & 0 & 0 & 0 & 0 & 0 & 0 & 0 & 0 & 1 & 0 \\ 1 & 0 & 0 & 0 & 0 & 0 & 0 & 0 & 0 & 0 & 1 \end{bmatrix} \begin{bmatrix} \mu \\ \lambda_1 \\ \lambda_2 \\ \lambda_3 \\ \lambda_4 \\ \lambda_5 \\ \lambda_6 \\ \lambda_7 \\ \lambda_8 \\ \lambda_9 \\ \lambda_{10} \end{bmatrix}
$$

which can be written as:

$$\boldsymbol{\ell} = \mathbf{X}\,\boldsymbol{\lambda} \tag{11.7}$$

where $\ell_{ij} = \ln(\hat{m}_{ij})$ have the familiar expression in (12.15) form and $\mathbf{X}$ is the design matrix consisting of 0s and 1s, which are derived from the indicator variables representing the levels of the factor variable S. For instance, columns 2 and 3 represent respectively, indicator variables Z_1 and Z_2.

When models S and QS are applied to the data in Table 11.1, we have the following expected values and the estimated local odds ratios under the symmetry model:

$$
\hat{m}_{ij} = \begin{bmatrix} 1520.0 & 250.0 & 120.5 & 51.0 \\ 250.0 & 1512.0 & 397.0 & 80.0 \\ 120.5 & 397.0 & 1772.0 & 192.0 \\ 51.0 & 80.0 & 192.0 & 492.0 \end{bmatrix} \quad \hat{\theta}_{ij} = \begin{bmatrix} 36.7718 & 0.5447 & 0.4761 \\ 0.5447 & 16.9994 & 0.5377 \\ 0.4761 & 0.5377 & 23.6497 \end{bmatrix}
$$

For this model, $G^2 = 19.25$ on 6 d.f, and for $i < j$,

$$\frac{\hat{\theta}_{12}}{\hat{\theta}_{21}} = 1$$

Clearly, the complete symmetry model (S) does not fit the data. We shall examine this further in chapter 11, section 11.2.5.

The QS model can also be characterized in terms of the symmetry of its odds ratios. It has, $\hat{\theta}_{ij} = \hat{\theta}_{ji}$, where θ is the local odds ratio. Because of this property, Goodman (1979a) has described the QS model as the *symmetric association model.* We illustrate this property below with the expected frequencies under the QS model for the data in Table 11.1, together with its estimated local odds ratios.

$$
\hat{m}_{ij} = \begin{bmatrix} 1520.000 & 263.380 & 133.584 & 59.036 \\ 236.620 & 1512.000 & 418.986 & 88.394 \\ 107.416 & 375.014 & 1772.000 & 201.570 \\ 42.964 & 71.605 & 182.431 & 492.000 \end{bmatrix} \quad \hat{\theta}_{ij} = \begin{bmatrix} 36.877 & 0.546 & 0.477 \\ 0.546 & 17.052 & 0.539 \\ 0.477 & 0.539 & 23.709 \end{bmatrix}
$$

From the above results, we see that for this model, $\hat{\theta}_{ij} = \hat{\theta}_{ji}$. Further, for example:

$$\left(\frac{\hat{m}_{13}}{\hat{m}_{31}}\right) = \left(\frac{\hat{m}_{14}}{\hat{m}_{41}}\right)\left(\frac{\hat{m}_{43}}{\hat{m}_{34}}\right) \quad \text{thus}$$
$$\left(\frac{133.584}{107.416}\right) = \left(\frac{59.036}{42.964}\right)\left(\frac{182.431}{201.570}\right) \quad = 1.2436$$

The above results will be true for all i and j. For this model, $G^2 = 7.27$ on $(I-1)(I-2)/2 = 3$ d.f. The model fits the data barely. Under the QS model,

$$\ln\left(\frac{\hat{m}_{ij}}{\hat{m}_{ji}}\right) = \hat{\alpha}_j - \hat{\alpha}_i \quad i < j \tag{11.8}$$

The expression in (11.8) therefore leads to the solutions: $\hat{\alpha}_2 = 1.1071$; $\hat{\alpha}_3 = 1.2180$; $\hat{\alpha}_4 = 1.3178$ with $\hat{\alpha}_1 = 1.0000$.

11.2.4 The Marginal Homogeneity Model (MHM)

When H_S, the symmetry model holds, we have $\sum_j \pi_{ij} = \sum_j \pi_{ji}$, that is, $\pi_{i+} = \pi_{+i}$. The latter is the formulation of the model of *unconditional marginal homogeneity* (UMH). The model assumes that the marginal totals are symmetric but the body of the table is not. This model is a *linear* model and it is therefore not log-linear. For the UMH model, the differences, $\sum_i m_{ik} - \sum_i m_{ki},\ k = 1, 2, \cdots, I$ are (Forthofer & Lehnen, 1981) compared with the null value of 0 in order to test UMH. We shall distinguish this from the conditional marginal homogencity (CMH) model test, which shall be described later in the next section.

We see that the model of complete symmetry implies marginal homogeneity. Specifically, if $I = 2$, both the S and UMH models are equivalent. But for $I > 2$, while S implies UMH, the converse is not always true. That is, marginal homogeneity does not imply symmetry (S). The UMH model is based on $(I-1)$ d.f. and is not log-linear. We give the implementation of this model in SAS software below together with a modified output.

```
set tab1;
proc catmod; weight count;
response marginals; model r*c=_response_/ml freq; repeated time 2; run;
```

```
            Analysis of Variance

Source         DF    Chi-Square    Pr > ChiSq
---------------------------------------------
Intercept       3      78744.17        <.0001
time            3         11.98        0.0075
Residual        0             .             .
```

The goodness-of-fit test statistic for the UMH model is provided in the SAS software output in the "TIME" line. In this case, the UMH model has $G^2 = 11.98$ on 3 degrees of freedom, which indicates that the UMH model does not fit.

Often used, however, is the corresponding conditional marginal homogeneity (CMH) model, which is related to both the symmetry and quasi-symmetry by the following relation:

$$S = QS \cap CMH \tag{11.9}$$

A conditional test of marginal homogeneity (CMH) assuming that QS is true is provided by examining the quantity $G^2(S) - G^2(QS)$. This conditional test for marginal homogeneity is based on $I(I-1)/2 - (I-1)(I-2)/2 = (I-1)$ degrees of freedom (d.f.). For this test to be valid, the QS model must hold true. In cases where the QS is not true, then the unconditional test for marginal homogeneity (that is, UMH) should be used. For the data in Table 11.1, since the model QS holds true, the conditional marginal homogeneity (CMH) test is therefore based on $G^2 = (19.25 - 7.27) = 11.98$ on $(6-3) = 3$ degrees of freedom. This model does not fit the data.

11.2.5 The Conditional Symmetry Model

We indicated earlier that the symmetry model does not fit the data in Table 11.1. To see, why the complete symmetry model does not fit, consider below the lower and upper off-diagonal triangles, $\triangle_1$ and $\triangle_2$ below for the data.

Right	Left(j)			
(i)	1	2	3	4
1				
2	234			
3	117	362		
4	36	82	179	

Lower-left $\triangle_1$ $i > j$
$n_1 = 1010$

Right	Left(j)			
(i)	1	2	3	4
1		266	124	66
2			432	78
3				205
4				

Upper-right $\triangle_2$ $i < j$
$n_2 = 1171$

The symmetry model in terms of the two triangles above, provides a test whether the probability of falling into cell (i, j) of $\triangle_1$ is the same as the probability of falling into cell (i, j) of $\triangle_2$. This model, however, does not take into account the fact that the overall observed subtotals in the two triangles are not always equal. In our example, these subtotals for example are 1010 and 1171, respectively. Thus, nonequality of these subtotals can seriously affect the probability of membership into each of the two triangles.

The conditional symmetry (CS) model, on the other hand, remedies this anomaly by testing whether the probability of falling into both cells in the two triangles are equal, after adjusting for the observed subtotals in the two triangles. That is, we are interested in testing whether the probability of falling in cell (i, j) is the same for both triangles, assuming that the probability of membership in the two triangles is equal. In this case, the expected values are now weighted by the proportion of cases in each triangle. Thus the CS model preserves the triangle totals.

The CS model, (McCullagh, 1978) can be formulated as:

$$\pi_{ij} = \gamma \pi_{ji} \qquad \text{for } (i < j) \tag{11.10}$$

or in logit form as

$$\log\left(\frac{m_{ij}}{m_{ji}}\right) = \eta \quad \text{for } i < j \tag{11.11}$$

The regression variable required to implement this model in SAS software is given below for a 4×4 table as:

$$\mathbf{CS} = \begin{bmatrix} 1 & 1 & 1 & 1 \\ 2 & 1 & 1 & 1 \\ 2 & 2 & 1 & 1 \\ 2 & 2 & 2 & 1 \end{bmatrix}$$

and in general for an $I \times I$ table,

$$C_{ij} = \begin{cases} 1 & \text{if } i \leq j \\ 2 & \text{if } i > j \end{cases}$$

The model is implemented in SAS software with the following:

```
set tab1;
proc genmod; class s; model count=s cs/dist=poi; run;
```

The implementation of this model implies that the CS model is a composite model (Lawal & Upton, 1990b; Lawal, 1996). That is, we can modify the nonstandard log-linear model above as:

$$l_{ij} = \mu + \lambda_k^{S_{ij}} + \eta^{CS}$$

and the generalized linear model in (12.15) as:

$$\ln(\mathbf{m}) = \mathbf{X}\boldsymbol{\lambda} + \mathbf{Z}\mathbf{u} = \mathbf{X}'\boldsymbol{\lambda}' \tag{11.12}$$

where $\mathbf{X}$, the design matrix in (12.15) is augmented by adding newly generated column(s) to represent either new factor variable(s) or regression variable(s). For example, the column vector $\mathbf{Z}$ has

$$\mathbf{Z} = [1\ 1\ 1\ 1\ 2\ 1\ 1\ 1\ 2\ 2\ 1\ 1\ 2\ 2\ 2\ 1]'$$

with corresponding parameter vector $\mathbf{u} = [\lambda^{CS}]'$. Thus the CS model has the design matrix $\mathbf{X}$ now modified to include the regression variable representing the CS component. Consequently, the new design matrix $\mathbf{X}'$ now has 12 columns.

The model when applied to the data in Table 11.1 gives the expected values:

$$\hat{m}_{ij} = \begin{bmatrix} 1520.000 & 268.455 & 129.395 & 54.765 \\ 231.542 & 1512.000 & 426.306 & 85.906 \\ 111.605 & 367.694 & 1772.000 & 206.173 \\ 47.235 & 74.095 & 177.827 & 492.000 \end{bmatrix} \quad \hat{\theta}_{ij} = \begin{bmatrix} 36.974 & 0.585 & 0.476 \\ 0.505 & 17.093 & 0.577 \\ 0.476 & 0.498 & 23.779 \end{bmatrix}$$

Under this model,

$$\ln\left(\frac{\hat{m}_{ij}}{\hat{m}_{ji}}\right) = \hat{\gamma} \tag{11.13}$$

Thus the expected value in cell $(2, 1)$ under the CS model is 231.542, which is $(231.542/1010 = 0.2292)$ of the $\triangle_1$ subtotal. Similarly, the expected value in cell $(1, 2)$ is also 268.455, which is $(268.455/1171 = 0.2292)$ of the $\triangle_2$ subtotal. We can therefore see that the (CS) model allows estimated probabilities to be the same for the two triangles, even though the subtotals are not the same.

The above leads to the solution:

$$\hat{\gamma} = \hat{\theta}_{ij}/\hat{\theta}_{ji} = 1.1594 \quad \text{for} \quad i < j$$

That is, the left eye is better than the right eye because $\hat{\gamma} > 1$.

When $\gamma = 1$, we have the model of complete symmetry (S). The CS model is based on $(I + 1)(I - 2)/2$ d.f. and has the likelihood equations:

$$\hat{m}_{ij} + \hat{m}_{ji} = f_{ij} + f_{ji} \quad \text{and}$$

$$\sum\sum_{i>j} \hat{m}_{ij} = \sum\sum_{i>j} f_{ij}$$

For this model $G^2 = 7.354$ on 5 d.f. The model fits the data in Table 11.1. We may note here that the complete symmetry model S implies either the CS model or the QS model or both.

11.2.6 The Quasi-Conditional Symmetry Model

The *quasi-conditional symmetry* model (QCS) has the formulation:

$$\pi_{ij} = \gamma\,\phi_j\,\pi_{ji} \qquad \text{for } (i < j) \tag{11.14}$$

where γ is unspecified.

A special case of (11.14) is the conditional symmetry model which has $\phi_j = 1$. For the QCS model:

$$\pi_{ij}\pi_{jk}\pi_{ki} = \gamma\pi_{ji}\pi_{kj}\pi_{ik} \quad (i < j < k)$$

That is, for this model, $\Omega_{(ij,jk)} = \Omega_{(ij,ik)} = \gamma$ for $1 \leq i < j < k \leq I$. The QCS model is sometimes described as the *extended quasi-symmetry* (EQS) model (Tomizawa, 1987). The model is based on $I(I-3)/2$ degrees of freedom. $\Omega_{(ij,st)}$ denotes the odds ratio for the 2×2 subtable formed from the i-th and j-th rows and the s-th and t-th columns, that is:

$$\Omega_{(ij,st)} = \frac{\Theta_{(ij,st)}}{\Theta_{(st,ij)}}, \quad \text{where}$$

$$\Theta_{(ij,st)} = \frac{\pi_{is}\pi_{jt}}{\pi_{js}\pi_{it}}$$

Estimates of $\Omega_{(ij,st)}$ under some model are obtained from its expected values $\hat{m}_{ij}$. Thus,

$$\hat{\Omega}_{(ij,st)} = \frac{\hat{m}_{is} \times \hat{m}_{jt} \times \hat{m}_{ti} \times \hat{m}_{sj}}{\hat{m}_{si} \times \hat{m}_{tj} \times \hat{m}_{it} \times \hat{m}_{js}}$$

Model QS in this context has $\Omega_{(ij,st)} = 1$ for $1 \leq i < j < k \leq I$. Further, for the local odds ratios formed from adjacent rows $(i, i+1)$ and adjacent columns $(j, j+1)$, $\theta_{ij} = \theta_{ji}$.

Model QCS is implemented with the following SAS software statements:

```
set tab1; proc genmod; class r c s;
model count=r c s cs/dist=poi; run;
```

The model when applied to the data in Table 11.1 has $G^2 = 6.823$ on 2 degrees of freedom. The model does not fit the data.

11.3 The Diagonal-Parameters Symmetry Models

This class of models are appropriate to square tables having ordinal classificatory variables.

11.3.1 The DPS Model

A decomposition of the (CS) model is Goodman's (1979b) *diagonal-parameters symmetry* model (DPS), which has the multiplicative form:

$$\pi_{ij} = \pi_{ji}\delta_k \quad i > j$$

where $k = i - j$ and the parameter δ_k represents the odds that an observation falls in cell (i, j), instead of in cell (j, i), $k = 1, 2, \cdots, (I-1)$. The CS model has $\delta_k = \gamma$ for all k.

Model DPS has the nonstandard log-linear model,

$$\ell_{ij} = \mu + \lambda_{ij}^S + \lambda_{ij}^D$$

and can be implemented in SAS software with the following statements:

```
set tab1; proc genmod; class s d;
model count=s d/dist=poi; run;
```

where **D** is the factor variable defined as:

$$\mathbf{D} = \begin{bmatrix} 7 & 1 & 2 & 3 \\ 4 & 7 & 1 & 2 \\ 5 & 4 & 7 & 1 \\ 6 & 5 & 4 & 7 \end{bmatrix}$$

When the model is applied to the data in Table 11.1, we have:

$$\hat{m}_{ij} = \begin{bmatrix} 1520.000 & 269.070 & 121.401 & 66.000 \\ 230.930 & 1512.000 & 427.284 & 80.599 \\ 119.599 & 366.716 & 1772.000 & 206.646 \\ 36.000 & 79.402 & 177.354 & 492.000 \end{bmatrix} \quad \hat{\theta}_{ij} = \begin{bmatrix} 36.987 & 0.626 & 0.347 \\ 0.468 & 17.099 & 0.618 \\ 0.719 & 0.462 & 23.788 \end{bmatrix}$$

The model has $G^2 = 0.498$ on $(I-2)(I-1)/2 = 3$ d.f. If we define τ_{j-i} as the log of parameter δ_{j-i}, that is, $\tau_{j-i} = \ln(\delta_{j-i})$, then maximum likelihood estimates of τ_{j-i} from this model are $\{\hat{\tau}_1 = 0.1529,\ \hat{\tau}_2 = 0.0150,\ \hat{\tau}_3 = 0.6061$. Consequently, we have

$$\hat{\delta}_1 = 1.1652 \qquad \hat{\delta}_2 = 1.0151 \qquad \hat{\delta}_3 = 1.8334$$

Because these parameter estimates are each greater than 1, we can therefore conclude that the left eye vision is better than the right eye vision. These results are consistent with those obtained in Goodman (1972). Because of our parameterization here, our values are the reciprocals of those in Goodman (1972).

The models based on the null hypotheses $H_{01}: \delta_1 = \delta_2$ and $H_{02}: \delta_2 = 1$ have respectively $G^2 = 2.03$, and $G^2 = 0.52$ on 4 degrees of freedom (Goodman, 1979c). Lawal (2001) has provided an alternative form of testing these and other similar null hypotheses for these kind of data. Both models based on $H-01$ and H_{02}, fit the data very well but then we would not encourage making statistical inference based on these models, since they are arrived at after the analysis of the data.

In terms of expected values under this model, therefore, we have:

$$\begin{aligned} \ln(\hat{m}_{12}) - \ln(\hat{m}_{21}) &= 0.1529 = \hat{\tau}_1 \\ \ln(\hat{m}_{13}) - \ln(\hat{m}_{31}) &= 0.0150 = \hat{\tau}_2 \\ \ln(\hat{m}_{14}) - \ln(\hat{m}_{41}) &= 0.6061 = \hat{\tau}_3 \end{aligned}$$

We may note here that $\ln(\hat{m}_{23}) - \ln(\hat{m}_{32}) = \ln(\hat{m}_{23}) - \ln(\hat{m}_{32}) = \hat{\tau}_1$ and $\ln(\hat{m}_{24}) - \ln(\hat{m}_{42}) = \hat{\tau}_2$.

With τ_{j-i} defined as the log of parameter δ_{j-i}, we can show that these log parameter estimates satisfy:

$$\hat{\tau}_{j-i} = -\sum_{t=1}^{j-i}\sum_{s=1}^{t} \ln(\hat{\Delta}_s), \quad s = 1, \cdots, (I-1) \tag{11.15}$$

where $\ln(\hat{\Delta}_1) = 0$.

Similarly, in terms of the log odds ratios, that is, Φ_{ij}, and using (11.15), we have

$$\begin{aligned} \ln(\hat{\Delta}_2) &= \hat{\Phi}_{12} - \hat{\Phi}_{21} = 0.2907 \quad = 2\hat{\tau}_1 - \hat{\tau}_2 \\ \ln(\hat{\Delta}_3) &= \hat{\Phi}_{13} - \hat{\Phi}_{31} = -0.7290 = -\hat{\tau}_3 + 2\hat{\tau}_2 + \hat{\tau}_1 \end{aligned}$$

Again, we note here that, $\ln(\hat{\Delta}_2) = \hat{\Phi}_{23} - \hat{\Phi}_{32} = 2\hat{\tau}_1 - \hat{\tau}_2$.

The DPS model for any 4×4 table for $1 \le i < j < k \le I$ has, for example:

$$\hat{\Omega}_{(12,23)} = \hat{\Omega}_{(23,34)} = \exp\{2\hat{\tau}_1 - \hat{\tau}_2\}$$
$$\hat{\Omega}_{(13,34)} = \hat{\Omega}_{(12,24)} = \exp\{-\hat{\tau}_3 + 2\hat{\tau}_2 + \hat{\tau}_1\}$$

11.3.2 The LDPS Model

Agresti (1983) considered a simpler version of the DPS model, the *linear diagonals-parameter symmetry* (LDPS) model, in which $\ln(\delta_k) = \tau_k$ have a linear pattern. The model has the multiplicative form:

$$\pi_{ij} = \pi_{ji}\, \delta^{j-i} \qquad j \ge i \tag{11.16}$$

where the log odds that an observation is a certain distance above the main diagonal, instead of the same distance below it, is assumed to depend linearly on the distance (Agresti, 1983). The model is also referred to by Tomizawa (1986) as simply the *linear diagonals-parameter model* (LDP) and (11.16) above leads to:

$$\ln\left(\frac{\pi_{ij}}{\pi_{ji}}\right) = \tau\,(j-i) \qquad \text{for } i < j \tag{11.17}$$

where $\tau = \ln(\delta)$. The nonstandard log-linear model formulation for this model is,

$$\ell_{ij} = \mu + \lambda_{ij}^S + \tau\,\lambda_{ij}^F$$

The LDPS model for the 4×4 table for $1 \le i < j < k \le I$ has:

$$\Omega_{(12,23)} = \Omega_{(23,34)} = 1$$
$$\Omega_{(13,34)} = \Omega_{(12,24)} = 1 \text{ and}$$
$$\Omega_{(12,34)} = 1$$

The model is based on $(I+1)(I-2)/2$ degrees of freedom and can be implemented with the SAS software statements:

```
set tab1; proc genmod;
class s; model count=s f/dist=poi; run;
```

where $\mathbf{F}$ is a regression variable defined as:

$$\mathbf{F} = \begin{bmatrix} 1 & 2 & 3 & 4 \\ 1 & 1 & 2 & 3 \\ 1 & 1 & 1 & 2 \\ 1 & 1 & 1 & 1 \end{bmatrix}$$

The expected values under this model for the data in Table 11.1 are given below:

$$\hat{m}_{ij} = \begin{bmatrix} 1520.000 & 263.370 & 133.352 & 59.121 \\ 236.630 & 1512.000 & 418.232 & 88.532 \\ 107.648 & 375.768 & 1772.000 & 202.268 \\ 42.879 & 71.468 & 181.732 & 492.000 \end{bmatrix}$$

For this model, $G^2 = 7.2804$ on $(I+1)(I-2)/2 = 5$ d.f., $\ln(\hat{f}) = 0.1071$ and

$$\ln\left(\frac{\hat{m}_{ij}}{\hat{m}_{ji}}\right) = \hat{\gamma}(j-i) = 0.1071\,(j-i) \quad \text{for} \quad i < j$$

where $\hat{\gamma} = \ln\left(\frac{\hat{m}_{12}}{\hat{m}_{21}}\right) = 0.1071$ for this data.

An extension of the LDPS model that employs the two-ratios parameter (Tomizawa, 1987) is the *2-ratios-parameter symmetry* (2RPS) model which is defined as:

$$\pi_{ij} = \pi_{ji}\eta\delta^{j-i} \quad \text{for } (i < j) \tag{11.18}$$

The MLE $\hat{m}_{ij}$ satisfies under this model the following equations:

$$\hat{m}_{ij} + \hat{m}_{ji} = f_{ij} + f_{ji},$$

$$\sum_{i=1}^{I}\sum_{j=1}^{I} j m_{ij} = \sum_{i=1}^{I}\sum_{j=1}^{I} j f_{ij} \quad \text{and}$$

$$\sum_{i<j} \hat{m}_{ij}\{I - 2(j-i)\} = \sum_{i<j} f_{ij}\{I - 2(j-i)\} \quad \text{for } 1 \leq (i,j) \leq I$$

The model in (11.18) reduces to models LDPS and CS when $\eta = 1$ and $\delta = 1$, respectively.

The model is based on $(I^2 - I - 4)/2$ degrees of freedom and can be implemented with the SAS software statements:

```
set tab1; proc genmod;
class s; model count=s f cs/dist=poi; run;
```

The model when applied to the data in Table 11.1 has $G^2 = 6.8252$ on 4 d.f., log-parameter estimates $\hat{f} = 0.0577$, $\hat{cs} = -0.0743$ and expected values:

$$\hat{m}_{ij} = \begin{bmatrix} 1520.000 & 266.470 & 131.890 & 57.274 \\ 233.530 & 1512.000 & 423.155 & 87.562 \\ 109.110 & 370.845 & 1772.000 & 204.649 \\ 44.726 & 72.438 & 179.351 & 492.000 \end{bmatrix} \quad \hat{\theta}_{ij} = \begin{bmatrix} 36.932 & 0.565 & 0.477 \\ 0.525 & 17.074 & 0.558 \\ 0.477 & 0.518 & 23.753 \end{bmatrix}$$

11.3.3 The QDPS Model

A further decomposition of the (DPS) model is the *quasi-diagonals parameter symmetry* (QDPS) model which takes the form:

$$\pi_{ij} = \phi_j\, \pi_{ji}\, \delta_k \qquad \text{for } (i < j) \tag{11.19}$$

where the parameter δ_k is the log odds that an observation falls in cell (i, j) satisfying $j - i = k$ instead of a cell (j, i) satisfying $j - i = -k, k = 1, 2, \cdots, I$. The *quasi-diagonals parameter symmetry* (QDPS) model satisfies the expression in (11.15) and has the nonstandard log-linear model form:

$$\ell_{ij} = \mu + \lambda_{ij}^{R} + \lambda_{ij}^{C} + \lambda_{ij}^{S} + \lambda_{ij}^{D}$$

The model is implemented in SAS software with the statement:

```
set tab1; proc genmod;
class r c s d; model count=r c s d/dist=poi; run;
```

For the *quasi-diagonals parameter symmetry* (QDPS) model, the expected values and the corresponding odds-ratios are:

$$\hat{m}_{ij} = \begin{bmatrix} 1520.000 & 267.923 & 122.077 & 66.000 \\ 232.077 & 1512.000 & 432.000 & 79.923 \\ 118.923 & 362.000 & 1772.000 & 203.077 \\ 36.000 & 80.077 & 180.923 & 492.000 \end{bmatrix} \quad \hat{\theta}_{ij} = \begin{bmatrix} 36.962 & 0.627 & 0.342 \\ 0.467 & 17.133 & 0.619 \\ 0.731 & 0.462 & 23.729 \end{bmatrix}$$

For this model, $G^2 = 0.2222$ on $(I-2)(I-3)/2 = 1$ d.f. and log estimates of D under this model are $\hat{\tau}_1 = -0.0567$ and $\hat{\tau}_2 = -0.4077$.

Following Tomizawa (1987), the log odds ratios satisfy the following:

$$\hat{\Phi}_{ij} - \hat{\Phi}_{ji} = \ln[\hat{\Delta}_{j-i+1}] \quad 1 \leq i < j \leq (I-1)$$

The above implies that for $i < j$, the log odds ratio $\hat{\Phi}_{ij}$ is $\hat{\Delta}_{j-i+1}$ more than the log odds ratio $\hat{\Phi}_{ji}$. That is, $\hat{\Phi}_{ij} - \hat{\Phi}_{ji}$ will be uniform for $i < j$ when $j - i$ is a constant. This implies that for the data in Table 11.1 in our example, the log odds ratios satisfy:

$$\begin{aligned} \ln(\hat{\Delta}_2) &= \hat{\Phi}_{12} - \hat{\Phi}_{21} = 0.2946 \quad = -\hat{\tau}_2 + 2\hat{\tau}_1 \\ \ln(\hat{\Delta}_3) &= \hat{\Phi}_{13} - \hat{\Phi}_{31} = -0.7596 = 2\hat{\tau}_2 - \hat{\tau}_1 \end{aligned}$$

with again, $\ln(\hat{\Delta}_2) = \hat{\Phi}_{23} - \hat{\Phi}_{32} = 0.2946$.

The model has for the 4×4 table the following:

$$\hat{\Omega}_{(12,23)} = \hat{\Delta}_2 \quad = \exp\{-\hat{\tau}_2 + 2\hat{\tau}_1\} = \hat{\Omega}_{(23,34)} \tag{11.20a}$$

$$\hat{\Omega}_{(12,34)} = \hat{\Delta}_3 \quad = \exp\{2\hat{\tau}_2 - \hat{\tau}_1\} \tag{11.20b}$$

$$\hat{\Omega}_{(13,34)} = \hat{\Delta}_2\,\hat{\Delta}_3 = \exp\{\hat{\tau}_2 + \hat{\tau}_1)\} \quad = \hat{\Omega}_{(12,24)} \tag{11.20c}$$

11.4 The Odds-Symmetry Models

Two extensions of the conditional symmetry model are the *odds-symmetry models* I and II (Tomizawa, 1985b) which are defined respectively as:

$$H_{OS1}: \quad \frac{\pi_{ij}}{\pi_{(i,j+1)}} = \frac{\pi_{ji}}{\pi_{(j+1,i)}} \qquad (i < j) \tag{11.21a}$$

$$H_{OS2}: \quad \frac{\pi_{(i-1,j)}}{\pi_{ij}} = \frac{\pi_{(j,i-1)}}{\pi_{ji}} \qquad (i < j) \tag{11.21b}$$

Following Tomizawa (1985a), model OS1 for instance indicates that

> the odds that the column value is in j instead of $j+1$ in row i in the upper-right triangle of the table is equal to the symmetric odds that the row value is in j instead of $j+1$ in column i in the lower-left triangle of the same table

Model OS2 can be similarly interpreted.

For unspecified parameters r_i and s_j, both models can be expressed respectively as:

$$H_{OS1}: \quad \frac{\pi_{ij}}{\pi_{ji}} = r_i \qquad (i < j) \tag{11.22a}$$

$$H_{OS2}: \quad \frac{\pi_{ij}}{\pi_{ji}} = s_j \qquad (i < j) \tag{11.22b}$$

The two models are log-linear and are easily programmed in SAS software.

A simple generalization of the odds-symmetry models is the *quasi-odds symmetry* (QOS) model, Tomizawa (1985b). Model (QOS) is equivalent to the *adjusted quasi-symmetry* (AQS) model described in Bishop *et al.* (1975) because the model adjusts to preserve the marginal totals of both the lower-left $\Delta_1,\ i > j$, and the upper-right $\Delta_2,\ i < j$, of observed frequency, rather than the marginal totals of the original contingency table. Under this model, expected values in cells $(1,2),(2,1),(I-1,I),(I,I-1)$ are exactly identical to the original observed counts. For the QOS model, $\Omega_{ij,jk} = \exp(\gamma_j)$ for $1 \leq i < j < k \leq I$ and this translates into the following:

$$\begin{aligned}
\Omega_{12,23} &= \Omega_{12,24} = \gamma_1 \\
\Omega_{13,34} &= \Omega_{23,34} = \gamma_2 \quad \text{and} \\
\Omega_{12,34} &= 1
\end{aligned}$$

Models OS1, OS2 are each based on $(I-1)(I-2)/2$ d.f. while model QOS is based on $(I-2)(I-3)/2$ d.f.

The factor variables for implementing models OS1 and OS2 (Lawal & Sundheim, 2000) are given respectively as:

$$\mathbf{OS1} = \begin{bmatrix} 7 & 1 & 1 & 1 \\ 4 & 7 & 2 & 2 \\ 4 & 5 & 7 & 3 \\ 4 & 5 & 6 & 7 \end{bmatrix}, \quad \mathbf{OS2} = \begin{bmatrix} 7 & 3 & 2 & 1 \\ 6 & 7 & 2 & 1 \\ 5 & 5 & 7 & 1 \\ 4 & 4 & 4 & 7 \end{bmatrix}$$

Models OS1, OS2, and QOS are implemented in SAS software with the following statements:

```
set tab1;
proc genmod; class s os1; model count=s os1/dist=poi; run; *** fits OS1***;
proc genmod; class s os2; model count=s os2/dist=poi; run; *** fits OS2***;
proc genmod; class r c s os1; model count=r c s os1/dist=poi; run; *** fits QOS***;
```

For the *odds-symmetry* I (OS1) model, the expected values and the corresponding odds ratios are:

$$\hat{m}_{ij} = \begin{bmatrix} 1520.000 & 270.463 & 130.363 & 55.174 \\ 229.537 & 1512.000 & 424.465 & 85.535 \\ 110.637 & 369.535 & 1772.000 & 205.000 \\ 46.826 & 74.465 & 179.000 & 492.000 \end{bmatrix} \quad \hat{\theta}_{ij} = \begin{bmatrix} 37.093 & 0.582 & 0.476 \\ 0.507 & 17.081 & 0.574 \\ 0.476 & 0.501 & 23.759 \end{bmatrix}$$

For this model, $G^2 = 7.2637$ on 3 d.f. and log estimates of parameters under this model are $\hat{\gamma}_1 = 0.1641, \hat{\gamma}_2 = 0.1386, \hat{\gamma}_3 = 0.1356$. Consequently, for $i < j$, we have:

$$\begin{aligned}
\ln\left(\frac{\hat{m}_{12}}{\hat{m}_{21}}\right) &= \ln\left(\frac{\hat{m}_{13}}{\hat{m}_{31}}\right) = \ln\left(\frac{\hat{m}_{14}}{\hat{m}_{41}}\right) = \ln(\hat{\gamma}_1) = 0.1641 \\
\ln\left(\frac{\hat{m}_{23}}{\hat{m}_{32}}\right) &= \ln\left(\frac{\hat{m}_{24}}{\hat{m}_{42}}\right) = \ln(\hat{\gamma}_2) = 0.1386 \\
\ln\left(\frac{\hat{m}_{34}}{\hat{m}_{43}}\right) &= \ln(\hat{\gamma}_3) = 0.1356
\end{aligned}$$

Further, the log odds ratios satisfy for $i < j$:

$$\begin{aligned}
\hat{\Phi}_{12} - \hat{\Phi}_{21} &= \ln(\hat{\gamma}_2) = 0.1386 \\
\hat{\Phi}_{23} - \hat{\Phi}_{32} &= \ln(\hat{\gamma}_3) = 0.1356 \\
\hat{\Phi}_{13} - \hat{\Phi}_{31} &= 0.000
\end{aligned}$$

Similarly, the odds-symmetry II model when applied to the data has a $G^2 = 7.2757$ on 3 d.f. and log estimates of parameters under this model are $\hat{\gamma}_1 = 0.1613, \hat{\gamma}_2 = 0.1491, \hat{\gamma}_3 = 0.1282$, again leading to:

$$\ln\left(\frac{\hat{m}_{14}}{\hat{m}_{41}}\right) = \ln\left(\frac{\hat{m}_{24}}{\hat{m}_{42}}\right) = \ln\left(\frac{\hat{m}_{34}}{\hat{m}_{43}}\right) = \ln(\hat{\gamma}_1) = 0.1613$$

$$\ln\left(\frac{\hat{m}_{13}}{\hat{m}_{31}}\right) = \ln\left(\frac{\hat{m}_{23}}{\hat{m}_{32}}\right) = \ln(\hat{\gamma}_2) = 0.1491$$

$$\ln\left(\frac{\hat{m}_{12}}{\hat{m}_{21}}\right) = \ln(\hat{\gamma}_3) = 0.1282$$

Again, the log odds ratios satisfy under this model:

$$\hat{\Phi}_{12} - \hat{\Phi}_{21} = \ln(\hat{\gamma}_3) = 0.1282$$
$$\hat{\Phi}_{23} - \hat{\Phi}_{32} = \ln(\hat{\gamma}_2) = 0.1491$$
$$\hat{\Phi}_{13} - \hat{\Phi}_{31} = 0.000$$

The estimated log odds ratios under the OS1 model can be summarized as follows:

$$\hat{\Phi}_{ij} - \hat{\Phi}_{ji} = \begin{cases} \hat{\gamma}_{i+1} & \text{if } |i-j| = 1 \\ 0 & \text{elsewhere} \end{cases}$$

Similarly, the estimated log-odds ratios under the OS2 models satisfy the following summary relation:

$$\hat{\Phi}_{ij} - \hat{\Phi}_{ji} = \begin{cases} \hat{\gamma}_{I-i} & \text{if } |i-j| = 1 \\ 0 & \text{elsewhere} \end{cases}$$

When the quasi odds-symmetry model is applied to the data in Table 11.1, we have $G^2 = 6.7934$ on 1 d.f. Obviously, this model does not fit the data. However, the expected values and corresponding estimated odds ratios under this model are given as:

$$\hat{m}_{ij} = \begin{bmatrix} 1520.000 & 266.000 & 132.814 & 57.186 \\ 234.000 & 1512.000 & 423.186 & 86.814 \\ 108.186 & 370.814 & 1772.000 & 205.000 \\ 44.814 & 73.186 & 179.000 & 492.000 \end{bmatrix} \quad \hat{\theta}_{ij} = \begin{bmatrix} 36.923 & 0.561 & 0.477 \\ 0.531 & 17.074 & 0.564 \\ 0.477 & 0.512 & 23.759 \end{bmatrix}$$

What are the expected values in the NW and SE corners? We observe that these values correspond exactly to the observed frequencies. These are as a result of the constraints imposed under this model. Under model QOS, the log parameter estimates are:

$$\hat{\gamma}_1 = 0.0000 \qquad \hat{\gamma}_2 = 0.0552 \qquad \hat{\gamma}_3 = 0.0970$$

And for $i < j$ we have,

$$\hat{\Phi}_{ij} - \hat{\Phi}_{ji} = \begin{cases} \hat{\gamma}_{i+1} & \text{if } |i-j| = 1 \\ 0 & \text{elsewhere} \end{cases}$$

In Table 11.7 are displayed the goodness-of fit statistic G^2 when the models described above are fitted to the selected 4×4 tables indicated.

Model	d.f.	Table 11.1 G^2	Table 11.3 G^2
S	6	19.249	65.081
QS	3	7.271	3.842
UMH	3	11.977	57.108
CMH	3	11.978	61.239
CS	5	7.354	46.379
QCS	2	6.823	3.577
DPS	3	0.498	35.780
LDPS	5	7.280	37.980
QDPS	1	0.222	3.160
2RPS	4	6.825	36.875
OS1	3	7.264	22.895
OS2	3	7.276	3.833
QOS	1	6.793	0.462

Table 11.7: Results of above models applied to the 4×4 data in Tables 11.1 to 11.3

11.5 Generalized Independence Models

For the $I \times I$ square contingency table, with row variable denoted by R and column variable denoted by C, let f_{ij} and $\hat{m}_{ij}$ denote, respectively, the observed and corresponding expected frequencies associated with the cell in row i and column j. We shall assume here that a multinomial sampling scheme applies to the $I \times I$ table, though other sampling schemes may also be assumed without loss of generality.

The log-linear model for an $I \times I$ table can be written in the form:

$$\ell_{ij} = \mu + \lambda_i^R + \lambda_j^C + \lambda_{ij}^{RC}$$

with the usual identifiability constraints imposed. In this setup, λ_{ij}^{RC} relate to the interaction term in the model. Goodman (1979b, 1985) has considered several models having different structures for λ_{ij}^{RC}. The model of independence (O), for instance, has $\hat{\lambda}_{ij}^{RC} = 0$ in the formulation above. A nonindependence model assumes the independence model as the baseline and tries to model the interaction structure λ_{ij}^{RC}. In this context, therefore, the independence model is often referred to as the null or baseline model. Often, λ_{ij}^{RC} is modeled as a function of the local odds ratios $\theta_{ij} = (\hat{m}_{ij}\hat{m}_{i+1,j+1})/(\hat{m}_{i,j+1}\hat{m}_{i+1,j})$ (Goodman, 1979b). Yamagushi (1990) has also modeled the interaction term in terms of $\Theta_{ij} = \ln(\theta_{ij}/\theta_{ji})$.

If we define $\Phi_{ij} = \ln(\theta_{ij})$, as the log-odds ratios, then for doubly classified $I \times I$ contingency tables having ordinal categories, the Φ has a diagonal pattern for most of the models that are being considered in this chapter section. For instance, the independence model (O) has $\hat{\Phi}_{ij} = 0$ for $(i, j) = 1, 2, \cdots, (I-1)$. The independence model applied to the data in Table 11.1 gives a $G^2 = 5704.27$ on 9 d.f. Clearly, this model does not fit the data. It is well known that the independence model is not adequate for the description of the observed frequencies in the general $I \times I$ ordered tables in which the classificatory variables are intimately related. For example, for the data in Table 11.1, $N = 7477$ and the diagonal cell counts are $f_{ii} = \{1520, 1512, 1772, 492\}$ for $i = 1, \cdots, 4$. The diagonal cells therefore account for about 71% of the entire data. Any model therefore that must explain the observed variations in the data must take cognizance of the diagonal cells. The corresponding expected values for the diagonal cells under the model of independence are: $\hat{m}_{ii} = \{503.976, 670.434, 823.484, 88.745\}$. We see that the diagonal counts are generally underestimated under this model. We may therefore consider a wide range of modifications to this baseline model (Goodman, 1972).

The inadequacy of the independence model is due primarily to the deflated main diagonal expected counts under the model. Thus, the introduction of a single *deflation factor*, say, ϵ, applied to the off-diagonal cells has the log-linear model formulation:

$$\ell_{ij} = \mu + \lambda_i^R + \lambda_j^C + \delta_{ij}\epsilon \tag{11.23}$$

where $\sum \lambda_i^A = \sum \lambda_j^C = 0$ and δ_{ij} is defined such that:

$$\delta_{ij} = \begin{cases} 1 & \text{if } i \neq j \\ 0 & \text{if } i = j \end{cases}$$

The model given by (11.23) is equivalent to the model obtained by the introduction of an *inflation factor*, say ϵ, applied to the main diagonal cells (Goodman, 1972). Again, in this case, the δ_{ij} takes the form:

$$\delta_{ij} = \begin{cases} 0 & \text{if } i \neq j \\ 1 & \text{if } i = j \end{cases}$$

The model described by either methods has been termed the *constant loyalty* or *uniform loyalty* model and Goodman refers to it in the context of social mobility as the *uniform inheritance* model, while Scheuren and Oh (1975) name it the *smoothed-quasi-independence* model. In the context of modeling agreement data, it has been described as the model of *exact agreement*. The model has one parameter more than the model of independence. Consequently, the model is based on $(I-1)^2 - 1 = I(I-2)$ degrees of freedom. Following Lawal and Upton (1990b), the model will be designated here as model (L).

Since we are going to employ the nonstandard log-linear formulations to fit all the models that will be discussed in this chapter, the non-standard log-linear form for model (L) is therefore given by:

$$\ell_{ij} = \mu + \lambda_i^R + \lambda_j^C + \phi L_{ij} \tag{11.24}$$

where **L** is a regression variable, defined as:

$$\mathbf{L} = \begin{bmatrix} 2 & 1 & 1 & 1 \\ 1 & 2 & 1 & 1 \\ 1 & 1 & 2 & 1 \\ 1 & 1 & 1 & 2 \end{bmatrix}$$

If the value of ϵ is not constant down the main diagonal, then we have the *variable loyalty* or the *non-uniform loyalty* model. This model is also well known as the *quasi-independence* model. The model is also designated as the model Q with ϵ replaced with ϵ_i in the log-linear model formulation in (11.23). The model has $\epsilon_i, i = 1, 2, \cdots, I$ extra parameters than the baseline (O) model. Hence it has $(I-1)^2 - I = (I^2 - 3I + 1)$ degrees of freedom. The model Q is more familiarly known as the *Mover-Stayer* model (Upton & Sarlvik, 1981) or the model of *quasi-perfect mobility* discussed in Bishop et al., (1975). The parameters ϵ_i were termed by Goodman (1969) as the *new index of immobility*. Both models when applied to the data in Table 11.1 has $G^2 = 492.465$ and 199.106 on 8 and 5 degrees of freedom respectively. Neither model fits the data.

11.5.1 The Uniform Association Model

The uniform association (U) model has the multiplicative form:

$$m_{ij} = \alpha_i \beta_j \theta^{ij}$$

and both the odds ratios and log odds ratios are expressed as:

$$\begin{aligned} \theta_{ij} &= \theta \quad \text{and} \\ \Phi_{ij} &= \phi \quad \text{for } (i,j) = 1, 2, \cdots, (I-1) \end{aligned}$$

The model has one more parameter (viz. θ) than the (O) model and thus has $I(I-2)$ degree of freedom. For this model all local odds ratios are equal and there is said to be a uniform local association for all cells in the square table. The model is sometimes referred to as *the uniform diagonals* model. This model is implemented in SAS software by defining a regression variable $U = i * j$. Models L, Q, and U are implemented in SAS software with the following statements respectively:

```
set tab1; u=r*c;
proc genmod; class r c; model count=r c L/dist=poi; run;
proc genmod; class r c q; model count=r c q/dist=poi; run;
proc genmod; class r c; model count=r c u/dist=poi; run;
```

Model U when implemented in SAS software to Table 11.1, gives a $G^2 = 1818.870$ on 8 d.f. and $\hat{\theta}_{ij} = 1.3582 \quad \forall \quad (i,j)$. Hence, the estimated log odd ratios under model U therefore becomes $\hat{\Phi}_{ij} = 0.3062$.

11.6 Diagonal Models

In this section, we consider the class of nonindependence models, which can be classified into three different groupings. These three groupings of diagonal models are described (Upton, 1985) below:

(i) The *principal* diagonal class models where $\Phi_{ij} = 0$ unless $i = j$.

(ii) The *diagonal band* class models where $\Phi_{ij} = 0$ unless $\mid i - j \mid \leq 1$

(iii) The *full diagonal* class models, which need not have zero terms but preserve features of the original structure of the simpler models.

We describe below some of the models that belong to the classes enumerated above.

11.6.1 The Principal Diagonal Models

Both the fixed-distance and variable distance models (Goodman, 1972, 1979b, and Haberman, 1978) belong to this class of models. The fixed distance model is equivalent (in the context of social mobility) to the vertical mobility model "V" in Hope (1982), which represents vertical distances traveled from the "origin" class to the "destination" class.

11.6.2 The Fixed and Variable Distance Models

The *fixed distance* model with constant or fixed distance parameter δ (that is, adjacent categories have constant distances between them) for an $I \times I$ table has the multiplicative form:

$$m_{ij} = \alpha_i \beta_j \delta^k, \quad \text{for } k =\mid i-j \mid$$

The model, following Lawal and Upton (1990b), is usually designated as model F. For this model, the structure of the log odds ratios is given by:

$$\hat{\Phi}_{ij} = \begin{cases} -\delta & \text{if } i = j \\ 0 & \text{if } i \neq j \end{cases} \tag{11.25}$$

We note here that the U model is equivalent to the above model if we replace $k =\mid i-j \mid$ by the product $i * j$.

The corresponding log-linear formulation for the model is as follows:

$$\ln(\hat{m}_{ij}) = \mu + \lambda_i^R + \lambda_j^C + \lambda_{ij}^D \quad (i, j = 1, 2, \cdots, I) \tag{11.26}$$

where

$$\lambda_{ij}^D = \delta \mid i - j \mid \quad \lambda_{ij}^D = \lambda_{ji}^D \quad \text{and} \quad \sum \lambda_i^R = \sum \lambda_j^C = 0$$

The model has one (δ) more parameter than the O model and hence is based on $I(I-2)$ degrees of freedom. Model F has the nonstandard log-linear formulation:

$$\ell_{ij} = \mu + \lambda_i^R + \lambda_j^C + \varphi F_{ij}$$

where **F** is the regression variable defined earlier under the (LDPS) model.

The *variable distance* model is also a member of the principal diagonal class models and has the multiplicative form:

$$m_{ij} = \alpha_i \beta_j \Lambda_{ij}$$

where

$$\Lambda_{ij} = \begin{cases} \displaystyle\prod_{k=i}^{j-1} \delta_k & \text{if } i < j \\ \displaystyle\prod_{k=j}^{i-1} \delta_k & \text{if } i > j \end{cases}$$

That is, δ_1, δ_2,$\cdots$,δ_{I-1} are the distances from categories 1 to 2, 2 to 3,$\cdots$, and $(I-1)$ to I respectively. That is, it assumes different intervals among the categories.

The structure of the log odds ratios under this model for some $\delta_k, 1 \leq k \leq (I-1)$ can be derived as:

$$\hat{\Phi}_{ij} = \begin{cases} -2\hat{\delta}_i & \text{if } i = j \\ 0 & \text{if } i \neq j \end{cases} \tag{11.27}$$

This variable distance model (V) is based on $(I-1)(I-2)$ degrees of freedom. The equivalent log-linear formulation of this model is similarly given by (12.14) except that λ_{ij}^D now takes the form, for some $\delta_k, 1 \leq k \leq (I-1)$:

$$\lambda_{ij}^D = \begin{cases} \displaystyle\sum_{k=i}^{j-1} \delta_k & (i < j) \\ 0 & (i = j) \\ \displaystyle\sum_{k=j}^{i-1} \delta_k & (j < i) \end{cases}$$

where the $\boldsymbol{\delta} = \{\delta_1, \delta_2, \cdots, \delta_{I-1}\}$ are the distances from categories 1 to 2, 2 to 3,$\cdots$, and $(I-1)$ to I, respectively.

Model (V) also have the nonstandard log-linear model formulation given by:

$$\ell_{ij} = \mu + \lambda_i^R + \lambda_j^C + \lambda_{ij}^{V_{1ij}} + \lambda_{ij}^{V_{2ij}} + \cdots + \lambda_{ij}^{V_{(I-1)ij}}$$

where the $\mathbf{V_1}, \cdots, \mathbf{V_{I-1}}$ are factor variables defined for a 4×4 table as:

$$\mathbf{V}_1 = \begin{bmatrix} 2 & 1 & 1 & 1 \\ 1 & 2 & 2 & 2 \\ 1 & 2 & 2 & 2 \\ 1 & 2 & 2 & 2 \end{bmatrix} \quad \mathbf{V}_2 = \begin{bmatrix} 2 & 2 & 1 & 1 \\ 2 & 2 & 1 & 1 \\ 1 & 1 & 2 & 2 \\ 1 & 1 & 2 & 2 \end{bmatrix} \quad \mathbf{V}_3 = \begin{bmatrix} 2 & 2 & 2 & 1 \\ 2 & 2 & 2 & 1 \\ 2 & 2 & 2 & 1 \\ 1 & 1 & 1 & 2 \end{bmatrix}$$

In general, for an $I \times I$ table, we need to construct $(I-1)$ such patterned factor variables to implement model V in SAS software. Both models F and V are implemented in SAS software with the following statements:

```
proc genmod; class r c v1-v3;
model count=r c f/dist=poi; run; /* fits fixed distance model*/;
model count=r c v1-v3/dist=poi; run; /* fits variable distance model*/;
```

The log-parameter estimates when PROC GENMOD in SAS® is applied to the data in Table 11.1 are given by: $\hat{f} = -2.6964$ for model F and $\hat{v}_1 = -1.5671; \hat{v}_2 = -1.1941; \hat{v}_3 = -1.3276$ for model V. Consequently, the estimated log odds ratios under model F and V using equations (11.25) and (11.27) become:

$$\hat{\Phi}_{ij}^F = \begin{bmatrix} 2.696 & 0.000 & 0.000 \\ 0.000 & 2.696 & 0.000 \\ 0.000 & 0.000 & 2.696 \end{bmatrix} \qquad \hat{\Phi}_{ij}^V = \begin{bmatrix} 1.567 & 0.000 & 0.000 \\ 0.000 & 1.194 & 0.000 \\ 0.000 & 0.000 & 1.328 \end{bmatrix}$$

The estimates of Φ satisfy the condition required for being classified as a principal diagonal class model.

Alternatively, if variables V_1, V_2, V_3 for model V are defined as quantitative variables in the CLASS statement of PROC GENMOD in SAS®, then the nonstandard log-linear model can be written as:

$$l_{ij} = \mu + \lambda_i^R + \lambda_j^C + \varphi_1\, V1_{ij} + \varphi_2\, V2_{ij} + \varphi_3\, V3_{ij}$$

and the structure of the odds ratios in this case takes the form:

$$\hat{\Phi}_{ij} = \begin{cases} \ln(\hat{\varphi}_i) * i & \text{if } i = j \\ 0 & \text{elsewhere} \end{cases}$$

for $i = 1, 2, \cdots, (I-1)$. In the above formulation, the log parameter estimates are given in this case as:

$$\ln(\hat{\varphi}_1) = 1.5671 \qquad \ln(\hat{\varphi}_2) = 0.5970 \qquad \ln(\hat{\varphi}_3) = 0.4425$$

Substitution of these values in the above expression for $\hat{\Phi}_{ij}$ leads to the same results obtained earlier.

Models F and V applied to the data in Table 11.1 give $G^2 = 255.7083$ on 8 d.f. and $G^2 = 204.6620$ on 6 d.f., respectively. Again, neither model fits the data.

11.6.3 The Diagonal Band Models

This class of models has

$$\Phi_{ij} = 0 \quad \text{unless} \quad |\, i - j \,| \leq 1$$

The uniform loyalty, the quasi-independence, and the triangles parameters models belong to this class of models. We discuss these models in what follows:

The Uniform Loyalty Model

The constant loyalty model (L) discussed earlier belongs to this class of models. In the language of mobility, the model differentiates those who do not change from those who do. That is, the model differentiates between the diagonal and the off-diagonal cells where the diagonal members are assumed to be homogeneous, that is they are all assumed to have the same probability of inheritance.

The model has the nonstandard log-linear model formulation in (11.24) and in general, for any square table, the structure of the log odds ratios under model L becomes (Goodman, 1969):

$$\Phi_{ij} = \begin{cases} 2\,\ln(\hat{\phi}) & (i = j) \\ -\ln(\hat{\phi}) & \mid i - j \mid = 1 \\ 0 & \text{elsewhere} \end{cases}$$

where $\ln(\hat{\phi})$ is the log estimate of parameter ϕ. When PROC GENMOD is applied to the data in Table 11.1, $\ln(\hat{\phi})$ is 1.9107 and the estimated log odds ratios become:

$$\hat{\Phi}^L_{ij} = \begin{bmatrix} 3.821 & -1.911 & 0.000 \\ -1.911 & 3.821 & -1.911 \\ 0.000 & -1.911 & 3.821 \end{bmatrix}$$

The Quasi-Independence Model

The quasi-independence (Q) described earlier and reformulated below (Tomizawa, 1992) belongs to this class of models.

$$m_{ij} = \alpha_i \beta_j \psi_{ij} \quad \text{where}$$
$$\psi_{ij} = 1 \quad \text{for } i \neq j$$

The corresponding log-linear model formulation is given by:

$$\ell_{ij} = \mu + \lambda_i^R + \lambda_j^C + \delta_{ij}\epsilon_i$$

where δ_{ij} is as defined previously. Specifically, model Q is sometimes described as the *symmetric diagonal band* model and as discussed earlier, the model is based on $(I^2 - 3I + 1)$ degrees of freedom.

The model has a nonstandard log-linear model:

$$\ell_{ij} = \mu + \lambda_i^R + \lambda_j^C + \lambda_{ij}^Q$$

where **Q** is a factor variable defined as:

$$\mathbf{Q} = \begin{bmatrix} 1 & 5 & 5 & 5 \\ 5 & 2 & 5 & 5 \\ 5 & 5 & 3 & 5 \\ 5 & 5 & 5 & 4 \end{bmatrix}$$

If we therefore let $\mathbf{q} = \{q_1, q_2, q_3, q_4, q_5\}$ be a vector of parameters (usually, $I + 1$), then it can be shown (Lawal, 2002b), that estimated log odds ratios under this model can be obtained from the following expression.

$$\hat{\Phi}_{ij} = \begin{cases} \hat{q}_i + \hat{q}_{i+1} & \text{if } i = j \\ -\hat{q}_j & \text{if } i < j \quad \text{and } \mid i - j \mid = 1 \\ 0 & \text{for } \mid i - j \mid > 1 \end{cases}$$

Note that $\hat{q}_{I+1} = 0$ and the estimates of the log-parameter vector $\mathbf{q}$ when the model is implemented in SAS software are given by:

$$\hat{\mathbf{q}} = \{2.9381,\ 1.2610,\ 1.5326,\ 2.4439,\ 0\}$$

and from the general expressions above, the estimated log odds ratios under model Q are calculated as:

$$\hat{\Phi}_{ij}^{Q} = \begin{bmatrix} 4.199 & -1.261 & 0.000 \\ -1.261 & 2.794 & -1.533 \\ 0.000 & -1.533 & 3.977 \end{bmatrix}$$

11.6.4 The Triangles Parameters Model

Goodman (1985) described the nonindependence triangle (T) model, where

$$m_{ij} = \alpha_i \beta_j \gamma_{ij}$$

where

$$\gamma_{ij} = \begin{cases} \tau_1 & \text{for } i > j \\ \tau_2 & \text{for } i < j \\ 1 & \text{for } i = j \end{cases}$$

The τ parameters pertain to the upper-right triangle and lower left triangles in the square table. The model has been described as the *asymmetric diagonal band* model (Goodman, 1972).

The model has a nonstandard log-linear model form:

$$\ell_{ij} = \mu + \lambda_i^R + \lambda_j^C + \lambda_{ij}^T$$

where $\mathbf{T}$ is a factor variable defined for a 4×4 table as:

$$\mathbf{T} = \begin{bmatrix} 1 & 1 & 1 & 1 \\ 2 & 1 & 1 & 1 \\ 2 & 2 & 1 & 1 \\ 2 & 2 & 2 & 1 \end{bmatrix}$$

The structure of the estimated log odds-ratios, $\hat{\Phi}_{ij}$ can be written more succinctly as:

$$\hat{\Phi}_{ij} = \begin{cases} \hat{\tau}_1 & \text{if } i < j \\ -(\hat{\tau}_1 + \hat{\tau}_2) & \text{if } i = j \\ \hat{\tau}_2 & \text{if } i > j \end{cases}$$

The log parameter estimates when this model is implemented in SAS software are: $\{\hat{\tau}_1 = -1.7996,\ \hat{\tau}_2 = -2.0240\}$. Hence, the estimated log odds ratios become using the above expressions:

$$\hat{\Phi}_{ij}^{T} = \begin{bmatrix} 3.824 & -1.800 & 0.000 \\ -2.024 & 3.824 & -1.800 \\ 0.000 & -2.024 & 3.824 \end{bmatrix}$$

Model T has 2 more parameters than the model O. Consequently, it is based on $(I^2 - 2I - 1)$ degrees of freedom.

The estimates of Φ for models L, T, and Q all satisfy the condition for being classified as a diagonal band class model, since $\hat{\Phi}$=0 for $\mid i - j \mid > 1$.

We can again implement these class of models in SAS software by the following statements:

```
set tab1;
proc genmod; class r c; model count=r c L/dist=poi;run; /* fits  model (L)*/;
proc genmod; class r c t; model count=r c t/dist=poi; run; /* fits  model (T)*/;
proc genmod; class r c q; model count=r c q/dist=poi; run;/* fits model (Q)*/;
```

Models T and Q have respectively $G^2 = 488.0967$ and $G^2 = 199.1062$ on 7 and 5 degrees of freedom. Again neither models fits the data in Table 11.1.

11.7 The Full Diagonal Models

Several models have been developed to take into account the diagonal symmetry of square tables with ordinal categories. We will assume here that changing status by one step in either direction on the scale has a different probability than that for two steps, etc, for greater distances. Goodman (1972) introduced the diagonal D model, which has

$$m_{ij} = \alpha_i \beta_j \delta_k \quad \text{for } i \neq j \tag{11.28}$$

where $k = i - j$. In this model, we are assuming that changing status are like a random walk (with drift) with steps taken with different probabilities in each direction. The model is sometimes called the asymmetric minor diagonal model.

If we consider the 2×2 subtable formed from adjacent rows $(i, i+1)$ and adjacent columns $(j, j+1)$, then the odds ratio for this subtable becomes using (11.28) (Goodman, 1969):

$$\theta_{ij} = \frac{(\alpha_i\ \beta_j\ \delta_{i-j})(\alpha_{i+1}\ \beta_{j+1}\delta_{i-j})}{(\alpha_{i+1}\ \beta_j\ \delta_{i-j+1})(\alpha_i\ \beta_{j+1}\ \delta_{i-j-1})}$$

which simplifies to

$$(\delta_k^2/(\delta_{k+1}\ \delta_{k-1}) \tag{11.29}$$

with $k = (i - j)$. Hence, taking logarithms,

$$\Phi_k = \ln\ (\delta_k^2/\delta_{k+1}\ \delta_{k-1}) = 2\ \ln\ (\delta_k) - \ln\ (\delta_{k+1}) - \ln\ (\delta_{k-1}) \tag{11.30}$$

where the δ's are the log-parameters of the model. There would be $(2I - 3)$ such distinct log odds ratios corresponding to the $(2I - 3)$ diagonals for an $I \times I$ table. These log-odds are conveniently labeled Φ_s, $s = -(I-2), \cdots, (I-2)$. Because there are $(2I - 3)$ parameters for this model, let us define a parameter vector $\mathbf{d}$ as: $\mathbf{d} = \{d_1, d_2, \cdots, d_{(I-1)}\}$, which for our 4×4 example becomes:

$$\mathbf{d} = \{d_1, d_2, d_3, d_4, d_5, d_6, d_7\}$$

The model has a nonstandard log-linear model:

$$\ell_{ij} = \mu + \lambda_i^R + \lambda_j^C + \lambda_{ij}^D$$

where $\mathbf{D}$ for the 4×4 Table 11.1 example, is the same $\mathbf{D}$ factor variable presented in our discussion of the (DPS) model. Although, the factor variable $\mathbf{D}$ implies that there are $(2I - 1)$ parameters to be estimated; however, two of these parameters are redundant, since there are only $(2I - 3)$ such distinct parameters (diagonals). Consequently, parameters d_6 and d_7 are set to zero in the SAS software implementation and the estimates of the distinct parameters of the model are:

$$\hat{d}_1 = -2.5734,\ \hat{d}_2 = -4.6413,\ \hat{d}_3 = -5.7242,\ \hat{d}_4 = -0.6090,\ \hat{d}_5 = -0.4042$$

with $\hat{d}_6 = \hat{d}_7 = 0.000$. Estimated log odds-ratios under model (D) satisfy:

$$\hat{\Phi}_{ij} = \begin{cases} -\{\hat{d}_1 + \hat{d}_I\} & \text{if } i = j \\ -\{\hat{d}_{k+1} - 2\hat{d}_k + \hat{d}_{k-1}\} & \text{for } i < j \text{ and } k = j - i \\ -\{\hat{d}_{(I+k')} - 2\hat{d}_{(I+k'-1)} + \hat{d}_{(I+k'-2)}\} & \text{for } i > j \text{ and } k' = i - j \end{cases}$$

where $k = 1, 2, \cdots, (I-2)$ and similarly $k' = 1, 2, \cdots, (I-2)$. In the application of the above expression for $\hat{\Phi}_{ij}$ for the $i > j$ case, we must assume that $\hat{d}_{I-1} = 0$ in this case for our expression to work.

Thus $k = (j-i) = 1, 2$ correspond to log parameter estimates $\{\hat{d}_1, \hat{d}_2, \hat{d}_3\} = \{-2.5734, -4.6413, -5.7242\}$ for $i < j$. Similarly, $k' = (i-j) = 1, 2$ correspond to parameters $\{\hat{d}_4, \hat{d}_5, \hat{d}_6\} = \{-0.6090, -0.4042, 0.000\}$ for $i > j$ respectively with $\hat{d}_7 = 0$. For this model therefore, estimated log odds ratios are given by:

$$\hat{\Phi}_{ij} = \begin{bmatrix} 3.182 & -0.506 & -0.980 \\ -0.814 & 3.182 & -0.506 \\ -0.200 & -0.814 & 3.182 \end{bmatrix}$$

For instance, $\hat{\Phi}_{21}$ and $\hat{\Phi}_{31}$ have $k' = 1$ and $k' = 2$ respectively. Hence,

$$\hat{\Phi}_{21} = -(\hat{d}_5 - 2\hat{d}_4 + \hat{d}_3) = -\{-0.4042 - 2(-0.6090) - 0.000\} = -0.814$$

$$\hat{\Phi}_{31} = -(\hat{d}_6 - 2\hat{d}_5 + \hat{d}_4) = -\{0.0 - 2(-0.4042) + (-0.6090)\} = -0.1995$$

In general, $\hat{d}_{2I-1} = \hat{d}_{2I-2} = 0$ under this model.

11.7.1 The Diagonals-Absolute Model

The diagonals-Absolute (DA) model has the multiplicative form:

$$m_{ij} = \alpha_i \beta_j \delta_k \quad \text{for } i \neq j \tag{11.31}$$

where $k =\mid i-j \mid$. The model has been referred to as the *symmetric minor diagonal* model (Lindsey, 1989; Fingleton, 1984) where the categories are assumed to be ordered and changing categories one step in either direction on the scale has a different probability than that for two steps, and so on, for greater distances.

The nonstandard log-linear model for this model is:

$$\ell_{ij} = \mu + \lambda_i^R + \lambda_j^C + \lambda_{ij}^{DA}$$

where **DA** is a factor variable defined as:

$$\mathbf{DA} = \begin{bmatrix} 4 & 1 & 2 & 3 \\ 1 & 4 & 1 & 2 \\ 2 & 1 & 4 & 1 \\ 3 & 2 & 1 & 4 \end{bmatrix}$$

For model (DA), there are I parameters to be estimated and if we denote such a parameter vector for a 4×4 table as: $\boldsymbol{\nu} = \{\nu_1, \nu_2, \nu_3, \nu_4\}'$, then the estimated log odds ratios Φ_{ij} are given by:

$$\hat{\boldsymbol{\Phi}}_k = \begin{cases} -2\hat{\nu}_1 & \text{if } k = 0 \\ 2\hat{\nu}_k - \hat{\nu}_{k+1} - \hat{\nu}_{k-1} & \text{if } k = 1, 2, \cdots, (I-2) \end{cases}$$

where $k = |i-j|$ and $\hat{\nu}_0 = \hat{\nu}_I = 0$. When this model is implemented in SAS software to the data in Table 11.1, the log-parameter estimates are:

$$\hat{\boldsymbol{\nu}} = \{\hat{\nu}_1, \hat{\nu}_2, \hat{\nu}_3, \hat{\nu}_4\} = \{-1.5895,\ -2.5286,\ -2.8288,\ 0.0000\}$$

The estimated log odd ratios under model DA therefore become:

$$\hat{\Phi}_{ij}^{DA} = \begin{bmatrix} 3.179 & -0.650 & -0.639 \\ -0.650 & 3.179 & -0.650 \\ -0.639 & -0.650 & 3.179 \end{bmatrix}$$

Models D, and DA have respectively when applied to the full table, $(I-2)^2$ and $(I-1)(I-2)$ degrees of freedom (Goodman, 1986a). The models are implemented in SAS software with the following statements.

```
proc genmod; class r c d; model count=r c d/dist=poi; run; /* fits  model (D)*/;
proc genmod; class r c da; model count=r c da/dist=poi; run; /* fits  model (DA)*/;
```

The *uniform association* model U described earlier is a member of this class of models as it has $\hat{\Phi}_{ij} = \hat{\phi}$ for all (i, j). For this model all local odds-ratios are equal and there is said to be a uniform local association for all cells in the square table. The model has $I(I-2)$ d.f. and is sometimes referred to as the *uniform diagonals* model.

11.7.2 Composite Models

Each of models O, U, L, F, V, T, D, and DA described above cannot alone completely explain the complex variation in most occupational mobility or similar data we are likely to come across. We will therefore consider second-order or third-order combinations of these models. These combinations of models or simply composite models generally fit better such data than the individual models. We examine some of these composite models that have received attention.

The composite model approach has been successfully employed by Goodman (1972, 1979a, 1986a,b, 1991), Upton (1985), Lawal and Upton (1990b), and Lawal (1992d) among others. For instance, the model DAT employed in Goodman (1986a) is a combination of models DA and T.

Model	d.f.	G^2	Model	d.f.	G^2
O	9	5704.27	Q	5	475.63
U	8	463.45	QU	4	43.36
F	8	27.47	QF	4	13.66
UF	7	27.25	LF	7	26.45
L	8	921.49	UL	7	52.37
V	6	19.22	QV	4	13.66
UV	5	17.11	VDA	4	6.07
LV	5	13.66	VT	4	12.54
DA	6	23.32	QDA	3	5.72
D	4	96.36	QD	1	2.02
VD	2	2.38	DAT	5	21.91
T	7	907.67	QT	4	466.39
UT	6	51.18	FT	6	25.18

Table 11.8: Results of analysis (up to first order only) for Table 11.1

Similarly, model LF, which is a combination of the loyalty and fixed distance models, has been described as the *loyalty-distance* model by Upton and Sarlvik (1981). They found that the model works very well with voting data. The model is based on $(I^2 - 3I + 1)$ d.f. Similarly, model VT is a combination of models V and T while model UV is a combination of U and V. Model UV for instance has been shown to be the most parsimonious model for analyzing the 5×5 British social mobility data (Lawal, 1992d) in Table 11.2.

Other composite models are models fitted to the off-diagonal cells $(i \neq j)$ of square tables which are prefixed here by Q. Some of these are the QO, QV, QUV, QFT, QVT, QT, and QU models. Thus model QUV is the composite model UV fitted to the off-diagonal cells. Model QO for instance is the quasi-independence model while model QV has been described by Goodman (1972) as the crossing-parameter model. We present in Tables 11.8 and 11.9, the results of the applications of all the models considered in this chapter to the 4×4 and 5×5 data in Tables 11.1 and 11.2, respectively.

Model	d.f.	G^2	Model	d.f.	G^2
O	16	792.1901	Q	11	235.782
U	15	79.4411	QU	10	14.079
F	15	93.0318	QF	10	17.791
UF	14	55.6923	LF	14	66.613
L	15	475.973	UL	14	54.241
V	12	38.196	QV	9	7.736
UV	11	10.762	VDA	9	10.448
LV	11	14.528	VT	10	12.222
DA	12	54.034	QDA	8	12.804
D	9	105.428	QD	5	9.486
VD	6	6.910	DAT	11	52.115
T	14	463.752	QT	10	227.933
UT	13	52.346	FT	13	64.393

Table 11.9: Results for the 5×5 British social mobility data in Table 11.2

We notice from the results that the fixed distance and variable distance models when fitted to the off-diagonal cells, that is, models, QF and QV are both equivalent for 4×4 tables even though both models are based respectively on $I(I-3)$ and $(I-2)^2$ degrees of freedom. When $I = 4$ both models therefore have the same number of degrees of freedom, hence the equivalence in this case. However, real differences between both models can be observed with tables in which $I \geq 5$. For 3×3 tables, while model QF does not exist, model QV, does exist and is based on 1 degree of freedom.

Parsimonious models for each of the tables discussed at the beginning of this chapter can be found from one or more combinations of the models discussed so far in this chapter. For instance, for the data in Table 11.1, while model VDA with a $G^2 = 6.07$ on 4 d.f. would be the most parsimonious under this group of models, we recall that both the conditional symmetry and LDPS models give $G^2 = 7.354, 7.2804$ on 5 degrees of freedom, respectively. Clearly, model LDPS is the most parsimonious for the data in Table 11.1.

We also give below some of the models that are equivalent to some of the models discussed earlier for 4×4 tables *only*.

$$\begin{aligned} \text{D} &\equiv \text{DPS} \\ \text{QD} &\equiv \text{QDPS} \\ \text{QDAT} &\equiv \text{QCS} \\ \text{QDA} &\equiv \text{QS} \end{aligned}$$

11.8 Classification of All the Above Models

All the symmetry-based and generalized independence models discussed so far in this chapter can be classified into three categories, namely:

(a) Nonindependence Models:

These group of models has the independence model (O) as its baseline model. Consequently, the independence model is often described for this group as the null model. Belonging to this group are models L, F, V, Q, T, D, and DA. Goodman (1986a) refers to this group of models as the nonindependence models. Each of the above models can be applied to the off-diagonal cells of the table. In such cases, the models will carry a prefix Q, the exception being model Q itself.

(b) Asymmetric Models:

A model that captures deviations from the symmetry model is described as the asymmetric model. Such model would have the symmetry model as the baseline model and, consequently, the symmetry model will be referred to in this case as the null (O) asymmetry model (Goodman, 1985). The symmetry model is therefore the null model for this group. Belonging to this group are models CS, LDPS, DPS, and 2RPS. They are composite model S+T, S+F, S+D, and S+T+F, respectively. Also belonging to this category are the odds-symmetry models I and II, that is, models OS1 and OS2. Both models are again composite models S+OS1 and S+OS2 respectively. In all these cases, the symmetry model (S) acts as the baseline model.

(c) Skew-Symmetric Models:

Yamagushi (1990) introduced the *skew-symmetry level* models, which are characterized by deviations from the (QS) model, and following Yamagushi (1990), the (QS) model will be described as the *null skew-symmetric* model (O). That is, the QS model is the baseline model for this category of models. Belonging to this group are models QS, QDPS, QCS, and QOS. Model QS, which is the null model for this group has been described by Goodman as the *RC asymmetry* model. Models QDPS, QCS, and QOS are composite models QS+D, QS+T, and QS+OS1, respectively. Yamagushi (1990) has described the QCS model, for instance, as the *uniform skew-symmetric level* model or as the triangles-parameter skew-symmetry (SP_{SK}) model. Similarly, model QOS is described as the *middle-value-effect skew-symmetry* model designated in Yamagushi (1990) as the M_{SK} model. The QDPS model on the other hand has been described as the *diagonals-parameter skew-symmetry* model which Yamagushi designates as the DP_{SK} model. We present the equivalent models as described in Yamagushi (1990).

$$\begin{aligned} \text{QCS} &\equiv (\text{SP}_{\text{SK}}) \\ \text{QDPS} &\equiv \text{DP}_{\text{SK}} \\ \text{QOS} &\equiv \text{M}_{\text{SK}} \end{aligned}$$

We may note here that the models in this category are equivalent to the symmetry + nonindependence models discussed in Goodman (1985).
Parsimonious models for the data in Tables 11.1, 11.2, and 11.3 are presented in Table 11.10 under the three different classifications presented above.

	Non-Ind			Asymmetric			Skew-symmetric		
Table	Model	d.f.	G^2	Model	d.f.	G^2	Model	d.f.	G^2
11.1	QDA	3	5.72	DPS	3	0.498	QDPS	1	0.222
11.2	UV	11	10.762	CS	9	10.346	QCS	5	2.697
11.3	LF	7	9.37	OS2	3	3.833	QS	3	3.842

Table 11.10: Parsimonious models for three set of data

Lawal and Sundheim (2002) have written a SAS® macro that will fit all the models discussed in the previous sections for any given $I \times I$ contingency table.

11.9 The Bradley-Terry Model

Consider a group of individuals being asked to respond and compare all possible pairs of I items, stating which is preferred, that is, to make comparisons between all possible pairs of such items. Such a situation could arise in a rating experiment, where, for instance, a cigarette smoker may be asked to rate I brands of cigarettes for taste. For given pairs of cigarette brands, a rater could probably state a preference after smoking them at the same occasion. Data collected this way is often referred to as *pairwise comparison* data. Such data results in square tables showing how many individuals prefer each brand of cigarettes as opposed to each other. The two variables of interest here can be described Lindsey (1989) as *prefer* and *not prefer*, each having I categories (the number of brands or items to be compared). We are generally interested in ranking the brands in order of "global preference" for the group of individuals. The ranks are obtained from the number of positive preferences expressed.

A model that is suitable for preference data is that proposed by Bradley and Terry (1952). The model assumes that each item or brand has a probability π_i of being preferred. Thus the probability that brand B_i is preferred to brand B_j is

$$\text{P}(B_i > B_j) = \Pi_{ij} = \frac{\pi_i}{\pi_i + \pi_j} \tag{11.32}$$

where Π_{ij} is the conditional probability that brand i is preferred to brand j, $\pi_i \geq 0$ $(i = 1, 2, \cdots, I)$, and $\sum_i \pi_i = 1$. Further, we shall assume that $\Pi_{ij} + \Pi_{ji} = 1$. The model assumes the independence of the ratings of the same pair by different judges and different pairs by the same judge.

If we let x_{ij} be the observed number of times that B_i is preferred to B_j in the comparisons, then, the expected frequencies under the Bradley-Terry model denoted here by $\hat{m}_{ij}$, is given by the following expression (Fienberg & Larntz, 1976):

$$\hat{m}_{ij} = \frac{n_{ij}\hat{\pi}_i}{\hat{\pi}_i + \hat{\pi}_j} \tag{11.33}$$

where $n_{ij} = x_{ij} + x_{ji}$. and the estimates of the expected frequencies must satisfy:

$$\hat{m}_{ij} + \hat{m}_{ji} = x_{ij} + x_{ji} = n_{ij} \quad \text{for } i \neq j \tag{11.34}$$

The above constraints suggest that the Bradley-Terry model can be implemented by fitting the model of quasi-symmetry (QS) to the generating $I \times I$ table having zeros on the main diagonal. Such a model would be based on $(I-1)(I-2)/2$ degrees of freedom. That is, the Bradley-Terry model is equivalent to a quasi-symmetry model fitted to the square table having zeros on its main diagonal. To implement fitting this model in SAS software, we can consider two approaches.

1. The first approach fits the B-T model by simply fitting the quasi-symmetry model to the $I \times I$ table, using the same approach we discussed earlier.

2. Alternatively, we can fit a model having a *Preference* factor variable and employ the equivalent S-factor design matrix discussed earlier for the 6×6 table. The advantage of this approach is that we can readily recognize the order of preference from estimates of the parameters of the Preference factor variable.

11.9.1 Example 11.1

The table below is from Andersen (1980). The table relates to preferences expressed for a series of six collective facilities in a Danish municipality. The data was originally analyzed in Lindsey (1989).

	Not preferred					
Preferred	1	2	3	4	5	6
1	-	29	25	22	17	9
2	49	-	35	34	16	14
3	50	42	-	40	22	15
4	54	43	37	-	33	16
5	61	61	54	44	-	27
6	69	64	63	62	51	-

Table 11.11: Preference for collective facilities in Denmark

Table 11.11 above is a 6×6 table having zeros on its main diagonal. Since the Bradley-Terry model is equivalent to the model of quasi-symmetry fitted to the above table, we can implement the B-T model with the following SAS software statements.

```
data terry;
do pref=1 to 6;
 do notp=1 to 6;
input count @@; output; end; end;
datalines;
0 29 25 22 17 9 ...69 64 63 62 51 0
;
data two; set terry; input bt s @@;
datalines;
0 1 1 2 2 3 3 4 4 5 5 6 ... 5 6 9 11 12 15 14 18 15 20 0 21
;
```

```
run;
***Fits QS Model****;
proc genmod; make 'obstats' out=aa; class  pref notp s;
model count=pref notp s/dist=poi link=log obstats; run;
***Fits PREF+SYMMETRY Model****;
proc genmod; where pref ne notp; make 'obstats' out=bb;
class  pref notp bt s;
model count=pref s/dist=poi link=log obstats; run;
```

where for a 6×6 table, we have the factor variables **S** and **BT** defined as:

$$\mathbf{S} = \begin{bmatrix} 1 & 2 & 3 & 4 & 5 & 6 \\ 2 & 7 & 8 & 9 & 10 & 11 \\ 3 & 8 & 12 & 13 & 14 & 15 \\ 4 & 9 & 13 & 16 & 17 & 18 \\ 5 & 10 & 14 & 17 & 19 & 20 \\ 6 & 11 & 15 & 18 & 20 & 21 \end{bmatrix}; \quad \mathbf{BT} = \begin{bmatrix} 0 & 1 & 2 & 3 & 4 & 5 \\ 1 & 0 & 6 & 7 & 8 & 9 \\ 2 & 6 & 0 & 10 & 11 & 12 \\ 3 & 7 & 10 & 0 & 13 & 14 \\ 4 & 8 & 11 & 13 & 0 & 15 \\ 5 & 9 & 12 & 14 & 15 & 0 \end{bmatrix}$$

Instead of the S factor variable, we could also use the BT symmetric factor variable generated above, which has zeros on its main diagonal and is constructed in such a way that, this factor variable has the number of levels that is equal to the number of possible pairings, that is, $I(I-1)/2$ which equals 15 in this case.

The QS model when applied to the data gives a $G^2 = 6.0721$ on 10 degrees of freedom. The {PREF+BT} model also gives $G^2 = 6.0721$ but with the same degrees of freedom. Both procedures lead to exactly the same results.

The Bradley-Terry model fits well the data but in order to determine the order of ranking or preference, we need to compute the estimated probabilities for each preference given by (11.33). The expected frequencies under these models are displayed in Table 11.12.

	NOT PREFERRED					
PREFERRED	1	2	3	4	5	6
1	-	30.071	25.007	23.272	15.016	8.634
2	47.929	-	34.158	31.798	21.203	12.912
3	49.993	42.842	-	36.093	24.531	15.541
4	52.728	45.202	40.907	-	27.005	17.158
5	62.984	55.797	51.469	49.995	-	26.754
6	69.366	65.088	62.459	60.842	51.246	-

Table 11.12: Expected values under the Bradley-Terry model

11.9.2 Computing Estimated Probabilities

From the constraints above, it is not too difficult to see that

$$\hat{\pi}_i = \frac{\hat{m}_{ij}\,\hat{\pi}_j}{\hat{m}_{ji}}$$

Therefore, if we let

$$\omega_i = \sum_{j \neq i}^{I} \frac{\hat{m}_{ji}}{\hat{m}_{ij}}$$

for $i = 1, 2, \cdots, I; j = 1, 2, \cdots, I$, then, the expected probabilities under the Bradley-Terry model are given by:

$$\hat{\pi}_i = \frac{1}{1+\omega_i} \tag{11.35}$$

For the data example above, using (11.35),we have the following expressions to estimate $\hat{\pi}_1$ for instance,

$$\hat{\pi}_2 = \hat{m}_{21}\hat{\pi}_1/\hat{m}_{12} \qquad \hat{\pi}_3 = \hat{m}_{31}\hat{\pi}_1/\hat{m}_{13} \qquad \hat{\pi}_4 = \hat{m}_{41}\hat{\pi}_1/\hat{m}_{14}$$

$$\hat{\pi}_5 = \hat{m}_{51}\hat{\pi}_1/\hat{m}_{15} \qquad \hat{\pi}_6 = \hat{m}_{61}\hat{\pi}_1/\hat{m}_{16}$$

Adding the above and noting that $\sum \hat{\pi}_i = 1$ leads to:

$$1 - \hat{\pi}_1 = (\frac{\hat{m}_{21}}{\hat{m}_{12}} + \frac{\hat{m}_{31}}{\hat{m}_{13}} + \frac{\hat{m}_{41}}{\hat{m}_{14}} + \frac{\hat{m}_{51}}{\hat{m}_{15}} + \frac{\hat{m}_{61}}{\hat{m}_{16}})\hat{\pi}_1$$

$$1 - \hat{\pi}_1 = (\frac{47.924}{30.071} + \frac{49.993}{25.007} + \frac{52.728}{23.272} + \frac{62.984}{15.016} + \frac{69.366}{8.634})\hat{\pi}_1$$

$$1 - \hat{\pi}_1 = (1.5939 + 1.9991 + 2.2657 + 4.1945 + 8.0342)\hat{\pi}_1$$

$$1 - \hat{\pi}_1 = (18.0875)\hat{\pi}_1$$

Hence, $\omega_1 = 18.0875$, and

$$\hat{\pi}_1 = \frac{1}{19.0875} = 0.0524$$

The table below gives the values of ω and the corresponding estimated probabilities.

i	ω_i	$\hat{\pi}_i$
1	18.0875	0.0524
2	10.9755	0.0835
3	8.5480	0.1047
4	7.4244	0.1187
5	3.5506	0.2198
6	1.3758	0.4209

From the magnitudes of the estimated probabilities in the above table, we can therefore conclude that the preferences are ranked in the order (6, 5, 4, 3, 2, 1), that is, in the same order as they are presented in Table 11.11 above, with facility 6 being the most preferred.

Alternatively, from the PREF+SYMMETRY fit, we have the following parameter estimates from SAS software for factor variable Preference (pref).

```
                                   Standard    Chi-
Parameter        DF  Estimate       Error    Square  Pr > ChiSq
---------------------------------------------------------------
Intercept         1    3.9366      0.1246    997.54      <.0001
pref         1    1   -2.0837      0.1652    159.13      <.0001
pref         2    1   -1.6175      0.1576    105.39      <.0001
pref         3    1   -1.3910      0.1554     80.17      <.0001
pref         4    1   -1.2658      0.1541     67.44      <.0001
pref         5    1   -0.6499      0.1519     18.31      <.0001
pref         6    0    0.0000      0.0000
```

Relative to category 6 of variable PREF, we see that the magnitudes of the parameters are in the order $(0, -0.65, -1.26, -1.39, -1.62, -2.08)$. These estimates retain the order as the data was presented again. This approach is much simpler since we do not need to compute probabilities to determine the order of preference or ranking.

The conditional probabilities Π_{ij} can be computed from either the expected frequencies or estimated probabilities. For example,

$$\hat{\Pi}_{ij} = \frac{\hat{\pi}_i}{\hat{\pi}_i + \hat{\pi}_j} = \frac{\hat{m}_{ij}}{\hat{m}_{ij} + \hat{m}_{ji}}$$

For the data in example 11.1, we have for instance,

$$\begin{aligned}\hat{\Pi}_{23} &= \frac{\hat{\pi}_2}{\hat{\pi}_2 + \hat{\pi}_3} = \frac{0.0835}{0.1882} = 0.444 \\ &= \frac{\hat{m}_{23}}{\hat{m}_{23} + \hat{m}_{32}} = \frac{34.158}{77.000} = 0.444\end{aligned}$$

Below are the computed conditional probabilities $\hat{\Pi}_{ij}$ for the data in Table 11.11.

$$\hat{\Pi}_{ij} = \begin{bmatrix} - & 0.3855 & 0.3334 & 0.3062 & 0.1925 & 0.1107 \\ 0.6145 & - & 0.4436 & 0.4130 & 0.2754 & 0.1655 \\ 0.6666 & 0.5564 & - & 0.4687 & 0.3228 & 0.1992 \\ 0.6938 & 0.5870 & 0.5313 & - & 0.3507 & 0.2200 \\ 0.8075 & 0.7246 & 0.6772 & 0.6493 & - & 0.3430 \\ 0.8893 & 0.8345 & 0.8008 & 0.7800 & 0.6570 & - \end{bmatrix}$$

11.9.3 Example 11.2

The following example relates to the matches for five professional women tennis players in 1988 (Agresti, 1990).

	Loser				
Winner	Graf	Navratilova	Sabatini	Evert	Shriver
Graf	-	1	3	2	2
Navratilova	0	-	3	3	2
Sabatini	2	0	-	2	1
Evert	0	2	1	-	0
Shriver	1	0	1	1	-

Table 11.13: Results of 1988 tennis tournament for five women players

The Bradley-Terry model fits the data well with G^2 value of 8.0071 on $(5-1)(5-2)/2 = 6$ degrees of freedom. The estimated probabilities are:

Player	$\hat{\pi}_i$
Graf	0.3477
Navratilova	0.3151
Sabatini	0.1264
Evert	0.0826
Shriver	0.1281

Based on the above estimates of the probabilities, there is a need to re-rank the players. The new ranking, which must take cognizance of the estimated probabilities would be: (Graf, Navratilova, Shriver, Sabatini, and Evert) in that order.

The corresponding RANK parameter estimates from SAS software output when the alternative method is employed is displayed in the following:

Parameter		DF	Estimate	Standard Error	Chi-Square	Pr > ChiSq
Intercept		1	-0.4977	1.0711	0.22	0.6422
rank	1	1	0.9988	0.8994	1.23	0.2668
rank	2	1	0.9003	0.9485	0.90	0.3425
rank	3	1	-0.0131	0.8766	0.00	0.9881
rank	4	1	-0.4386	0.9788	0.20	0.6541
rank	5	0	0.0000	0.0000		

The above log estimates relate to (Graf, Nav, Sabb, Evert, Shriver) = (0.999, 0.900, −0.013,−0.439, 0). Based on the magnitude of these estimates, obviously, there is a need to re-rank these estimates in order of magnitude. Consequently, the new ranking would be in the order (0.999, 0.900, 0, −0.013, −0.439), that is, in the order (Graf, Navratilova, Shriver, Sabbatini, Evert). This ranking is consistent again with that obtained from the estimated probabilities.

The estimated conditional probabilities based on the original ranking are given as:

$$\hat{\Pi}_{ij} = \begin{bmatrix} - & 0.5246 & 0.7334 & 0.8080 & 0.7308 \\ 0.4754 & - & 0.7137 & 0.7923 & 0.7110 \\ 0.2666 & 0.2863 & - & 0.6048 & 0.4967 \\ 0.1920 & 0.2077 & 0.3952 & - & 0.3921 \\ 0.2692 & 0.2890 & 0.5033 & 0.6079 & - \end{bmatrix}$$

Thus the predicted probability that Graf would beat Navratilova in a match in 1988 is 0.5246, while the probability that Navratilova will defeat Graf in the same match is $1 - 0.5246 = 0.4754$. Similar interpretations could be given to the other estimated predicted probabilities. These models are implemented in SAS software with S and RANK defined for the 5×5 appropriately as in previous sections.

Lawal (2002a) has applied the Bradley-Terry model to the 1984-1993 10-year season results from the 14 teams in the American Professional Baseball, comprising of teams from both Eastern and Western divisions. The model incorporates home field advantage into the Bradley-Terry model for each of the 10 years. Logit and Poisson regression approaches were employed to model the usual "home field" advantage. The results extend those provided in Agresti (1990).

11.10 Measures of Agreement

We consider here the case in which a sample of N individuals or subjects is rated independently by the same two raters (A and B), to one of I nominal or ordinal categories. The responses f_{ij} of the two raters can then be used to construct a two-way $I \times I$ contingency table with the main diagonal cells representing the agreement between the two raters. The f_{ij} relate to the number of subjects jointly classified into category i by rater A and category j by rater B. Thus maximum agreement occurs when both raters give the same categorical response. For nominal scale of measurement, a table may display association (which is the dependency of one categorical level in A on another in B) and yet low or high agreement. For example, if on an ordinal scale of measurements objects are consistently rated one level higher than rater B, then we would expect the association to be strong but agreement to be weak.

Several authors have argued the case for the often-called *chance-agreement* effect, where for example two raters A and B employ different set of criteria for classifying objects. In such a case, the observed agreement will be said to be primarily due to chance. We shall be concerned here with observed agreement for the *beyond-chance* situations.We discuss below two measures of agreement.

11.10.1 Two Measures of Agreement

The first measure of agreement is the *kappa* measure of agreement, proposed by Cohen (1960), which measures the proportion of agreement between two raters. The *kappa* is an adjustment for agreement by chance, as defined under independence. The measure is defined in terms of observed frequencies as:

$$\hat{\kappa} = \frac{N \sum f_{ii} - \sum_i f_{i+} f_{+i}}{N^2 - \sum_i f_{i+} f_{+i}} \tag{11.36}$$

As pointed out by Tanner and Young (1985), the numerator in the expression for kappa above can be rewritten as:

$$\frac{1}{N} \sum_{i=1}^{a} (f_{ii} - \hat{m}_{ii}) \quad \text{where} \quad \hat{m}_{ii} = \frac{f_{i+} f_{+i}}{N}$$

which shows that $\hat{\kappa}$ is based on the sum of differences between the observed and expected cell counts on the main diagonal of the table with the expected counts being those obtained under the model of independence. Thus $\hat{\kappa}$ employs the model of independence as a baseline for measuring the beyond chance agreement between two raters. Thus an observed cell count will be considered discrepant if it is significantly different from the corresponding expected cell count under the model of independence.

κ ranges from $-\infty \leq \kappa \leq 1$. It is 0 when the observed and expected by chance alone amounts of agreement are equal, and it equals 1 when there is complete agreement between the raters. Kappa will be positive if the observed agreement is greater than chance agreement and will be negative if the observed agreement is less than the chance agreement. The asymptotic variance of kappa is given (Fleiss et al., 1969) as:

$$\frac{\omega + \omega^2 - \sum_i p_{i+} p_{+j} (p_{i+} + p_{+j})}{N(1 - \omega^2)}$$

where $\omega = \sum_i p_{i+} p_{+i}$, $p_{i+} = \sum_j p_{ij}$, and $p_{ij} = f_{ij}/N$.

Jolayemi (1990a) also proposed a measure of agreement, denoted here by τ, where $-1 < \tau < 1$, and defined as:

$$\hat{\tau} = \sqrt{\hat{\lambda}} \quad \text{where}$$
$$\hat{\lambda} = \frac{X^2}{(I-1)N} \tag{11.37}$$

λ is an R^2-type statistic, and X^2 is the value of Pearson's goodness-of-fit test statistic under the model of independence. He classified the agreement as being poor or almost perfect as follows:

$$|\hat{\tau}| = \begin{cases} 0.00 - 0.20 & \text{Poor} \\ 0.21 - 0.40 & \text{Slight} \\ 0.41 - 0.60 & \text{Moderate} \\ 0.61 - 0.81 & \text{Substantial} \\ > 0.81 & \text{Almost perfect} \end{cases}$$

The τ measure of agreement has been demonstrated to be better than the κ measure for not too large sample sizes.

11.10.2 Example 11.3

The data in Table 11.14 below relate self-reporting of crimes as reported in San Jose, CA, from the Law Enforcement Assistance Administration (1972). The data were reported in Bishop et al. (1975). The 5×5 table is a cross-classification of original police descriptions of a sample of crimes versus the victims' categorization of the crimes based on recall

Police	Victim's recall					
categorization	Assault	Burglary	Larceny	Robbery	Rape	Totals
Assault	33	0	0	5	1	39
Burglary	0	91	2	0	0	93
Larceny	0	12	56	0	0	68
Robbery	0	0	6	54	0	60
Rape	5	0	0	0	25	30
Totals	38	103	64	59	26	290

Table 11.14: Law Enforcement Assistance Administration data.

For the above data: $\sum_i f_{i+}f_{+i} = [39 \times 38 + \cdots + 26 \times 30] = 19733$. Hence,

$$\hat{\kappa} = \frac{75110 - 19733}{290^2 - 19733} = 0.8603$$

Similarly, a model of independence fitted to the above data yields an $X^2 = 872.6969$, and hence,

$$\hat{\lambda} = \frac{872.6969}{4 \times 290} = 0.7523$$

From the above, $\hat{\tau} = \sqrt{0.7523} = 0.8674$. This estimated value of τ agrees very closely with the κ estimate.

There is therefore strong evidence of a very strong agreement (almost perfect) between the characterization of the crimes by the police and the victims. In order words, there is a strong agreement between the police and victims in their characterizations of the crimes than if the characterizations were independent. The characterizations are dependent, since the model of independence yields an $X^2 = 872.6969$ on 16 degrees of freedom. The SAS software implementation of the above result is presented below together with a relevant output.

```
data agree;
do pol=1 to 5; do vic=1 to 5;
input L Q sym cs count @@; output; end; end;
datalines;
2 2 1 0 33.... 2 6 15 0 25
;
proc freq; weight count; tables pol*vic/agree; run;

                        Kappa Statistics
Statistic           Value       ASE      95% Confidence Limits
--------------------------------------------------------------
Simple Kappa       0.8603    0.0236        0.8140       0.9066
Weighted Kappa     0.8465    0.0312        0.7854       0.9076
Sample Size = 290
```

Other models that are considered for the data in order to explain fully the dependent structure in the data are the symmetry model, the conditional symmetry model and the unconditional marginal homogeneity model (UMH). The symmetry model gives a G^2 value of 26.0851 on 10 d.f. The conditional symmetry also gives a G^2 value of 18.5134 on 9 d.f., while the UMH gives a value of 22.4182 on 4 d.f. The quasi-symmetry model is very difficult to fit to this data because of so many zeros. The data clearly provide strong evidence against marginal homogeneity.

Apart from the independence model, which assumes independent ratings by the two raters, the model below measures the structure of the overall agreement in the table and has the log-linear model formulation:

$$\ln(m_{ij}) = \mu + \lambda_i^A + \lambda_j^B + \delta_{ij} I(i = j) \tag{11.38}$$

where $I(i = j)$ is an indicator function. The case in which $\delta_{ij} = \delta$ has (11.38) being described as the *homogeneous agreement* model (HA), which measures the overall agreement in the table. The model has two components, the first three terms on the RHS representing chance and the last term representing agreement. The model is very sensitive only to discrepancies that may be present on the main diagonal. The model is sensitive only to discrepancies that may be present on the main diagonal. Melia and Diener-West describe the parameter δ in model (11.38) as the *exact agreement term*, and the model has been referred to as the *exact agreement* model.

This model when implemented is equivalent to the *uniform loyalty* model (Upton & Sarlvik, 1981) or the *equal weight agreement* model (Tanner & Young, 1985) and has $G^2 = 76.7793$ on 15 degrees of freedom. The estimate of parameter δ is 3.570, with asymptotic standard error (a.s.e. = 0.2124) indicating very strong agreement along the main diagonals of the table.

A model similar to the above but that have the $\delta_{ij} = \delta_i$ is the familiar *quasi-independence* or *nonuniform loyalty* model (Q) discussed earlier. This model when applied to data has $G^2 = 67.5917$ on 11 degrees of freedom. The quasi-independence model when used in the context of agreement of raters can help us assess patterns of agreement (Tanner & Young, 1985). Tanner and Young describe this model as the *differential weight agreement* model. For the data in Table 11.14, neither model HA or Q fits the data.

11.10.3 The Case of Ordinal Categories

While models HA and Q may be appropriate for situations in which the categories are nominal, they are unsuitable for situations in which the response variables are ordinal, as, for example, in a rating experiment. This is because, while both models may account for the high concentration of observations on the main diagonal, they do not account for the off-diagonal cells where disagreement between the raters are manifested. Further, with ordered categories, high ratings by rater A are almost always accompanied by high ratings by rater B . Similarly, low ratings by A are also accompanied by low ratings by B. In this case, Agresti (1990) suggests that we might consider the beyond-chance agreement model as consisting of two components. The first component relates to the linear-by-linear association between the raters, and a second component that reflects agreement in excess of what we would normally expect by chance from the linear-by-linear baseline model.

(a) The first component which is the linear-by-linear association component between the raters is modeled as:

$$\ln(m_{ij}) = \mu + \lambda_i^A + \lambda_j^B + \beta u_i u_j \tag{11.39}$$

where $\{u_i\}$ are ordered fixed scores assigned to the levels of the ordinal scale such that $u_1 < u_2 < \cdots < u_I$. With integer scores, then, $u_i = i$, and the model in (11.39) becomes the *linear-by-linear association* (LL) model in Goodman (1979a). The model is based on $I(I-2)$ degrees of freedom. The model is equivalent to the uniform association model (U) when integer scores are employed. As Agresti (1996) observed, the parameter β in equation (11.39) relates to the direction and strength of the association between A and B. The model in (11.39) reduces to the independence model when $\beta = 0$, and if $\beta > 0$, then the association is positive, that is, A increases as B increases. Similarly, when $\beta < 0$, then the association is negative, and A decreases as B increases. Further, the association is stronger as $|\beta|$ increases.

(b) The second component reflects agreement in excess of what we would normally expect by chance from the linear-by-linear baseline model. Agresti (1990) suggests a log-linear model of the form

$$\ln(m_{ij}) = \mu + \lambda_i^A + \lambda_j^B + \beta u_i u_j + \delta I(i = j) \tag{11.40}$$

This model which is based on $(I^2 - 2I - 1)$ degrees of freedom is referred to as the *parsimonious quasi-symmetry* (PQS) model. A generalization of the above model occurs when we do not have homogeneity along the diagonals, in which case $\delta = \delta_i$ for $i = 1, 2, \cdots, I$. In this case, the log-linear model formulation becomes

$$\ln(m_{ij}) = \mu + \lambda_i^A + \lambda_j^B + \beta u_i u_j + \delta_{ij} I(i = j) \tag{11.41}$$

The model expressed by (11.41) has been described by Goodman (1979a) as the *quasi-uniform association* (QUA) model and it is based on $I(I-3)$ degrees of freedom.

Another model that is often employed is the *ordinal quasi-symmetry (OQS)* model, which has the log-linear model formulation (Agresti, 1996):

$$\ln(m_{ij}) = \mu + \lambda_i^A + \lambda_j^B + \lambda_{ij}^{AB} + \beta u_i \tag{11.42}$$

where $\lambda_{ij}^{AB} = \lambda_{ji}^{BA}$ for all i and j. For this model,

$$\lambda_i^A - \lambda_j^B = \beta u_i$$

It is obvious that this model is equivalent to the quasi-symmetry model when $\beta = 0$. When $\beta > 0$, then the responses are more likely to be at the low end of the ordinal scale for the column variable than for the row variable, and that when $\beta < 0$, the mean response will be higher for the column variable (Agresti, 1996).

The structure of the estimated log odds ratios formed from adjacent rows and column categories (that is, local odds ratios) under models (11.39) and (11.40) are given respectively by:

$$\hat{\Phi}_{ij} = \hat{\beta}, \quad \text{for } (i, j) = 1, 2, \cdots, I - 1 \tag{11.43}$$

and

$$\hat{\Phi}_{ij} = \begin{cases} 2\hat{\delta} + \hat{\beta} & \text{if } (i = j) \\ -\hat{\delta} + \hat{\beta} & \text{if } \mid i - j \mid = 1 \\ \hat{\beta} & \text{elsewhere} \end{cases} \tag{11.44}$$

On the other hand, the odds ratios from 2×2 subtable formed from rows i and j and columns i and j is defined as:

$$\theta_{ij} = \frac{m_{ii}m_{jj}}{m_{ij}m_{ji}} \quad \text{for } i \neq j$$

In this case, the corresponding log odds, given in terms of parameter estimates for models (11.39), and (11.40), respectively, are:

$$\hat{\Phi}_{ij} = (u_i - u_j)^2\hat{\beta} \quad \text{and} \tag{11.45a}$$

$$\hat{\Phi}_{ij} = (u_i - u_j)^2\hat{\beta} + 2\hat{\delta} \tag{11.45b}$$

where $(i, j) = 1, 2, \cdots, (I - 1)$. Following Darrock and McCloud (1986), two categories i and j, are said to be indistinguishable if $\theta_{ij} = 1$ and $\theta_{ik} = \theta_{jk}$ for all $k \neq i, j$. Consequently, the index of distinguishability which is based on the odds-ratios is:

$$\nu_{ij} = 1 - \theta_{ij}^{-1} \tag{11.46}$$

Clearly, ν_{ij} is maximum when there is perfect agreement. For most cases $0 \leq \nu_{ij} \leq 1$.

We may observe here that there is a relationship between the expected values under model (OQS) as expressed in (11.42) and to those obtained from the *the linear-diagonal paramcters model* (LDPS) expressed in (11.17). Consequently, for 4×4 tables, models OQS and LDPS are equivalent. That is,

$$\text{OQS} \equiv \text{LDPS}$$

Application of the LDPS model therefore, gives direct estimate of the β parameter in model OQS. The model can also be implemented in SAS software by the use of PROC LOGISTIC.

11.10.4 Example 11.4

As an example, we consider the data below which arose from diagnosis of multiple sclerosis, reported in Weslund and Kurland (1953). Sixty-nine New Orleans patients were examined by two neurologists, one from New Orleans, and the other from Winnipeg. The two neurologists classified each patient into one of the following classes: (1) certain multiple sclerosis; (2) probable multiple sclerosis; (3) possible multiple sclerosis (odds 50:50); and (4) doubtful, unlikely, or definitely not multiple sclerosis.

Employing the methodology described in the previous sections, the complete symmetry (S) when applied to the above data gives $G^2 = 11.9483$ on 6 d.f. The conditional symmetry (CS) model also gives a G^2 value of 6.3575 on 5 d.f., while the quasi-symmetry (QS) model gives a $G^2 = 2.0367$ on 3 d.f. Since the categories are considered ordered, then a conditional marginal symmetry model is obtained (since the conditional symmetry model holds) as: $G^2 = 11.9483 - 6.3575 = 5.5908$ on 1 d.f. The conditional test is valid because the CS model holds. The symmetry, the conditional symmetry and the QS models all fit these data. The marginal homogeneity hypothesis, however, is not tenable for the data. The symmetry model

New Orleans neurologist	Winnipeg neurologist 1	2	3	4	Totals
1	5	3	0	0	8
2	3	11	4	0	18
3	2	13	3	4	22
4	1	2	4	14	21
Totals	11	29	11	18	69

Table 11.15: Diagnostic classification regarding multiple sclerosis for the New Orleans patients

indicates that mis- classification by either pathologist were compensating misclassification (n_{ij} approximately equals to n_{ji}) for all $i \neq j$. Also QS fitting the data similarly indicates that the association is roughly symmetric.

Let us now focus on implementing the log-linear models discussed in the previous sections to the data above. These models can be implemented in SAS software with the following statements.

```
data agree;
do new=1 to 4; do win=1 to 4; input L Q sym cs count @@;
ul=new; vl=win; output; end; end;
datalines;
2 2 1 0 5 ... 2 5 10 0 14
;
****Fit exact agreement (HA) model****;
proc genmod; class new win s q; model count=new win L/dist=poi link=log; run;
****Fit QUA model****;
proc genmod; class new win s q; model count=new win Q/dist=poi link=log;
****Fit linear-by-linear (U) model****;
proc genmod; class new win s q; model count=new win ul*vl/dist=poi link=log;
****Fit PQS model****;
model count=new win ul*vl L/dist=poi link=log; model count=ul S/dist=poi link=log;
****Fit OQS model****;
run;
```

When the models discussed above are each applied to the data in Table 11.15, we display the results of these fits in Table 11.16.

Models	d.f.	G^2	X^2
O	9	46.2621	44.0662
HA	8	29.2252	26.2960
Q	5	10.1855	8.1690
U	8	8.8430	10.4662
PQS	7	8.8367	10.3516
QUA	4	4.0184	4.2060
OQS	5	4.2904	3.7510
DA	6	6.4601	6.9508

Table 11.16: Results of fitting the models to the data in Table 11.15

The exact agreement model (HA) fits poorly from the results in Table 11.16. The parsimonious quasi-symmetry (PQS) or the "linear-by-linear+exact agreement" model fits the data well with the following log parameter estimates:

Parameter	d.f.	Estimate	s.e.	Wald 95% Confidence limits		Chi-square	Pr > χ^2
β	1	1.0412	0.2971	0.4589	1.6234	12.28	0.0005
δ	1	0.0277	0.3487	-0.6557	0.7111	0.01	0.9367

We notice that while the effect of the β parameter is significant, that of the covariate is not. The model has a G^2 value of 8.8367 on 7 d.f. Since the last term is not significant, we can drop this term from the model and refit the reduced model. The reduced model , which is the uniform association model (U), gives a G^2 value of 8.8434 on 8 d.f. (this is the most parsimonious model) with $\hat{\beta} = 1.0556$ (a.s.e. = 0.2360), and $\hat{\theta} = e^{1.0556} = 2.874$, for adjacent rows and column categories, indicating that the odds that the diagnosis of New Orleans neurologist is $i+1$ rather then i is estimated to be 2.874 higher when the diagnosis of the Winnipeg neurologist is $j+1$ than when it is j for all cases in which $\mid i-j \mid = 1$. Similarly, the odds that the diagnosis of New Orleans neurologist is $i+2$ rather then i is estimated to be $(2.874)^2 = 8.260$ higher when the diagnosis of the Winnipeg neurologist is $j+1$ than when it is j in this case. Lawal (2003b) discusses the case when the δ parameter is significant (see exercise 6 at the end of this chapter).

Another approach to measuring agreement when categories are ordinal is to obtain a different measure of agreement different from κ. The κ that we discussed earlier is most useful for nominal categories. An alternative measure with ordinal categories is the *weighted kappa*, κ_w, which is defined (Spitzer et al., 1967) by:

$$\kappa_w = \frac{\sum\sum \omega_{ij}\pi_{ij} - \sum\sum \omega_{ij}\pi_{i+}\pi_{+j}}{1 - \sum\sum \omega_{ij}\pi_{i+}\pi_{+j}} \tag{11.47}$$

where $0 \leq \omega_{ij} = 1 - (i-j)^2/(I-1)^2 \leq 1$. The following results are obtained when this is implemented in SAS software.

```
set agree;
proc freq; weight count; tables new*win/chisq agree; run;

   Test of Symmetry

Statistic (S)     9.7647
DF                     6
Pr > S            0.1349

                    Kappa Statistics
Statistic          Value        ASE      95% Confidence Limits
--------------------------------------------------------------
Simple Kappa      0.2965     0.0785        0.1427       0.4504
Weighted Kappa    0.4773     0.0730        0.3341       0.6204
Sample Size = 69
```

For the above data, $\hat{\kappa}_w = 0.4773$. Under the model of independence, $X^2 = 44.0662$ and $G^2 = 46.2641$. Hence $\hat{\tau} = \sqrt{44.0662/(3 \times 69)} = 0.4614$. The measure of agreement based on τ indicates that the agreement between the two neurologists is moderate.

11.11 Multirater Case

We now extend the theory developed in the earlier sections to the case in which we have multiple raters. The simplest of this is the case involving three raters.

The data in Table 11.17 below relate to the degree of necrosis for 612 tumors as cross-classified by three raters with grading scale: 1, none; 2, $<10\%$; 3, $\geq 10\%$. The data, a $2 \times 3 \times 3$ contingency table are again taken from Melia and Diener-West (1994).

Rater A	Rater B	Rater C 1	Rater C 2	Rater C 3
1	1	315	105	13
	2	14	22	3
	3	3	0	1
2	1	33	16	2
	2	8	16	7
	3	0	2	4
3	1	5	6	1
	2	1	3	4
	3	0	4	24

Table 11.17: Degree of necrosis of tumor in 612 eyes cross-classified by three raters

The PQS model, in (11.40) for example, can be extended to the case where we have three raters. In this case, we would have:

$$\begin{aligned}\ln(m_{ijk}) &= \mu + \lambda_i^A + \lambda_j^B + \lambda_k^C \\ &+ l_1 I(i=j) + l_2 I(i=k) + l_3 I(j=k) + l_4 I(i=j=k) \\ &+ \beta_1 u_i u_j + \beta_2 u_i u_k + \beta_3 u_j u_k + \beta_4 u_i u_j u_k\end{aligned} \tag{11.48}$$

The l_1, l_2, l_3 pertain to exact agreement between pairs of raters A and B; A and C; and B and C, respectively, beyond that due to the linear-by-linear association while l_4 describes the additional exact agreement among all the three raters. The l's are created by the following expressions.

$$l_1 = \begin{cases} 2 & \text{if } i = j \\ 0 & \text{otherwise} \end{cases} \quad l_2 = \begin{cases} 2 & \text{if } i = k \\ 0 & \text{otherwise} \end{cases}$$

$$l_3 = \begin{cases} 2 & \text{if } j = k \\ 0 & \text{otherwise} \end{cases} \quad l_4 = \begin{cases} 2 & \text{if } i = j = k \\ 0 & \text{otherwise} \end{cases}$$

We have adopted the scores $u_i = i - 2$, that is, scores centered at zero for these data set so that we can compare our parameter estimates with those obtained in Melia and Diener-West (1994).

We present below the log-linear formulations of the models described in Melia and Diener-West (1994).

$$\ln(m_{ijk}) = \mu + \lambda_i^A + \lambda_j^B + \lambda_k^C \tag{M1}$$

$$\ln(m_{ijk}) = \mu + \lambda_i^A + \lambda_j^B + \lambda_k^C + l_1 I(i=j) + l_2 I(i=k) + l_3 I(j=k) + l_4 I(i=j=k) \tag{M2}$$

$$\ln(m_{ijk}) = \mu + \lambda_i^A + \lambda_j^B + \lambda_k^C + \beta_1 u_i u_j + \beta_2 u_i u_k + \beta_3 u_j u_k + l_1 I(i=j) + l_2 I(i=k) + l_3 I(j=k) + l_4 I(i=j=k) \tag{M3a}$$

$$\ln(m_{ijk}) = \mu + \lambda_i^A + \lambda_j^B + \lambda_k^C + \beta_1 u_i u_j + \beta_2 u_i u_k + \beta_3 u_j u_k + \beta_4 u_i u_j u_k \tag{M3b}$$

$$\ln(m_{ijk}) = \mu + \lambda_i^A + \lambda_j^B + \lambda_k^C + \beta_1 u_i u_j + \beta_2 u_i u_k + \beta_3 u_j u_k + l_1 I(i=j) + l_2 I(i=k) + l_3 I(j=k) \tag{M5a}$$

$$\ln(m_{ijk}) = \mu + \lambda_i^A + \lambda_j^B + \lambda_k^C + \beta_1 u_i u_j + \beta_2 u_i u_k + \beta_3 u_j u_k + l_{i,j,k} I(i,j,k) \tag{M5b}$$

$$\ln(m_{ijk}) = \mu + \lambda_i^A + \lambda_j^B + \lambda_k^C + \beta_1 u_i u_j + \beta_2 u_i u_k + \beta_3 u_j u_k + \beta_4 u_i u_j u_k + l_1 I(i=j) + l_2 I(i=k) + l_3 I(j=k) \tag{M5c}$$

$$\ln(m_{ijk}) = \mu + \lambda_i^A + \lambda_j^B + \lambda_k^C + \beta_1 u_i u_j + \beta_2 u_i u_k + \beta_3 u_j u_k + \beta_4 u_i u_j u_k + l_1 I(i=j) + l_2 I(i=k) + l_3 I(j=k) + l_4 I(i=j=k) \tag{M5d}$$

Model (M5b), which assumes homogeneity in pairwise l, has $l(i,j,k)$ defined in this case by:

$$l(i,j,k) = l_1(i=j) + l_2(i=k) + l_3(j=k)$$

11.11.1 Results

In Table 11.18 are the results of employing the models in (M1) to (M5d).

Model	d.f.	Deviance	AIC
M1	20	384.3596	-
M2	16	114.0314	-
M3			
(a)	17	28.3525	-5.6
(b)	16	23.0422	-9.0
M5			
(a)	14	19.1596	-8.8
(b)	15	15.5831	-14.4
(c)	13	9.6941	-16.31
(d)	12	9.5375	-14.40

Table 11.18: Comparison of log-linear models for the data in Table 11.17

The results in Table 11.18 are consistent with those in Melia and Diener-West (1994). The column labeled AIC refers to the Akaike information criterion for selecting the most parsimonious models among those models that fit the data. These values are presented for only models that fit our data. It is obvious from the values of the AIC that model (M5c) fits best. That is, model (M5c) is the most parsimonious model for the data in Table 11.17. Relevant parameter estimates under this model are presented in Table 11.19.

Parameter	d.f.	Estimate	Standard error	Wald 95% confidence limits		Chi-square	Pr > Chisq
β_1	1	0.5689	0.2310	0.1161	1.0217	6.06	0.0138
β_2	1	0.9383	0.2294	0.4887	1.3880	16.73	<0.0001
β_3	1	1.0040	0.2334	0.5465	1.4614	18.50	<0.0001
β_4	1	0.4997	0.1695	0.1675	0.8318	8.69	0.0032
l_1	1	0.6613	0.2093	0.2511	1.0715	9.98	0.0016
l_2	1	-0.0837	0.1891	-0.4543	0.2868	0.20	0.6578
l_3	1	0.3627	0.1930	-0.0157	0.7410	3.53	0.0603

Table 11.19: Analysis of parameter estimates

The SAS software implementation of these models discussed above is carried out with the following SAS software program.

```
data agree;
do A=-1 to 1 BY 1; do B=-1 to 1 BY 1;  do C=-1 to 1 BY 1;
input COUNT @@;
IF A EQ B THEN D1=2;       ELSE D1=1;
IF A EQ C THEN D2=2;       ELSE D2=1;
IF B EQ C THEN D3=2;       ELSE D3=1;
IF A EQ B EQ C THEN D4=2; ELSE D4=1;
D=D1+D2+D3; U12=A*B; U13=A*C; U23=B*C; U123=A*B*C;
output; end; end; end;
datalines;
315 105 13 14 22 3 3 0 1 33 16 2 8 16 7 0 2 4 5 6 1 1 3 4 0 4 24
;
run;
        proc genmod; class A B C;
(i)     model count=A B C/dist=poi; run;
(ii)    model count=A B C D1-D4/dist=poi; run;
(iii)   model count=A B C U12 U13 U23/dist=poi; run;
(iv)    model count=A B C U12 U13 U23 U123/dist=poi; run;
(v)     model count=A B C U12 U13 U23 D1-D3/dist=poi; run;
(vi)    model count=A B C U12 U13 U23 D/dist=poi; run;
(vii)   model count=A B C U12 U13 U23 U123 D1-D3/dist=poi; run;
(viii)  model count=A B C U12 U13 U23 U123 D1-D4/dist=poi; run;
```

** We represent the l_1, l_2, l_3 and l_4 in the text respectively by D1-D4 in the SAS software implementation above.
Because the effect of exact agreement as measured by l_1 between raters A and B is highly significant, we can therefore conclude that our chosen model indicates that there is strong agreement beyond that would be expected by chance in the form of the contributions of $\beta_1, \beta_2, \beta_3$ and β_4, the linear-by-linear associations parameters. The results also indicate that β_4, the three-way linear-by-linear association parameter is highly significant. Following Melia and Diener-West (1994), we present below parameter estimates of the linear-by-linear association for each pair of raters by the level of the third rater. That is, for any pair $p = 1, 2, 3$, we calculate the parameter estimate to be equal to:

$$\hat{\beta}_p + u_q \hat{\beta}_4$$

where $u_1 = -1$, $u_2 = 0$, and $u_3 = 1$ are respectively the centered scores derived from $u_i = i - 2$. For the pair of raters {A,C} for example, corresponding to $p = 2$ in the table below and the third level of rater C, we have, since $q = 3$ in this case,

$$\hat{\beta}_2 + (1 * \hat{\beta}_4) = 0.9383 + (1 * 0.4997) = 1.4380.$$

These results are displayed in Table 11.20.
As observed in Melia and Diener-West (1994), increase in the degree of necrosis increases the strength of the linear-by-linear association between any pairs of raters, as the level of the third rater also increases. This indicates that raters tend to agree

p	Rater pair	Level of third rater, q		
		1	2	3
1	{A, B}	0.0692	0.5689	1.0686
2	{A, C}	0.4386	0.9383	1.4380
3	{B, C }	0.5043	1.0040	1.5037

Table 11.20: Estimates of linear-by-linear association for each pair of raters by level of third rater

about which cases have a large degree of necrosis than they about which cases have little or no necrosis. However, this degree of agreement depends on which pair of raters being considered. While the linear-by-linear association is weaker between raters A and B than between A and C or B and C, raters A and B have the strongest additional exact agreement, $\hat{l}_1 = 0.661$. There is no significant additional agreement between raters B and C with $\hat{l}_3 = 0.3627$ and also no significant additional exact agreement between A and C since $\hat{l}_2 = -0.0837$ is not significant both at $\alpha = 0.05$.

The above findings that raters A and B have strongest exact agreement while raters A and C and raters B and C have strongest linear-by-linear association suggest as Melia and Diener-West put it, that "rater C's evaluations may be shifted with respect to those of raters A and B." This is exemplified by the marginal distributions. For instance, rater C is less likely to assign cases to "no necrosis," 379 of 612 (or 62%), than are raters A and B; 78% (476 of 612) and 81% (496 of 612), respectively, and correspondingly, more likely to assign a higher degree of necrosis, 59 of 612 for rater C as compared to 48 and 38 out of 612 for raters A and B respectively.

11.12 Exercises

1. For the data in Table 10.21 in chapter 10,

   (a) Fit the symmetry model to these data.

   (b) Fit the quasi-symmetry model to the data and use it to test for marginal homogeneity.

   (c) Fit the QI, CS, and diagonal parameter models to the above data.

2. For the data in Table 10.22 in chapter 10,

   (a) Fit the models of independence, QI, F, V, T, Q, D, and DA to these data.

   (a) Fit symmetry models to these data.

   (c) Fit skew-symmetry models to the above data and discuss the most parsimonious model.

3. The following data is supplied by E. Jensen of Faellesforeningen for Danmarks Brugsforeninger, Copenhagen: 15 persons examined all possible pairings of 4 different samples for taste, resulting in the following preference table (David, 1988, pp. 115 to 116).

	A_1	A_2	A_3	A_4
A_1	-	3	2	2
A_2	12	-	11	3
A_3	13	4	-	5
A_4	13	12	10	-

Analyze the data assuming the Bradley-Terry model and also test the goodness of fit of your model. Give a ranking for the four samples and estimate the probabilities π_i.

4. The data below relate to 44 Caucasian women from North Carolina who were under 30 and married to their first husbands. Women were asked to respond for pairs of numbers x and y between 0 and 6 with $x < y$. The question asked was, "given a choice of having, during your entire lifetime, either x or y children, which would you choose?" The data are summarized below (Imrey, et al., 1976).

Alternative	Preferred number of children						
choice	0	1	2	3	4	5	6
0	-	17	22	22	15	26	25
1	2	-	19	13	10	9	11
2	1	0	-	11	11	6	6
3	3	1	7	-	6	2	6
4	1	10	12	13	-	4	0
5	1	11	18	15	17	-	11
6	2	13	20	22	14	12	-

Table 11.21: Family size preference

(a) Test whether the Bradley-Terry model fits.
(b) Estimate the probabilities π_i.

5. The data below relate to two pathologist classifying each of 118 slides in terms of carcinoma in situ of the uterine cervix (Landis & Koch, 1977a)based on the most involved lesion. The classification is into one of the ordered categories (1 = negative, 2 = atypical squamous hyperplasia, 3 = carcinoma in situ, 4 = squamous carcinoma with early stromal invasion, 5 = invasive carcinoma) resulting into the 5×5 table below:

	Pathologist B				
Pathologist A	1	2	3	4	5
1	22	2	2	0	0
2	5	7	14	0	0
3	0	2	36	0	0
4	0	1	14	7	0
5	0	0	3	0	3

Obtain estimates of κ and τ for the above data. Find a parsimonious log-linear model of the form discussed in this chapter for these data and interpret your parameter estimates.

6. The data in Table 11.22, are from the Collaborative Ocular Melanoma Study (COMS) (1989) and were analyzed in Melia and Diener-West (1994).

	Rater B				
Rater A	1	2	3	4	5
1	291	74	1	1	1
2	186	256	7	7	3
3	2	4	0	2	0
4	3	10	1	14	2
5	1	7	1	8	3

Table 11.22: Scleral extension in 885 eyes: cross-classified by two raters (Melia & Diener-West, 1994)

They came from multi-center clinical trials investigating the treatment of choroidal melanoma, a very rare cancer of the eye. A detailed description of the data is available in Melia and Diener-West (1994). The data are a summary of the classification by two raters A and B (pathologists) of the extent of scleral extension of choroidal melanoma in 885 eyes. The category grading scale is: (1) none or innermost layers; (2) within sclera, but does not extend to scleral surface; (3) extends to scleral surface; (4) extrascleral extension without transection; (5) extrascleral extension with presumed residual tumor in orbit. The categories are assumed ordered, and agreement among raters has implications for the grading system reliability of the histopathological features believed to be important prognostic indicators for the disease. Melia and Diener-West have already analyzed these data. Show that the PQS model is the most parsimonious model for this data.

7. The data below refer to 264 marriages in Surinam (Speckman, 1965). Here husbands and wives are categorized in terms of four religious groups: C = Christian, M = Moslems, S = Sanatin Dharm, A = Arya Samaj. S and A are two Hindustan religious groups.

	Wives				
Husbands	C	M	S	A	Total
C	17	1	4	3	25
M	1	66	4	2	73
S	5	4	96	14	119
A	4	2	18	23	47
Total	27	73	122	42	264

Table 11.23: Marriage in Surinam (Speckmann, 1965)

Fit the models of symmetry, conditional symmetry, quasi-symmetry, and quasi-independence to this data. Do any of the distance models fit these data? Also fit the various diagonal models to the data.

8. The table below relates mother's education to father's education (Mullins & Sites, 1984) for a sample of eminent Black Americans (defined as persons having biographical sketch in the publication *Who's Who Among Black Amer-*

icans). Fit the symmetry, conditional symmetry, marginal homogeneity, and quasi-symmetry to the data and interpret the data.

Mother's education	Father's education			
	8th Grade or less	Part High school	High school	College
8th Grade or less	81	3	9	11
Part high school	14	8	9	6
High school	43	7	43	18
College	21	6	24	87

9. For the two data sets below, fit appropriate models to these data and interpret.

Danish occupational mobility data (Svalastoga, 1959)

Father's Status	**Son's Status**					Total
	(1)	(2)	(3)	(4)	(5)	
(1)	18	17	16	4	2	57
(2)	24	105	109	59	21	318
(3)	23	84	289	217	95	708
(4)	8	49	175	384	198	814
(5)	6	8	69	201	246	530
Total	79	263	658	865	562	2427

Origin religion	**Current religion**						Total
	(1)	(2)	(3)	(4)	(5)	(6)	
(1)	123	2	0	0	1	48	174
(2)	10	420	9	1	4	217	661
(3)	2	21	102	1	5	54	185
(4)	0	8	2	15	0	6	31
(5)	0	4	0	0	7	5	16
(6)	1	3	0	1	1	62	68
Total	136	458	113	18	18	392	1135

Table 11.24: **Religious Mobility**: Cross-classification of origin religion by current religion

where: (1) Catholic, (2) Anglican, (3) Mainline Protestant, (4) Fundamentalist Protestant, (5) Other protestant, (6) None.

10. Caussinus (1965) presented the data in the table below, which relate to a cross-classification of individuals by their social and professional status in 1954 and for the same individuals in 1962.

	Status in 1962					
1954	1	2	3	4	5	6
1	187	13	17	11	3	1
2	4	191	4	9	22	1
3	22	8	182	20	14	3
4	6	6	10	323	7	4
5	1	3	4	2	126	17
6	0	2	2	5	1	153

(a) Fit the symmetry model to the data. Does this model fit the data?

(b) Fit both the QS and QI models to the data. Test for marginal homogeneity.

(c) Fit nonindependence models to the data. Which is the most parsimonious model?

Chapter 12

Analysis of Repeated Measures Data

12.1 Introduction

Data arising from repeated measures either by design or by epidemiologic (observational) designs often occur when observations on subjects or objects are taken over several times or occasions. Such data are often described as longitudinal data.The theory on the analysis of repeated measures data is not new, especially when the response variable if of the continous type. Classical multivariate methodology exists for this analysis. In recent years, because of the inherent correlation between the outcome observations over time (it is sometimes believed that observations closer together are often more correlated than those farther apart) several methods of analysis have been developed. The mixed effects models for example, have been developed in recent years to take care of this problem, and the SAS® book SAS® System for Mixed Models by Littel et al., (1996) provides excellent examples and various correlation structures for analyzing longitudinal data when the outcome variable is continuous.

Given that we are concerned in this chapter with the situation in which the outcome variable is categorical, that is, longitudinal data with categorical outcome variable, the logistic regression for a binary outcome for example, assumes that observations are independent across time. But it is not uncommon in longitudinal data to at least imagine, for example, if observations were taken at say, time 1, time 2 up to time 6 then observations at times 1 and 2 are more likely to have higher correlations than say, at times 1 and 6.

Recent advances in the methodology of analyzing repeated measured data with categorical outcome variable (Liang & Zeger, 1986) have made it possible to actually model the covariance structure of the repeated observations. We shall examine this with examples in a later section. We will, however, begin our discussion in this chapter with the following binary correlated data.

12.2 Logit Models for Repeated Binary Response

Consider the case when the response variable is binary, that is, $I = 2$ categories, and let the $k = 1, 2, \cdots, p$ denote the relevant populations. If we assume that there are t repeated occassions, then the logit of the response can be modeled for the j-th occassion as:

$$L(j;k) = \ln\left[\psi_1(j;k)/\psi_2(j;k)\right]$$

where $\psi_i(j,k)$ represents the i-th response probability at the j-th occassion in population k. The above can be succintly written in the form:

$$\begin{aligned} L(j;k) &= \alpha + \beta_k^P + \beta_j^T + \beta_3^{P*T} \quad \text{that is,} \\ \text{logit}_{jk} &= \boldsymbol{\beta}\mathbf{X} \end{aligned} \tag{12.1}$$

where P refers to the population, T to the occassion and $P * T$ to the interaction between the population and occassion (*if it exists*). The usual constraints are needed for proper identifiability of the parameters.

12.2.1 Example 12.1

The table below is from Woolson and Clarke (1984) and relates to the longitudinal studies of coronary risk factors in schoolchildren. A sample of 11 to 13 year-old children in 1977 were classified by gender and by relative weight (obese, not obese) in 1977, 1979, and 1981. .

	Responses								
Gender	NNN	NNO	NON	NOO	ONN	ONO	OON	OOO	Total
Male	119	7	8	3	13	4	11	16	181
Female	129	8	7	9	6	2	7	14	182

Table 12.1: Classification of children by gender and relative weight

Here NNN indicates *not obese* in 1977, 1979 and 1981, and NON similarly indicates *not obese* in 1977, *obese* in 1979, and *not obese* in 1981. Similar interpretations can be obtained for the remaining six outcomes. Here, there are two subpopulations and the response variable obese, not obese is observed at $T = 3$ occasions (1977, 1979, and 1981). We thus have a 2×2^3 contingency table. We intend to model the nonobsese response category.

If we let 1977, 1979, and 1981 occasions be represented by $j = 1, 2,$ and 3 respectively and $k = 1$ for male and $k = 2$ for female similarly, then the logit can be appropriately represented by $L(j,k)$. Since the occasions are ordered, we may employ the linear occasion effects $\{\beta_l^T = \beta\nu_l\}$, for fixed scores $\{\nu_l\}$. There are 6 marginal distributions, since there are two gender levels and 3 occasions. The following model is first employed for our analysis:

$$L(j,k) = \alpha + \beta_k^G + \beta\nu_j + \gamma_k\nu_l \tag{12.2}$$

If the marginal distribution for 1997 response (occasion 1) is identical for males and female respondents, then the method of Lipsitz (1988) allows us to fit this *population-averaged* or *marginal* model. In this case, models are specified in terms of marginal parameters such as those describing the mean response at time 1, time 2, or time 3. Alternatively, we can fit the cumulative logit model to the data. The results of both considerations are displayed respectively below. The models assume that interaction is present that leads to a saturated model. Let us first consider the marginal model to the data in Table 12.1. We give below the relevant SAS software instructions (together with the corresponding output) for implementing the model in (12.2) where year1, year2, and year3 refer respectively to 1977, 1979, and 1981.

```
data binary;
input gender$ year1$ year2$ year3$ count @@;
datalines;
m n n n 119 m n n o 7 ... f o o n 7 f o o o 14
;
proc catmod order=data; weight count;
response marginals;   /* use marginal homogeneity model*/
model year1*year2*year3=gender|_response_; repeated year; run;
```

```
                    Population Profiles

               Sample    gender    Sample Size
               ------------------------------
                  1        m            181
                  2        f            182

                    Response Profiles

            Response    year1    year2    year3
            -----------------------------------
                1         n        n        n
                2         n        n        o
                3         n        o        n
                4         n        o        o
                5         o        n        n
                6         o        n        o
                7         o        o        n
                8         o        o        o

         Function    Response                  Design Matrix
Sample    Number     Function      1       2       3       4       5       6
----------------------------------------------------------------------------
   1         1        0.75691      1       1       1       0       1       0
             2        0.79006      1       1       0       1       0       1
             3        0.83425      1       1      -1      -1      -1      -1

   2         1        0.84066      1      -1       1       0      -1       0
             2        0.79670      1      -1       0       1       0      -1
             3        0.81868      1      -1      -1      -1       1       1

                       Analysis of Variance
          Source                  DF    Chi-Square    Pr > ChiSq
          ------------------------------------------------------
          Intercept                1       2267.75        <.0001
          gender                   1          0.54        0.4613
          year                     2          2.87        0.2382
          gender*year              2          5.93        0.0516
          Residual                 0           .             .

             Analysis of Weighted Least Squares Estimates

                                            Standard        Chi-
Effect               Parameter   Estimate      Error      Square    Pr > ChiSq
------------------------------------------------------------------------------
Intercept                1         0.8062     0.0169     2267.75        <.0001
gender                   2        -0.0125     0.0169        0.54        0.4613
year                     3       -0.00743     0.0121        0.37        0.5410
                         4        -0.0128     0.0111        1.33        0.2496
gender*year              5        -0.0294     0.0121        5.86        0.0155
                         6        0.00915     0.0111        0.67        0.4117
```

The above response functions are obtained from Table 12.2, which are generated in SAS software by PROC FREQ.

Sample	Response function			Total
	year1	year2	year3	
1 (Men)	137	143	151	181
2 (Females)	153	145	149	182

Table 12.2: Frequencies of all no obese

where, for instance, 0.75691 is computed from 137/181 and so on. The model has six parameters representing respectively, the overall effect (1), a gender effect (2), two years effects (3, 4) and the two gender∗year interaction effects (5, 6). The numbers in the parentheses represent corrresponding columns in the design matrix columns as well as parameters in the analysis of WLS estimates in the output.

Similarly, we have for the cumulative logit fit the following select SAS software program and corresponding output.

```
set binary;
proc catmod order=data; weight count;
response logits;  /* fits cummulative logit model */
model year1*year2*year3= gender|_response_; repeated year;
run;
```

```
          Function      Response                        Design Matrix
Sample     Number       Function         1         2         3         4         5         6
-------------------------------------------------------------------------------------------
     1          1        1.13579         1         1         1         0         1         0
                2        1.32526         1         1         0         1         0         1
                3        1.61608         1         1        -1        -1        -1        -1

     2          1        1.66314         1        -1         1         0        -1         0
                2        1.36582         1        -1         0         1         0        -1
                3        1.50744         1        -1        -1        -1         1         1

                             Analysis of Variance

                Source                 DF   Chi-Square    Pr > ChiSq
                --------------------------------------------------
                Intercept               1       172.28        <.0001
                gender                  1         0.49        0.4840
                year                    2         2.71        0.2577
                gender*year             2         5.70        0.0580
                Residual                0          .             .

                  Analysis of Weighted Least Squares Estimates

                                                  Standard          Chi-
Effect                   Parameter   Estimate      Error          Square    Pr > ChiSq
-------------------------------------------------------------------------------------
Intercept                        1     1.4356      0.1094         172.28        <.0001
gender                           2    -0.0765      0.1094           0.49        0.4840
year                             3    -0.0361      0.0785           0.21        0.6453
                                 4    -0.0901      0.0698           1.66        0.1971
gender*year                      5    -0.1871      0.0785           5.68        0.0171
                                 6     0.0563      0.0698           0.65        0.4203
```

The results from both analyses are very similar. While the overall effects each of gender, year, and gender∗year terms are not significant based on the pvalues from the Wald test, examination of the parameter estimates indicates that for 1997 there is a significant interaction presence between gender and year with pvalues of 0.0155 and 0.0171, respectively, from both models. We can therefore fit unsaturated model to this data set, bearing in mind that we must include the effects of the first intercept, gender (male) effect and linear effect of year and the first interaction term. In other words, we need the design vectors of parameters 1, 2, 3, and the product of

2 and 3 in our model statement. These correspond respectively to design vectors 1, 2, 3, and 5 respectively. The reduced model is implemented in SAS software with the following statements and a partial output.

```
set binary;
proc catmod order=data; weight count; population gender; response logits;
model year1*year2*year3=(1  1  1  1,
                         1  1  0  0,
                         1  1 -1 -1,
                         1 -1  1 -1,
                         1 -1  0  0,
                         1 -1 -1  1)
                        (1='Intercept',
                         2='Gender Male',
                         3='Year 1977',
                         4= 'Gender Year 1')/ pred freq;  run;
```

Analysis of Variance

Source	DF	Chi-Square	Pr > ChiSq
Intercept	1	172.04	<.0001
Gender Male	1	0.47	0.4918
Year 1977	1	1.29	0.2562
Gender Year 1	1	4.43	0.0353
Residual	2	2.50	0.2861

Analysis of Weighted Least Squares Estimates

Effect	Parameter	Estimate	Standard Error	Chi-Square	Pr > ChiSq
Model	1	1.4345	0.1094	172.04	<.0001
	2	-0.0752	0.1094	0.47	0.4918
	3	-0.0820	0.0722	1.29	0.2562
	4	-0.1520	0.0722	4.43	0.0353

The weighted least squares analysis from the SAS software output gives $W_Q = 2.50$ on 2 d.f. The model fits the data with a pvalue of 0.2861. The parameter estimates are also displayed. Since the interaction term is significant (pvalue = 0.0353), for males therefore the odds are $e^{2*(-0.0752)} = 0.86$, indicating that they are 14% less likely to be classified as nonobese than being obese in 1977 among boys. Put another way, the odds are $1/0.86 = 1.16$, that is, 16% times higher to be classified as obese than nonobese among boys in 1977. Among girls (females), the odds is $e^{2*(-0.0752-0.1520)} = 0.64$, indicating that females are 36% less likely to be classified as nonobese than being obese or are ($1/.64 = 1.58$), 58% more likely to be classified as obese in 1977, a slightly higher odds than the boys. For the year 1977 the response to not being obese has the odds $e^{(-0.082)} = 0.92$ against being obese, that is, in 1977 girls are 9% more likely to be obese than not being obese. For the boys, we have the odds to be $e^{(-.0820-2*(-.1520))} = 1.25$, that is, boys are 25% more likely to be classified not being obese than being obese 1n 1977.

12.2.2 Alternative Analysis

An alternative analysis of the data in example 12.1 is to fit a Rasch model to this data set. The Rasch model, originally developed by Rasch (1961), is a logistic item response model that describes subject i's response to item k. If this response is denoted by y_{ik}, then the Rasch model is formulated as:

$$P(y_{ik}|k_i) = \frac{\exp(\alpha_i + \boldsymbol{x}'_{ik}\boldsymbol{\beta})}{1 + \exp(\alpha_i + \boldsymbol{x}'_{ik}\boldsymbol{\beta})} \tag{12.3}$$

Andersen (1970, 1973) proposed a conditional argument for estimating the parameter $\boldsymbol{\beta}$ and Tjur (1982) showed that the conditional model can be fitted as a log-linear model. To implement this, each margin R_k of each item will be fitted together with a factor variable giving the total score of successes (see Lindsey, 1995). In order to implement this model, we define binary variables R1, R2, R3 for years 1977, 1979, and 1981 respectively, which takes the value 1 if respondent replies not obsese (that is, N) and 0 otherwise. That is,

$$R_i = \begin{cases} 1 & \text{if not obese (N)} \\ 0 & \text{if obese (O)} \end{cases}$$

For the three years, we next count how many 1's (maximum of 3) are recorded for each year. We can implement this in SAS software with the following:

```
set binary;
if year1 eq 'n' then r1=1;else r1=0;
if year2 eq 'n' then r2=1;else r2=0;
if year3 eq 'n' then r3=1;else r3=0;
total=r1+r2+r3;
datalines;
m n n n 119 1 1 1 3 ... f o o o 14 0 0 0 0
;
run;
```

gender	year1	year2	year3	r1	r2	r3	total	count
m	n	n	n	1	1	1	3	119
m	n	n	o	1	1	0	2	7
m	n	o	n	1	0	1	2	8
m	n	o	o	1	0	0	1	3
m	o	n	n	0	1	1	2	13
m	o	n	o	0	1	0	1	4
m	o	o	n	0	0	1	1	11
m	o	o	o	0	0	0	0	16
f	n	n	n	1	1	1	3	129
f	n	n	o	1	1	0	2	8
f	n	o	n	1	0	1	2	7
f	n	o	o	1	0	0	1	9
f	o	n	n	0	1	1	2	6
f	o	n	o	0	1	0	1	2
f	o	o	n	0	0	1	1	7
f	o	o	o	0	0	0	0	14

In the above table, we see that for the (NNN) combination $(R1, R2, R3) = (1, 1, 1)$ and yields a total of 3 positive (1's) responses. The Rasch model now can be implemented in GENMOD with binary explanatory variables R1, R2, R3, factor variable TOTAL with 4 levels and any other covariates (in this case gender). The following models are employed with their appropriate interpretations.

(i) {R1, R2, R3, TOTAL, GENDER}: Model 1

This model implies that the responses are the same for each gender. The model gives a deviance of 11.9916 on 11 d.f. This model fits, indicating that the no obesity classification of the school children is uniform across gender.

(ii) { (R1, R2, R3, TOTAL)* GENDER }: Model 2

This model has the dependence of gender incorporated into the model. The model gives a deviance of 4.3312 on 4 degrees of freedom. This model also fits well but is too structured. We relax some of these restrictions in the next two models.

(iii) {(R1, R2, R3)* GENDER, TOTAL}: Model 3

This model tries to answer that the "no obese" classification is the same for the two sexes. This hypothesis tests if the total number of "N" varies from males and females. Notice the absence of the gender∗total interaction term in this model. Again this model, gives a deviance of 5.1658 on 6 d.f. This model fits indicating that the classificatory variable does not vary among the sexes.

(iv) {R1, R2, R3, TOTAL*GENDER}: Model 4

This last model, in the language of item response methodology, tests if the item characteristic curve is the same across gender. Notice again the absence of the interaction terms between gender and the item responses. This model gives a deviance of 10.4542 on 6 d.f. The model fits but is not as good as model 3. The responses seem to have the same frequency of occurrence of "no obesity" for the two sexes.

(v) {TOTAL, GENDER, TOTAL*GENDER}: Model 5

Model 5 examines if the probability of "no obese" responses are the same. This model has a deviance value of 13.4024 on 8 d.f. Again this model fits but we consider model 3, the most parsimonious.

The above models are implemented in GENMOD with the following model statements (put together for brevity only).

```
proc genmod;
class gender total;
model count=r1 r2 r3 total gender /dist=poi type3; run;/* fits model 1 */
model count=r1|gender r2|gender r3|gender total|gender/dist=p; run;/* fits model 2 */
model count=r1|gender r2|gender r3|gender total/dist=poi type3; run; /* fits model 3 */
model count=r1 r2 r3 total|gender /dist=poi type3; run;/* fits model 4 */
model count=total|gender /dist=poi type3;run; /* fits model 5 */
```

Our chosen model here is model 3, and we display below partial output of the parameter estimates under this model from GENMOD. Once again, the effect of gender is not significant ($p = 0.6030$). While the total number of "no obsese" recorded are significantly different (with positive NO's being the largest frequency), the estimates below (based on the magnitudes of $r1, r2, r3$) indicate that 1977 recorded the expected classification "no obese" most often from the model, followed by 1981 and lastly 1979. Only the 1977 gender interaction $r1$∗gender is significant ($p = 0.0121$), indicating that the distribution of no responses is more frequent in 1977 among girls than boys since the parameter estimate is negative.

Analysis Of Parameter Estimates

Parameter		DF	Estimate	Standard Error	Wald 95% Confidence Limits		Chi-Square	Pr > ChiSq
Intercept		1	2.6303	0.2394	2.1611	3.0995	120.73	<.0001
r1		1	-0.6345	0.3349	-1.2909	0.0219	3.59	0.0582
gender	m	1	0.1499	0.2881	-0.4148	0.7146	0.27	0.6030
r1*gender	m	1	-0.8367	0.3335	-1.4904	-0.1830	6.29	0.0121
r2		1	-1.2353	0.3480	-1.9173	-0.5532	12.60	0.0004
r2*gender	m	1	0.1711	0.3223	-0.4607	0.8028	0.28	0.5956
r3		1	-0.9426	0.3428	-1.6144	-0.2707	7.56	0.0060
r3*gender	m	1	0.4584	0.3425	-0.2128	1.1297	1.79	0.1807
total	3	1	5.0306	0.6334	3.7890	6.2721	63.07	<.0001
total	2	1	1.2843	0.4084	0.4838	2.0848	9.89	0.0017

The results of the analysis described above are displayed in Table 12.3.

Model	d.f.	G^2	pvalue
I	9	11.9916	0.2138
II	4	4.3312	0.3630
III	6	5.1658	0.5227
IV	6	10.4542	0.1068
V	8	13.4024	0.0987

Table 12.3: Results of analysis

12.3 Generalized Estimating Equations

For longitudinal data, observations of subjects at multiple points in time are usually correlated. Basically, the generalized estimating equations (GEE) comprise the following:

(a) n subjects, indexed by $i = 1, 2, \cdots, n$, and all observations arising from the multiple measurements or observations on a single subject are usually referred to as a cluster. Such multiple measurements on the same subject, whether under different conditions or over time, introduces within-subject correlation. Generalized estimating equations (GEE) procedure allows the within-subject correlation to be worked into the parameter estimation procedure. We would therefore expect n clusters from n subjects. The GEE analysis is based on first collapsing across subjects, and then model the marginal parameters. This approach yields the population-averaged regression parameters.

(b) T_i measurements over time on the ith subject ($t = 1, 2, \cdots, T,\ i = 1, 2, \cdots, n$).

(c) y_{it} is the observed response on subject i at time t (or measurement t), where, for a binary response outcome,

$$y_{it} = \begin{cases} 1 & \text{if subject has response} \\ 0 & \text{otherwise} \end{cases}$$

(d) $\mathbf{x}_{it}$ is the set of p covariates measured on subject i at time t, where the p covariates are indexed by $k = 1, 2, \cdots, p$. Both time-varying (within-subject) and time-stationary (between-subject) covariates could be included here.

A typical data structure for longitudinal data is presented in Table 12.4 where each of $\mathbf{x}_{ij}$ and $\mathbf{y}_i$ are t-dimensional.

Subjects	Covariates					Response
1	$\mathbf{x}_{11}$	$\mathbf{x}_{12}$	$\cdots$	$\mathbf{x}_{1,p-1}$	$\mathbf{x}_{1p}$	$\mathbf{y}_1$
2	$\mathbf{x}_{21}$	$\mathbf{x}_{22}$	$\cdots$	$\mathbf{x}_{2,p-1}$	$\mathbf{x}_{2t}$	$\mathbf{y}_2$
$\vdots$	$\vdots$	$\vdots$	$\vdots$	$\vdots$	$\vdots$	$\vdots$
n	$\mathbf{x}_{n1}$	$\mathbf{x}_{n2}$	$\cdots$	$\mathbf{x}_{n,p-1}$	$\mathbf{x}_{nt}$	$\mathbf{y}_n$

Table 12.4: Typical data structure

The GEE approach collapses first across subjects, and then models the marginal parameters, yielding population-averaged regression parameters. A natural approach to the analysis of such data is to model the *marginal* distributions of the observed response at each measurement times as a function of the covariates. Such a model allows us to look forward to future applications of results from it. The marginal models are sometimes referred to as *cross-sectional models*.

The marginal density of y_{it}, which is Bernoulli, is given by

$$f(y_{it} \mid \mathbf{x}_{it}) = \pi_{it}^{y_{it}}(1-\pi_{it})^{(1-y_{it})} \tag{12.4}$$

where, following Williamson et al., (1999), we shall assume that

$$\pi_{it} = \pi_{it}(\boldsymbol{\beta}) = \text{pr}(y_{it}=1|\mathbf{x}_{it},\boldsymbol{\beta}) = \left(\frac{e^{\boldsymbol{\beta}_0+\boldsymbol{\beta}_1'\mathbf{x}_{it}}}{1+e^{\boldsymbol{\beta}_0+\boldsymbol{\beta}_1'\mathbf{x}_{it}}}\right) \tag{12.5}$$

Although we have used a link function in (12.5) above, in general if we assume a GLM model for y_{it}, then,

$$E(y_{it}) = \mu_{it} \qquad g(\mu_{it}) = \mathbf{x}_{it}'\boldsymbol{\beta} \tag{12.6}$$

where g is the link function. Liang and Zeger (1986) show that consistent estimates of the marginal model parameters can be obtained from the solutions of the estimating equations by treating the correlation parameters as nuisance parameters. This procedure models the within-subject correlation structure, which in turn increases the efficiency of the estimators of the βs.

$$\sum_{i=1}^{n}\mathbf{D}_i'\mathbf{V}_i^{-1}(\mathbf{Y}_i-\boldsymbol{\mu}_i) = \sum_{i=1}^{n}\mathbf{D}_i'\mathbf{A}_i^{1/2}\mathbf{R}_i(\boldsymbol{\alpha})\mathbf{A}_i^{1/2}(\mathbf{Y}_i-\boldsymbol{\mu}_i) = 0 \tag{12.7}$$

where

(i) $\mathbf{D}_i'$ is a $T_i \times p$ matrix whose $(t,k)^{th}$ component is $\partial\mu_{it}/\partial\beta_k$.

(ii) $\mathbf{A}_i$ is a $T_i \times T_i$ symmetric positive definite diagonal matrix representing the variance of $\mathbf{Y}_i$ which are given by $\text{Var}(Y_{it}|\mathbf{x}_{it}) = \pi_{it}(1-\pi_{it})$.

(iii) $\mathbf{R}_i(\boldsymbol{\alpha})$ is a $T_i \times T_i$ *working* correlation matrix which depends on the $p \times 1$ vector of unknown parameters and $\mathbf{V}_i$ is the corresponding $T_i \times T_i$ *working* covariance matrix.

(iv) $\boldsymbol{\alpha} = (\alpha_1, \alpha_2, \cdots, \alpha_p)$ are parameters describing the within-subject correlation.

(v) $(\mathbf{Y}_i - \boldsymbol{\mu}_i)$ is a $T_i \times 1$ residual vector, measuring deviations of observed responses of the ith subject from its mean.

The maximum likelihood estimate, $\hat{\beta}$, is the solution to the estimating equations using iterative procedure usually the Gauss-Newton procedure. If $\mathbf{R}_i$ is set to be the identity matrix, then the estimating equations become those that would apply if the measurements were independent. In this case, all the observations both within and among subjects are assumed to be independent, and the GEE in (12.7) reduces to

$$\sum_{i=1}^{n}\mathbf{D}_i'\mathbf{V}_i^{-1}(\mathbf{Y}_i-\boldsymbol{\mu}_i) = 0 \tag{12.8}$$

We notice that the approach requires the specification of the first and second moments of the vector of correlated binary responses for each individual. Prentice (1988) has also adopted this procedure. Before we consider our first example here, we may note the following concerning the GEE.

- When employing the GEE method, a working correlation must be employed or one must be chosen to validate the marginal expectation (Pepe & Anderson, 1994). The choice of the working correlation structure (since this is often not known) is often left to the analyst. We shall discuss possible choices of the working correlation structures that are available to the analyst in the next section.

- The estimator of the covariance matrix of $\boldsymbol{\beta}$ is robust. That is, the GEE method has the property of being a consistent estimator of the covariance matrix of the estimators of $\boldsymbol{\beta}$, even if the working correlation matrix is misspecified (Chang, 2000).

- Diagnostics tests should be carried out to ascertain if the final model from the GEE fits the data as accurately as possible. In this case, it has been found that the usual residual plots could be misleading and a non-parametric method "the Wald-Wolfowitz" run test (Chang, 2000) would be very appropriate in this case.

It is also important that we define the concept of time-varying and time-stationary covariates,

- A time-stationary covariate is a between-subject variate that would be repeated in each of the T_i measurements for the i-th subject. An example here would be the variable gender.

- A time-varying is a within subject variate that assumes different values for each of the T_i measurements on the i-th subject. Examples are income which varies over time and age.

12.4 Example 12.2: Mother Rat Data

The following example by courtesy of Stuart Lipsitz relate to the effect of some possibly toxic substance on the offsprings of $n = 46$ pregnant rats who are given different doses of a toxic substance. The outcome for a baby (offspring) in the litter is whether he/she has a birth defect. The data also contains information about the weight of the offspring (lighter babies tend to have more birth defects, since one possible defect is a missing limb) and sex of the offspring. We wish to model therefore the probability of a defect as a function of dose, sex, and weight. Since births are in litters (offsprings from the same parents), we can therefore consider data arising from this study to be clustered (litters).

Let the response variable be Y_{ij} from dose i on the j-th subject be defined as:

$$Y_{ij} = \begin{cases} 1 & \text{if defect is found} \\ 0 & \text{otherwise} \end{cases}$$

Suppose the probability (in the j-th cluster) of a birth defect depends on the dose and sex, weight, that is,

$$\begin{aligned} \pi_{ij} &= \text{pr}[\text{Y}_{ij} = 1|\text{dose}_i, \text{sex}_{ij}, \text{weight}_{ij}] \\ &= \frac{\exp\left(\beta_0 + \beta_1 \text{dose}_i + \beta_2 \,\text{sex}_{ij} + \beta_3 \,\text{weight}_{ij}\right)}{1 + \exp\left(\beta_0 + \beta_1 \,\text{dose}_i + \beta_2 \,\text{sex}_{ij} + \beta_3 \,\text{weight}_{ij}\right)} \end{aligned} \tag{12.9}$$

where dose_i is a cluster-level covariate since it is the same for all offspring in the litter (we only need the subscript i).

Also,

$$\text{sex}_{ij} = \begin{cases} 1 & \text{if male} \\ 0 & \text{if female} \end{cases}$$

is a within-cluster covariate since it can be different for different offspring in the litter, and weight is similarly a within-cluster covariate.

The model in (12.9) can be written as:

$$\log\left(\frac{\pi_{ij}}{1-\pi_{ij}}\right) = \boldsymbol{\beta}\,\mathbf{X}, \quad \text{that is,} \tag{12.10}$$

$$\text{logit}_{ij} = \beta_0 + \beta_1\,\text{dose}_i + \beta_2\,\text{sex}_{ij} + \beta_3\,\text{weight}_{ij}$$

The GEE procedure uses iterative *generalized* least squares with the weight matrix, **W** with nonzero off-diagonal elements that are functions of the correlations among the observations. Using the correlations among the Pearson's based residuals, the matrix of correlation **W** is reestimated at each iteration, until converge is attained. For our data, we are interested in estimating $\boldsymbol{\beta} = [\beta_0, \beta_1, \beta_2, \beta_3]$. We present a sample of the mother rat data below for the first 2 of the 46 clusters. The entire data is presented in appendix G.1.

```
CLUSTER  DOSE    WEIGHT   SEX    DEFECT (DEFECT: 1=Yes, 0=No)
  49    0.000    0.989    F       0
  49    0.000    0.898    M       0
  49    0.000    0.945    M       0
  49    0.000    0.899    M       0
  49    0.000    0.933    F       0
  49    0.000    0.842    F       0
  49    0.000    0.896    F       0
  49    0.000    1.006    M       0
  49    0.000    1.115    M       0
  49    0.000    1.007    F       0
  49    0.000    0.958    F       0
  49    0.000    0.999    M       0
  49    0.000    0.909    F       0
  49    0.000    0.848    F       0
  49    0.000    0.999    F       0
  53    0.100    0.751    F       0
  53    0.100    0.902    F       0
  53    0.100    0.875    F       0
  53    0.100    0.964    M       0
  53    0.100    0.973    M       0
  53    0.100    0.965    M       0
  53    0.100    0.925    M       0
  53    0.100    0.936    M       0
  53    0.100    1.012    M       0
  53    0.100    0.858    M       0
  53    0.100    0.816    F       0
  53    0.100    1.007    M       0
 ..................................
```

12.4.1 Correlation Structure

GEE uses several correlation structures to model the correlation matrix, among the observations within each cluster. We shall first examine two of these structures for now and extend the results to other structures later in the chapter.

1. If we assume the *compound symmetry* correlation structure, which SAS software calls **EXCH** for *exchangeable*, the structure assumes equal correlations within subjects at all time points in the model. That is,

$$\text{Corr}(Y_{ij}, Y_{ik}) = \rho, \quad \Longrightarrow \quad \text{Corr}(Y_{ij}, Y_{ik}) = \begin{bmatrix} 1 & \rho & \rho & \rho \\ \rho & 1 & \rho & \rho \\ \rho & \rho & 1 & \rho \\ \rho & \rho & \rho & 1 \end{bmatrix}$$

2. Also, if we assume that observations in a cluster are independent (that is, all correlations are assumed zero), which is the usual assumption in logistic regression, then this will be equivalent to specifying that the correlation structure is **IND** *independent* in SAS software.

We implement the GEE in GENMOD with in the following sections for the data in our example above.

12.4.2 GEE with Exchangeable Structure

The following SAS software program is employed to fit the GEE with EXCH covariance structure.

```
data rat;
input cluster dose weight sex $ defect @@;
49 0.000 0.989 F 0 49 0.000 0.898 M 0 ...199 0.000 0.791 M 0 199 0.000 0.961 F 0
;
proc genmod data=rat descending;  class cluster sex ;  /* must use class for cluster id */
  model defect =  dose weight sex /
        link=logit dist=binomial; /* logistic regression */
                                   /* binomial distribution */
                                  /* binomial = bernoulli */
                                  /* when binomial sample */
                                  /* size = 1             */
  repeated subject=cluster / type=EXCH corrw ;
                                  /* subject = cluster id;*/
                                  /* type = correlation;  */
                                  /* corrw prints out     */
                                  /* correlation matrix   */
  run;
```

In the above, the GEE is invoked with the *repeated* statement. The *type=EXCH* asks for the exchangeable correlation structure, while *corrw* asks for the printing of the working correlation matrix. The **descending** specifies that the "defect = 1" be modeled. We present below a selected output from implementing the above program in SAS software.

```
                    GEE Model Information

      Correlation Structure                         Exchangeable
      Subject Effect                            cluster (46 levels)
      Number of Clusters                                      46
      Correlation Matrix Dimension                            16
      Maximum Cluster Size                                    16
      Minimum Cluster Size                                     2

                Analysis Of GEE Parameter Estimates
                 Empirical Standard Error Estimates

                                    95% Confidence
                          Standard        Limits
Parameter     Estimate      Error     Lower     Upper        Z Pr > |Z|
-----------------------------------------------------------------------
Intercept       3.2840     1.7096   -0.0667    6.6347     1.92   0.0547
dose           35.8357     6.0009   24.0742   47.5972     5.97   <.0001
weight         -7.4991     1.9348  -11.2912   -3.7071    -3.88   0.0001
sex        F   -0.4600     0.2811   -1.0109    0.0910    -1.64   0.1018
sex        M    0.0000     0.0000    0.0000    0.0000      .        .
```

In the selected output above, the GEE model information tells us that there are 46 clusters in the data, with each cluster ranging from 2 to 16 (that is, the number of litters). This implies that while some clusters have sixteen offspring, some have just two offspring. There would therefore be a 16×16 working correlation matrix. Since the correlation structure assumes equal correlations between pairs of observations within each cluster, we have therefore presented the first seven columns of row 1 for this matrix below. Essentially, $\hat{\rho} = 0.1384$.

```
                         Working Correlation Matrix

              COL1      COL2      COL3      COL4      COL5      COL6      COL7
  ROW1      1.0000    0.1384    0.1384    0.1384    0.1384    0.1384    0.1384
```

The estimate of the intracluster correlation coefficient using weighted (GEE) is 0.1384 (SAS software doesn't print out the pvalue, but a test would probably reject $H_0 : \rho = 0$). We observe here that the intracluster correlation is probably significant, and we would expect its estimated value to be smaller for a model without the covariates and should increase with a model involving one or more covariates that are essential for a better explanation of the variability within the clusters.

Naive Estimate Under Independence

A similar model employing the independence covariance structure (the usual logistic model under the scoring algorithm) is implemented with the following SAS software program with the corresponding selected output.

```
set rat;
proc genmod data=rat;  class cluster sex ;  model defect =  dose weight sex /
   link=logit dist=binomial; repeated subject=cluster / type=IND ; /* naive independ */
  run;
                        Analysis Of GEE Parameter Estimates
                        Empirical Standard Error Estimates

                              Empirical  95% Confidence Limits
Parameter    Estimate       Std Err          Lower        Upper          Z  Pr>|Z|
---------------------------------------------------------------------------------
INTERCEPT      2.5701        1.7347        -0.8299       5.9700   1.4816  0.1385
DOSE          34.3778        6.1975        22.2310      46.5246   5.5471  0.0000
WEIGHT        -6.7969        2.0081       -10.7327      -2.8610   -3.385  0.0007
SEX        F  -0.2840        0.3148        -0.9010       0.3330   -.9021  0.3670
SEX        M   0.0000        0.0000         0.0000       0.0000   0.0000  0.0000
Scale          0.9200             .              .            .        .       .
```

12.4.3 Comparing Estimates

The parameter estimates from the GEE with exchangeable correlation and independent structures are displayed below.

Parameter	EXCH Estimate	IND Estimate
INTERCEPT	3.2840	2.5701
DOSE	35.836*	34.378*
SEX F	-0.4600	-0.2840
WEIGHT	-7.499*	-6.797*

* implies significance at .05 using robust variance

The following observations are presented:

- The estimates from both methods are similar (since both are asymptotically unbiased).
- The biggest difference is in SEX effect, which is not significant using either estimate. We estimate that the odds of a birth defect increases by a factor of

$$\exp(-\hat{\beta}_2) \approx \exp(.46) = 1.58$$

  for male versus female offspring. The other two covariates appear to significantly predict birth defects.
- From this output, we see that male offspring and offspring whose mother had higher doses tend to have increased odds of birth defects, and offspring who weigh more have a decreased odds of a birth defect.
- We estimate that the odds of a birth defect increases by a factor of

$$\exp(\hat{\beta}_3 * -.1) \approx \exp(-7.5 * -.1) = 2.12$$

  for an offspring that weighs .1 kg less.
- Similarly, we estimate that the odds of a birth defect increases by a factor of

$$\exp(\hat{\beta}_1 * .15) \approx \exp(35.84 * .15) = 216.16$$

  for those whose mother has a high dose (.15) versus no dose (0).

All the odds above are computed using the parameters estimates from the exchangeable model.

12.4.4 Naive Standard Errors

By default, SAS® PROC GENMOD prints out the *robust* (or empirical) standard errors of the parameter estimates under GEE. We can therefore obtain printed values of the "naive" or sometimes called "model-based" standard errors by including the **modelse** option in the **repeated** statement. The model based-parameter parameter standard errors for the GEE under EXCH and IND correlation structures are respectively given below.

```
proc genmod data=rat;
class cluster sex ;
model defect =  dose weight sex /
  link=logit dist=binomial;  repeated subject=cluster / type=EXCH modelse;
run;
```

Analysis Of GEE Parameter Estimates
Model-Based Standard Error Estimates
(EXCH STRUCTURE)

Parameter		Estimate	Standard Error	95% Confidence Limits Lower	Upper	Z	Pr > |Z|
Intercept		3.2840	1.5677	0.2113	6.3567	2.09	0.0362
dose		35.8357	6.2331	23.6190	48.0524	5.75	<.0001
weight		-7.4991	1.7409	-10.9113	-4.0870	-4.31	<.0001
sex	F	-0.4600	0.3412	-1.1287	0.2087	-1.35	0.1776
sex	M	0.0000	0.0000	0.0000	0.0000	.	.
Scale		1.0000	.	.	.	.	.

```
              Model-Based Standard Error Estimates
                       (IND STRUCTURE)

                                       95% Confidence
                           Standard        Limits
Parameter     Estimate      Error     Lower      Upper          Z Pr > |Z|
---------------------------------------------------------------------------
Intercept       2.5701     1.4737   -0.3183     5.4584      1.74    0.0812
dose           34.3778     4.3212   25.9083    42.8473      7.96    <.0001
weight         -6.7969     1.5775   -9.8886    -3.7051     -4.31    <.0001
sex         F  -0.2840     0.3498   -0.9695     0.4016     -0.81    0.4169
sex         M   0.0000     0.0000    0.0000     0.0000       .        .
Scale           1.0000       .         .          .          .        .
```

We notice immediately that while the parameter estimates are identical to the earlier cases respectively, the standard errors are now different. The standard errors presented in the IND case now equals those that would normally be obtained from a logistic regression using reweighted least squares, that is, from the WALD approach.

12.4.5 Efficiency

We now examine the efficiency of the GEE versus the ordinary logistic regression (GEE independence). The table below gives the estimate efficiency for the independence GEE (ordinary logistic regression) versus the weighted GEE.

	VARIANCE		Efficiency (%)
PARAMETER	EXCH	IND	[Var(EXCH)/Var(IND)]
INTERCEP	2.9227	3.0092	97.13
DOSE	36.0108	38.4090	93.76
SEX	0.0790	0.0991	79.72
WEIGHT	3.7435	4.0325	92.83

- Now, we see that, for the cluster-level covariate (dose, and actually the intercept), ordinary logistic regression is almost as efficient as weighted logistic regression.
- Ordinary logistic regression appears to be pretty efficient for estimating the effects of weight, a within-cluster covariate.
- However, we see that it appears to be very inefficient (79.72%) for estimating the effect of sex, a within-cluster covariate.
- In general, ordinary logistic regression can be very efficient for estimating the effects of cluster-level covariates, but can be inefficient for the effects of within cluster covariates.

12.4.6 Hypothesis Involving Parameters

We can also use PROC GENMOD to make joint tests about the β's. For example, suppose we want to test that none of the covariates in the model below are important:

$$\text{logit}(\pi_{ij}) = \beta_0 + \beta_1 \,\text{dose}_i + \beta_2 \,\text{sex}_{ij} + \beta_3 \,\text{weight}_{ij}$$

That is, we wish to test the hypothesis:

$$H_0 : \beta_1 = 0, \quad \beta_2 = 0, \quad \beta_3 = 0$$

We can write this null as a contrast of the elements of the parameter vector $\boldsymbol{\beta} = [\beta_0, \beta_1, \beta_2, \beta_3]'$. In particular, we can use

$$H_0 : C\boldsymbol{\beta} = \mathbf{0}$$

or

$$H_0 : \begin{bmatrix} 0 & 1 & 0 & 0 \\ 0 & 0 & 1 & 0 \\ 0 & 0 & 0 & 1 \end{bmatrix} \begin{bmatrix} \beta_0 \\ \beta_1 \\ \beta_2 \\ \beta_3 \end{bmatrix} = \begin{bmatrix} 0 \\ 0 \\ 0 \end{bmatrix}$$

Since there is no likelihood involved with the estimating equations, the statistic calculated for a contrast is a Wald statistic (a multivariate generalization of the estimate divided by its standard error). The contrast statement is implemented in GENMOD for the EXCH model for instance by:

```
 set rat; if sex='F' then sexg=1; if sex='M' then sexg=0;
run;
proc genmod data=rat;
  class cluster sex;  model defect =  dose weight sexg /
    link=logit dist=binomial; CONTRAST 'no effects'
                         dose 1 weight 0  sexg 0,
                         dose 0 weight 1  sexg 0,
                         dose 0 weight 0  sexg 1  / wald;
CONTRAST 'ALTER' dose 1, weight 1, sexg 1/wald;
   repeated subject=cluster / type=EXCH ;  run;
```

```
                 Contrast Results for GEE Analysis

                               Chi-
Contrast             DF      Square    Pr > ChiSq    Type
---------------------------------------------------------
NO EFFECT             3       43.77        <.0001    Wald
ALTER                 3       43.77        <.0001    Wald
```

The contrast test gives a $Q_W = 43.77$ On 3 d.f. with $p =< 0.0001$. A similar result was obtained from the GEE with IND correlation structure. Here again, $Q_W = 37.34$ on 3 d.f. with $p =< 0.0001$. Both tests tell us that there are covariate effects. The hypothesis above can alternatively be constructed within SAS software as indicated with the "ALTER" contrast. Both contrast formulations lead to the same result as expected.

12.5 The Six Cities Longitudinal Study Data

Now, we apply the above methods to the Six Cities Study (Ware et al., 1984) on the health effects of pollution. We analyze below the data from only two of the cities. Children in Kingston, Harriman, and Portage were examined for wheezing at each of ages 7 through 10 years of age. The mothers' smoking habits were also recorded at the start of the study.

The response of interest at age t ($t = 7, 8, 9, 10$) is the wheeze status of the child, where

$$\text{Status} = \begin{cases} 0 & \text{if no wheeze} \\ 1 & \text{if wheeze} \end{cases}$$

The covariates are city_i, the child's city of residence (city_i equals 1 if the child lived in Kingston–Harriman, the more polluted city, and 0 if the child lived in Portage); smoke_{ij}, the maternal cigarette smoking at that age in packs per day and the child's age in years ($t_{i1} = 7, t_{i2} = 8, t_{i3} = 9, t_{i4} = 10$). The complete data are presented in appendix G.2.

Interest centers on whether age has an effect (because as the child gets older, we would expect he/she to get stronger physically, and consequently wheeze less). Since age has four levels and equidistance, we sought to find whether this effect is linear, quadratic, or cubic. We therefore created linear, quadratic, and cubic orthogonal polynomials in age, say $(\text{agel}_{ij}, \text{ageq}_{ij}, \text{agec}_{ij})$. The coefficients of these polynomials are displayed below. In particular, we have the following orthogonal polynomials,

	Age			
	7	8	9	10
agel	-3	-1	1	3
ageq	1	-1	-1	1
agec	-1	3	-3	1

Let the probability of wheeze be modeled as

$$\pi_{ij} = \text{pr}[\text{WHEEZE}_{ij} = \text{YES}|\mathbf{x}_{ij}]$$

Then, the model becomes

$$\log\left(\frac{\pi_{ij}}{1-\pi_{ij}}\right) = \beta_0 + \beta_1 \,\text{city}_i + \beta_2 \,\text{smoke}_{ij} + \beta_3 \,\text{agel}_{ij} + \beta_4 \,\text{ageq}_{ij} + \beta_5 \,\text{agec}_{ij} \tag{12.11}$$

or as

$$\text{logit}[\pi_{ij}] = \beta_0 + \beta_1 \,\text{city}_i + \beta_2 \,\text{smoke}_{ij} + \beta_3 \,\text{agel}_{ij} + \beta_4 \,\text{ageq}_{ij} + \beta_5 \,\text{agec}_{ij} \tag{12.12}$$

The model is implemented in SAS software with the following statements (with the necessary data transformation statements). The orthogonal components of the age effects are defined as *tl*, *tq*, and *tc* in the SAS software statements.

```
DATA ONE(KEEP=ID City TIME Smoke WHEEZE);
infile 'six.dat';
INPUT ID City S1 S2 S3 S4 WHEEZE1 WHEEZE2 WHEEZE3 WHEEZE4;
WHEEZE=WHEEZE1; TIME=1; Smoke=S1; OUTPUT;
WHEEZE=WHEEZE2; TIME=2; Smoke=S2; OUTPUT;
WHEEZE=WHEEZE3; TIME=3; Smoke=S3; OUTPUT;
WHEEZE=WHEEZE4; TIME=4; Smoke=S4; OUTPUT; RUN;
/* FORMING ORTHOGONAL POLYNOMIALS */
DATA TWO; SET ONE;
if time=1 then do; tl=-3; tq=1;  tc=-1; end;
if time=2 then do; tl=-1; tq=-1; tc=3;  end;
if time=3 then do; tl=1;  tq=-1; tc=-3; end;
if time=4 then do; tl=3;  tq=1;  tc=1;  end; run;
proc genmod data=two;
  class time id;
  model outcome = city smoke tl tq tc/dist=b link=logit type3;
  repeated subject=id/ type=EXCH corrw WITHINSUBJECT=time modelse;
                                /*WITHINSUBJECT is time variable*/;   run;
```

The data for this example are presented in appendix G.2; even though it has a lot of missing values, the sample size is still sufficient to estimate an unstructured correlation matrix, which is the most general.

Although we have specified the exchangeable correlation structure in the model statement, we now consider other correlation structures that have gained wide acceptance in the GEE theory.

1. **Autoregressive**

   If we assume that the correlation structure has a first-order autoregressive model, AR(1), then the structure is of the form:

   $$\rho_{ijk} = \rho^{|j-k|} \qquad 0 < \rho < 1$$

   Here, the correlation between two observations c times apart is ρ^{c-1}. That is, for $T = 4$, we have

   $$\text{Corr}(y_{ij}, y_{ik}) = \begin{bmatrix} 1 & \rho & \rho^2 & \rho^3 \\ \rho & 1 & \rho & \rho^2 \\ \rho^2 & \rho & 1 & \rho \\ \rho^3 & \rho^2 & \rho & 1 \end{bmatrix}$$

   That is, adjacent observations have higher correlations than noadjacent ones.

2. **MDEP (M)**

   The **m**-dependent structure with $m = 1, 2, \cdots$, has the correlation structure of the following form:

   $$\text{Corr}(Y_{ij}, Y_{i,j+\tau}) = \rho_\tau$$

   unless $\tau > M$, in which

   $$\text{Corr}(Y_{ij}, Y_{i,j+\tau}) = 0.$$

   The 1-dependent structure has for instance

   $$\text{Corr}(y_{ij}, y_{i,j+t}) = \begin{bmatrix} 1 & \rho & 0 & 0 \\ \rho & 1 & \rho & 0 \\ 0 & \rho & 1 & \rho \\ 0 & 0 & \rho & 1 \end{bmatrix}$$

   which indicates that correlations between adjacent observations are nonzero and equal.

   The corresponding 2-dependent structure has

   $$\text{Corr}(y_{ij}, y_{i,j+t}) = \begin{bmatrix} 1 & \rho_1 & \rho_2 & 0 \\ \rho_1 & 1 & \rho_1 & \rho_2 \\ \rho_2 & \rho_1 & 1 & \rho_1 \\ 0 & \rho_2 & \rho_1 & 1 \end{bmatrix}$$

   which again indicate that observations one time period and two time periods apart have respectively nonzero and equal correlations ρ_1 and ρ_2, respectively.

3. **Unstructured**

   Here the correlation matrix is unstructured and has the form

   $$\rho_{ijk} = \rho_{jk}$$

   which again indicates that all correlations are to be estimated independently from the data.

The above correlation structures are specified in the REPEATED statement option as type = **AR(1)**, type = **MDEP(2)**, and type = **UN**, respectively. The type = IND still refers to the situation where we are assuming that the observations are independently distributed. We present below the analysis of GEE parameter estimates with empirical standard error results when each of these correlation structures is invoked on the data in this example.

```
Correlation Structure: Independence

                     Standard   95% Confidence
Parameter Estimate     Error        Limits            Z Pr > |Z|
-----------------------------------------------------------------
Intercept   -1.6131    0.1595  -1.9256  -1.3005  -10.12   <.0001
City         0.5177    0.1916   0.1420   0.8933    2.70   0.0069
Smoke        0.0128    0.0076  -0.0021   0.0278    1.68   0.0929
tl          -0.0531    0.0250  -0.1021  -0.0041   -2.12   0.0338
tq           0.0072    0.0513  -0.0934   0.1078    0.14   0.8884
tc          -0.0227    0.0206  -0.0632   0.0177   -1.10   0.2708


Correlation Structure: Exchangeable

                     Standard   95% Confidence
Parameter Estimate     Error        Limits            Z Pr > |Z|
-----------------------------------------------------------------
Intercept   -1.6305    0.1535  -1.9313  -1.3298  -10.63   <.0001
City         0.5357    0.1891   0.1650   0.9064    2.83   0.0046
Smoke        0.0124    0.0069  -0.0011   0.0259    1.80   0.0715
tl          -0.0492    0.0247  -0.0977  -0.0007   -1.99   0.0470
tq          -0.0033    0.0503  -0.1019   0.0954   -0.06   0.9482
tc          -0.0220    0.0204  -0.0620   0.0180   -1.08   0.2816


Correlation Structure: AR(1)

                     Standard   95% Confidence
Parameter Estimate     Error        Limits            Z Pr > |Z|
-----------------------------------------------------------------
Intercept   -1.6406    0.1555  -1.9454  -1.3357  -10.55   <.0001
City         0.5478    0.1899   0.1755   0.9200    2.88   0.0039
Smoke        0.0132    0.0072  -0.0008   0.0273    1.85   0.0650
tl          -0.0477    0.0244  -0.0955   0.0001   -1.95   0.0507
tq           0.0076    0.0503  -0.0910   0.1062    0.15   0.8802
tc          -0.0254    0.0208  -0.0662   0.0153   -1.22   0.2213


Correlation Structure: MDEP(1)

                     Standard   95% Confidence
Parameter Estimate     Error        Limits            Z Pr > |Z|
-----------------------------------------------------------------
Intercept   -1.6238    0.1561  -1.9298  -1.3178  -10.40   <.0001
City         0.5334    0.1906   0.1599   0.9070    2.80   0.0051
Smoke        0.0127    0.0073  -0.0016   0.0270    1.74   0.0820
tl          -0.0429    0.0244  -0.0908   0.0049   -1.76   0.0787
tq           0.0231    0.0501  -0.0752   0.1213    0.46   0.6452
tc          -0.0274    0.0214  -0.0693   0.0145   -1.28   0.1997

Correlation Structure: MDEP(2)

                     Standard   95% Confidence
Parameter Estimate     Error        Limits            Z Pr > |Z|
-----------------------------------------------------------------
Intercept   -1.6600    0.1594  -1.9724  -1.3477  -10.42   <.0001
City         0.5742    0.1948   0.1924   0.9559    2.95   0.0032
Smoke        0.0136    0.0075  -0.0011   0.0283    1.82   0.0691
tl          -0.0512    0.0247  -0.0995  -0.0028   -2.07   0.0382
tq          -0.0041    0.0515  -0.1049   0.0968   -0.08   0.9371
tc          -0.0245    0.0207  -0.0650   0.0160   -1.19   0.2353
```

```
Correlation Structure: MDEP(3)

                    Standard    95% Confidence
Parameter Estimate    Error         Limits              Z Pr > |Z|
-------------------------------------------------------------------
Intercept  -1.6320   0.1534  -1.9327  -1.3313  -10.64   <.0001
City        0.5368   0.1889   0.1666   0.9069    2.84   0.0045
Smoke       0.0127   0.0069  -0.0008   0.0261    1.84   0.0662
tl         -0.0478   0.0245  -0.0959   0.0003   -1.95   0.0516
tq          0.0022   0.0501  -0.0960   0.1003    0.04   0.9656
tc         -0.0235   0.0206  -0.0638   0.0168   -1.14   0.2536

Correlation Structure: Unstructured

                    Standard    95% Confidence
Parameter Estimate    Error         Limits              Z Pr > |Z|
-------------------------------------------------------------------
Intercept  -1.6304   0.1534  -1.9311  -1.3298  -10.63   <.0001
City        0.5366   0.1888   0.1665   0.9068    2.84   0.0045
Smoke       0.0124   0.0069  -0.0011   0.0260    1.80   0.0711
tl         -0.0478   0.0245  -0.0959   0.0003   -1.95   0.0512
tq          0.0025   0.0500  -0.0955   0.1006    0.05   0.9597
tc         -0.0235   0.0206  -0.0639   0.0168   -1.14   0.2530
```

12.5.1 Comparing Parameter Estimates

If we look at the estimates of $\boldsymbol{\beta}$ using the different correlation models, we see that the estimates of $\boldsymbol{\beta}$ are similar using all of the correlation structures. This is to be expected since they are all (asympotically) unbiased. In particular, city is significant, and smoking is marginally significant. The linear age effect is also marginally significant.

Suppose we want to test for no age effect in the model defined in (12.11), that is, we wish to test

$$H_0 : \beta_3 = \beta_4 = \beta_5 = 0$$

we can construct this contrast as

$$H_0 : \begin{bmatrix} 0 & 0 & 0 & 1 & 0 & 0 \\ 0 & 0 & 0 & 0 & 1 & 0 \\ 0 & 0 & 0 & 0 & 0 & 1 \end{bmatrix} \begin{bmatrix} \beta_0 \\ \beta_1 \\ \beta_2 \\ \beta_3 \\ \beta_4 \\ \beta_5 \end{bmatrix} = \begin{bmatrix} 0 \\ 0 \\ 0 \end{bmatrix}$$

This contrast is not significant using any of the correlation models when implemented in SAS software. These results are presented in Table 12.5 for the various correlation structures considered.

CORR. MODEL	d.f.	Q_W	pvalue
IND	3	6.23	0.1008
EXCH	3	5.66	0.1292
AR(1)	3	5.90	0.1165
MDEP(1)	3	5.40	0.1447
MDEP(2)	3	6.31	0.0974
MDEP(3)	3	5.63	0.1310
UNSTR	3	5.65	0.1300

Table 12.5: Test for no age effects

The results in Table 12.5 indicate that irrespective of the correlation structures employed, we would fail to reject the null hypotheis. Thus indicating that there is no age effect. The test is implemented in SAS software by the contrast statement **'no time effects' tl 1, tq 1, tc 1/e wald**. This relevant SAS software statement is presented under the Estimated Relative Efficiency section.

Alternatively, we could have conducted the above test by looking at the log-likelihood. When the age effects (that is, the full model) are in the model for the exchangeable structure for instance, the log-likelihhod is −726.5196. When the effects are removed from the model (leading to a reduced model), the new log-likelihood is −728.4838. Hence, a test of hypotheses:

$$H_0 : \beta_2 = \beta_3 = \beta_4 = 0 \quad \text{versus} \quad H_A : \text{at least one} \quad \beta \neq 0$$

equals

$$\begin{aligned} -2 * (\text{log-likelihood}_{\text{reduced}} - \text{log-likelihood}_{\text{Full}}) &= -2\{-728.4838 + 726.5196\} \\ &= 3.9284 \end{aligned}$$

or we could have used the differences in the deviance values which are respectively for the reduced and full model equal 1456.9677 (on 1384 d.f.) and 1453.0392 (on 1381 d.f.). Once again, the test is based on

$$\begin{aligned} (\text{Deviance}_{Reduced} - \text{Deviance}_{Full})/\phi &= (1456.9677 - 1453.0392)/1.00 \\ &= 3.9285 \end{aligned}$$

Because the cubic component of the age effect is not significant from all the models, we therefore removed this effect from all subsequent models. We give below the estimated correlation matrices for the responses ages 7, 8, 9, and 10 for the different models.

12.5.2 Correlation Matrices

```
Exchangeable  Correlation Matrix

              Col1          Col2          Col3          Col4
Row1        1.0000        0.4289        0.4289        0.4289
Row2        0.4289        1.0000        0.4289        0.4289
Row3        0.4289        0.4289        1.0000        0.4289
Row4        0.4289        0.4289        0.4289        1.0000

AR(1)  Correlation Matrix

              Col1          Col2          Col3          Col4
Row1        1.0000        0.4814        0.2317        0.1115
Row2        0.4814        1.0000        0.4814        0.2317
Row3        0.2317        0.4814        1.0000        0.4814
Row4        0.1115        0.2317        0.4814        1.0000

MDEP(1)
              Col1          Col2          Col3          Col4
Row1        1.0000        0.4824        0.0000        0.0000
Row2        0.4824        1.0000        0.4824        0.0000
Row3        0.0000        0.4824        1.0000        0.4824
Row4        0.0000        0.0000        0.4824        1.0000

MDEP(2)
              Col1          Col2          Col3          Col4
Row1        1.0000        0.4807        0.3686        0.0000
Row2        0.4807        1.0000        0.4807        0.3686
Row3        0.3686        0.4807        1.0000        0.4807
Row4        0.0000        0.3686        0.4807        1.0000
```

```
MDEP(3)
              Col1          Col2          Col3          Col4
Row1        1.0000        0.4806        0.3681        0.4000
Row2        0.4806        1.0000        0.4806        0.3681
Row3        0.3681        0.4806        1.0000        0.4806
Row4        0.4000        0.3681        0.4806        1.0000

Unstructured Correlation Matrix

              Col1          Col2          Col3          Col4
Row1        1.0000        0.4713        0.3698        0.3998
Row2        0.4713        1.0000        0.4980        0.3738
Row3        0.3698        0.4980        1.0000        0.4906
Row4        0.3998        0.3738        0.4906        1.0000
```

If one looks at the unstructured correlation matrix, observations closest in time are the most highly correlated. Further, the correlation matrix from the unstructured model looks pretty similar to the MDEP(3) correlation matrix, meaning the MDEP(3) correlation matrix is probably a good fit. What is obvious from this analysis, and which is most important, is that the observations look far from independent, and, we would like to see how much efficiency we gain in estimating $\boldsymbol{\beta}$ when using more complex correlation structures over independence (ordinary logistic regression).

12.5.3 Interpretation of Parameter Estimates

Since $\hat{\beta}_1$ is approximately 0.5 using all of the correlation structures, the more polluted city (Kingston-Harriman) tends to increase the odds of wheezing by a factor of

$$\exp(\hat{\boldsymbol{\beta}}_1) \approx \exp(.5) = 1.65$$

(given the age and maternal smoking level).
Since $\hat{\beta}_2$ is approximately .013 using all of the correlation structures, an increase in two packs (40 cigarettes) per day smoked by the mother tends to increase the odds of the child wheezing by a factor of

$$\exp[\hat{\boldsymbol{\beta}}_2 \cdot 40)] \approx \exp[.013 \cdot 40] = 1.68$$

(given the age and city).

12.5.4 Confidence Intervals

A 95% confidence interval for the log-odds ratio can be calculated via,

$$\hat{\boldsymbol{\beta}}_\ell \pm 1.96\sqrt{\widehat{Var}(\hat{\boldsymbol{\beta}}_\ell)}$$

where $\widehat{Var}(\hat{\boldsymbol{\beta}}_\ell)$ comes from the output (and it is better to use the robust variance). Thus, a 95% C.I. for the odds ratio is calculated by exponentiating the endpoints of this confidence interval,

$$\exp\left[\hat{\boldsymbol{\beta}}_\ell \pm 1.96\sqrt{\widehat{Var}(\hat{\boldsymbol{\beta}}_\ell)}\right]$$

Suppose we use the exchangeable correlation model for confidence intervals (with the robust variance):

```
Correlation Structure: Exchangeable

PARAMETER ESTIMATES with robust variance

                    Empirical Standard Error Estimates

                          Standard   95% Confidence
          Parameter Estimate    Error        Limits              Z Pr > |Z|
          -------------------------------------------------------------
          Intercept  -1.6305   0.1535  -1.9313  -1.3298  -10.63   <.0001
          City        0.5357   0.1891   0.1650   0.9064    2.83   0.0046
          Smoke       0.0124   0.0069  -0.0011   0.0259    1.80   0.0715
          tl         -0.0492   0.0247  -0.0977  -0.0007   -1.99   0.0470
          tq         -0.0033   0.0503  -0.1019   0.0954   -0.06   0.9482
          tc         -0.0220   0.0204  -0.0620   0.0180   -1.08   0.2816
```

Then a 95% confidence interval for the city odds ratio is

$$\exp[0.536 \pm 1.96(0.189)] = [1.18, 2.48]$$

Similarly, the 95% confidence interval for an increase in two packs (40 cigarettes) per day smoked by the mother is

$$\begin{aligned}\exp\left\{40\left[\hat{\beta}_3 \pm 1.96\sqrt{\widehat{Var}(\hat{\beta}_3)}\right]\right\} &= \exp\{40[0.0124 \pm 1.96(0.00689)]\} \\ &= [0.96, 2.82]\end{aligned} \tag{12.13}$$

12.5.5 Estimated Relative efficiency

To get a rough idea of the asymptotic efficiency of the estimates, we can compare the robust variance estimators of $\hat{\beta}$ under two different correlation models. First, we give parameter estimates and the robust standard errors under the different models for the case excluding both the quadratic and cubic age effects (since the preceeding analysis indicate that both effects are not significant at $\alpha = 0.05$) under the IND, EXCH, AR(1), MDEP(3) and UNSTR correlation structures. Partial SAS software outputs for these implementation are presented below.

```
Independence:

proc genmod data=two ;
 class id time;
 model wheeze = city smoke tl tq/ link=logit dist=binomial;
 contrast 'no time effects' tl 1/ e wald;
  repeated subject=id  /type=IND corrw WITHINSUBJECT=time modelse;
                        /* WITHINSUBJECT is time variable */
run;


                     Analysis Of GEE Parameter Estimates
                      Empirical Standard Error Estimates

                          Standard   95% Confidence
          Parameter Estimate    Error        Limits              Z Pr > |Z|
          -------------------------------------------------------------
          Intercept  -1.6137   0.1596  -1.9266  -1.3008  -10.11   <.0001
          City        0.5181   0.1917   0.1425   0.8938    2.70   0.0069
          Smoke       0.0128   0.0076  -0.0021   0.0278    1.68   0.0920
          tl         -0.0530   0.0251  -0.1023  -0.0038   -2.11   0.0347
```

Analysis Of GEE Parameter Estimates
Model-Based Standard Error Estimates

Parameter	Estimate	Standard Error	95% Confidence Limits		Z	Pr > |Z|
Intercept	-1.6137	0.1076	-1.8245	-1.4029	-15.00	<.0001
City	0.5181	0.1318	0.2597	0.7765	3.93	<.0001
Smoke	0.0128	0.0057	0.0017	0.0240	2.25	0.0244
tl	-0.0530	0.0293	-0.1105	0.0044	-1.81	0.0702

Contrast Results for GEE Analysis

Contrast	DF	Chi-Square	Pr > ChiSq	Type
no linear effect	1	4.46	0.0347	Wald

Exchangeable:

Analysis Of GEE Parameter Estimates
Empirical Standard Error Estimates

Parameter	Estimate	Standard Error	95% Confidence Limits		Z	Pr > |Z|
Intercept	-1.6308	0.1537	-1.9321	-1.3295	-10.61	<.0001
City	0.5367	0.1891	0.1661	0.9073	2.84	0.0045
Smoke	0.0124	0.0069	-0.0011	0.0258	1.80	0.0724
tl	-0.0488	0.0248	-0.0974	-0.0003	-1.97	0.0485

Analysis Of GEE Parameter Estimates
Model-Based Standard Error Estimates

Parameter	Estimate	Standard Error	95% Confidence Limits		Z	Pr > |Z|
Intercept	-1.6308	0.1516	-1.9279	-1.3337	-10.76	<.0001
City	0.5367	0.1895	0.1652	0.9082	2.83	0.0046
Smoke	0.0124	0.0073	-0.0019	0.0266	1.70	0.0895
tl	-0.0488	0.0229	-0.0937	-0.0040	-2.13	0.0329

AR(1):

Analysis Of GEE Parameter Estimates
Empirical Standard Error Estimates

Parameter	Estimate	Standard Error	95% Confidence Limits		Z	Pr > |Z|
Intercept	-1.6399	0.1555	-1.9446	-1.3351	-10.55	<.0001
City	0.5496	0.1901	0.1771	0.9222	2.89	0.0038
Smoke	0.0132	0.0072	-0.0009	0.0272	1.84	0.0661
tl	-0.0526	0.0244	-0.1005	-0.0047	-2.15	0.0312

Analysis Of GEE Parameter Estimates
Model-Based Standard Error Estimates

Parameter	Estimate	Standard Error	95% Confidence Limits		Z	Pr > |Z|
Intercept	-1.6399	0.1439	-1.9219	-1.3578	-11.40	<.0001
City	0.5496	0.1784	0.2000	0.8993	3.08	0.0021
Smoke	0.0132	0.0070	-0.0005	0.0269	1.88	0.0595
tl	-0.0526	0.0291	-0.1096	0.0043	-1.81	0.0701

MDEP(3):

Analysis Of GEE Parameter Estimates
Empirical Standard Error Estimates

Parameter	Estimate	Standard Error	95% Confidence Limits		Z	Pr > |Z|
Intercept	-1.6328	0.1536	-1.9338	-1.3317	-10.63	<.0001
City	0.5383	0.1889	0.1681	0.9085	2.85	0.0044
Smoke	0.0126	0.0069	-0.0009	0.0261	1.83	0.0671
tl	-0.0506	0.0245	-0.0986	-0.0026	-2.07	0.0388

Analysis Of GEE Parameter Estimates
Model-Based Standard Error Estimates

Parameter	Estimate	Standard Error	95% Confidence Limits		Z	Pr > |Z|
Intercept	-1.6328	0.1520	-1.9306	-1.3349	-10.74	<.0001
City	0.5383	0.1898	0.1662	0.9103	2.84	0.0046
Smoke	0.0126	0.0073	-0.0017	0.0269	1.73	0.0831
tl	-0.0506	0.0243	-0.0982	-0.0030	-2.08	0.0374

Unstructured:

Analysis Of GEE Parameter Estimates
Empirical Standard Error Estimates

Parameter	Estimate	Standard Error	95% Confidence Limits		Z	Pr > |Z|
Intercept	-1.6328	0.1536	-1.9339	-1.3317	-10.63	<.0001
City	0.5386	0.1889	0.1683	0.9089	2.85	0.0044
Smoke	0.0124	0.0069	-0.0012	0.0259	1.79	0.0729
tl	-0.0501	0.0245	-0.0981	-0.0021	-2.05	0.0408

Analysis Of GEE Parameter Estimates
Model-Based Standard Error Estimates

Parameter	Estimate	Standard Error	95% Confidence Limits		Z	Pr > |Z|
Intercept	-1.6328	0.1522	-1.9310	-1.3346	-10.73	<.0001
City	0.5386	0.1901	0.1661	0.9112	2.83	0.0046
Smoke	0.0124	0.0073	-0.0019	0.0267	1.70	0.0890
tl	-0.0501	0.0243	-0.0977	-0.0025	-2.06	0.0390

From the outputs above, the estimated (robust) standard errors (empirical s.e) of $\hat{\boldsymbol{\beta}}$ under exchangeability, AR(1), MDEP(3), and unstructured are very similar under the reduced model:

$$\text{logit}[\pi_{ij}] = \beta_0 + \beta_1 \,\text{city}_i + \beta_2 \,\text{smoke}_{ij} + \beta_3 \,\text{agel}_{ij} \tag{12.14}$$

Hence, we appear to gain the greatest efficiency for estimating $\boldsymbol{\beta}$ by assuming an exchangeable model instead of independence, and we do not appear to gain much efficiency by assuming AR(1), MDEP(M), or unstructured instead of exchangeable. As long as we use some correlation model other than independence, the simplest of which is exchangeable, we gain efficiency. The estimated relative efficiency of the elements of $\hat{\boldsymbol{\beta}}$ for an independence model versus an exchangeable model is given in Table 12.6.

The efficiencies are very high, with the smallest being for the maternal smoking effect. The estimates under the naive assumption of independence appear very efficient for all effects, except for the within-cluster (time-varying) covariate Maternal Smoking. This result is very similar to the result we found for the teratology data set in example 12.2. In that data set, the within-cluster covariate was the sex of

	Standard errors		Efficiency (%)
Parameters	EXCH	IND	[Var(EXCH)/Var(IND)]
Intercept	0.1537	0.1596	92.74
City	0.1891	0.1917	97.31
Smoke	0.0069	0.0076	82.43
tl	0.0248	0.0251	97.62

Table 12.6: Relative efficiency of IND versus EXCH models

the offspring, and the efficiency of the independence (ordinary logistic regression) estimates was only 79.72%. As a general rule, therefore, ordinary logistic regression gives high efficiency for the time-stationary (cluster-level) covariates, and low efficiency for the time-varying (within-cluster) covariates.

12.5.6 "Naive" Versus Robust Variance Estimate

Thus far, we have not discussed using the "naive" variance estimate. One question often asked is, when can I use the naive estimate of variance (the one that assumes you have modeled the correlation structure correctly)? The general rule is that for the robust estimator to be a good estimate, the number of clusters (n) should be large. On the other hand, if you have specified the correlation model correctly, the naive variance estimator will be a good estimate as long as $\sum_{i=1}^{n} K_i$ is large. This can occur if either the number of subjects within the cluster is large (K_i is large) or the number of clusters (n) is large, or both are intermediate. Lipsitz (1999) suggested that, for instance, that when n is small, say $n < 25$, then you should carefully model your correlation so that you can use the naive variance estimate (the robust is not a good estimate in this case).

However, if only $n = 20$ children had been available, with wheeze measurements each day for a year ($K_i = 365$ days), then one would be better off assuming that the correlation model is correct, and use the "naive" variance estimate. For the Six Cities Study example in this section, the number of clusters $n = 412$ is large enough to use the robust variance estimator (you would like to have at least $n = 25$).

One may also wish to calculate the relative bias of the "naive" estimator by looking at

$$\text{REL BIAS} = \frac{\text{SE(naive)} - \text{SE(robust)}}{\text{SE(robust)}}$$

This is necessary because with large n, the "robust" variance estimators are correct, and the naive are only correct if we have modeled the correlation correctly. Thus we can get an idea of the relative bias of the naive estimator by looking at the relative bias under different correlation structures. We present these results for both the independence and exchangeable correlation structures below under the reduced model which excludes the quadratic and cubic age effects. First we present that of the independence correlation structure:

(a) Ordinary Logistic Regression

```
Correlation Structure: Independence

PARAMETER ESTIMATES with naive variance (Model-Based Estimates)

                    Standard   95% Confidence
Parameter Estimate    Error        Limits              Z Pr > |Z|
-------------------------------------------------------------
Intercept  -1.6137   0.1076  -1.8245  -1.4029  -15.00   <.0001
City        0.5181   0.1318   0.2597   0.7765    3.93   <.0001
Smoke       0.0128   0.0057   0.0017   0.0240    2.25   0.0244**
tl         -0.0530   0.0293  -0.1105   0.0044   -1.81   0.0702

PARAMETER ESTIMATES with robust variance (Empirical Estimates)

                    Standard   95% Confidence
Parameter Estimate    Error        Limits              Z Pr > |Z|
-------------------------------------------------------------
Intercept  -1.6137   0.1596  -1.9266  -1.3008  -10.11   <.0001
City        0.5181   0.1917   0.1425   0.8938    2.70   0.0069
Smoke       0.0128   0.0076  -0.0021   0.0278    1.68   0.0920
tl         -0.0530   0.0251  -0.1023  -0.0038   -2.11   0.0347
```

The biases of the standard errors of the parameters are summarized in Table 12.7.

Parameter	SE(NAIVE)	SE(ROBUST)	Relative bias
INTERCEP	0.1076	0.1596	-0.3258
City	0.1318	0.1917	-0.3125
Smoke	0.0057	0.0076	-0.2500
tl	0.0293	0.0251	0.1673

Table 12.7: Relative bias of standard errors under independence model

Here, we see that the relative biases are mostly greater than 25%. A "myth" that has prevailed is that, since ordinary logistic regression treats all observations within clusters as independent, we are in effect assuming that we have more information than we actually do, and that we will therefore always underestimate the true variance. We see here that that is not always true, i.e., the naive variance does not always underestimate the true variance: It depends on what parameter you are estimating. We demonstrate this with a simple theoretical justification below:

Suppose we have two correlated means, $\bar{Y}_1$ and $\bar{Y}_2$, where $Cov(\bar{Y}_1, \bar{Y}_2) > 0$, and also suppose we look at $(\bar{Y}_1 - \bar{Y}_2)$, then the naive variance under independence is

$$\text{Var}\,(\bar{Y}_1 - \bar{Y}_2) = \text{Var}\,(\bar{Y}_1) + Var(\bar{Y}_2)$$

which is larger than the true variance,

$$\text{Var}\,(\bar{Y}_1 - \bar{Y}_2) = \text{Var}\,(\bar{Y}_1) + \text{Var}\,(\bar{Y}_2) - 2\text{Cov}\,(\bar{Y}_1, \bar{Y}_2)$$

since $\text{Cov}\,(\bar{Y}_1, \bar{Y}_2) > 0$.

On the other hand, suppose we consider instead, $(\bar{Y}_1 + \bar{Y}_2)$. The naive variance under independence is again,

$$\text{Var}\,(\bar{Y}_1 + \bar{Y}_2) = \text{Var}\,(\bar{Y}_1) + \text{Var}\,(\bar{Y}_2)$$

which this time is smaller than the true variance,

$$\text{Var}\left(\bar{Y}_1 - \bar{Y}_2\right) = \text{Var}\left(\bar{Y}_1\right) + \text{Var}\left(\bar{Y}_2\right) + 2\text{Cov}\left(\bar{Y}_1, \bar{Y}_2\right)$$

12.5.7 Effect of Bias on P Values

From the results from both the "naive" (model-based) and "robust" (empirical) estimates, smoking appears significant using the "naive" estimate, but not significant when the robust estimate is employed under the independence model. We present again the results below.

Hence, we should always use the robust variance when using ordinary logistic regression.

(b) Bias of Naive Variance in the Exchangeable Model

Results under this model are similarly displayed below.

```
Correlation Structure: Exchangeable

PARAMETER ESTIMATES with naive variance

                    Standard   95% Confidence
Parameter Estimate    Error        Limits              Z Pr > |Z|
------------------------------------------------------------------
Intercept  -1.6308   0.1516  -1.9279  -1.3337  -10.76   <.0001
City        0.5367   0.1895   0.1652   0.9082    2.83   0.0046
Smoke       0.0124   0.0073  -0.0019   0.0266    1.70   0.0895
tl         -0.0488   0.0229  -0.0937  -0.0040   -2.13   0.0329

PARAMETER ESTIMATES with robust variance

                    Standard   95% Confidence
Parameter Estimate    Error        Limits              Z Pr > |Z|
------------------------------------------------------------------
Intercept  -1.6308   0.1537  -1.9321  -1.3295  -10.61   <.0001
City        0.5367   0.1891   0.1661   0.9073    2.84   0.0045
Smoke       0.0124   0.0069  -0.0011   0.0258    1.80   0.0724
tl         -0.0488   0.0248  -0.0974  -0.0003   -1.97   0.0485
```

The corresponding summary table for the biases is similarly presented in Table 12.8. The relative biases of the "naive" standard errors from the exchangeable model are all less than 10%.

Variable	SE(NAIVE)	SE(ROBUST)	Relative bias
Intercept	0.1516	0.1537	-0.0137
City	0.1895	0.1891	0.0021
Smoke	0.0073	0.0069	0.0580
tl	0.0229	0.0248	-0.0766

Table 12.8: Relative bias of standard errors under exchangeable model

For the exchangeable model, unlike the independence model, smoking is not significant using either the "naive" estimate and the robust estimate under the reduced model (pvalue >0.05 in each case).

The point is that the robust estimate should definitely be used if you are using the idependence correlation model and it is still a good idea to use it with the other correlation models, but probably not 100% necessary, especally when n is small (since the robust estimate needs n to be large for it to be good).

12.5.8 Parsimonious Model

For the data in Example 12.3, both the quadratic and cubic effects of age are not significant as shown previously; hence the reduced model is given by the expression in (12.14). This model only incorporates the linear effect of age in the model.

The robust parameter estimates under this model with the exchangeable correlation structure are again displayed below.

```
proc genmod data=two;
class id time;
model wheeze=city smoke tl/dist=bin link=logit;
repeated subject=id/type=exch corrw within=time modelse;
run;
```

Analysis Of GEE Parameter Estimates
Empirical Standard Error Estimates

Parameter	Estimate	Standard Error	95% Confidence Limits		Z	Pr > |Z|
Intercept	1.6308	0.1537	1.3295	1.9321	10.61	<.0001
City	-0.5367	0.1891	-0.9073	-0.1661	-2.84	0.0045
Smoke	-0.0124	0.0069	-0.0258	0.0011	-1.80	0.0724
tl	0.0488	0.0248	0.0003	0.0974	1.97	0.0485

Both the linear effect of age and city effect are significant with an estimated correlation parameter $\hat{\rho} = 0.4287$. From the output above, we can conclude the odds of wheezing are $\exp(0.5367) = 1.71$ times higher in the more polluted city (Kingston-Harriman) than in the less polluted city of Portage. Similarly, for a unit increase in age of the child, the odds of wheezing increase by $\exp(0.0488) = 1.05$, while the odds increase by $\exp(2 * 0.0488) = 1.10$ for two-unit increase in age (say from 7 to 9 years, for instance).

12.6 Analysis of Nonbinary Response Data

The examples of the GEE approach in the previous sections deal mainly with cases in which the outcome variable is binary or dichotomous. We consider in this section the case in which the outcome variable is not binary. The following example from Thall and Vail (1990) relates to data arising from a clinical trial of 59 epileptics. The complete data are presented in appendix G.3. However, we present next the first and last 5 observations from the complete data.

id	y1	y2	y3	y4	trt	base	age
1	5	3	3	3	0	11	31
2	3	5	3	3	0	11	30
3	2	4	0	5	0	6	25
4	4	4	1	4	0	8	36
5	7	18	9	21	0	66	22

....

55	3	5	4	3	1	16	32
56	1	23	19	8	1	22	26
57	2	3	0	1	1	25	21
58	0	0	0	0	1	13	36
59	1	4	3	2	1	12	37

The data relate to patients suffering from simple or complex partial seizures who were randomized to receive either the antiepileptic drug *progabide* or placebo, as an adjuvant to chemotherapy. At each of four successive postrandomization clinic visits ($y1, y2, y3, y4$), the number of seizures occurring over the previous 2 weeks was reported. The data above further displays the age (in years) of the epileptic, the treatment assigned and the 8-week baseline seizure counts.

Following Thall and Vail (1990), the following covariates were employed: the log of the baseline seizure rates, obtained as the log of $\frac{1}{4}$ of the 8-week seizure count, log of age, the treatment, and Visit$_4$ for the fourth clinic visit. However, before we adopt the model proposed by Thall and Vail (1990), let us consider the four 2-week visits as a covariate having four equally spaced levels. This leads us to further consider employing the linear, quadratic and cubic effects of the covariate visits in our model. We give below the SAS software program for the data step together with a sample output of the first twenty observations for the first five epileptics.

```
data epilepsy;
infile 'c:\classdata\cl11\lepsyii.txt';
input id y1-y4 trt base age;
datalines;
run;
data new; set epilepsy; by id;
array xx[*] y1-y4; do visit=1 to 4; y=xx[visit]; output; end;
drop y1-y4; run;
DATA TWO; SET new;
***FORMING ORTHOGONAL CONTRASTS ***;
/* Statements here similar to in previous example;
..............
logage=log(age); logbase4=log(base/4); v4=(visit=4); run;
proc print; run;
```

Obs	id	trt	base	age	visit	y	wkl	wkq	wkc	logage	logbase4	v4
1	1	0	11	31	1	5	-3	1	-1	3.43399	1.01160	0
2	1	0	11	31	2	3	-1	-1	3	3.43399	1.01160	0
3	1	0	11	31	3	3	1	-1	-3	3.43399	1.01160	0
4	1	0	11	31	4	3	3	1	1	3.43399	1.01160	1
5	2	0	11	30	1	3	-3	1	-1	3.40120	1.01160	0
6	2	0	11	30	2	5	-1	-1	3	3.40120	1.01160	0
7	2	0	11	30	3	3	1	-1	-3	3.40120	1.01160	0
8	2	0	11	30	4	3	3	1	1	3.40120	1.01160	1
9	3	0	6	25	1	2	-3	1	-1	3.21888	0.40547	0
10	3	0	6	25	2	4	-1	-1	3	3.21888	0.40547	0
11	3	0	6	25	3	0	1	-1	-3	3.21888	0.40547	0
12	3	0	6	25	4	5	3	1	1	3.21888	0.40547	1
13	4	0	8	36	1	4	-3	1	-1	3.58352	0.69315	0
14	4	0	8	36	2	4	-1	-1	3	3.58352	0.69315	0
15	4	0	8	36	3	1	1	-1	-3	3.58352	0.69315	0
16	4	0	8	36	4	4	3	1	1	3.58352	0.69315	1
17	5	0	66	22	1	7	-3	1	-1	3.09104	2.80336	0
18	5	0	66	22	2	18	-1	-1	3	3.09104	2.80336	0
19	5	0	66	22	3	9	1	-1	-3	3.09104	2.80336	0
20	5	0	66	22	4	21	3	1	1	3.09104	2.80336	1

....

We propose a model of the form:

$$\begin{aligned}\ell_{ij} &= \beta_0 + \beta_1 \,\text{trt}_i + \beta_2 \,\text{base4}_{ij} + \beta_3 \,\text{trt*base4}_{ij} + \beta_4 \,\text{age}_i \\ &\quad + \beta_5 \,\text{wkl}_{ij} + \beta_6 \,\text{wkq}_{ij} + \beta_7 \,\text{wkc}_{ij}\end{aligned} \tag{12.15}$$

where

$$\text{trt} = \begin{cases} 1 & \text{if progabide} \\ 0 & \text{if placebo} \end{cases}$$

and ℓ_{ij} are the log of counts, trt$*$base4 is the interaction between treatment group and log base4 as defined previously, and wkl, wkq, and wkc are the orthogonal linear, quadratic, and cubic components of the weeks of visits, respectively. We present below the results of employing the usual log-linear model to the epileptic data and those from the GEE using the independent (IND) and exchangeable correlation structures. The results from the other correlation structures are very similar.

```
data prelim;
set two;
proc genmod data=prelim;
title 'Epileptic seizure data: Standard Poisson regression model';
model y=trt logbase4 trt*logbase4 logage wkl wkq wkc/ dist=poisson type3;
run;
proc genmod data=prelim;
title 'Epileptic seizure data: Marginal model with independent correlation
structure';
class id visit;
model y=trt logbase4  trt*logbase4 logage wkl wkq wkc/ dist=poisson  type3;
contrast 'linear' wkl 1, wkq 1, wkc 1/e wald;
repeated subject=id/type=EXCH corrw within=visit modelse;
run;
```

LOG-LINEAR MODEL

Analysis Of Parameter Estimates

Parameter	DF	Estimate	Standard Error	Wald 95% Confidence Limits Lower	Upper	Chi-Square	Pr > ChiSq
Intercept	1	-2.7661	0.4074	-3.5647	-1.9676	46.09	<.0001
trt	1	-1.3386	0.1568	-1.6459	-1.0314	72.90	<.0001**
logbase4	1	0.9486	0.0436	0.8632	1.0341	473.46	<.0001**
trt*logbase4	1	0.5615	0.0635	0.4370	0.6860	78.16	<.0001**
logage	1	0.8876	0.1165	0.6593	1.1159	58.05	<.0001**
wkl	1	-0.0301	0.0102	-0.0502	-0.0101	8.66	0.0033**
wkq	1	-0.0180	0.0227	-0.0625	0.0266	0.63	0.4290
wkc	1	-0.0111	0.0101	-0.0308	0.0087	1.20	0.2724
Scale	0	1.0000	0.0000	1.0000	1.0000		

GEE with IND correlation structure:

Analysis Of GEE Parameter Estimates
Empirical Standard Error Estimates

Parameter	Estimate	Standard Error	95% Confidence Limits Lower	Upper	Z	Pr > |Z|
Intercept	-2.7661	0.9393	-4.6072	-0.9250	-2.94	0.0032
trt	-1.3386	0.4255	-2.1726	-0.5047	-3.15	0.0017
logbase4	0.9486	0.0965	0.7595	1.1377	9.83	<.0001
trt*logbase4	0.5615	0.1739	0.2207	0.9024	3.23	0.0012
logage	0.8876	0.2727	0.3530	1.4222	3.25	0.0011
wkl	-0.0301	0.0170	-0.0634	0.0031	-1.78	0.0755
wkq	-0.0180	0.0438	-0.1037	0.0678	-0.41	0.6814
wkc	-0.0111	0.0196	-0.0494	0.0273	-0.57	0.5716

```
GEE with EXCH structure:

                       Analysis Of GEE Parameter Estimates
                       Empirical Standard Error Estimates

                                              95% Confidence
                                  Standard        Limits
        Parameter     Estimate     Error     Lower      Upper        Z Pr > |Z|
        ----------------------------------------------------------------------
        Intercept      -2.8434    0.9512   -4.7077   -0.9792    -2.99    0.0028
        trt            -1.3528    0.4287   -2.1930   -0.5125    -3.16    0.0016**
        logbase4        0.9489    0.0985    0.7558    1.1420     9.63    <.0001**
        trt*logbase4    0.5696    0.1744    0.2277    0.9115     3.27    0.0011**
        logage          0.9097    0.2752    0.3704    1.4491     3.31    0.0009**
        wkl            -0.0301    0.0170   -0.0634    0.0031    -1.78    0.0755
        wkq            -0.0180    0.0438   -0.1037    0.0678    -0.41    0.6814
        wkc            -0.0111    0.0196   -0.0494    0.0273    -0.57    0.5716
```

The results indicate that the linear, quadratic, and cubic effect of weeks of visits are not significant based on the type 3 analysis. Hence with a hypothesis of no weeks of visits effect in (12.15), that is, testing that

$$H_0 : \beta_5 = \beta_6 = \beta_7 = 0,$$

we can construct the following contrast:

$$H_0 : \begin{bmatrix} 0 & 0 & 0 & 0 & 0 & 1 & 0 & 0 \\ 0 & 0 & 0 & 0 & 0 & 0 & 1 & 0 \\ 0 & 0 & 0 & 0 & 0 & 0 & 0 & 1 \end{bmatrix} \begin{bmatrix} \beta_0 \\ \beta_1 \\ \beta_2 \\ \beta_3 \\ \beta_4 \\ \beta_5 \\ \beta_6 \\ \beta_7 \end{bmatrix} = \begin{bmatrix} 0 \\ 0 \\ 0 \end{bmatrix}$$

This contrast is not significant using any of the correlation structures when implemented in SAS software. The results for the exchangeable correlation structure for instance give a Wald's test value of 6.70 on 3 degrees of freedom with a corresponding nonsignificant pvalue of $p = 0.0820$. Refitting the model with no weeks of visits effect, we have for the reduced model

$$\ell_{ij} = \beta_0 + \beta_1 \,\text{trt}_i + \beta_2 \,\text{base4}_{ij} + \beta_3 \,\text{trt*base4}_{ij} + \beta_4 \,\text{age}_i \tag{12.16}$$

and the following parameter estimates from the various correlation structures.

```
Parameter      POISSON      EXCH       AR(1)      1-DEP      2-DEP      3-DEP      UNSTR
----------------------------------------------------------------------------------------
Intercept      -2.7634    -2.7634    -3.0533    -3.0533    -3.2516    -3.0529    -3.0932
trt            -1.3356    -1.3356    -1.4864    -1.4864    -1.5876    -1.4862    -1.5211
logbase4        0.9486     0.9486     0.9413     0.9413     0.9364     0.9413     0.9320
trt*logbase4    0.5615     0.5615     0.6200     0.6200     0.6600     0.6199     0.6359
logage          0.8876     0.8876     0.9790     0.9790     1.0413     0.9788     0.9973
----------------------------------------------------------------------------------------
```

It is clear that not much is gained in terms of parameter estimates between the log-linear, independent, and exchangeable correlation structures. Similarly, the parameter estimates from both the AR(1) and 1-dependent correlation structures are exactly the same except for the different standard errors. The 2-dependent, the 3-dependent, and the unstructured correlation structures returned different parameter estimates as shown. For all correlation structures employed for this example, we notice that the standard errors produced under the GEE are much greater than those produced from the usual log-linear regression. The standard errors produced under the log-linear models assume that the observations are independent from visits to visits, which is not necessaily the case. Hence, these standard errors may be misleading and the appropriate model would be one that assumes some correlations between observations relating to visit1 to visit4. Under the GEE models all the parameters of the models are highly significant.

12.6.1 Interpretation of Parameter Estimates

Since $\hat{\beta}_4$ is approximately 0.98 using all the correlation structures, we see that the number of epileptic attacks increases by a factor of

$$\exp\{(\hat{\beta}_4)\} \approx \exp\{(0.98)\} = 2.66$$

that is, approximately by 3 attacks given the levels of the other covariates for the 8-week period for a unit increase in age. Since there is significant interaction effect of treatment and logbase4, we see that the negative value of the estimate for β_2 indicates that the seizure count is significantly lower after 8 weeks for the treatment group compared to the placebo group and that, further, this decrease is affected by the baseline seizure counts for each group.

12.6.2 Thall and Vail Model

Thall & Vail (1990) proposed the following model for the epileptic data in this example:

$$\ell_{ij} = \beta_0 + \beta_1 \,\text{trt}_i + \beta_2 \,\text{base4}_{ij} + \beta_3 \,\text{trt*base4}_{ij} + \beta_4 \,\text{age}_i + \beta_5 v4_{ij} \tag{12.17}$$

where

$$\text{trt} = \begin{cases} 1 & \text{if progabide} \\ 0 & \text{if placebo} \end{cases} \quad \text{and} \quad v4 = \begin{cases} 1 & \text{if visit=4} \\ 0 & \text{otherwise} \end{cases}$$

The values of $v4$ and trt were displayed earlier for the first 5 subjects.

The model in (12.17) is implemented under various correlation structures, and the parameter estimates under both the IND and EXCH correlation structures are displayed below along with the accompanying SAS software program.

```
data two; set epilepsy;
proc genmod data=two;
title 'Epileptic seizure data: Marginal model with INDEPENDENT correlation
structure';
class id visit;
model y=trt logbase4  trt*logbase4 logage v4/ dist=poisson type3;
repeated subject=id/type=IND corrw within=visit modelse;
run;
proc genmod data=two;
title 'Epileptic seizure data: Marginal model with exchangeable correlation
structure';
class id visit;
model y=trt logbase4  trt*logbase4 logage v4/ dist=poisson type3;
repeated subject=id/type=EXCH corrw within=visit modelse;
run;
```

IND:

```
                  Analysis Of GEE Parameter Estimates
                  Empirical Standard Error Estimates

                                      95% Confidence
                           Standard        Limits
Parameter     Estimate      Error     Lower     Upper        Z Pr > |Z|
-----------------------------------------------------------------------
Intercept      -2.7258    0.9382  -4.5646  -0.8870    -2.91    0.0037
trt            -1.3386    0.4255  -2.1726  -0.5047    -3.15    0.0017
logbase4        0.9486    0.0965   0.7595   1.1377     9.83    <.0001
trt*logbase4    0.5615    0.1739   0.2207   0.9024     3.23    0.0012
logage          0.8876    0.2727   0.3530   1.4222     3.25    0.0011
v4             -0.1598    0.0651  -0.2874  -0.0321    -2.45    0.0142
```

EXCH:

```
                  Analysis Of GEE Parameter Estimates
                  Empirical Standard Error Estimates

                                      95% Confidence
                           Standard        Limits
Parameter     Estimate      Error     Lower     Upper        Z Pr > |Z|
-----------------------------------------------------------------------
Intercept      -2.7599    0.9491  -4.6202  -0.8996    -2.91    0.0036
trt            -1.3361    0.4293  -2.1775  -0.4946    -3.11    0.0019
logbase4        0.9495    0.0987   0.7561   1.1428     9.62    <.0001
trt*logbase4    0.5625    0.1749   0.2197   0.9053     3.22    0.0013
logage          0.8965    0.2751   0.3574   1.4356     3.26    0.0011
v4             -0.1598    0.0651  -0.2874  -0.0321    -2.45    0.0142
```

The overdispersion parameter for the above model is 4.4139 with an estimated correlation coefficient under the exchangeable structure of $\hat{\rho} = 0.3543$. The type 3 GEE analysis also indicate that trt, logbase4, logage, and v4 are significant at the 5% point with pvalues of $0.0171, 0.0038, 0.0111$, and 0.0417, respectively.

12.6.3 Diggle, Liang, and Zeger Model

Diggle, Liang, and Zeger (1995) proposed the following Poisson regression model for the epileptic data:

$$\ell_{ij} = \ln t_{ij} + \beta_0 + \beta_1 x_{i1} + \beta_2 \operatorname{trt}_i + \beta_3 x_{i1} * \operatorname{trt}_i, \quad \begin{matrix} j = 0, 1, 2, 3, 4 \\ i = 1, 2, \cdots, 59 \end{matrix} \tag{12.18}$$

where the covariates are defined as:

$$x_{i1} = \begin{cases} 1 & \text{if visit=1,2,3 or 4} \\ 0 & \text{if baseline} \end{cases} \qquad \operatorname{trt} = \begin{cases} 1 & \text{if progabide} \\ 0 & \text{if placebo.} \end{cases}$$

and

$$t_{ij} = \begin{cases} 8 & \text{if } j = 0 \\ 2 & \text{if } j = 1, 2, 3, \text{or } 4 \end{cases}$$

The t_{ij} are included in the model to account for the different observation periods. Notice here that there is no age effect in the model. Diggle et al. (1995) conclude that patients in the two treatment groups "appear to be comparable in terms of baseline age and eight-week baseline seizure counts".

Again, the results of implementing this model for the exchangeable correlation structure are displayed below, where we have included the data transformation for the first two patients in the partial output.

```
data new1; set new; output;
if visit=1 then do; y=base; visit=0; output; end; run;
data new2; set new1; if visit=0 then do; x1=0; ltime=log(8); end;
else do; x1=1; ltime=log(2); end; x1trt=x1*trt; run;
proc genmod data=new2;
class id;
model y=x1 trt x1*trt/ dist=poisson offset=ltime type3;
repeated subject=id/type=EXCH corrw modelse; run;
```

Obs	id	trt	base	age	intercpt	visit	y	x1	ltime
1	1	0	11	31	1	1	5	1	0.69315
2	1	0	11	31	1	0	11	0	2.07944
3	1	0	11	31	1	2	3	1	0.69315
4	1	0	11	31	1	3	3	1	0.69315
5	1	0	11	31	1	4	3	1	0.69315
6	2	0	11	30	1	1	3	1	0.69315
7	2	0	11	30	1	0	11	0	2.07944
8	2	0	11	30	1	2	5	1	0.69315
9	2	0	11	30	1	3	3	1	0.69315
10	2	0	11	30	1	4	3	1	0.69315

....

Analysis Of GEE Parameter Estimates
Empirical Standard Error Estimates

Parameter	Estimate	Standard Error	95% Confidence Limits		Z	Pr > |Z|
Intercept	1.3476	0.1574	1.0392	1.6560	8.56	<.0001
x1	0.1087	0.1156	-0.1179	0.3354	0.94	0.3472
trt	0.0265	0.2219	-0.4083	0.4613	0.12	0.9049
x1*trt	-0.1016	0.2134	-0.5198	0.3166	-0.48	0.6339

Results from this model suggest that there is very little difference between the treatment and placebo groups in the change of seizure counts before and after treatment, since the estimated coefficient of β_3 is nonsignificant in the above analysis. The overdispersion parameter for this model is 19.69 with an estimate of ρ being 0.7712. The overdispersion parameter for this model is relatively higher (19.62) here as compared to that obtained from the Thall and Vail model.

12.6.4 Effect of Removing Patient 49

We observe that patient with *id* number 49 (equivalent to patient 207) in Thall and Vail (1990) and Diggle et al. (1995) has unusual and extremely high seizure count of 151 in 8 weeks at baseline and 302 counts in 8 weeks. We examine the effect of dropping this patient on our analyses below.

Again, the model incorporating the linear, quadratic, and cubic components of weeks of visit in the model has the hypothesis of no time effects, providing a Wald statistic of $Q_W = 4.15$ on 3 d.f. with a pvalue of 0.2457, again indicating that this hypothesis is tenable.

When patient 49 is deleted from the data, with the analysis employing the Thall and Vail (1990) model in (12.17), we have the following parameter estimates under the exchangeable correlation structure.

```
proc genmod data=two;
 where id ne 49;
   class id visit;
model y=trt logbase4  trt*logbase4 logage v4/ dist=poisson type3;
repeated subject=id/type=EXCH corrw within=visit;
run;
```

Analysis Of GEE Parameter Estimates
Empirical Standard Error Estimates

Parameter	Estimate	Standard Error	95% Confidence Limits		Z	Pr > |Z|
Intercept	-2.3223	0.8734	-4.0342	-0.6103	-2.66	0.0078
trt	-0.5160	0.4178	-1.3349	0.3029	-1.24	0.2168
logbase4	0.9500	0.0983	0.7573	1.1426	9.67	<.0001
trt*logbase4	0.1375	0.1946	-0.2439	0.5189	0.71	0.4797
logage	0.7662	0.2535	0.2694	1.2630	3.02	0.0025
v4	-0.1464	0.0758	-0.2949	0.0022	-1.93	0.0534

The model indicates again that the treatment has very little effect on seizure counts after randomization of the subjects to both groups. The negative value of the treatment coefficient, however, indicates that there is some reduction in the seizure counts for the progabide group, although not that significant. The overdispersion parameter for this model is 4.1496 with estimated correlation coefficient of 0.3353.

Similarly, when we employed the Diggle et al. (1995) model in (12.18), we obtain the following results, again based on the exchangeable correlation structure.

```
proc genmod data=new2;
where id ne 49;
class id;
model y=x1 trt x1*trt/ dist=poisson offset=ltime type3;
repeated subject=id/type=EXCH corrw;
run;
```

Analysis Of GEE Parameter Estimates
Empirical Standard Error Estimates

Parameter	Estimate	Standard Error	95% Confidence Limits		Z	Pr > |Z|
Intercept	1.3476	0.1574	1.0392	1.6560	8.56	<.0001
x1	0.1087	0.1156	-0.1179	0.3354	0.94	0.3472
trt	-0.1080	0.1937	-0.4876	0.2716	-0.56	0.5770
x1*trt	-0.2995	0.1709	-0.6345	0.0354	-1.75	0.0797

The model indicates again that the treatment has very little effect on seizure counts after randomization of the subjects to both groups. The negative value of the treatment coefficient however indicates that there is some reduction in the seizure counts for the progabide group, although this effect is not significant. The overdispersion parameter for this model is 10.5308 with estimated correlation coefficient of 0.5932.

12.7 Exercises

1. Ware, et al. (1984) analyzed the wheezing data in Table 12.9 from a six-city study on the health effects of pollution. We give below the data from only one of the cities. Children in Steubenville, OH, were examined for wheezing at each of ages 7 through 10 years of age. The mothers' smoking habits were also recorded at the start of the study.

Mother smokes					Nonsmoking mother				
			Age 10					Age 10	
Age 7	Age 8	Age 9	No	Yes	Age 7	Age 8	Age 9	No	Yes
No	No	No	118	6	No	No	No	237	10
		Yes	8	2			Yes	15	4
	Yes	No	11	1		Yes	No	16	2
		Yes	6	4			Yes	7	3
Yes	No	No	7	3	Yes	No	No	24	3
		Yes	3	1			Yes	3	2
	Yes	No	4	2		Yes	No	6	2
		Yes	4	7			Yes	5	11

Table 12.9: Wheezing data

Analyze the above data and interpret your results. Is there an age effect on the child's wheezing? What effect does mother smoking status has on the child's wheezing?

2. Responses to three questions on abortion in surveys conducted over three years (Haberman, 1978, p. 482) are given in Table 12.10.

	Year		
Response	1972	1973	1974
YYY	334	428	413
YYN	34	29	29
YNY	12	13	16
YNN	15	17	18
NYY	53	42	60
NYN	63	53	57
NNY	43	31	37
NNN	501	453	430

Table 12.10: Abortion survey data

The questions were: Should a pregnant woman be able to obtain a legal abortion if
(1) she is married and does not want more children;
(ii) the family has very low income and cannot afford any more children;
(iii) she is not married and does not want to marry the man?

Fit a Rasch model to the above data.

3. Data on responses (C: correct, W: wrong) to four questions from the arithmetic reasoning test on the Armed Services Vocational Aptitude Battery, with samples from White and Black males and females (Lindsey, 1995), are reproduced in Table 12.11.

	White		Black	
Response	M	F	M	F
CCCC	86	42	2	4
WCCC	1	7	3	0
CWCC	19	6	1	2
WWCC	2	2	3	3
CCWC	11	15	9	5
WCWC	3	5	5	5
CWWC	6	8	10	10
WWWC	5	8	5	8
CCCW	23	20	10	8
WCCW	6	11	4	6
CWCW	7	9	8	11
WWCW	12	14	15	7
CCWW	21	18	7	19
WCWW	16	20	16	14
CWWW	22	23	15	14
WWWW	23	20	27	29

Table 12.11: Arithmetic reasoning data

Fit a Rasch model to the above data and test if your results are the same for each of the four groups.

4. The data below is a subset of data from Woolson and Clarke (1984). The variables are respectively sex (0 = female), y1–y3 (1 = obese), and count of the number of people with each pattern. Thus, this data set contains records on 1014 children who were 7-9 years old in 1977 (the first year measurements were obtained). Measures of obesity were obtained in three survey years: 1977, 1979, and 1981 (y1, y2, y3, respectively).

```
0 1 1 1 20
0 1 1 0 7
0 1 1 . 11
0 1 0 1 9
0 1 0 0 8
0 1 0 . 1
0 1 . 1 3
0 1 . 0 1
0 1 . . 7
0 0 1 1 8
0 0 1 0 8
0 0 1 . 3
0 0 0 1 15
0 0 0 0 150
0 0 0 . 38
0 0 . 1 6
0 0 . 0 16
0 0 . . 45
0 . 1 1 13
0 . 1 0 3
0 . 1 . 4
0 . 0 1 2
0 . 0 0 42
0 . 0 . 33
0 . . 1 14
0 . . 0 55
1 1 1 1 21
1 1 1 0 6
1 1 1 . 11
1 1 0 1 6
1 1 0 0 2
1 1 0 . 1
1 1 . 1 4
1 1 . 0 0
1 1 . . 7
1 0 1 1 19
1 0 1 0 13
1 0 1 . 3
```

```
1 0 0 1 14
1 0 0 0 154
1 0 0 . 25
1 0 . 1 3
1 0 . 0 16
1 0 . . 47
1 . 1 1 8
1 . 1 0 1
1 . 1 . 5
1 . 0 1 4
1 . 0 0 47
1 . 0 . 23
1 . . 1 13
1 . . 0 39
```

Analyze the above data.

5. The following data is from a longitudinal study of dental growth in children (Potthoff & Roy, 1964). Measured was the distance from the center of the pituitary gland to the maxillary fissure for children at ages 8, 10, 12, and 14. The data are presented below.

	Age (in years): **Girls**					Age in years): **Boys**			
Girls	8	10	12	14	Boys	8	10	12	14
1	21	20	21.5	23	1	26	25	29	31
2	21	21.5	24	25.5	2	21.5	22.5	23	26.5
3	20.5	24	24.5	26	3	23	22.5	24	27.5
4	23.5	24.5	25	26.5	4	25.5	27.5	26.5	27
5	21.5	23	22.5	23.5	5	20	23.5	22.5	26
6	20	21	21	22.5	6	24.5	25.5	27	28.5
7	21.5	22.5	23	25	7	22	22	24.5	26.5
8	23	23	23.5	24	8	24	21.5	24.5	25.5
9	20	21	22	21.5	9	23	20.5	31	26
10	16.5	19	19	19.5	10	27.5	28	31	31.5
11	24.5	25	28	28	11	23	23	23.5	25
					12	21.5	23.5	24	28
					13	17	24.5	26	29.5
					14	22.5	25.5	25.5	26
					15	23	24.5	26	30
					16	22	21.5	23.5	25

(a) Perform an analysis of the above data assuming an exchangeable structure.

(b) Repeat (a) using an AR(1) structure.

(c) Compare and contrast the two analyses. Which seems best?

Appendices

All appendices for this book can be found in the CD-Rom enclosed with this text. A separate table of contents and indexes are prepared for this chapter.

Table of the Chi-Squared Distribution

d.f.	Right-Tail Probability						
	0.250	0.100	0.050	0.025	0.010	0.005	0.001
1	1.32	2.71	3.84	5.02	6.63	7.88	7.88
2	2.77	4.61	5.99	7.38	9.21	10.60	10.60
3	4.11	6.25	7.81	9.35	11.34	12.84	12.84
4	5.39	7.78	9.49	11.14	13.28	14.86	14.86
5	6.63	9.24	11.07	12.83	15.09	16.75	16.75
6	7.84	10.64	12.59	14.45	16.81	18.55	18.55
7	9.04	12.02	14.07	16.01	18.48	20.28	20.28
8	10.22	13.36	15.51	17.53	20.09	21.95	21.95
9	11.39	14.68	16.92	19.02	21.67	23.59	23.59
10	12.55	15.99	18.31	20.48	23.21	25.19	25.19
11	13.70	17.28	19.68	21.92	24.72	26.76	26.76
12	14.85	18.55	21.03	23.34	26.22	28.30	28.30
13	15.98	19.81	22.36	24.74	27.69	29.82	29.82
14	17.12	21.06	23.68	26.12	29.14	31.32	31.32
15	18.25	22.31	25.00	27.49	30.58	32.80	32.80
16	19.37	23.54	26.30	28.85	32.00	34.27	34.27
17	20.49	24.77	27.59	30.19	33.41	35.72	35.72
18	21.60	25.99	28.87	31.53	34.81	37.16	37.16
19	22.72	27.20	30.14	32.85	36.19	38.58	38.58
20	23.83	28.41	31.41	34.17	37.57	40.00	40.00
21	24.93	29.62	32.67	35.48	38.93	41.40	41.40
22	26.04	30.81	33.92	36.78	40.29	42.80	42.80
23	27.14	32.01	35.17	38.08	41.64	44.18	44.18
24	28.24	33.20	36.42	39.36	42.98	45.56	45.56
25	29.34	34.38	37.65	40.65	44.31	46.93	46.93
26	30.43	35.56	38.89	41.92	45.64	48.29	48.29
27	31.53	36.74	40.11	43.19	46.96	49.64	49.64
28	32.62	37.92	41.34	44.46	48.28	50.99	50.99
29	33.71	39.09	42.56	45.72	49.59	52.34	52.34
30	34.80	40.26	43.77	46.98	50.89	53.67	53.67
40	45.62	51.81	55.76	59.34	63.69	66.77	66.77
50	56.33	63.17	67.50	71.42	76.15	79.49	79.49
60	66.98	74.40	79.08	83.30	88.38	91.95	91.95
70	77.58	85.53	90.53	95.02	100.43	104.21	104.21
80	88.13	96.58	101.88	106.63	112.33	116.32	116.32
90	98.65	107.57	113.15	118.14	124.12	128.30	128.30
100	109.14	118.50	124.34	129.56	135.81	140.17	140.17
110	119.61	129.39	135.48	140.92	147.41	151.95	151.95
120	130.05	140.23	146.57	152.21	158.95	163.65	163.65

Bibliography

Abramowitz, M., & Stegun, I.A. (1970). *Handbook of Mathematical Functions with Formulas, Graphs and mathematical Tables.* U.S. Govt. Printing Office, Washington.

Agresti, A. (1983). A simple diagonals-parameter symmetry and quasi-symmetry model. *Statist. Prob. Letters,* 1, 313-316.

Agresti, A. (1984). *Analysis of Ordinal Categorical Data.* John Wiley and Sons, New York.

Agresti, A. (1990). *Categorical Data Analysis* Wiley, New York.

Agresti, A. (1992). A survey of exact inference for contingency tables (with discussion). *Statistical Science,* 7: 131-177.

Agresti, A. (1996). *An Introduction to Categorical Data Analysis.* John Wiley and Sons, New York.

Agresti, A., Mehta, C.R., & Patel, N.R. (1990). Exact inference for contingency tables with ordered categories. *JASA*, 85: 453-458.

Agresti, A., & Wackerly, D. (1977). Some exact conditional tests of independence for $R \times C$ cross-classification tables. *Psychometrika*, 42: 111-125.

Agresti A., Wackerly D., & Boyett J.M. (1979). Exact conditional tests for cross-classification: approximation of attained significance levels. *Psychometrika, 44:* 75-83.

Aitkin, M. (1979). A simultaneous test procedure for contingency tables. *Applied Statistics*, 28, 233-242.

Aitken, M., Anderson, D., Francis, B. & Hinde, J. (1989). *Statistical Modelling in GLIM.* Clarendon Press, Oxford.

Aicken, M. (1983). *Linear Statistical Analysis of Discrete Data* J. Wiley & Sons, New York.

Akaike, H. (1974). A new look at the statistical model identification. *IEEE Trans. Autom. Contr.* AC-19, 716-723.

Alba, R.D. (1987). Interpreting the parameters of log-linear models. *Sociological Methods and Research,* 16, 45-77.

Amemiya, T. (1981). Qualitative response models: a survey. *J. of Econ. Literature, 19*, 1483-1536.

Andersen, E. B. (1970) Asymptotic properties of conditional maximum likelihood estimators. *Jour. of the Royal Statistical Society, B32*, 283-301.

Andersen, E. B. (1973). *Conditional Inference and Models for Measuring.* Copenhagen: Mentalhygieniejuisk Forlag.

Andersen, E.B. (1980). *Discrete Statistical Models with Social Science Applications.* North-Holland, Amsterdam.

Anderson, E.B. (1997) *Introduction to the Statistical Analysis of Categorical Data.*, Springer-Verlag, Berlin.

Anderson, T.W., & Goodman, L.A. (1957). Statistical inference about Markov chains. *Ann. Math. Statist.*,28, 89-110.

Bachman, J.G., Johnson, L.D., & O'Malley, P.M. (1980). *Monitoring the future: Questionnaire responses from the nation's high school seniors*, 1979 and 1980. Ann Arbor, MI: Survey Research Center.

Baker, R.J. (1977). Exact distributions derived from two-way tables. *Applied Statistics*, 26: 199-206, Corr. 27: 109.

Barnard, G.A. (1989). On alleged gains in power from lower P-values. *Statist. Med.*, 8: 1469-1477.

Barnard, G.A. (1990) Comment *Statist. Med., 9,* 373-375.

Becker, M.P., and Clogg, C.C. (1989). Analysis of sets of two-way contingency tables using association models. *J. Amer. Statist. Assoc.*, 84, 142-151.

Berchtold, H. (1972). vertrauensgrenzen und vergleich zweier wahrschein-lichkeiten. Z. *Wahrscheinlichkeitstheorie und Verw. Gebiete, 22*, 112-119.

Berkson, J. (1953). A statistically precise and relatively simple method of estimating the bio-assay with quantal response based on the logistic function. *J. Amer. Statist. Assoc.*, 41, 569-99.

Berkson, J. (1955). Maximum likelihood and minimum χ^2 estimate of the logistic function. *J. Amer. Statist. Assoc.*, 50, 130-162.

Berstock, D. A., Villers, C., & Latto, C. (1978). Diverticular disease and carcinoma of the colon. Unpublished manuscript.

Best, D.J., & Roberts, D.E. (1975). Algorithm AS 91: The percentage points of the χ^2 distribution. *Applied Statist.*, 24, 385-388.

Bhapkar, V.P. (1966). A note on the equivalence of two test criteria for hypotheses in categorical data. *J. Amer. Statist. Assoc.*, 61, 228-235.

Bhapkar, V.P. (1979). On tests of marginal symmetry and quasi-symmetry in two and three-dimensional contingency tables. *Biometrics*, 35, 417-426.

Birch, M.W. (1963). Maximum likelihood in three-way contingency tables. *J. Roy. Statist. Soc.*, B25, 220-233.

Birch, M.W. (1964). The detection of partial association, I: The 2 × 2 case. *J. Roy. Statist. Soc.*, B26, 313-324.

Bishop, Y.M.M, Fienberg, S.E., & Holland, P.W. (1975). *Discrete Multivariate Analysis*. MIT Press.

Bliss, C.I. (1935). The calculation of the dosage-mortality curve. *Ann. Appl. Biol., 22*, 134-167.

Bortkewicz, L. von (1898). *Das Gesetz der Kleinen Zahlen*. Leipzig: Teubner.

Bradley, J. V. (1968). *Distribution Free Statistical Tests*. Englewood Cliffs, NJ: Prentice Hall.

Bradley, R.A., & Terry, M.E. (1952). Rank analysis of incomplete block designs I. The method of paired comparisons. *Biometrika*, 39, 324-345.

Breslow, N.E., & Day, N.E. (1980). *Statistical Methods in Cancer Research*. Lyon: International Agency for Research on Cancer.

Brown, M.B. (1976). Screening effects in multidimensional contingency tables. *Applied Statistics*, 25, 37-46.

Burstein, H. (1981). Binomial 2×2 test for independent samples with independent proportions. *Comm. Statist., A10, 1,* 11-29.

Caussinus, H. (1965). Contributions a l'analyse statistique des tableaux de correlation. *Ann. Fac. Sci. Univ. Toulouse*, 29, 77-182.

Chamberlain, G. (1980). Analysis of covariance with qualitative data. *Review of Economic Statistics,* 48, 225-238.

Chang, Y. (2000). Residuals analysis of the generalized linear models for longitudinal data. *Statist. Med., 19*, 1277-1293.

Chatterji, S.D. (1963). Some elementary characterizations of the Poisson distribution. *Amer. Math. Monthly, 70*, 958-964.

Christensen, R. (1987). *Plane Answers to Complex Questions: The Theory of Linear Models.* Springer-Verlag, New York.

Christensen, R. (1990). *Log-Linear Models.* Springer-Verlag, New York.

Clarke, S. R., & Norman, J.M. (1995). Home ground advantage of individual clubs in English soccer. *The Statistician,* **44,** 509–521.

Clayton, M.K., Geisser, S and Jennings, D.E. (1986). A comparison of several model selection procedures. In *Bayesian Inference and Decission Techniques,* edited by P. Goel and A. Zellner. Amsterdam: North Holland.

Clogg, C.C. (1978). Adjustments of rates using multiplicative models. *Demography, 15,* 523-539.

Clogg, C.C. (1982). Some models for the analysis of association in multiway cross-classifications having ordered categories. *Jour. Amer. Statist. Assoc.*, 77, 803-815.

Clogg, C.C. (1995). Latent Class Models. in *Handbook of Statistical Modeling for the Social and Behavioral Sciences*, eds Arminger, G.E, Clogg, C.C. & Sobel, M. Plenum Press, New York. pp. 311-359.

Clogg, C.C., & Eliason, S. R. (1987). Some common problems in log- linear analysis. *Sociol. Meth. & Research*, 16, 8-44.

Clogg. C.C. & Eliason, S.R. (1988). A flexible procedure for adjusting rates and proportions, including statistical methods for group comparisons. *Amer. Sociol. Rev.*, 53, 267-283.

Clogg, C.C., Eliason, S.R., & Grego, J.M. (1990). Models for the analysis of change in discrete variables.in *Statistical Methods in Longitudinal Research*, 2, Time series and categorical longitudinal data.

Clogg, C.C., & Goodman, L.A. (1984). Latent structure analysis of a set of multidimensional contingency tables. *Jour. Amer. Statist. Assoc.*, 79, 762-771.

Clogg, C.C., & Shihadeh, E.S. (1994). *Statistical Models for Ordinal Variables*, Thousand Oaks, CA:Sage.

Clogg, C.C. & Shockey, J.W. (1988). Multivariate Analysis of discrete Data. *in Handbook of Multivariate Experimental Psychology*, eds J.R. Nesselroade and R.B. Cattell, New York:Plenum Press, pp. 337-365.

Clogg, C.C, Shockey, J.W., & Eliason, S.R. (1990). A general statistical framework for adjustment of rates.*Sociol. Meth. & Res.*, 19, 156-195

Cochran, W.G. (1942). The X^2 correction for continuity. *Iowa State College Journal of Science*, 16, 421-436.

Cochran, W.G. (1952). The χ^2 test of goodness-of-fit. *Ann. Math. Statist.*, 23, 315-345.

Cochran, W.G. (1954). Some methods of strengthening the common χ^2 tests. *Biometrika*, 10, 417-451.

Cohen, J. (1960). A coefficient of Agreement for nominal scales. *Educ. Psychol. Meas.*, 20, 37-46.

Collet, D. (1991). *Modeling Binary Data.* Chapman & Hall, London.

Conover, W.J. (1974). Some reasons for not using the Yates continuity correction on 2×2 contingency tables (with comments). *J. Amer. Statist. Assoc.*, 69, 374-382.

Correa, D., Pickle, L.W., Fontham, E., Lin, Y., & Haenszel, W. (1983). Passive smoking and lung cancer *Lancet 8350*, 595-597.

Cox, C.P. (1987). *A Handbook of Statistical Methods.* Wiley, New York.

Cox, D.R. (1970a). *The Analysis of Binary Data.* Methuen, London.

Cox, D.R. (1970b). The continuity correction. *Biometrika, 57,* 217-219.

Cox, D.R., & Snell, E.J. (1989). *Analysis of Binary Data,* Second Edition. Chapman & Hall, London.

Cox, M.A.A., & Plackett, R.L. (1980). Small samples in contingency tables. *Biometrika*, 67: 1-14.

Cramer, H. (1946). *Mathematical Methods of Statistics.* Princeton, NJ: Princenton University Press.

Cressie, W., & Read, T.R.C. (1984). Multinomial goodness-of-fit tests. *J. R. Statist. Soc., B*, 46, 440-464.

Cressie, W., & Read, T.R.C. (1988). *Goodness-of-fit statistics for discrete multivariate data.* Springer-Verlag, New York.

Darroch, J. N., & McCloud, P. I. (1986). Category distinguishability and observed agreement. *Australian Journal of Statistics*, 28, 371-388.

David, H.A. (1988). *The Method of Paired Comparisons.* Oxford: Oxford University Press.

Davidson, R.R., & Beaver, R.J. (1977). On extending the Bradley-Terry model to incorporate within-pair order effects. *Biometrics,* **33,** 693–702.

Deming,W.E., & Stephan, F.F. (1940). On a least squares adjustment of a sampled frequency table when the expected marginal totals are known. *Ann. Math. Statist.*, 11, 427-444.

Demo, D.H., & Parker, F.F. (1987). Academic achievement and self-esteem among black and white college students. *J. Social Psychol.*, 127, 345-355.

DiFrancisco, W., & Critelman, Z. (1984). Soviet political culture and covert participation in policy implementation. *Amer. Polit. Sci. Rev.*, 78, 603-621.

Diggle, P. J., Liang, K., & Zeger, S. L. (1995) *Analysis of Longitudinal Data.* Oxford Science Publications, Oxford.

Dobson, A.J. (1990). *An Introduction to Generalized Linear Models*, Chapman and Hall, London.

Dozzi, M., & Riedwyl, H. (1984). Small sample properties of asymptotic tests for two binomial proportions. *Biom. J., 26*, 505-516.

Duncan, O.D., & McRae, J.A. Jr. (1979). Multiway contingency analysis with a scaled response or factor. In *Sociological Methodology.* San Francisco: Jossey-Bass.

Duncan, O.D., Schuman, H., & Duncan, B. (1973). *Social Change in a Metropolitan Community.* New York: Russell Sage Foundation.

Edwards, A.W.F. (1963). The measure of association in a 2×2 table. *J. Roy. Statist. Soc.*, A126, 109-114.

Eliason, S. (1990). *Categorical Data Analysis System*, Version 3.50. Dept. of Sociology, University of Iowa, Ames.

Elson, R.C., & Johnson, W.D. (1994). *Essentials of Biostatistics*, F.A. Davis, Philadelphia.

Everitt, B.S. (1977). *The Analysis of Contingency Tables.* London: Chapman and Hall.

Farewell, V. T. (1982). A note on regression analysis of ordinal data with variability of classification. *Biometrika*, 69: 27-32.

Fienberg, S.E. (1980). *The Analysis of Cross-Classified Categorical Data.* New York: John Wiley.

Fienberg, S.E., & Larntz, K. (1976). Loglinear representation for paired and multiple comparison models. *Biometrika*, 63, 245-254.

Fingleton, B. (1984). *Models of Category Counts.* Cambridge University Press, Cambridge.

Finney, D.J. (1941). The estimation from individual records of the relationship between dose and quantal response. *Biometrika, 34*, 320-334.

Finney, D.J. (1971). *Probit Analysis*, 3rd edition, Cambridge University Press.

Firth D., & Treat, B. R. (1988). Square contingency tables and GLIM. *GLIM Newletter*,16, 16-20.

Fisher, R. A. (1924). The condition under which χ^2 measures the discrepancy between observation and hypothesis. *Jour. R. Statist. Soc., 87*, 442-450.

Fisher, R.A. (1935). The case of zero survivors (Appendix to Bliss, C.I. (1935)). *Ann. Appl. Biol., 22*, 164-165.

Fisher, R.A. (1970). *Statistical Methods for Research Workers.* 14th edition, Edinburgh:Oliver & Boyd. Oliver & Boyd, Edinburgh (14th edition).

Fisher, R. A. (1973). *Statistical Methods and Scientific Inference.* New York: Hafner.

Fleiss, J.L. (1981). *Statistical Methods for Rates and Proportions*, second edition. New York: Wiley.

Fleiss, J.L., Cohen, J., and Everitt, B. S. (1969). Large-sample standard errors of kappa and weighted kappa. *Psychol. Bull.*, 72, 323-327.

Forthofer, R. N., & Lehmen, R.G. (1981). *Public Program Analysis: A New Categorical Data Approach.* Belmont, CA: Lifetime Learning Publications.

Freeman, D.H. (1987). *Applied Categorical Data Analysis.* Marcel Dekker, Inc., New York.

Freeman, G.H., & Halton, J.H. (1951). Note on an exact treatment of contingency, goodness of fit and other problems. *Biometrika*, 38: 141-149.

Freeman, M.F., & Tukey, J.W. (1950). Transformations related to the angular and the square root. *Ann. Math. Statist.*, 27, 601-611.

Friedl, H. (1995). A note on generating the factors for symmetry parameters in log-linear models. *GLIM Newletter*, 24, 33-36.

Gabriel, K.R. (1969). Simultaneous test procedures: Some theory of multiple comparisons. *Ann. Math. Statist.*, 40, 224-250.

Gail, M.H., & Mantel, N. (1977). Counting the number of $r \times c$ contingency tables with fixed margins. *J. Amer. Statist. Assoc.*, 72: 859-862.

Gibbons, J.D., & Pratt, J.W. (1975). P-values: Interpretation and methodology. *Amer. Statist., 29*, 20-25.

Gilula, Z. (1986). Grouping and association in contingency tables: An exploratory canonical correlation approach. *J. Amer. Statist. Assoc.*, 81, 773-779.

Gilula, Z., Krieger, A.M., & Ritov, Y. (1988). Ordinal association in contingency tables: Some interpretive aspects. *Jour. Amer. Statist. Assoc.*, 83, 540-545.

Glass, D.V. (1954). *Social Mobility in Britain.* Glenco, III: Free Press.

Goldstein, A. (1965). *Biostatistics: An Introductory Text.* New York: Macmillan.

Good, I.J. (1976). On the application of symmetric Dirichlet distributions and their mixtures to contingency tables. *Ann. Math. Statist.*, 4: 1159-1189.

Good, I.J., Gover, T.N., & Mitchell, G.J. (1970). Exact distributions for X^2 and for the likelihood-ratio statistic for the equi-probable multinomial distribution. *J. Amer. Statist. Assoc.*, 65, 267-283.

Goodman, L.A. (1962). Statistical methods for analyzing processes of change. *Amer. J. Sociol.*, 68, 57-78.

Goodman, L.A. (1964). Interaction in multi-dimensional contingency tables. *Ann. Math. Statist. 35*, 632-646.

Goodman, L.A. (1968). The analysis of cross-classified data: independence, quasi-independence, and interactions in contingency tables with or without missing entries. *J. Amer. Statist. Assoc., 63*, 1091-1131.

Goodman, L.A. (1969). On partioning chi-square and detecting partial association in multiple classifications. *J. Roy. Statist. Soc.,* B31, 486-498.

Goodman, .A. (1970). The multivariate analysis of qualitative data: interactions among multiple classifications. *J. Amer. Statist. Assoc.,* 65, 226-256.

Goodman, L.A. (1971a). The analysis of multidimensional contingency tables: stepwise procedures and direct estimation methods for building models for multiple classifications. *Technometrics, 13*, 33-61.

Goodman, L.A. (1971b). The partioning of chi-square, the analysis of marginal contingency tables, and the estimation of expected frequencies in multidimensional contingency tables. *J. Amer. Statist. Assoc., 66*, 339-344.

Goodman, L.A. (1972). Some multiplicative models for the analysis of cross-classified data, in L. Le Cam et al., *Proceedings of the Sixth Berkeley Symposium on Mathematical Statistics and Probability*, 649-696. Berkeley: University of California Press.

Goodman, L.A. (1978). *Analyzing Qualitative/Categorical Data*. Cambridge, MA: Abt Books.

Goodman, L.A. (1979a). Simple models for the analysis of Association in cross-classifications having ordered categories. *Jour. Amer. Statist. Assoc.*, 74, 537-552.

Goodman, L.A. (1979b). Multiplicative models for the analysis of occupational mobility tables and other kinds of cross-classification tables. *Amer. Jour. Sociology,* 84, 804-829.

Goodman, L. A. (1979c). Multiplicative models for square contingency tables with ordered categories. *Biometrika*, 66, 413-418.

Goodman, L.A. (1981). Criteria for determining whether certain categories in a cross-classification table should be combined with special reference to occupational categories in an occupational mobility table. *Amer. J. Sociology*, 87, 612-650.

Goodman, L.A. (1984). *The Analysis of Cross-Classified Data Having Ordered Categories*, Cambridge, MA: Harvard University Press.

Goodman, L.A. (1985). The analysis of cross-classified data having ordered and/or unordered categories: Association models, correlation models and symmetry models for contingency tables with or without missing entries. *The Annals of Statistics*, 13, 10-69.

Goodman, L.A. (1986a). Some useful extensions of the usual correspondence analysis approach and the usual log-linear models approach in the analysis of contingency tables. *Internat. Statist. Rev.*, 54, 243-309.

Goodman, L.A (1986b). Correspondence analysis and log-linear models *Int. Statist. Rev.*, 54, 243-309.

Goodman, L. A. (1991). Measures, models, and graphical displays in the analysis of cross-classified data. *J. Amer. Statist. Assoc.*, 86, 1085-1111. (with comments)

Goodman, L.A. (1996). A single general method for the analysis of cross-classified data: Reconciliation and synthesis of some methods of Pearson, Yule, and Fisher, and also some methods of correspondence analysis and association analysis. *J. Amer. Statist. Assoc.*, 91, 408-428.

Goodman, L.A., & Kruskal, W. H. (1954). Measures of association for cross classifications. *J. Amer. Statist. Assoc.*, 49, 732-764

Green, P.J. (1984). Iteratively re weighted least squares for maximum likelihood estimation and some robust and resistant alternatives (with discussion). *J. R. Statist. Soc., B*, 46, 149-192.

Greenland, S. (1991). On the logical justification of conditional tests for two-by-two contingency tables. *Amer. Statist., 45,* 248-251.

Grizzle, J.E. (1967). Continuity correction in the chi-square test for 2×2 tables. *Amer. Statist.*, 21, 28-32.

Grizzle, J.E., Starmer, C.F., & Koch, G.G. (1969). Analysis of categorical data by linear models. *Biometrics*, 25, 489-504.

Guttman, L. (1968). A general nonmetric technique for finding the smallest coordinate space for a configuration of points. *Psychometrika*, 33, 469-506.

Haberman, S.J. (1973). The analysis of residuals in cross- classification tables. *Biometrics*, 29: 205-220.

Haberman, S.J. (1978). *Analysis of Qualitative Data, Volumes 1 & 2.* Academic Press: New York.

Haberman, S. (1981). Tests for independence in two-way contingency tables based on canonical correlation and on a linear-by-linear interactions. *The Annals of Statistics*, 9, 1178-1186.

Haldane, J.B.S. (1937). The exact value of the moments of the distribution of χ^2, used as atest of goodness-of-fit when expectations are small. *Biometrika*, 29, 133-143; Corr. 13, 220.

Hansteen, V., Moinichen, E., Lorentsen, E., Andersen, A., Strom, O., Soiland, K., Dyrbekk, D., Refsum, A.M., Tromsdal, A., Knudsen, K., Eika, C., Bakken, J., Smith, P., & Hoff, P.I. (1982). One year's treatment with propranolol after myocardial infarction: preliminary report of Norwegian multicentre trial. *Brit. Med. J. (Clin Res Ed)*, 284(6310), 155-60.

Hedlund, R.D. (1978). Cross-over voting in a 1976 presidential primary. *Public Opinion Quart.*, 41, 498-514.

Hildebrand, D., & Ott, L. (1991). *Statistical Thinking for Managers.* Third edition. PWS Kent:MA.

Hill, I.D. (1988). Discussion on the paper by William Rice. *Biometrics,44,* 14-16.

Hirji, K. F., Tan, S.J., & Elashoff, R.M. (1991). A quasi-exact test for comparing two binomial proportions. *Statist. Med.*, 10, 1137-1153.

Hogg, R. V., & Tanis, E. A. (1997). *Probability and Statistical Inference*, Fifth Edition, Prentice Hall. Upper Saddle River, NJ.

Holloway, J.W. (1989). A comparison of the toxicity of the pyrethroid trans-cypermethrin, with and without the synergist piperonyl butoxide, to adult moths from two strains of *Heliothis virescens.* Final year dissertation, University of Reading, UK.

Hommel, G. (1978). Tail probabilities for contingency tables with small expectations. *JASA*, 23: 764-766.

Hope, K. (1982). Vertical and non vertical class mobility in three countries. *Amer. Sociol. Review*, 47, 99-113.

Hubert, J.J. (1992). *Bioassay.* Kendall/Hunt Publishing Company: Dubuque, Iowa.

Imrey, P.B., Johnson, W.D., & Koch, G.G. (1976). An incomplete contingency table approach to paired comparison experiments. *Jour. Amer. Statist. Assoc.*, 71, 614-623.

Iverson, G.R. (1979). Decomposing chi-square: A forgotten technique. *Sociol. Methods Res.*, 8: 143-157.

Jarret, R.G. (1979). A note on the intervals between coal-mining disasters. *Biometrika*, 66, 191-193.

Johnson, N.L., & Kotz, S. (1982). *Distributions in Statistics: Discrete Distributions.* Boston: Houghton Mifflin.

Jolayemi, E.T. (1990a). On the measure of agreement between two raters. *Biom. J*, 32, 87-93.

Jolayemi, E.T. (1990b). The model selection for one-dimensional multinomials. *Biometrical Journal*, 32, 827-834.

Kateri, M. (1993). Fitting asymmetry and non-independence models with GLIM. *GLIM Newsletter*, 22, 46-50.

Kaufman, R.L. and Schervish, P.G. (1986). Using adjusted cross tabulations to interpret log-linear relationships. *Amer. Soc. Rev.*, 51, 717-733.

Kaufman, R.L, & Schervish, P.G. (1987). Variations on a theme: More uses of odds ratios to interpret log-linear models. *Sociological Methods and Research.* 16, 218-255.

Kelsey, J.L., & Hardy, R. J. (1975). Driving of motor vehicles as a risk factor for acute herniated lumbar intervertebral disc. *Amer. Journal of Epidemiology*, 102, 63-73.

Kendall, M.G. (1945). The treatment of ties in rank problems. *Biometrika*, 33, 239-251.

Kendall, M. G. (1952). *The Advanced Theory of Statistics*, Vol. 1, 5th edition. Griffin, London.

Keil, J.E., Sutherland, S.E., Hames, C., Lackland, D., Gazes, P., Knapp, R., & Tyroler, H. (1995). Coronary disease mortality and risk factors in black and white men. *Archives of Internal Medicine*, 155, 1521-1527.

Kleinbaum, D.G., Kupper, L.L., Muller, K.E., & Nizam, A. (1998). *Applied Regression Analysis and Other Multivariable Methods.* Duxbury Press, Pacific Grove, CA.

Klotz, J., & Teng, J. (1977) One-way layout for counts and the exact enumeration of the Kruskal-Wallis H distribution with ties. *J. Amer. Statist. Assoc.*, 72, 165-169.

Koch, G.G., Amara, I.A., Davis, G.W. & Gillings, D.B. (1982). A review of some statistical methods for covariance analysis of categorical data. *Biometrics, 38*, 563-595.

Koch, G.G., & Bhapkar, V.P. (1982). Chi-square tests. In *Encyclopedia of Statistical Sciences*, Jonson, N.L., and Kotz, S. (eds), 442-457. Wiley, New York.

Koch, G.G., Imrey, P.B., Singer, J.M., Atkinson, S.S., & Stokes, M.E. (1985). *Lecture Notes for Analysis of Categorical Data.* Montreal: Les Presses de L'Université de Montréal.

Koch, G.G & Stokes, M.E. (1981). Chi-square tests: numerical examples. In *Encyclopedia of Statistical Sciences*, Johnson, N.L., & Kotz, S. (eds). Wiley, New York.

Koehler, K. (1986). Goodness-of-fit tests for log-linear models in sparse contingency tables. *JASA*, 81: 483-493.

Koehler, K.J., & Larntz, K. (1980). Empirical investigation of goodness-of-fit statistics for sparse multinomials. *J. Amer. Statist. Assoc.*, 75, 336-344.

Kreiner, S. (1992). Exact inference in multidimensional tables. Comment on Agresti (1992). *Statistical Science* 7: 163-165.

Kudo, A. & Tarumi, T. (1978). 2×2 Tables emerging out of different chance mechanisms. *Commun. Statist.*, 977-986.

Kullback, S. (1959). *Information Theory and Statistics.* Wiley, New York.

Kutylowski, A. J. (1989). Analysis of symmetric cross-classifications. In *Lecture Notes in Statistics*, 57, 188-197.

Lancaster, H. O. (1949a). The derivation and partition of χ^2 in certain discrete distributions. *Biometrika, 36*, 117-122.

Lancaster, H.O. (1949b). Statistical control of counting experiments. *Biometrika, 39,* 419-422.

Lancaster, H.O. (1961). Significance tests in discrete distributions *J. Amer. Statist. Assoc., 56,* 223-234.

Landis, J.R., & Koch, G.G. (1977a). The measurement of observer agreement for categorical data. *Biometrics*, 33, 159-174.

Landis, J.R., & Koch, G.G. (1977b). An application of hierarchical Kappa-type statistics in the assessment of majority agreement among multiple observers. *Biometrics*, 33, 363-374.

Larntz, K. (1978). Small-sample comparison of exact levels for chi- squared goodness-of-fit statistics. *JASA*, 73: 253-263.

Lawal, H.B. (1980). Tables of percentage points of Pearson's goodness-of-fit statistic for use with small expectations. *Appl. Statist.*, 29, 292-298.

Lawal, H.B. (1984). Comparisons of X^2, Y^2, Freeman-Tukey and William's improved G^2 test statistics in small samples of one-way multinomials. *Biometrika*, 71, 415-458.

Lawal, H.B. (1986). Review of approximations to the null distribution of Pearson's X^2 test statistic. *Nig. Jour. Scientific Research*, 1, 65-69.

Lawal, H.B. (1989a). Comparing power behaviors for some goodness-of-fit test statistics in sparse multinomials. *Biom. Journal.*, 31, 297-306.

Lawal, H.B. (1989b). On the X^2 statistic for testing independence in two-way contingency tables *AMSE Review*, 12: 37-51.

Lawal, H.B. (1992a). A modified X^2 test when some cells have small expectations in the multinomial distribution. *J. Statist. Comput. Simul.*, 40, 15-27.

Lawal, H.B. (1992b). Using discrete distributions to compute powers for goodness-of-fit test statistics in a one-way multinomial setting. *Biometrical Journal*, 34, 429-435.

Lawal, H.B. (1992c). Approximating the non-null distribution of the likelihood ratio test statistic. *Biometrical Journal*, 34, 473-483.

Lawal, H.B. (1992d). Parsimonious uniform-distance association models for the occupational mobility table data. *J. Japan Statist. Soc.*, 22, 123-134.

Lawal, H.B. (1993a). Association, symmetry, and diagonal models for occupational mobility and other similar square contingency tables having ordered categorical variables. *Biometrical Journal*, 35, 193-206.

Lawal, H.B. (1993b). Comparisons of some chi-squared goodness-of-fit test statistics in sparse one-way multinomials. *Biometrical Journal*, 35, 589-599.

Lawal, H.B. (1996). Using SAS Genmod Procedure to fit Diagonal class Models to Square Contingency Tables Having Ordered Categories *Proceedings of the MidWest SAS Users Group 96*, 149-160.

Lawal, H.B. (2001). Modeling symmetry models in square contingency tables. *Jour. Statist. Comp. Simul.*, 71, 59-83.

Lawal, H.B. (2002a). Modeling the 1984-1993 American League Baseball results as dependent categorical data. *Jour. Applied Probability*, 27, 53-66.

Lawal, H. B. (2002b). The structure of the log odds-ratios in non-independence and symmetry diagonal models for square contingency tables. *Jour. of Quality & Quantity, 36(3)*, 197-220.

Lawal, H. B. (2003a). Implementing point-symmetry models for square tables having ordered categories in SAS. To appear in *Journal of Italian Statistical Society.*

Lawal, H.B. (2003b) Application of non standard log-linear models to symmetry and diagonals parameter models in square contingency tables. Submitted, *South African Statist. Jour.*.

Lawal, H.B., & Upton, G.J.G (1980). An approximation to the distribution of the X^2 goodness-of-fit statistic for use with small expectations. *Biometrika, 67,* 447-453.

Lawal, H.B., & Upton, G.J.G. (1984).On the use of X^2 as a test of independence in contingency tables with small cell expectations. *Australian J. Statist.*, 26, 75-85.

Lawal, H.B., & Upton, G.J.G. (1990a). Comparisons of some chi- squared tests for the test of independence in sparse two-way contingency tables. *Biom. J.*, 32: 59-72.

Lawal, H.B. & Upton, G.J.G. (1990b). Alternative interaction structures in square contingency tables having ordered classificatory variables. *Quality and Quantity*, 24, 107-127.

Lawal, H.B., & Upton, G.J.G. (1995). A Computer algorithm for fitting models to $N \times N$ contingency tables having ordered categories. *Comm. Statist.*, Simula, 24(3), 793-805.

Lawal, H.B., & Sundheim, R. (2002). Generating factor variables for asymmetry, non-independence and skew-symmetry models in square contingency tables using a SAS macro. *Journal of Statistics and Computing*, Vol. 7, (8), 1-23.

Le, C.T. (1992). *Fundamentals of biostatistical inference*; New York: Marcel Dekker.

Lehmann, E.L. (1966). Some concepts of dependence. *Ann. Math. Statist.*, 37, 1137-1153.

Liang, K.Y., & Zeger, S.L. (1986). Longitudinal data analysis using generalized linear models. *Biometrika, 73,* 13-22.

Lindgren, B.W. (1968). *Statistical Theory.* Macmillan, London.

Lindsey, D.V. (1964). The Bayesian analysis of contingency tables. *Ann. Math. Statist.*, 35: 1622-1643.

Lindsey, J.K. (1989). *The Analysis of Categorical Data Using GLIM.* John Wiley, New York.

Lindsey, J.K. (1992). Fitting distributions in GLIM as log-linear models. *GLIM Newsletter*, 21, 9-12.

Lindsey, J.K. (1995). *Modelling Frequency and Count Data.* Oxford Science Publications, Oxford.

Lipsitz, S. (1988). Methods for analyzing repeated categorical outcomes. Unpublished PhD dissertation, Department of Biostatistics, Harvard University.

Lipsitz, S. (1999). Lecture Notes on the GEE for BMTRY 726. Department of Biometry and Epidemiology, MUSC, Charleston, SC.

Littel, R.C., Milliken, G.A., Stroup, W.W., & Wolfinger, R.D. (1996). *SAS® System for Mixed Models.* SAS Institute, Inc., Cary, NC.

Lord, F. (1953). On the statistical treatment of Footbal Numbers. *Amer. Psychologist*, 8, 750-751.

Lunneborg, C. E. (1994). *Modeling Experimental and Observational Data.* Duxbury. Belmont, CA.

Mantel, N. (1963). Chi-square tests with one degree of freedom: Extensions of the Mantel-Haenszel procedure. *J. Amer. Statist. Assoc.*, 58, 690-700.

Mantel, N., & Greenhouse, S.W. (1968). What is the continuity correction? *Amer. Statist., 22,* 27-30.

Mantel, N., & Haenszel, W. (1959). Statistical aspects of the analysis of data from retrospective studies of disease. *J. Nat. Cancer Inst.*, 22, 719-748.

Maxwell, A.E. (1961). *Analyzing Qualitative Data*, Methuen, London.

McCullagh, P. (1978). A class of parametric models for the analysis of square contingency tables with ordered categories. *Biometrika 65*, 413-418.

McCullagh, P. (1980). Regression models for ordinal data (with discussion). *J. R. Statist. Soc.*, B 42, 109-142.

McCullagh, P., & Nelder, J.A. (1989). *Generalized Linear Models.* 2nd edition, Chapman and Hall, London.

Mead, R., & Curnow, R.N. (1983). *Statistical Methods in Agriculture and Experimental Biology.* Chapman and Hall, London.

Mecklenburg, R.S., Benson, E.A., Benson, J.W., Fredlung, P.N., Guinn, T., Metz, R.J., Nielsen, R. L., & Sannar, C.A. (1984). Acute complications associated with insulin pump therapy: Report of experience with 161 patients. *JAMA*, 252(23), 3265-3269.

Mehta, C.R., & Hilton, J.F. (1993). Exact power of conditional and unconditional tests: Going beyond the 2 × 2 contingency table. *Amer. Statistician, 47*, 91-98.

Mehta, C.R., & Patel, N.R.(1983). A network algorithm for performing Fisher's exact test in $r \times c$ contingency tables. *Jour. Amer. Statist. Assoc.* 78, 427-434.

Mehta, C.R., Patel, N.R.(1990). Exact significance testing by the method of control variates. *COMPSTAT*,9, 141- 144.

Melia, B.M., & Diener-West M. (1994). Modeling inter rater agreement for pathologic features of choroidal melanoma. In *Case Studies in Biometry*, edited by Nicolas Lange *et al.*. Wiley and Sons.

Mendenhall, W. (1968). *Introduction to Linear Models and the Design and Analysis of Experiments.* Belmont, CA: Wadsworth.

Morgan, B.J.T. (1992). *Analysis of Quantal Response Data.* London:Chapman and Hall.

Morgan, S.P., & Teachman, J.D. (1988). Logistic regression: Description, examples, and comparisons. *J. Marriage Family*, 50, 929-936.

Mosteller, F., & Tukey, J. W. (1977). *Data Analysis and Regression.* Boston: Addison-Wesley.

Mullins, E.J. & Sites, P. (1984). The origins of contemporary eminent black Americans: A three-generation analysis of social origin. *Amer. Sociol. Rev.*, 49, 672-685.

Nass, C.A.G. (1959). The X^2 test for small expectations in contingency tables with special reference to accidents and absenteeism. *Biometrika, 46*, 365-85.

Nelder, J.A., & Wedderburn, R.W.M. (1972). Generalized linear models. *J. R. Statist. Soc., A*, 135, 370-384.

Neyman, J. (1949). Contribution to the theory of the χ^2 test. *Proceedings of the First Berkeley Symposium on Mathematical Statistics and Probability*, 239-273.

Ott, L. (1984). *An Introduction to Statistical Methods and Data Analysis.* Boston: Duxbury Press.

Pagano, M. & Gauvreau, K. (1993). *Principles of Biostatistics.* Duxbury Press, Belmont, CA.

Paul, J.S, Ronald, S.R., & Ronald, W.M. (1979). Exact and approximate distribution of the chi-square statistic for equi-probability. *Comm. Statist.*, B8(2), 131-147.

Pearson, E.S. (1947). The choice of a statistical test illustrated on the interpretation of data classified in 2 × 2 tables. *Biometrika*, 34, 139-167.

Pearson, K. (1900). On a criterion that a given system of deviations from the probable in the case of a correlated system of variables is such that it can be reasonably supposed to have arisen from random sampling. *Philo. Mag.*, Series 5, 50: 157-175.

Pearson, K. (1904). Mathematical contributions to the theory of evolution XIII: On the theory of contingency and its relation to association and normal correlation. *Draper's Co. Research Memoirs. Biometric Series.*

Pepe, M.S., & Anderson, G.L. (1994). A cautionary note on inference for marginal regression models with longitudinal data and general correlated response data. *Comm. Statist.-Simula, 23(4)*, 939-951.

Pirie, W.R, & Hamden, M.A. (1972). Some revised continuity corrections for discrete distribution. *Biometrics, 28*, 693-701.

Plackett, R.L. (1981). *The Analysis of Categorical Data.* Griffin, London.

Potthoff, R.F., & Roy, S.N. (1964) A generalized multivariate analysis of variance model useful especially for growth curve models. *Biometrika*, 51, 665-680.

Pregibon, D. (1981). Logistic regression diagnostics. *Ann. Statist., 9*, 705-724.

Prentice, R. L. (1988). Correlated binary regression with covariates specific to each binary observations. *Biometrics, 44*, 1033-48.

Prentice, R. L., & Zhao, L. P. (1991). Estimating equations for parameters in means and covariances of multivariate discrete and continuous responses. *Biometrics, 47*, 825-39.

Radelet, M. (1981). Racial characteristics and the imposition of the death penalty. *Amer. Sociol. Rev.*, 46, 918-927.

Radlow, R., & Alf, E.F. (1975). An alternate multinomial assessment of the accuracy of the χ^2 test of goodness of fit. *JASA*, 70: 811-813.

Raftery, A.E. (1986). A note on Bayes factors for log-linear contingency table models with vague prior information. *Jour. R. Statist. Soc.*, 48, 249-250.

Randolf, R.H., & Wolfe, D.A. (1979). *Introduction to the Series of Nonparametric Statistics.* John Wiley, New York.

Rasch, G. (1961). On general laws and the meaning of measurement in psychology, pp. 321-333. In *Proc. 4th Berkeley Symp. Math. Statist. Probab.*, vol 4, ed. J. Newman. University of CA Press, Berkeley.

Rayner, J.C.W., & Best, D.J. (1982). The choice of class probabilities and number of classes for the simple X^2 goodness-of-fit test. *Sankhya*, B44, 28-38.

Read, T.R.C., & Cressie, W.(1988). *Goodness-of-Fit Statistics for Discrete Multivariate Data.* Springer-Verlag, New York.

Reynolds, H.T. (1977). *The Analysis of Cross-Classifications.* Collier Macmillan, London.

Rice, W.R. (1988). A new probability model for determining exact P values for 2 × 2 contingency tables when comparing binomial proportions. *Biometrics, 44,* 1-22.

Roscoe, J.T., & Byars, J.A. (1971). An investigation of the restraints with respect to sample size commonly imposed on the use of the chi- square statistic. *J. Amer. Statist. Assoc.*, 66, 755-759.

Rosner, B. (2000). *Fundamentals of Biostatistics.* Duxbury, Belmont, CA.

Sachs, L. (1986). Alternatives to the chi-square test of homogeneity in 2 × 2 tables and to Fisher's exact test. *Biometrical Journal, 28,* 975-979.

Sakamoto, Y, Ishiguro, M., & Kitagawa, G. (1986). *Akaike Information Criterion Statistics.* KTL Scientific Publishers, MA.

Samuels, M. L., & Witmer, J. A. (1999). *Statistics for the Life Sciences.* Prentice Hall, Upper Saddle River, NJ.

Santner, T., & Duffy, D.E. (1989). *The Statistical Analysis of Discrete Data.* New York: Springer-Verlag.

Schork, M.A., & Remington, R.D. (2000). *Statistics with applications to the Biological and Health Sciences.* Prentice Hall, Englewood Cliffs, NJ.

Schouten, F.J., & Loch Oh, H. (1975). A data analysis approach to fitting square tables. *Comm. Statist.*, 4, 595-615.

Schouten, F.J., Molenaar, I.W, Van Strik, R., & Boomsma, A. (1980). Comparing two independent binomial proportions by a modified chi- square test. *Biom. J. 22,* 241-248.

Schwarz, G. (1978). Estimating the dimensions of a model. *Annals of Statistics*, 6, 461-464.

Senchaudhuri, P., Cyrus, R.M., & Patel, N.R. (1993). Estimating exact P-values by the method of control variates, or Monte Carlo Rescue. *ASA, San Francisco Conference*

Senie, R.T, Rosen, P.P., Lesser, M.L., & Kinne, D.W. (1981). Breast self-examination and medical examination related to breast cancer status. *Amer. J. of Public Health*, 71, 583-590.

Shapiro, S., Slone, D., Rosenberg, L., Kaufman, D., Stolley, P.D., & Miettinen, O.S. (1979). Oral contraceptive use in relation to myocardial infarction. *Lancet, i:* 743-746.

Simpson, E.H. (1951). The interpretation of interaction in contingency tables. *J. Roy. Statist. Soc.*, B13: 238-241.

Slaton, T.L., Piegorsch, W.W., and Durham, S.D. (2000). Estimation and testing with overdispersed proportions using the beta-logistic regression model of Heckman and Willis. *Biometrics*, 56, 125-133.

Slomczynski, K.M., & Krauze, T.K. (1987). Cross-National similarity in social mobility patterns: A direct test of the Featherman-Jones-Hauser hypothesis. *Amer. Sociological Review,* 52, 598-611.

Slutsky, E.E. (1925). Uber Stochastische asymptoten und Grenzwerte. *Metron, 5*, 1-90.

Snedecor, G.W., & Cochran, G.W. (1973). *Statistical Methods.* Iowa State University Press, Ames.

Sobel, M.E., Hout, M., & Duncan, O.T. (1985). Exchange, structure and symmetry in occupational mobility. *Amer. Jour. Sociology*, 91, 359-372.

Somers, R.H. (1962). A new asymmetric measure of association for ordinal variables *Amer. Sociol. Rev.*, 27, 799-811.

Speckman, J.J. (1965). *Marriage and kinship among the East Indians in Suriname.* Assen, Netherlands: Van Gorcum.

Spitzer, R. L., Cohen, J., Fleiss, J.L, and Endicott, J. (1967). Quantification of agreement in psychiatric diagnosis. *Arch. Gen. Psychiatry*, 17, 83-87.

Stanley, G., Appadu, B., Mead, M.R. & Rowbotham, D.J. (1996). Dose requirements, efficacy and side effects of morphine and pethidine delivered by patient controlled analgesia after gynaecological surgery. *Brit. J. Anaesth.*, 76, 484-486.

StatXact (1991). *StatXact: Statistical Software for Exact Nonparametric Inference.* Cytel Software, Cambridge, MA.

Stern, E., Misczyncki, M., Greenland, S., Damus, K., & Coulson, A. (1977). Pap testing and hysterectomy prevalence: a survey of communities with high and low cervical cancer rates. *Amer. J. of Epidem.*, 106, 296-305.

Stevens, S. S. (1946). On the theory of scale measurements. *Science*, 103, 677-680.

Stevens S. S.(1951). Mathematics, measurement, and psychophysics, pp. 1-49. In *Handbook of Experimental Psychology*, ed. by S.S. Stevens. New York: John Wiley.

Stevens, S.S. (1968). Measurement, statistics and the schematic view. *Science, 161,* 849-854.

Stokes, M.E., Davis C. S., & Koch, G. G. (1995). *Categorical Data Analysis Using the SAS® System.* SAS Institute, Inc., Cary, NC.

Strand, A.L. (1930). Measuring the toxicity of insect fumigants. *Industrial and Engineering Chemistry: analytical edition*, 2, 4-8.

Stuart, A. (1955). A test for homogeneity of the marginal distributions in a two-way classification. *Biometrika*, 42, 412-416.

Sugiura, N., & Otake, M. (1974). An extension of the Mantel-Haenszel procedure to a k $2 \times c$ contingency tables and the relation to the logit models. *Comm. Statist., 3(9)*, 829-842.

Svalastoga, K. (1959). *Prestige, class, and mobility.* Heinemann, London.

Tanner, M.A., & Young, M.A. (1985). Modeling agreement among raters. *J. Amer. Statist. Assoc.*, 80, 175-180.

Tate, M.W., & Hyer, L.A. (1973). Inaccuracy of the X^2 test of goodness-of- fit when expected frequencies are small. *J. Amer. Statist. Assoc.*, 68, 836-841.

Thall, P.F., & Vail, S. C. (1990). Some covariance models for longitudinal count with overdispersion. *Biometrics, 46*, 657-71.

Tjur, T. (1982) A connection between Rasch's item analysis model and a multiplicative Poisson model. *Scandinavian Journal of Statistics, 9*, 23-30.

Tomizawa, S. (1985a). Decompositions for odds-symmetry models in a square contingency table with ordered categories. *J. Japan Statist. Soc.*, 15, 151-159.

Tomizawa, S. (1985b). Analysis of data in square contingency tables with ordered categories using the conditional symmetry and its decomposed models. *Environmental Health Perspectives*, 63, 235-239.

Tomizawa, S. (1986). A decomposition for the inclined point-symmetry model in a square contingency table. *Biom. J.* 28, 371-380.

Tomizawa, S. (1987). Decompositions for 2-Ratios-Parameter symmetry model in square contingency tables with ordered categories. *Biom. J.*, 29, 45-55.

Tomizawa, S. (1992). Quasi-diagonals-parameter symmetry model for square contingency tables with ordered categories. *Calcutta Statist. Assoc. Bull.*, 39, 53-61.

Truett, J., Cornfield, J. & Kannel, W. (1967). A multivariate analysis of the risk of coronary heart disease in Framingham. *J. Chronic Diseases*, 20, 511-524.

Tukey, J. W. (1961). Data analysis and behavioral science or learning to bear the quantitative man's burden by shunning badmandments. In *The Collected Works of John W. Tukey, Vol III (1986)*, ed. L.V. Jones. Belmont, CA: Wadsworth, pp. 391-484.

Upton, G.J.G. (1978). *The Analysis of Cross-Tabulated Data.* John Wiley, Chichester.

Upton, G.J.G. (1982). A comparison of alternative tests for the 2 × 2 comparative trial. *J. Roy. Statist. Soc., Ser. A, 145,* 86-105.

Upton, G.J.G. (1985). A survey of log-linear models for ordinal variables in an $I \times J$ contingency tables. *Guru Nanak Journal of Sociology*,6, 1-18.

Upton, G.J.G. (1991). The explanatory analysis of survey data using log-linear models. *The Statistician*, 40, 169-182.

Upton, G.J.G. (1992). Fisher's exact test *J. Roy. Statist. Soc., Ser. A, 155,* 395-402.

Upton, G.J.G., & Sarlvik, B. (1981). A loyalty-distance model for voting change. *J. R. Statist. Soc., A*, 144, 247-259.

Upton, G.J.G. & Fingleton, (1985). *Spatial Data Analysis by Example.* Vol. I. John Wiley, New York.

Velleman, P.F., & Wilkinson, L. (1993). Nominal, ordinal, interval and ratio typologies are misleading. *American Statistician*, 47, 65-72.

von Eye, A., & Indurkhya, A. (2000). Log-linear representations of the Mantel-Haenszel and Breslow-Day tests. *Methods of Psychol. Research*, 5, No. 4, 13-30.

von Eye, A. & Niedermeier, K. E. (1999). *Statistical Analysis of Longitudinal Categorical Data in the Social and Behavioral Sciences.* Lawrence Erlbaum Associates, Mahwah, NJ.

von Eye, A., & Spiel, C. (1996). Standard and non-standard log-linear symmetry models for measuring change in categorical variables. *American Statistician*, 50, 300-305.

Wald, A. (1943). Tests of statistical hypotheses concerning several parameters when the number of observations is large. *Amer. Math. Soc. Transactions, 54*, 426-482.

Ware, J.H., Dockery, D.W., Spiro, A. III, Speizer, F.E., and Ferris, B.G. Jr. (1984). Passive smoking, gas cooking and respiratory health in children living in six cities. *Amer. Review of Respiratory Diseases*, 129, 366-374.

Ware, J.H., Lipsitz, S., & Speizer, F. E. (1988). Issues in the analysis of repeated categorical outcomes. *Statistics in Medicine, 7*, 95-107.

West, E.N., & Kempthorne, O. (1972). A comparison of the χ^2 and likelihood ratio tests for composite alternatives. *J. Statist. Comput. Simul.*, 1, 1-33.

Whittaker, J. (1990). *Graphical Models in Applied Multivariate Statistics.* New York: Wily.

Wilks, S. S. (1938). The large-sample distribution of the likelihood ratio for testing composite hypotheses. *Ann. Math. Statist. 9,* 60-62.

Williams, D.A. (1975). The analysis of binary responses from toxicological experiments involving reproduction and teratogenicity. *Biometrics*, 31, 949-952.

Williams, D.A. (1982). Extra-binomial variation in logistic linear models. *Applied Statist.*, 31, 144-148.

Williams, E.J. (1952). Use of scores for the analysis of association in contingency tables, *Biometrika*, 39, 274-289.

Williamson, J., Lipsitz, S., & Kim, K. M. (1999). GEECAT and GEEGOR: Computer programs for the analysis of correlated categorical response data. *Computer Methods and Programs in Biomedicine, 58* 25-34.

Woodward, G., Lange, S.W., Nelson, K.W., & Calvert, H.O. (1941). The acute oral toxicity of acetic, chloracetic, dichloracetic and trichloracetic acids. *J. of Industrial Hygience and Toxicology*, 23, 78-81.

Woolf, B. (1955). On estimating the relation between blood group and disease. *Ann. Human Genet., 19*, 251-253.

Woolson, R. F., & Clarke, W. R. (1984). Analysis of categorical incomplete longitudinal data. *J. Royal Statistical Society, A147:* 87-99.

Yamaguchi, K. (1987). Models for comparing mobility tables: Toward parsimony and substance. *Amer. Sociol. Rev.*, 32, 482-494.

Yamaguchi, K. (1900). Some models for the analysis of asymmetric association in square contingency tables with ordered categories. In *Sociol. Method.*, ed. C.C. Clogg. Oxford: Basil Blackwell, 181-212.

Yarnold, J.K. (1970). The minimum expectation in χ^2 goodness-of-fit tests and the accuracy of approximations for the null distribution. *J. Amer. Statist. Assoc.*, 65, 865-886.

Yates, F. (1934). Contingency tables involving small numbers and the χ^2 test. *J. R. Statist. Soc., Suppl., 1,* 217-235.

Yule, G.U. (1900). On the association of attributes in Statistics. *Phil. Trans.*, A, 194, 257-319.

Yule, G.U. (1912). On the methods of measuring association between two attributes (with discussion) *J. Roy. Statist. Soc.*, 75, 579-642.

Zeger, S.L. (1988). A regression model for time series counts. *Biometrika*, 75, 621-629.

Zelterman, D. (1987). Goodness-of-fit tests for large sparse multinomial distributions. *J. Amer. Statist. Assoc.*, 82, 624-629.

Zelterman, D., Chan, S.F., & Mielke, P.W. (1993). Exact tests of significance in higher dimensional tables. *ASA, San Francisco Conference.*

Subject Index